Chemistry and Application of H-Phosphonates

Phosphorus is ubiquitous in our world. It is the key element in the genetic tape that guides the reproduction of all species; it is essential to all forms of life; it is one of the major elements found in many solid-state metal phosphide materials such as GaP and InP.

Chemistry and Application of H-Phosphonates

Kolio D. Troev

Institute of Polymers
Bulgarian Academy of Sciences
Sofia 1113, Bulgaria

Amsterdam • Boston • Heidelberg • London • New York • Oxford
Paris • San Diego • San Francisco • Singapore • Sydney • Tokyo

ELSEVIER

Elsevier
Radarweg 29, PO Box 211, 1000 AE Amsterdam, The Netherlands
The Boulevard, Langford Lane, Kidlington, Oxford OX5 1GB, UK

First edition 2006

Copyright © 2006 Elsevier B.V. All rights reserved

No part of this publication may be reproduced, stored in a retrieval system or transmitted in any form or by any means electronic, mechanical, photocopying, recording or otherwise without the prior written permission of the publisher

Permissions may be sought directly from Elsevier's Science & Technology Rights Department in Oxford, UK: phone (+44) (0) 1865 843830; fax (+44) (0) 1865 853333; email: permissions@elsevier.com. Alternatively you can submit your request online by visiting the Elsevier web site at http://elsevier.com/locate/permissions, and selecting *Obtaining permission to use Elsevier material*

Notice
No responsibility is assumed by the publisher for any injury and/or damage to persons or property as a matter of products liability, negligence or otherwise, or from any use or operation of any methods, products, instructions or ideas contained in the material herein. Because of rapid advances in the medical sciences, in particular, independent verification of diagnoses and drug dosages should be made

Library of Congress Cataloging-in-Publication Data
A catalog record for this book is available from the Library of Congress

British Library Cataloguing in Publication Data
A catalogue record for this book is available from the British Library

ISBN-13: 978-0-444-52737-0
ISBN-10: 0-444-52737-0

For information on all Elsevier publications
visit our website at books.elsevier.com

Printed and bound in The Netherlands

06 07 08 09 10 10 9 8 7 6 5 4 3 2 1

Working together to grow
libraries in developing countries

www.elsevier.com | www.bookaid.org | www.sabre.org

ELSEVIER   BOOK AID International   Sabre Foundation

The author gratefully acknowledges the assistance and helpful suggestions offered by
Dr. Emil Georgiev, GE, USA.

Preface

The interest in the chemistry and application of the diesters of H-phosphonic acid has increased dramatically over the recent years. Although closely related to the esters of phosphoric acid, the diesters of H-phosphonic acid have quite a rich and distinctive chemistry as compared to their phosphate analogues, which is based on the presence of a number of functional groups in their molecules (P–OR, P–H, and P=O), especially, the highly reactive P–H group. This group can be converted into a number of derivatives, like P–OH, P–NR, P–OR, and P–SH. These transformations proceed in mild conditions with practically quantitative yield. The accumulation of sufficient experimental data in the last few decades has not only provided an extended picture of the application range of this class of compounds, but has also allowed for some theoretical and mechanistic studies aimed at gaining a deeper understanding of most of the reaction mechanisms involving the diesters of H-phosphonic acid.

The rich chemistry, low cost, and easy availability of these esters of phosphonic acid makes them an excellent choice as a synthone in a number of practically important reactions. These esters are intermediates in the synthesis of an important classes of compounds such as

 (i) α-aminophosphonic acids;
 (ii) bisphosphonates;
(iii) epoxyphosphonates;
(iv) α-hydroxyphosphonates;
 (v) phosphoramides;
(vi) poly(alkylene H-phosphonate)s; and
(vii) nucleoside H-phosphonates.

α-Aminophosphonic acids are an important class of biologically active compounds, which have received an increasing amount of attention because they are considered to be structural analogues of the corresponding α-amino acids. The utilities of α-amino phosphonates as peptide mimics, haptens of catalytic antibodies, enzyme inhibitors, inhibitors of cancers, tumors, and viruses, antibiotics, and pharmacological agents are well documented. Bisphosphonates are drugs that have been widely used in different bone diseases, and have recently been used successfully against many parasites. Poly(alkylene H-phosphonate)s are promising, biodegradable, water soluble, new polymer carriers of drugs. The P–H group in the repeating units of poly(alkylene H-phosphonate)s determines a variety of chemical functionalities. Immobilization of the widely utilized low molecular–mass biologically active substances onto the poly(alkylene H-phosphonate)s enables the formation of new type of polymer prodrugs with improved properties—greater effectiveness, lowered toxicity, higher resistance, and controlled release. Nucleoside H-phosphonates seem to be the most attractive candidates as starting materials in the chemical synthesis of

DNA and RNA fragments. The 5'-hydrogen phosphonate-3'-azido-2',3'-dideoxythymidine is one of the most significant anti-HIV prodrugs, which is currently in clinical trials.

The book contains chapters dealing with physical and spectral properties (^{1}H, ^{13}C, ^{31}P and ^{17}O NMR data); characteristic reactions; important classes of compounds based on the esters of H-phosphonic acid; their application as physiologically active substances, flame retardants, catalysts, heat, and light stabilizers; lubricants; scale inhibitors; polymer carriers of drugs; preparation of diesters of H-phosphonic acid, and general procedures for conducting the most important reactions.

Kolio D. Troev

About the Author

Kolio Dimov Troev was born in Rupkite, district Chirpan, Bulgaria, in 1944. He did his undergraduate work at the Higher Institute of Chemical Technology, Sofia, and received his doctorate in the field of organophosphorus chemistry in 1974 from the Institute of Organic Chemistry, Bulgarian Academy of Sciences, with Prof. Georgy Borissov. In 1985, he received a Doctor of Science degree from the Institute of Polymers, where he worked. In 1988, he became Professor of Chemistry at the same Institute. Since 1989, he has been the head of the laboratory 'Phosphorus-containing monomers and polymers', which he established in 1989. His research interests have been the areas of organophosphorus chemistry, especially esters of H-phosphonic acid. He is the author of more than 110 papers in this field published in the *Phosphorus, Sulfur, Silicon and Related Elements, Heteroatom Chemistry, Journal of the American Chemical Society, European Polymer Journal, Polymer, Bioorganic & Medicinal Chemistry, Journal of Medicinal Chemistry, Macromolecular Rapid Communication, Polymer Degradation and Stability,* and the *Journal of Polymer Science, Part A: Polymer Chemistry*. Since 2003, he is a director of the Institute of Polymers, Bulgarian Academy of Sciences.

He and his wife, Krassimira, have a daughter, who is a lawyer, and a son, who is an economist.

Contents

Preface vii
About the Author .... ix-x

1 Methods for Preparation and Physical Properties of H-Phosphonates 1
    1.1   From Phosphorus Trichloride and Alcohols 1
    1.2   From H-Phosphonic Acid 3
    1.3   From White Phosphorus and Alcohols 4
    1.4   Others 4
    1.5   Physical Properties 4
    1.6   Thermal Stability 5
    Appendix 8

2 Structure and Spectral Characteristics of H-Phosphonates 11
    2.1   Electronic Structure of the Phosphorus Atom 11
    2.2   Nature of Chemical Bonds in the Phosphoryl Group 12
    2.3   Molecular Structure of Dimethyl H-Phosphonate 13
    2.4   Tautomerization of Dialkyl H-Phosphonates 15
    2.5   Spectral Characteristics 18
            2.5.1   Infrared spectra 18
            2.5.2 ^{1}H NMR spectra 18
            2.5.3 ^{31}P NMR spectra 19
            2.5.4 ^{13}C NMR spectra 20
            2.5.5 ^{17}O NMR spectra 20
    Appendix 21

3 Reactivity of H-Phosphonates 23
    3.1   Acidity of H-Phosphonate Diesters 23
    3.2   Disproportionation 25
    3.3   Reduction 25
    3.4   Hydrolysis 25
    3.5   Acidolysis 29
    3.6   Substitution Reactions at the Phosphorus Atom 29
            3.6.1   Transesterification reaction 29
            3.6.2   Side reactions 32
            3.6.3   Synthetic applications 33
    3.7   Characteristic Reactions of P–H Group 39
            3.7.1   Oxidation of P–H group 39
            3.7.2   Addition Reactions 53
    3.8   Reactions with the Participation of the α-Carbon Atom 77
            3.8.1   Reaction of dialkyl H-phosphonates with amines 77

		3.8.2	Reaction of diaryl H-phosphonates with amines	82
		3.8.3	Reaction with amides and urethanes	85
		3.8.4	Reaction with hydrogen halides	87
		3.8.5	Reaction with metal salts	88
		3.8.6	Reaction with ammonium halides	91
		3.8.7	Reaction with chlorosilanes	92
	Appendix			102

4 Important Classes of Compounds — 107

	4.1	Aminophosphonates and Aminophosphonic Acids		107
		4.1.1	Methods for the preparation of aminophosphonates	108
		4.1.2	Hydrolytic cleavage of phosphorus–carbon bond	138
		4.1.3	Synthetic application	141
	4.2	Bisphosphonates		144
		4.2.1	Methods for preparation of bisphosphonates	145
	4.3	Nucleoside H-Phosphonates		159
		4.3.1	Methods of preparation	160
		4.3.2	Reactivity of nucleoside H-phosphonates	166
		4.3.3	Synthetic application	174
	4.4	Dialkyl Epoxyalkylphosphonates		180
		4.4.1	1,2-Epoxyalkylphosphonates	180
		4.4.2	2,3-Epoxyalkylphosphonates	185
		4.4.3	Reactivity of 1,2-epoxyalkylphosphonates	185
	4.5	Poly(Alkylene H-Phosphonate)s		188
		4.5.1	Methods of preparation	189
		4.5.2	Reactivity of poly(alkylene H-phosphonate)s	195
		4.5.3	Application of poly(alkylene H-phosphonate)s	201
	4.6	Metal Salts of Dialkyl H-Phosphonates		210
		4.6.1	Phosphite-type metal salts	211
		4.6.2	Phosphonate-type metal salts	220
	4.7	Complexes of Dialkyl H-Phosphonates		222
		4.7.1	Molecular complexes with Lewis acids	223
		4.7.2	Complexes with some main group and f-elements	224
		4.7.3	Complexes with transition metals	226
		4.7.4	Complexes containing P(III)-type ligands	227
	Appendix			245

5 Application of H-Phosphonate Diesters and their Derivatives — 253

	5.1	Physiologically Active Substances		253
		5.1.1	Fungicides and bactericides	254
		5.1.2	Herbicides	256
		5.1.3	Plant-growth regulators	258
		5.1.4	Insecticides	260
		5.1.5	Drugs suppressing the growth of cancer, tumor, virus, or parasites	260
		5.1.6	Anti-HIV prodrugs	261
		5.1.7	Antiresorption drugs, carriers of radioactive metals	263

5.2	Polymer Additives	264
	5.2.1 Flame retardants and antioxidants	264
5.3	Degrading and Alkylating Agents of Polymers	271
5.4	Heat, Light, and UV Stabilizers	275
5.5	Catalysts	276
5.6	Corrosion Inhibitors	278
5.7	Scale Inhibitors	279
5.8	Lubricants	280

Subject Index 285

– 1 –

Methods for Preparation and Physical Properties of H-Phosphonates

The diesters of H-phosphonic acid can be obtained by several synthetic procedures. This chapter outlines the most commonly used approaches in that respect.

1.1. FROM PHOSPHORUS TRICHLORIDE AND ALCOHOLS

Dialkyl H-phosphonates are produced in the USA, Japan, Germany and other countries on an industrial scale primarily from phosphorus trichloride and alcohols. (The general procedure for the preparation of dialkyl H- phosphonates is given in the Appendix).

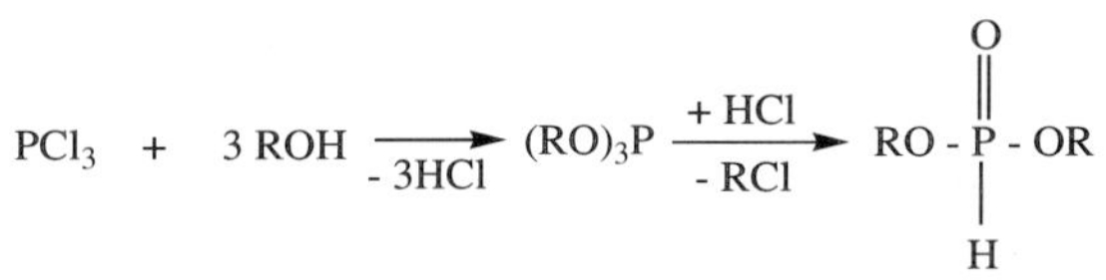

$$PCl_3 + 3\,ROH \xrightarrow{-3HCl} (RO)_3P \xrightarrow[-RCl]{+HCl} RO-\underset{\underset{H}{|}}{\overset{\overset{O}{\|}}{P}}-OR$$

Addition of an alcohol to the phosphorus trichloride at about 0 °C leads to rapid stepwise alkoxylation of phosphorus, followed by dealkylation of the trialkyl phosphite to dialkyl H-phosphonate [1,2]. The initial procedures for this reaction, which include cooling of the reaction mixture, have been modified many times [2–6]. It has been established that the cooling step is not necessary in the case of large alkoxy substituents with more than four carbon atoms in the chain [7]. Another study reveals that when some water is added to the reaction mixture together with the alcohol, the yield of dialkyl phosphonates increases significantly [8–11]. Another approach for synthesis of dialkyl H-phosphonates has been proposed, which does not include cooling of the reaction mixture but uses a solvent instead [12]. Methods have also been developed for hydrogen chloride elimination from the reaction mixture [13,14] by washing it with water. Mixed dialkyl H-phosphonates can be obtained when the above reaction is carried out with an equimolar mixture of two different alcohols [15].

The synthesis of higher dialkyl H-phosphonate homologues usually includes the initial treatment of phosphorus trichloride with methyl alcohol, and then the transesterification of so-formed dimethyl H-phosphonate with higher alcohols [16,17].

$$CH_3O-\underset{\underset{H}{|}}{\overset{\overset{O}{\|}}{P}}-OCH_3 + 2ROH \rightleftharpoons RO-\underset{\underset{H}{|}}{\overset{\overset{O}{\|}}{P}}-OR + 2CH_3OH$$

The hydrolysis of cyclic chlorophosphites [18–24], obtained by reacting phosphorus trichloride and diols, is one of the best methods for the preparation of cyclic H-phosphonates (see Appendix).

$$PCl_3 + HO-R-OH \xrightarrow[-2HCl]{0-5\,^\circ C} R\underset{O}{\overset{O}{\diagdown}}P-Cl \xrightarrow[-HCl]{+H_2O} R\underset{O}{\overset{O}{\diagdown}}\underset{H}{\overset{O}{P}}$$

A general method for the synthesis of cyclic H-phosphonates starting with phosphorus trichloride and aliphatic glycols has been described by Lucas (see Appendix) [25]. 4-Methyl-2-oxo-2-hydro-1,3,2-dioxaphospholane was obtained in two stages:

(a) during the first stage, 1,2-propanediol reacts with PCl$_3$ yielding 2-chloro-4-methyl-1,3,2-dioxaphospholane;
(b) at the second stage, 2-chloro-4-methyl-1,3,2-dioxaphospholane was hydrolyzed to give 4-methyl-2-oxo-2-hydro-1,3,2-dioxaphospholane [25].

$$\underset{CH_2-OH}{\overset{CH_3}{\underset{|}{\overset{|}{CH-OH}}}} \xrightarrow[-2HCl]{+PCl_3} \underset{H_2C-O}{\overset{CH_3}{\underset{|}{\overset{|}{HC-O}}}}\!\!\!\!\!\!\!\!>P-Cl \xrightarrow[-HCl]{+H_2O} \underset{H_2C-O}{\overset{CH_3}{\underset{|}{\overset{|}{HC-O}}}}\!\!\!\!\!\!\!\!>\!\!\!\underset{H}{\overset{O}{P}}$$

The hydrolysis was carried out in CH$_2$Cl$_2$ solution with a mixture of water and 1,4-dioxane. It was essential to use slightly less than the stoichiometric amount of water (1: 0.8); otherwise premature, undesirable polymerization occurred [26].

2-Oxo-2-hydro-1,3,2-dioxaphosphorinane or 4-methyl-2-oxo-2-hydro-1,3,2-dioxaphosphorinane were obtained followed the same procedure starting 1,3-propanediol or 1,3-butanediol and PCl$_3$. The hydrolysis was carried out in the presence of triethylamine.

$$H_2C\!\!\underset{CH_2OH}{\overset{CH_2OH}{\diagup\!\!\!\diagdown}} \xrightarrow[-2HCl]{+PCl_3} \bigcirc\!\!\!\!P-Cl \xrightarrow[-HCl]{+H_2O} \bigcirc\!\!\!\!\overset{O}{P}\!-\!H$$

1.2. FROM H-PHOSPHONIC ACID

Saks et al. [27] used for the first time H-phosphonic acid for the preparation of dialkyl H-phosphonates. There are a few patents devoted to the preparation of dialkyl H-phosphonates via direct esterification of H-phosphonic acid with alcohols [28–31]. The process involves heating under reflux a mixture of H-phosphonic acid, an alcohol in excess of that required

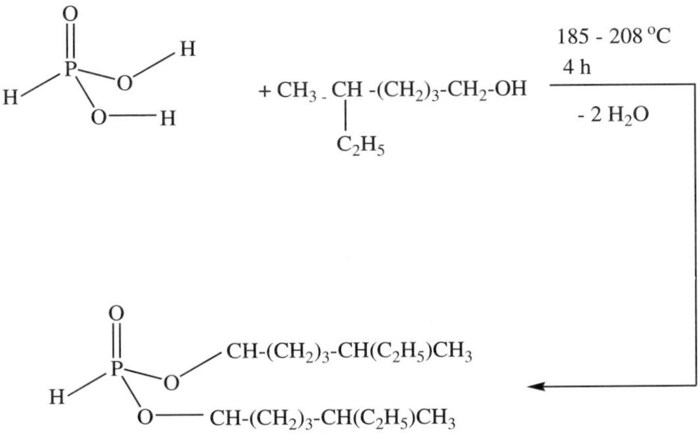

$$\text{H-P(O)(OH)}_2 + CH_3\text{-}CH(C_2H_5)\text{-}(CH_2)_3\text{-}CH_2\text{-}OH \xrightarrow[-2 H_2O]{185 - 208 \,°C,\ 4h} \text{H-P(O)(O-CH(CH}_2)_3\text{-CH(C}_2H_5)CH_3)_2$$

stoichiometrically to form the dialkyl H-phosphonate, and a substantial proportion of an inert solvent such as toluene. The water formed during the esterification is removed continuously. An improved process for the preparation of dialkyl H-phosphonates by means of refluxing H-phosphonic acid with alcohols having at least four carbon atoms, in an excess of at least 45% over the stoichiometrical amount, under azeotropic separation of the reaction water is described in [32] (see Appendix). In comparison to known processes, the dialkyl H-phosphonates are obtained according to the instant process in higher yield and with higher purity. Mixtures of two different alcohols are also used in this process. It has been shown that the yield of dialkyl H-phosphonates increases when the synthesis is carried out in the presence of sulfonic acid [29] or trialkyl phosphates [33,34].

Diphenyl H-phosphonate is obtained by the treatment of H-phosphonic acid with a twofold excess of triphenyl phosphite [35–37].

$$H_3PO_3 + 2\ P(OC_6H_5)_3 \longrightarrow 3\ (C_6H_5O)_2P(O)H$$

Another approach employs treatment of H-phosphonic acid or its monoalkyl ester with a carboxylic acid anhydride and an alcohol at 20–50 °C. Dimethyl H-phosphonate is obtained according to this procedure in quantitative yield based on the starting phosphonic acid (See Appendix) [38].

$$(HO)_2P(O)H + CH_3OH + (CH_3CO)_2O \longrightarrow (CH_3O)_2P(O)H$$

1.3. FROM WHITE PHOSPHORUS AND ALCOHOLS

This method for preparation of dialkyl H-phosphonates involves oxidation of white phosphorus followed by treatment of the intermediate with alcohols [39].

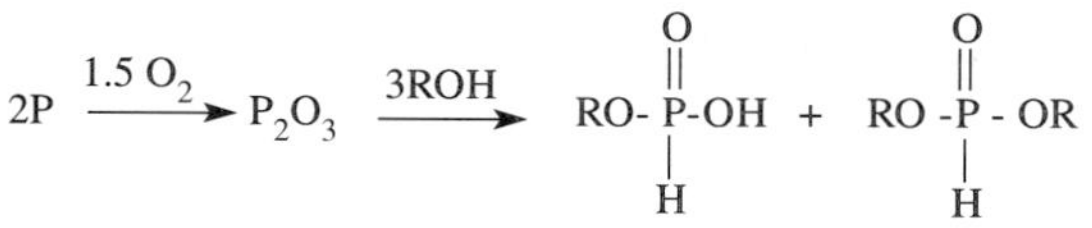

$$2P \xrightarrow{1.5\ O_2} P_2O_3 \xrightarrow{3ROH} RO-\underset{H}{\underset{|}{P}}(=O)-OH \ + \ RO-\underset{H}{\underset{|}{P}}(=O)-OR$$

1.4. OTHERS

Asymmetric dialkyl H-phosphonates are formed by heating equimolar mixtures of two types symmetric dialkyl H-phosphonates [40,41].

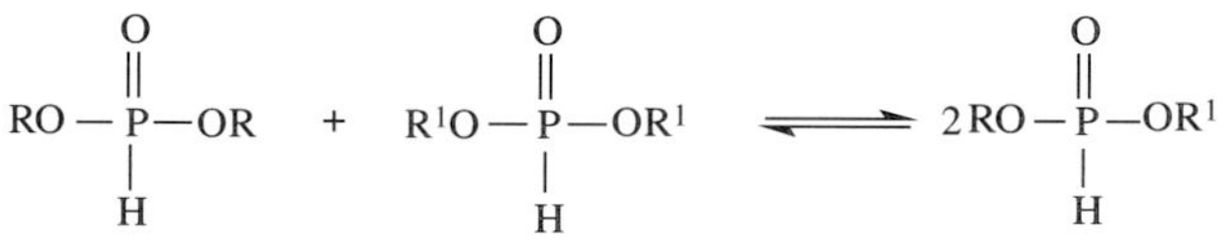

$$RO-\underset{H}{\underset{|}{P}}(=O)-OR \ + \ R^IO-\underset{H}{\underset{|}{P}}(=O)-OR^I \ \rightleftharpoons \ 2\,RO-\underset{H}{\underset{|}{P}}(=O)-OR^I$$

It has been shown [42] that mixed (asymmetric) dialkyl H-phosphonates can also be obtained in good yields under phase transfer conditions:

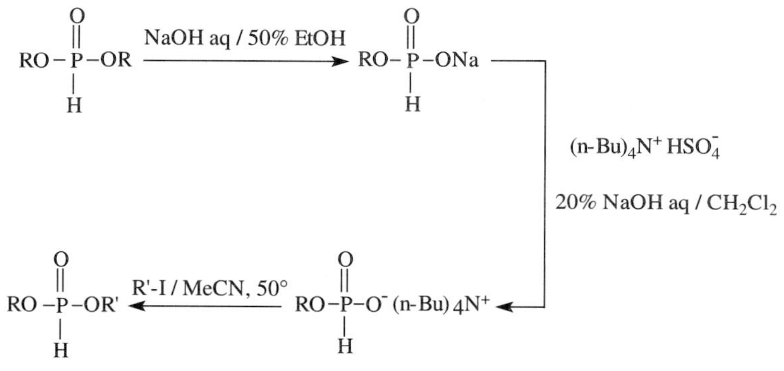

$$RO-\underset{H}{\underset{|}{P}}(=O)-OR \xrightarrow{NaOH\ aq\ /\ 50\%\ EtOH} RO-\underset{H}{\underset{|}{P}}(=O)-ONa$$

$$(n\text{-}Bu)_4N^+\ HSO_4^-$$
$$20\%\ NaOH\ aq\ /\ CH_2Cl_2$$

$$RO-\underset{H}{\underset{|}{P}}(=O)-OR' \xleftarrow{R'\text{-}I\ /\ MeCN,\ 50°} RO-\underset{H}{\underset{|}{P}}(=O)-O^-\ (n\text{-}Bu)_4N^+$$

The mixed dialkyl H-phosphonates are of particular interest in organophosphorus chemistry because of the presence of a chiral center at the phosphorus atom.

Other methods for synthesis of dialkyl H-phosphonates have also been reported [43], but the above mentioned are the most frequently used and are of major practical importance.

1.5. PHYSICAL PROPERTIES

Dialkyl H-phosphonates are liquids under normal conditions, soluble in alcohols, diethyl ether, acetone, chloroform, tetrahydrofurane, benzene, and other common

1.6 Thermal Stability

organic solvents. Table 1.1. summarizes some characteristic physical constants: boiling point, refraction, dipole moment, and density for a number of diesters of H-phosphonic acid.

1.6. THERMAL STABILITY

Under normal conditions, dialkyl phosphonates are stable compounds. At elevated temperatures (above 160 °C), they begin to decompose. Dimethyl H-phosphonate is the most unstable homologue in that respect. At a temperature of 173 °C, it pyrolyzes to monomethyl H-phosphonate and dimethyl methylphosphonate [55,56].

$$(CH_3O)_2P(O)H \xrightarrow{\Delta} CH_3OP(O)(OH)H + (CH_3O)_2P(O)CH_3$$

Moreover, the monomethyl ester of H-phosphonic acid undergoes further rearrangement, yielding the monomethyl ester of the methyl phosphonic acid.

$$2CH_3OP(O)(OH)H \xrightarrow{\Delta} H_3PO_3 + CH_3OP(O)(OH)CH_3$$

Another decomposition product observed in these studies [55,56] is tetramethyl pyrophosphonate. Its formation is probably due to the condensation of two molecules of the monomethyl ester of methylphosphonic acid.

$$2CH_3OP(O)(OH)CH_3 \xrightarrow[-H_2O]{\Delta} CH_3O-\underset{\underset{CH_3}{|}}{\overset{\overset{O}{\|}}{P}}-O-\underset{\underset{CH_3}{|}}{\overset{\overset{O}{\|}}{P}}-OCH_3$$

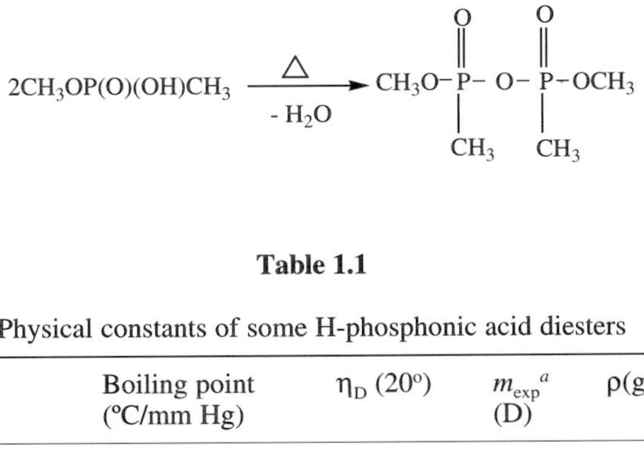

Table 1.1

Physical constants of some H-phosphonic acid diesters

Compound	Boiling point (°C/mm Hg)	η_D (20°)	m_{exp}^a (D)	ρ (g/cm^3)	Reference
$(CH_3O)_2P(O)H$	56–58/10	1.4036	2.98	1.1944	[44,45]
$(C_2H_5O)_2P(O)H$	68.7–70/10	1.4080	3.11	1.0756	[44]
$(C_3H_7O)_2P(O)H$	87.0/6	1.4183		1.0179	[44]
$(i-C_3H_7O)_2P(O)H$	80–81/16	1.4090	3.14	0.9981	[46]
$(C_4H_9O)_2P(O)H$	115/10	1.4240		0.9898	[46]
$(i-C_4H_9O)_2P(O)H$	105/9	1.4210		0.9766	[46]
$(C_2H_5O)(C_6H_{13}O)P(O)H$	104–105/3.5	1.4268		0.9883	[47]
$(C_5H_{11}O)_2P(O)H$	102–105/1–2	1.4306			[48]
$(ClCH_2CH_2O)_2P(O)H$	119–120/3.5–4	1.4708		1.4025	[49,50]
$(C_6H_5O)_2P(O)H$	100/0.008	1.5570		1.2268	[51–53]

^{a}Dipole moments taken from [54].

Furthermore, the generated phosphonic acid further decomposes to phosphoric acid and phosphine.

$$4\ H_3PO_3 \longrightarrow 3\ H_3PO_4 + PH_3$$

The oxidation of the phosphine that is generated in the above reaction takes place as a radical chain process, and at a certain PH_3/O_2, ratio the mixture may ignite.

Methyl phosphonic acid is obtained in a high degree of purity and yield approaching 100% of theory by pyrolysis of dimethyl H-phosphonate in liquid phase [57–59]. The condensation of the latter to pyromethyl phosphonic acid takes place at about 250–270 °C.

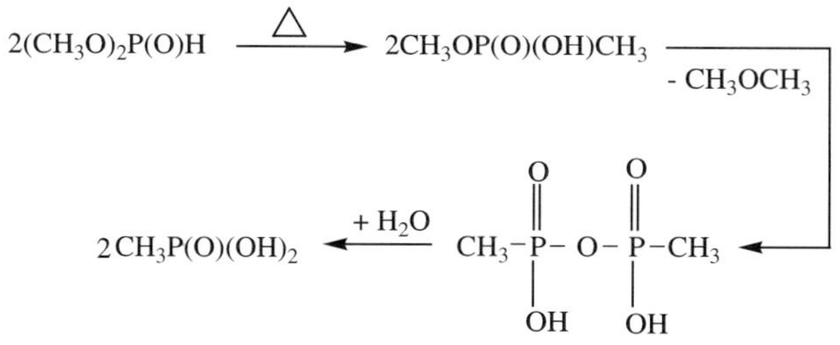

$$2(CH_3O)_2P(O)H \xrightarrow{\Delta} 2CH_3OP(O)(OH)CH_3 \xrightarrow{-CH_3OCH_3}$$

$$2\,CH_3P(O)(OH)_2 \xleftarrow{+H_2O} CH_3\!-\!\underset{OH}{\overset{O}{\overset{\|}{P}}}\!-\!O\!-\!\underset{OH}{\overset{O}{\overset{\|}{P}}}\!-\!CH_3$$

These reactions occur rapidly upon addition of dimethyl H-phosphonate to a reaction medium having a temperature of about 290–300 °C. The reaction can be carried out rapidly and in good yield when a high-boiling heavy paraffin oil (such as Nujol) is employed.

Higher dialkyl H-phosphonate homologues such as butyl and amyl usually decompose at higher temperatures.

REFERENCES

1. W. Gerrard, *J. Chem. Soc.*, **1944**, 85.
2. W. Gerrard, M. J. D. Isaacs, G. Machell, K. B. Smith, P. L. Wyvill, *J. Chem. Soc.*, **1953**, 1920.
3. H. McCombie, B. Saunders, G. Stacey, *J. Chem. Soc.*, **1945**, 380.
4. U.S. Pat. 2,409,039 (**1946**); C. A., 41, 1233b (**1947**).
5. U.S. pat. 2,494,862 (**1950**); C. A., 45, 304d (**1950**).
6. E. Gefter, M. Kabachnik, *Izv. Akad. Nauk SSSR, Ser. Khim*, **1957**, 194.
7. C. Campbell, D. Chadwick, S. Kaufman, *Ing. Eng. Chem.*, **1957**, 49, 1871.
8. U.S. Pat. 2,692,890 (**1954**); C. A., **49**, 125291i (**1955**).
9. Brit. Pat. 730,957 (**1957**); *Ref. Zh. Khim.*, **1955**, 16, 54883.
10. C. Y. Tsai, K. L. Chen, *Rohsuch Tung Pao*, **1964**, 11, 1003.
11. USSR Pat. 167848 (**1965**); *Ref. Zh. Khim.*, **1966**, 13K, 113.
12. Ts. Tsokov, St. Gaitandzhiev, *Izv. Inst. Khim. Sredstva Selsko Stopanstvo, Acad.Selsk. Nauki*, **1962**, 1, 25.
13. Czech. Pat. 109145 (**1963**); C. A., 60, 105490 (**1963**).
14. J. A. Mendelbaum, A. L. Itskova, N. N. Melnikov, *Khimia Organitsheskih Ssoedinenii Fosfora*, Nauka, Leningrad, **1967**, 228.
15. Ger. Pat. 1,078,558 (**1961**); C. A., 55, 14308d (**1961**).

References

16. USSR Pat. 127649 (**1960**), *Ref.Zh. Khim.*, **1961**, 13, 38.
17. K. Petrov, E. Nifantiev, R. Goltsova, D. Stegolev, B. Bushmin, *Zh. Obshch. Khim.*, **1962**, 32, 3723.
18. E. Nifantief, E. Milliaresy, N. Ruchkina, *Zh. Obshch. Khim.*, **1967**, 37, 1105.
19. A. Zwierzak, *Canad. J. Chem.*, **1967**, 45, 2501.
20. M. Mikolajczyk, *Chem. Commun.*, **1969**, 1221.
21. B. Costisella, H. Cross, *J. Prakt. Chem,* **1972**, 314, 532.
22. C. Bodkin, P. Simpson, *J. Chem. Soc., Perkin Trans. II*, **1973**, 5, 676.
23. D. Predvoditelev, T. Chukbar, J. Zeleneva, E. Nifantief, *Zh. Obshch .Khim.*, **1981**, 17, 1305.
24. E. Nifantief, P. Koroteev, N. Pugsashova, *Zh. Obshch. Khim.*, **1981**, 51, 1990.
25. H. J. Lucas, F. W. Mitchell, C. N. Scully, *J. Am. Chem. Soc.*, **1950**, 72, 5491.
26. E. Nifantief, J. S. Nasonovskij, A. A. Borisenko, *J. Gen. Chem. USSR (Engl. Transl.)*, **1971**, 41, 1885.
27. A. N. Saks, N. K. Lebitskii, *Zh. Rusk. Phyz. Khim. Obshtestv.*, **1903**, 25, 211.
28. (a) Brit. Pat. 699,154 (**1953**); C.A. 49, 6301 (**1955**); (b) U.S. Pat. 2,670,368 (**1954**); C. A. 49, 2483 (**1955**); (c) Brit. Pat. 1,298,156 (**1972**); (d) Ger. Pat. 1,668,031 (**1973**).
29. K. Petrov, E. Nifantief, R. Goltsova, M. Belovetsev, S. Korneev, *Zh. Obshch. Khim.*, **1962**, 32, 1277.
30. E. Nifantief, Y. Uldasheva, I. Nasonobskii, *Zh. Prikl. Khim.*, **1969**, 42, 2590.
31. U. Titarenko, L. Vasjakina, J. Uljanova, L. Sohadze, V. Tserbak, *Zh. Obshch. Khim.*, **1972**, 45, 2090.
32. U.S. Pat. 3,725,515 (**1973**); *Ref. Zh. Khim.*, **1977**, 3, 107.
33. Ger. Pat. 1,078,136 (**1960**); C. A., 56, 3423c (**1962**).
34. U.S. Pat. 3,036,109 (**1962**); C. A., 57, 13612g (**1962**).
35. E. Walsh, *J. Am. Chem. Soc.*, **1959**, 81, 3023.
36. Fr. Pat. 1,170,515 (**1959**).
37. Brit. Pat. 835,785 (**1959**); C. A., 55, 456i (**1959**).
38. Ger. Pat. DE 4,121,696 (**1991**); C. A., 118, 234241m (**1993**).
39. U.S. Pat. 2,661,364 (**1954**); C. A., 49, 1774b (**1954**).
40. Brit. Pat. 841671 (**1960**); C.A., 55, 3433g (**1961**).
41. Ger. Pat. 1059425 (**1959**); C. A., 56, 11446d (**1962**).
42. M. Kluba, A. Zwierzak, *Synthesis*, **1978**, 135.
43. H. D. Block, *Houben & Weil's "Methoden der Organischen Cemie"*, Bd. E1, G. Thieme, Stuttgart, **1982**, 322–338.
44. A. Arbuzov, P. Rakov, *Izv. Akad. Nauk SSSR, Ser. Khim.*, **1950**, 237.
45. O. Foss, *Acta Chem. Scand.*, **1947**, 1, 8.
46. B. Arbuzov, V. Vinogradova, *Izv. Akad. Nauk SSSR, Ser. Khim.*, **1947**, 617.
47. B. Arbuzov, V.Vinogradova, *Dokladi. Akad. Nauk SSSR* , **1952**, 83, 79.
48. A. de Roos, H. Toet, *Rec. Chim.*, **1958**, 77, 946.
49. E. Gefter, M. Kabachnik, *Izv. Akad. Nauk SSSR, Ser. Khim.*, **1957**, 194.
50. U.S. Pat. 2,494,862 (**1950**); C. A., 45, 304d (**1950**).
51. E. Walsh, *J. Am. Chem. Soc.*, **1959**, 81, 3023.
52. Fr. Pat. 1,170,515 (**1959**).
53. Brit. Pat. 835785 (**1959**).
54. T. Gramstad, *Acta Chem. Scand.*, **1992**, 46, 1087.
55. L. Beach, R. Drogin, J. Shewmaker, *Prod. Res.& Dev.*, **1963**, 2, 145.
56. U.S. Pat. 3,093,673 (**1963**); C. A., **1963**, 59, 11563d.
57. U.S. Pat. 2,951,863 (**1960**).
58. U.S. Pat. 4,129,588 (**1978**).
59. U.S. Pat. 3,089,889 (**1963**); C. A., **1963**, 59, 11562d.

APPENDIX

General procedure for the preparation of dialkyl H-phosphonates from phosphorus trichloride (ref: U.S. Pat. 2,582,817 (**1952**)*.*

Dialkyl H-phosphonates $(RO)_2P(O)H$ with R containing 1–3 C atoms are prepared in good yields by reacting PCl_3 and the corresponding alcohol. The reactants are brought in contact with each other, and an inert organic refrigerant in a spray chamber at substantially atmospheric pressure; the vapors of the refrigerant serve to remove RCl and HCl formed. For the production of dimethyl H-phosphonate, methyl chloride is a satisfactory refrigerant, keeping the reaction zone at about –11 to –8 °C. Usually, an ROH/PCl_3 ratio of 3:1 can be used, although higher ratios are possible. Yields of the dimethyl H-phosphonate reach 93%.

General procedure for the preparation of dialkyl H-phosphonates by direct esterification of H-phosphonic acid (ref: U.S. Pat. 3,725,515, **1973***).*

2-Ethylhexanol-1 (0.8 kg) was refluxed while stirring. A mixture of 0.5 kg of H-phosphonic acid and 1.5 kg of 2-ethylhexanol-1 (in a total excess of 45%) was added dropwise thereto within 4 h. The reaction temperature rose from 185 to 208 °C. Water (250 mL) was separated using a water separator. One gram of the cooled reaction mixture consumed 0.9 mL or 0.1 N sodium hydroxide solution. After the subsequent distillation, 93 g of octene (corresponding to a conversion rate of 2-ethylhexanol-1 used at only 5%), unaltered excess 2-ethylhexanol-1, 150 g of di-(2-ethylhexyl)ether (corresponding to a conversion rate of 2-ethylhexanol-1 used at only 7.5%), and 1.68 kg of di-(2-ethylhexyl) H-phosphonate (boiling point 130–135 °C/0.1 mm Hg) were obtained in a yield of 90% of the theory.

General procedure for the preparation of dialkyl H-phosphonates by direct esterification of H-phosphonic acid in the presence of carboxylic acid anhydride (ref: DE 4,121,696, **1991***).*

A reactor was charged with phosphonic acid (0.5 mol) and acetic anhydride (1.5 mol) was added dropwise; then methanol (1.1 mol) was added such that the temperature of the reaction mixture remained below 50 °C. Conversion of phosphonic acid into dimethyl H-phosphonate was 100%.

General procedure for the preparation of 2-hydroxy-2-oxo-1,3,2-dioxaphospholanes or 2-hydroxy-2-oxo-1,3,2-dioxaphosphorinanes and their derivatives (ref: Can. J. Chem., **1959***, 37, 1498).*

By transesterification: Diethyl H-phosphonate (27.6 g, 0.2 mol) and 1,2- or 1,3-glycol (0.2 mol) were placed in a round-bottom flask connected to a Claisen head fitted with a downward condenser and a receiver for vacuum distillation. The resulting solution was heated to 130 °C at 120–160 mm pressure under nitrogen. Ethanol evolution ceased after about 3 h of heating at the above temperature. The remaining crude product was fractionated at 2–3 mm. The cyclic hydrogen phosphonates were obtained as very viscous, colorless liquids.

By hydrolysis of cyclic chlorophosphites (ref: Can. J. Chem., **1967***, 45, 2501; J. Am. Chem. Soc.,* **1950***, 72, 5491)*

First stage: A solution of the corresponding glycol (0.2 mol) and triethylamine (40.4 g, 0.4 mol) in benzene (150 ml) was added dropwise, with efficient stirring and cooling, to a solution of phosphorus trichloride (27.4 g, 0.2 mol) in benzene (200 ml) at 5–10 °C. The mixture was kept for 1 h at room temperature and then filtered. The triethylamine hydrochloride was washed with benzene (2 × 100 ml). Evaporation of the filtrate and flash distillation of the residue in vacuo yielded the corresponding cyclic chlorophosphite.

Second stage: A mixture of water (3.6 g, 0.2 mol), triethylamine (20.2 g, 0.2 mol), and tetrahydrofuran (10 ml) was added dropwise, with efficient stirring and cooling to a solution of cyclic chlorophosphite (0.2 mol) in benzene (250 ml) at 0–5 °C. The mixture was kept for 1 h at room temperature and then filtered. The triethylamine hydrochloride was washed with benzene (2 × 100 ml). The solvent was removed under reduced pressure and the residue distilled to give the corresponding cyclic H-phosphonate.

– 2 –

Structure and Spectral Characteristics of H-Phosphonates

2.1 ELECTRONIC STRUCTURE OF THE PHOSPHORUS ATOM

In dialkyl H-phosphonates, the phosphorus atom is in the sp^3 hybrid state. To obtain this state, the phosphorus atom has to render a 3s electron to the oxygen atom of the phosphoryl group and as a result, the phosphorus becomes positively charged and the oxygen, negatively charged.

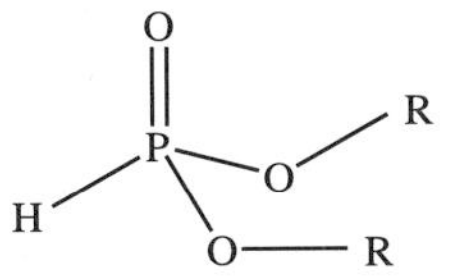

$$P(3s^2 3p^3) \xrightarrow{-e^-} \overset{+}{P}(3s^1 3p^3) \longrightarrow P(sp^3)$$

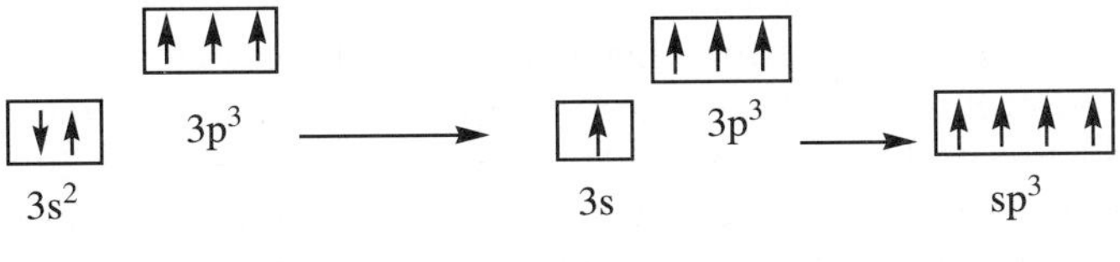

In this hybrid state, the phosphorus atom forms four σ bonds and it is positively charged. This is its first special feature. The sp^3 hybrid state defines the fourth valence of the phosphorus atom, but in dialkyl H-phosphonates it is of fifth valence. In the nontransition elements including phosphorus, s and p atomic orbitals do not form hybridized orbitals with hole exponents, that is, valence states such as sp, sp^2, and sp^3 are formal and arbitrary to a great extent. This is the second special feature of the phosphorus atom in dialkyl H-phosphonates, which can be explained by the type of bonds in the phosphoryl group (P=O).

2.2 NATURE OF CHEMICAL BONDS IN THE PHOSPHORYL GROUP

The chemical bonds in the phosphoryl group of the dialkyl H-phosphonates are one σ bond and one π bond, formed at the expense of the phosphorus free d orbitals and one of the free 2p electron pairs of the oxygen.

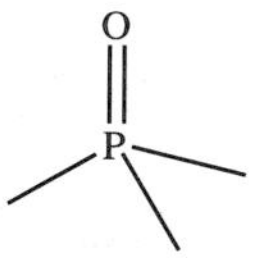

The type of the chemical bond in the phosphoryl group depends on the type of substituents at the phosphorus atom [1]. The nature of the chemical bonds in the phosphoryl group of the phosphonous oxides is studied and the following structures are proposed [2].

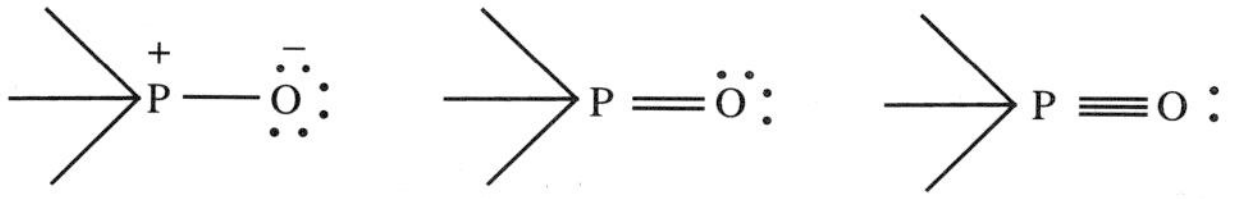

        I   II   III

Structure **I** contains one σ bond between the phosphorus and oxygen, which has three free electron pairs; the phosphorus free d-orbitals do not participate in chemical bond formation. Structure **II** is constructed from one σ bond and one π bond, where the oxygen atom has two free electron pairs. One of the free electron pairs of the oxygen atom participates in the formation of one π bond. Structure **III** includes one σ bond and two π bonds, and this oxygen atom has one free electron pair. Quantum-mechanical investigations have shown that in the formation of a bond between phosphorus and oxygen, two phosphorus d orbitals must participate in the formation of the phosphoryl group because of the symmetry of the two electron pairs of oxygen. For the phosphoryl group in the dialkyl H-phosphonates, it is assumed that the chemical bond includes structures **I** and **III** [3]. The presence of the positive charge at the phosphorus atom gives rise to a deformation of its 3d orbitals [4]. The alkoxy groups as acceptor substituents will result in additional reduction of the phosphorus electron density, so the possibility of a p_π–d_π conjunction increases. As a result, the content of structure **III** will increase. The σ bond is formed by recovery of sp^3 phosphorus hybrid orbital with a p_x-oxygen orbital. Both π bonds are formed with the participation of phosphorus 3d orbitals and p_z and p_y orbitals of the phosphoryl oxygen atom. The interatomic distance between phosphorus and oxygen atoms in the phosphoryl group is shorter (0.1248 Å) than the interatomic distance between phosphorus and oxygen atoms in the alkoxy group (P–O–C). This is an indication of a higher degree of p_π–d_π interaction between phosphorus and oxygen in the phosphoryl group. The infrared absorption band of the phosphoryl group of dialkyl H-phosphonates lies in the range 1260–1310 cm^{-1}, and the band of the phosphonous oxides is in the range 1170–1176 cm^{-1}. The absorption bands show that in the

2.3 Molecular Structure of Dimethyl H-Phosphonate

phosphonous oxides, the bond length of the P=O group is longer than that of the dialkyl H-phosphonates. The difference stems from the different degrees of p_π–d_π interaction of the oxygen and phosphorus atoms in the P=O group in the dialkyl H-phosphonates [5]. The higher electronegativity of the substituents on the phosphorus atom in dialkyl H-phosphonates compared to those in alkylphosphonous oxides results in greater reduction of the electron density of the phosphorus atom; this is a major reason for the higher degree of participation of the phosphorus $3d$ orbitals in the formation of the P=O bond.

2.3 MOLECULAR STRUCTURE OF DIMETHYL H-PHOSPHONATE

There are no published experimental geometries for the H-phosphonate diesters. Dimethyl H-phosphonate, however, was the subject of some *ab-initio* computational studies [1,6]. *Ab initio* conformational analyses that were carried out for dimethyl H-phosphonate with the 6–31+G* basis set [7] predict that there are four stable conformers of this compound [6] with the different conformers being the result of rotations about the C–O bonds in the molecule (Figure 2.1). The relative energies of these conformers are given in Table 2.1.

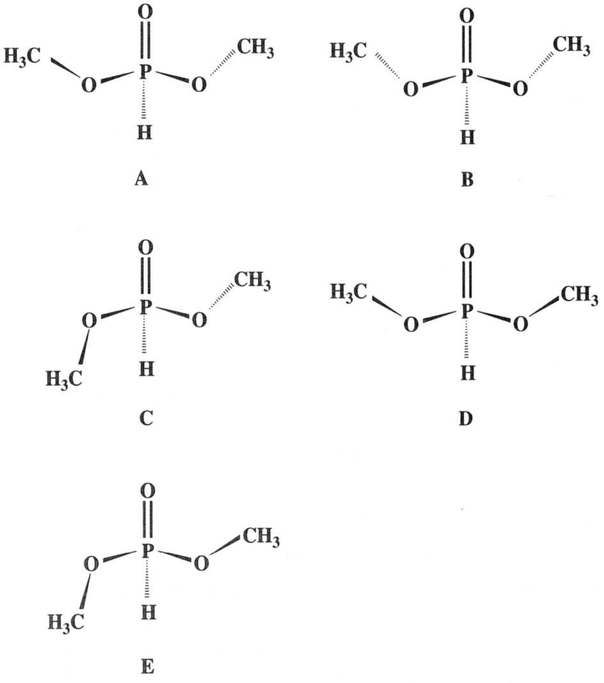

Figure 2.1 Conformers of dimethyl H-phosphonate, computed with the HF/6–31+G* basis set. **A** is the global minimum and **D** is a transition structure.

Table 2.1

Energies[a] of the dimethyl H-phosphonate conformers [6]

Conformer[b]	HF	MP2[c]	ZPE[d]	RE[e]	N[f]
A	−645.17830	−646.09095	61.5	0.0	0
B	−645.17575	−646.08742	61.2	1.9	0
C	−645.17464	−646.08756	61.4	2.0	0
D	−645.17750	−646.08980	61.3	0.5	1
E	−645.17516	−646.08824	61.4	1.6	0

[a]Energy is expressed in hartrees.
[b]Structures optimized using HF/6−31+G* basis set.
[c]MP2 energies calculated on HF/6−31+G* geometries.
[d]Zero-point vibrational energies are in kcal/mol, scaled by 0.9.
[e]Relative MP2 energies including ZPE corrections of the different conformers in kcal/mol.
[f]Number of negative eigenvalues, note that $n = 1$ for conformer D corresponding to a transition structure.

Table 2.2

Selected optimized structural parameters for dimethyl H-phosphonate, computed with the HF/6−31+G* basis [2]. $H^1P(O^1)(O^2C^1H_3)(O^3C^2H_3)$

Parameter	Configuration A	Configuration B	Configuration C	Configuration D
$P=O^1$	1.4564	1.4564	1.4509	1.4511
$P-O^2$	1.5848	1.5748	1.5848	1.5832
$P-O^3$	1.5723	1.5748	1.5728	1.5818
$P-H^1$	1.3771	1.3825	1.3846	1.3804
O^2-C^1	1.4216	1.4222	1.4226	1.4244
O^3-C^2	1.4225	1.4223	1.4178	1.4162
$< O^2PO^1$	114.06	116.52	115.79	115.80
$< O^3PO^1$	117.38	116.53	115.37	113.25
$< H^1PO^1$	115.33	113.09	114.14	116.59
$< C^1O^2P$	122.41	122.20	122.42	122.29
$< C^2O^3P$	122.45	122.20	123.81	123.03
tor. $O^3PO^2O^1$	120.63	118.64	119.94	122.27
tor. $H^1PO^1O^2$	238.37	239.32	240.14	242.69
tor. $C^1O^2PO^1$	−29.89	−31.33	−30.66	30.28
tor. $C^2O^3PO^1$	−48.57	30.79	184.51	179.97

Note: Bond lengths are in angstroms and angles in degrees.

The above data indicate that conformer A is the energy minimum structure and conformer D, because of the presence of a single imaginary frequency, corresponds to a transition structure. Selected optimized parameters for all stable conformers of dimethyl H-phosphonate are listed in Table 2.2 [6].

The charge distribution in the molecule of the most stable conformer of dimethyl H-phosphonate, obtained with Mulliken population analysis on HF/6−31+G*//HF/6−31+G* level is the following:

$$\begin{array}{c} \text{O} \;\; -0.920 \\ \| \\ \text{--C--O--P 2.170 O--C--} \\ | \\ \text{H} \\ -0.071 \end{array}$$

with C charges of 0.144.

Although the atomic charges obtained with Mulliken population analysis are basis-set dependant, these calculations clearly indicate the following trends: the electron density is the lowest at the phosphorus atom and the highest at the phosphoryl oxygen. This is due to the strongly polar character of the phosphoryl group, which is responsible to a great extent for the reactivity of this class of compounds. At the same time, the above results indicate the lack of a positive charge at the hydrogen atom of the P–H group. The last result is consistent with some early considerations [8,9] concerning the low P–H acidity of dialkyl H-phosphonates.

2.4 TAUTOMERIZATION OF DIALKYL H-PHOSPHONATES

Although it was accepted that the diesters of H-phosphonic acid existed primarily in tautomeric form **2** [10, 11], in which the phosphorus atom is trivalent, it is now well established that these compounds exist in form **1**, in which the phosphorus is tetravalent. The ^{31}P{H} NMR spectra

$$\underset{\mathbf{1}}{\overset{\text{O}}{\underset{\text{H}}{\overset{\|}{\text{P}}}}\!\!\!\!\begin{array}{c}\text{R}\\\text{O}\\\text{O}\!\!-\!\!\text{R}\end{array}} \;\; \overset{\text{B:}}{\rightleftharpoons} \;\; \underset{\mathbf{2}}{\overset{\text{RO}}{\underset{\text{RO}}{\overset{\cdot\cdot}{\text{P}}}}\!\!-\!\!\text{OH}}$$

of all diesters of H-phosphonic acid show signals in the range 0–15 ppm, which is characteristic of a tetravalent phosphorus atom with a P–H bond. There is no evidence for a trivalent phosphorus atom. There are no signals in the range 130–135 ppm, characteristic of trialkyl phosphites. The infrared spectra of all diesters of H-phosphonic acid show strong absorptions corresponding to P=O and P–H bond.

Although there is no direct experimental evidence for the existence of the three-coordinated phosphite form, the tautomerization of dialkyl H-phosphonates to dialkyl phosphites has often been invoked to explain the mechanism of other reactions of dialkyl H-phosphonates, which result in a P–H bond cleavage. Dimethyl phosphite, although never experimentally observed, has been studied recently by *ab-initio* computations. Conformational analysis for this compound [6] results in four stable conformers (Figure 2.2). Conformer

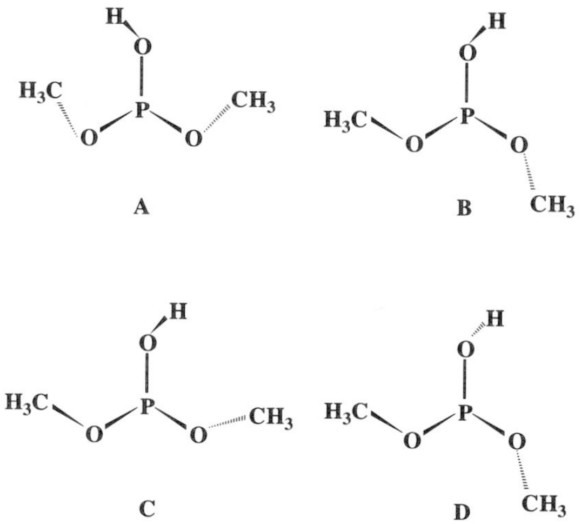

Figure 2.2 Conformers of dimethyl phosphite computed with the HF/6-31+G* basis set. **B** is the global minimum structure.

Table 2.3

Selected optimized structural parameters[a] for the most stable conformer of dimethyl phosphite. $(H_3C^1O^2)(H_3C^2O^3)PO^1H^1$

Bond lengths (Å)	Bond angles (deg.)	Dihedral angles (deg.)
P-O^1 1.6233	< O^{1}PO^2 94.85	tor. O^{3}PO^{1}O^2 103.13
P-O^2 1.6407	< O^{1}PO^3 103.18	tor. H^{1}O^{1}PO^2 −13.60
P-O^3 1.6154	< C^{1}O^{2}P 120.19	tor. C^{1}O^{2}PO^1 −171.38
O^1-H^1 0.9514	< C^{2}O^{3}P 125.44	tor. C^{2}O^{3}PO^1 −42.95
O^2-C^1 1.4105	< H^{1}O^{1}P 114.35	
O^3-C^2 1.4118		

[a]Geometry optimized with the HF/6–31+G* basis set.

B is the global minimum structure whose optimized structural parameters are summarized in Table 2.3.

The energy profile of the noncatalyzed tautomerization of dimethyl phosphonate to dimethyl phosphite [6], which includes the corresponding transition structure, is shown in Figure 2.3. The predicted tautomerization energy of 7.9 kcal/mol for dimethyl H-phosphonate is close to the value 9.8 kcal/mol reported by Guthrie [12] for the isomerization of diethyl H-phosphonate. This value of 9.8 kcal/mol was calculated from thermodynamic data for an aqueous solution at 25 °C, which may explain the difference in the two calculated values. This result suggests that both species involved in this tautomerization equilibrium have similar solvation energies, thereby maintaining the same energy difference both in the gas phase and in solution. The three-centered transition structure reveals a single imaginary frequency at 2180 cm^{-1}, which corresponds mainly to the P–H stretch. In

2.4 Tautomerization of Dialkyl H-Phosphonates

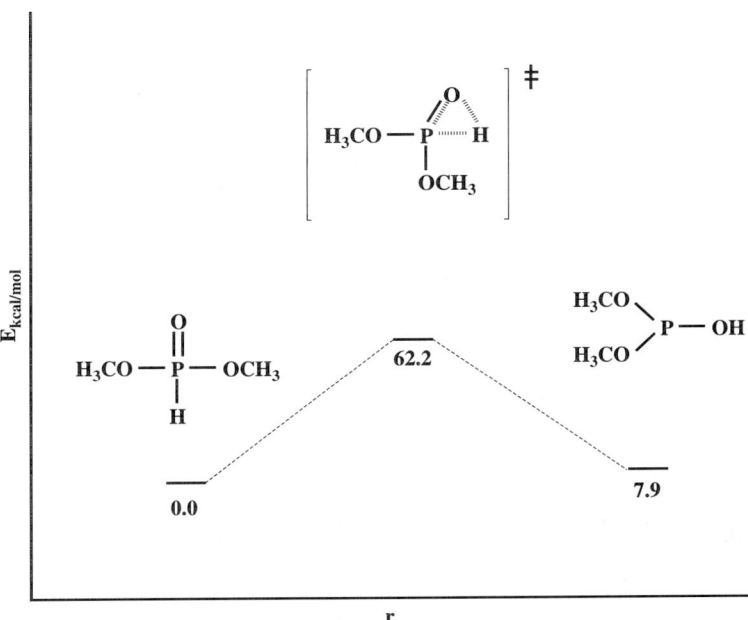

Figure 2.3 MP2 / 6–31+G* // HF / 6–31+G* energy profile of the noncatalyzed tautomerization of dimethyl H-phosphonate to dimethyl phosphite.

this transition structure, the P–H, P–O (phosphoryl), and O–H (to be formed) bond lengths have intermediate values, compared to those in the two tautomers. The computed activation energy (62.2 kcal/mol) must be significantly high in absolute value for the noncatalyzed process to be likely to take place.

These computational results provide indirect support to the early considerations that the tautomerization of dialkyl H-phosphonates is a catalytic process that is accelerated in

the presence of bases and acids [13]. One of the fundamental reasons for the assumption that in a basic medium the dialkyl H-phosphonates exist in the tricoordinate phosphite form is their chemical behavior in base-promoted reactions such as transesterification [14,15] and Atherton–Todd [16–18]. It is assumed that the first step of these reactions involves proton abstraction from dialkyl H-phosphonates:

$$(RO)_2P(O)H + :B \longrightarrow [(RO)_2PO]^- [HB]^+$$

There is currently no proof for the existence of dialkyl H-phosphonates in phosphite form in basic medium. Studies of the dialkyl H-phosphonate structure in a basic medium (e.g., amines) by ^{1}H and ^{31}P NMR and IR spectroscopy have shown that dialkyl H-phosphonates exist entirely in their four-coordinate phosphonate form [19–22]. An alkylation of the amine and not a deprotonation of dialkyl H-phosphonates occurs:

$$(RO)_2P(O)H + NR^1_3 \longrightarrow RO\text{-}\underset{H}{\underset{|}{P}}(=O)\text{-}O^- \overset{+}{N}R^1_3 \text{ (R)}$$

In the ^{31}P NMR spectrum of the methylammonium salt of the monomethyl ester of H-phosphonic acid, the phosphorus signal appears at 4.55 ppm as a doublet of quartets with $^1J(P,H) = 600$ Hz indicating phosphonate structure. Kers et al. [23] confirm that in the reaction mixture of diethyl H-phosphonate and pyridine or pyridine with 2 equiv. of tertiary base (TEA), there were no signals that could be assigned to triethyl phosphite. The results obtained show that the basic activation in these reactions involves at the first step formation of the monoalkyl H-phosphonate anion $[(RO)P(O)(H)O]^-$, which further plays the key role in interactions where the phosphorus atom acts as a nucleophile.

2.5 SPECTRAL CHARACTERISTICS

2.5.1 Infrared spectra

The IR spectra of diesters of H-phosphonic acid contain characteristic bands attributed to the P=O, P–H, and P–O–C vibrations that are used in the structural characterization of this type of compound. The positions of the P–H and P=O absorption bands depend largely on the type of the substituents R. More detailed studies of the vibrational spectra and force constants for some dialkyl H-phosphonate representatives have been reported recently [24].

The vibrational frequencies that correspond to these absorption bands for H-phosphonic acid diesters as well as the characteristic absorption bands of some of their representatives are summarized in Table A1.

2.5.2 ^{1}H NMR spectra

NMR spectroscopy is one of the most powerful tools for structural characterization of dialkyl H-phosphonates, since the latter are not suitable for X-ray analysis. The most

2.5 Spectral Characteristics

characteristic feature in the ^{1}H NMR spectra of H-phosphonate diesters is the ^1J(PH) coupling constant. Depending on the type of J(PH) coupling, H-phosphonate diesters have two different types of hydrogen atoms in their molecules: the first one is connected directly to the phosphorus atom giving rise to the ^1J(PH) coupling. The other type is ^3J(PH) spin-coupled with the phosphorus atom through the oxygen heteroatoms. In other words, the ^1J(PH) and ^3J(PH) coupling constants are the ones for which significant spin–spin interaction between the phosphorus nucleus and protons through up to three σ bonds has been observed. The coupling constant ^1J(PH) in dimethyl H-phosphonate varies in a wide range—more than 100 Hz depending on the type of the solvent and on the temperature of the sample [25,26]. In polar organic solvents such as pyridine, acetone, dimethylformamide, and acetonitrile as well as in nonpolar ones such as cyclohexane and tetrachloroethane, a decrease in the ^1J(PH) coupling constant with 2–3 Hz is observed compared to pure dimethyl H-phosphonate. The maximum value of ^1J(PH) = 734.3 Hz has been reported when trifluoroacetic acid is used as a solvent. An increase in ^1J(PH) has also been observed in aqueous solution (723.8 Hz). A temperature change from −80 to 120 °C of the solution of dimethyl H-phosphonate in toluene results in a decrease of the ^1J(PH) coupling to 8.7 Hz. The value of ^1J(PH) in diphenyl H-phosphonate is considerably larger (734 Hz) than in diethyl H-phosphonate (691 Hz), in spite of the insignificant difference in the induction effects of both phenoxy and ethoxy groups [27].

The following ranges for the coupling constants have been observed for H-phosphonate diesters: ^1J(PH) from 670 to 740 Hz; ^3J(PH) from 6.5 to 12.0 Hz. The ^{1}H NMR data for some representative dialkyl H-phosphonates are given in Table A2.

2.5.3 ^{31}P NMR spectra

^{31}P NMR spectroscopy is the most precise method for determining the structure of the phosphorus-containing compounds. Chemical shifts for ^{31}P depend on: (i) imbalance of σ–bonds, caused by the difference in electronegativity of the atoms and by the effect of the free electron pairs, (ii) degree of occupation of phosphorus d orbitals, and (iii) deviation from geometric symmetry. In addition, the following factors influence the phosphorus chemical shifts: (i) degree of ionization of the bonds [28], (ii) formation of complexes with cations [29], (iii) valence angle O–P–O [30], (iv) torsion angle [31], (v) temperature [32], and (vi) solvent [33].

The phosphorus atom of the diesters of H-phosphonates is bonded with the hydrogen of the P–H group. The signal of the phosphorus atom is split into two due to the interaction between the phosphorus nucleus and proton of the P–H group. These two signals are further split by the interaction of the phosphorus nucleus and protons of the α–carbon atoms. As a result, the signal of the phosphorus nucleus in dimethyl H-phosphonate appears as a doublet of septets and in diethyl H-phosphonate as a doublet of quintets. The typical range for the ^{31}P chemical shifts in the diesters of H-phosphonic acid is 0–15 ppm referenced to 85% H_3PO_4 [34–41]. As in the case of their proton spectra, the ^{31}P NMR spectra of these compounds reveal two types of P–H coupling constants corresponding to a direct ^1J(PH) coupling and to a ^3J(PH) coupling through the oxygen heteroatom (see A3).

2.5.4 ^{13}C NMR spectra

A characteristic feature of the carbon NMR spectra of dialkyl H-phosphonates is the J(PC) coupling through the oxygen heteroatom. These coupling constants have values in the range 5–8 Hz and can be observed through up to three bonds: ^2J(PC) and ^3J(PC) (see A4).

2.5.5 ^{17}O NMR spectra

The ^{17}O NMR spectra for a number of phosphonates have been recently reported [42]. A characteristic feature in these spectra is the relatively sharp signal for the phosphoryl oxygen (P=O), whereas the signals of the bridging oxygens (–O–P–O–) are broader and rarely show the doublet due to the ^1J(PO) coupling. The following values have been reported for diethyl H-phosphonate: δ 106.5 (d, ^1J(PO) = 170 Hz, P=O), δ 70.2 (s, –O–P–O). The signal line width at half height for the phosphonates reported ranges from 90 to 150 Hz (less than 50 Hz for diethyl H-phosphonate) and from 200 to 250 Hz for the phosphoryl and the bridging oxygens, respectively.

REFERENCES

1. J. Van Wazer, C. Ewig, *J. Am. Chem. Soc.*, **1986**, 108, 4354.
2. M. Schmidt, S. Yabushita, M. Gordon, *J. Am. Chem. Soc.*, **1984**, 108, 382.
3. H. Jaffe, *J. Phys. Chem.*, **1954**, 58, 185.
4. R. Gillespie, *J. Chem. Soc.*, **1952**, 1002.
5. E. Wegner, *J. Am. Chem. Soc.*, **1963**, 85, 161.
6. E. M. Georgiev, J. Kaneti, K. Troev, D. Max Roundhill, *J. Am. Chem. Soc.*, **1993**, 115,10964.
7. W. J. Hehre, L. Radom, P. von Rague Schleyer, J. Pople, *Ab initio Molecular Orbital Theory*, Wiley, New York (**1986**).
8. J. David, H. Hallam, *J. Chem. Soc.*, **1966**, 1103.
9. R. Wolf, D. Houll, F. Mathis, *Spectrochim. Acta*, **1967**, 23A, 1641.
10. A. Arbuzov, *Zh. Russ. Phys. Khim. Soc.*, **1906**, 38, 691.
11. F. Milobendzki, *Chem. Ber.*, **1912**, 45, 298.
12. J. P. Guthrie, *Can. J. Chem.*, **1979**, 57, 236.
13. P. R. Hammond, *J. Chem. Soc.*, **1962**, 1365.
14. M. Kabachnik, Y. Polikarpov, *Dokl. AN SSSR*, **1957**, 115, 512.
15. M. Imaev, V. Masslennikov, V. Gorina, O. Krasheninikova, *Zh. Obsh. Khim.*, **1965**, 35, 75.
16. L. Risel, E. Herrmann, J. Steinbach, *Phosphorus& Sulfur*, **1983**, 18, 253.
17. A. Kong, R. Engel, *Bull. Chem. Soc. Jpn.*, **1985**, 58, 3671.
18. E. Nifantiev, V. Blagoweshtenskij, A. Sokurenko, S. Skljaskil, *Zh. Obsh. Khim.*, **1973**, 44, 108.
19. K. Troev, G. Borissov, *Phosphorus& Sulfur*, **1987**, 29, 129.
20. K. Troev, D. M. Roundhill, *Phosphorus& Sulfur*, **1988**, 37, 243.
21. K. Troev, D. M. Roundhill, *Phosphorus& Sulfur*, **1988**, 37, 247.
22. K. Troev, E. M. G. Kirilov, D. M. Roundhill, *Bull. Chem. Soc. Jpn.*, **1990,** 63, 1284.
23. A. Kers, I. Kers, J. Stawinski, M. Sobkowski, A. Kraszewski, *Tetrahedron*, **1996**, 52, 9931.
24. S. A. Katcyba, N. I. Monakhova, L. Kh. Ashrafulina, R. P. Shagidulin, *J. Mol. Str.*, **1992**, 269, 1–21.

25. S. Smith, R. Cox, *J. Chem. Phys.*, **1966**, 45, 2848.
26. S. Smith, A. Ihring, *J. Chem. Phys.*, **1967**, 46, 1181.
27. E. Fluk, H. Binder, *Z. Naturforsch.*, **1967**, B22, 805.
28. K. Moedritzer, *Inorg. Chem.*, **1967**, 6, 936.
29. A. Costello, T. Glonek, J. Van Wazer, *Inorg. Chem.*, **1976**, 15, 972.
30. D. Gorenstein, *J. Am. Chem. Soc.*, **1975**, 97, 898.
31. D. Gorenstein, D. Kar, *Biochem. Biophys. Res. Commun.*, **1975**, 65, 1073.
32. D. Gorenstein, J. Finday, R. Motti, B. Luxon, D. Kar, *Biochemistry*, **1976**, 15, 3796.
33. D. Lenar, D. Kearus, *J. Am. Chem. Soc.*, **1980**, 102, 7611.
34. C. Dungan, V. Mark, J. Van Wazer. In: *Topics in Phosphorus Chemistry* (eds. M. Grayson & E. Griffih), New York, London, Sidney, **1967,** Vol. 5.
35. K. Moedritzer, *J. Inorg. Nucl. Chem.*, **1961**, 22, 19.
36. J. Van Wazer, C. Callis., J. Shoolery, R. Jones, *J. Am. Chem. Soc.*, **1956**, 78, 5715.
37. C. Callis, J. van Wazer, J. Shoolery, W. Anderson, *J. Am. Chem. Soc.*, **1959**, 79, 2719.
38. H. Finegold, *Ann. New York Acad. Sci.*, **1958**, 70, 875.
39. R. Jones, A. Katritzky, *Angew. Chem.*, **1962**, 74, 60.
40. N. Muller, P. Lanterbur, J. Goldenson, *J. Am. Chem. Soc.*, **1956**, 78, 3557.
41. V. Mark, J. Van Wazer, *J. Org. Chem.*, **1964**, 24, 1006.
42. H. Dahn, V. V. Taan, M. N. U. Truong, *Magn. Res. Chem.,* **1992**, 30, 1089.

APPENDIX

Table A1

Infrared spectroscopic data for some H-phosphonic acid diesters $(RO)_2P(O)H$

R	Group	Stretching vibration ν (cm^{-1})	R	Group	Stretching vibration ν (cm^{-1})
Alkyl	P–H	2350–2440	CH_3	P–O–CH_3	1190
Alkyl	P=O	1260–1310	C_2H_5	P–O–C_2H_5	1160
Alkyl	P–O–C	1030–1090	C_6H_5	P–O–C_6H_5	1190–1240

Table A2

^{1}H NMR data for some H-phosphonate diesters $(R^1O)(R^2O)P(O)H$ in $CDCl_3$

R^1	R^2	δ (ppm); J (Hz)
CH_3	CH_3	3.51 (d, 6H, ^{3}J(H,H) = 11.95 Hz, CH_3); 6.49 (d, 1H, ^{1}J(P,H) = 697.2 0 Hz, P–H);
CH_3CH_2	CH_3CH_2	0.94 (t, 6H, ^{3}J(H,H) =7.05Hz, CH_3); 4.01–4.08 (m, 4H, POCH$_2$); 6.38 (d, 1H, ^{1}J(P,H) = 691.1 Hz, P–H)
a b c d $CH_3CH_2CH_2CH_2$	a b c d $CH_3CH_2CH_2CH_2$	0.95 (t, 6H, ^{3}J(H,H) = 7.32 Hz, a–CH_3); 1.42 (q, 4H, ^{3}J(H,H) = 7.43 Hz, b–CH_2); 1.67 (q, 4H, ^{3}J(H,H) = 6.98 Hz, c–CH_2); 4.08–4.11 (m, 4H, POCH$_2$); 6.80 (d, 1H, ^{1}J(P,H) = 695 Hz, P–H)

Table A3

^{31}P NMR data of some H-phosphonic acid diesters

Compound	δ (ppm)	^{n}J(P,H) (Hz)
(CH$_3$O)$_2$P(O)H	11.61	^{1}J(P,H) = 697.22 ^{3}J(P,H) = 11.92
(C$_2$H$_5$O)$_2$P(O)H	9.8	^{1}J(P,H) = 686 ^{3}J(P,H) = 9.8
(C$_3$H$_7$O)$_2$P(O)H	7.41	^{1}J(P,H) = 685
[(CH$_3$)$_2$CHO]$_2$P(O)H	3.38	^{1}J(P,H) = 670
(C$_4$H$_9$O)$_2$P(O)H	7.63	^{1}J(P,H) = 685.4 ^{3}J(P,H) = 9.2
[(CH$_3$)$_3$CO]$_2$P(O)H	3.21	^{1}J(P,H) = 682
(C$_6$H$_5$O)$_2$P(O)H	0.95	^{1}J(P,H) = 741
(p–CH$_3$–C$_6$H$_4$O)$_2$P(O)H	1.3	^{1}J(P,H) = 740

Table A4

^{13}C{H} NMR data for some phosphonic acid diesters

Compound	Assignment δ (ppm); J (Hz)
(CH$_3$O)$_2$P(O)H	51.07 [d, ^{2}J(P,C) = 5.8 Hz, CH$_3$]
(CH$_3$CH$_2$O)$_2$P(O)H	16.67 [d, ^{3}J(P,C) = 6.0, CH$_3$] 61.72 [d, ^{2}J(P,C) = 6.0, CH$_2$]
a b c d (CH$_3$CH$_2$CH$_2$CH$_2$O)$_2$P(O)H	12.95 [s, a–CH$_3$] 18.26 [s, b–CH$_2$] 32.0 [d, ^{3}J(P,C) = 5.9, c–CH$_2$] 65.06 [d, ^{2}J(P,C) = 5.8, d–CH$_2$]

– 3 –

Reactivity of H-Phosphonates

The diesters of H-phosphonic acid occupy a major position in organophosphorus chemistry since they are frequently intermediates in the synthesis of a variety of bioactive products including aminophosphosphonates, aminophosphonic acids, P–C phosphonates, hydroxyalkyl phosphonates, phosphates, amidophosphates, nucleoside H-phosphonates, poly(alkylene H-phosphonate)s and poly(alkylene phosphate)s, phosphorus-containing polyesters, polyurethanes, etc. The strongly polar character of the phosphoryl group of the H-phosphonates is responsible to a great extent for the reactivity of this class of compounds. The versatility of these compounds is determined by the presence of two types of reaction centers in their molecule: the phosphorus atom and the α-carbon atom of the alkoxy groups, and of three functional groups—alkoxy, P–H, and P=O. This fact uniquely defines the chemical reactivity of dialkyl phosphonates and their usefulness in various synthetic applications.

3.1 ACIDITY OF H-PHOSPHONATE DIESTERS

H-phosphonate diesters are tautomeric systems in which the phosphite–phosphonate equilibrium is practically entirely shifted to the four-coordinated phosphonate form [1,2].

$$\text{RO}-\underset{\underset{H}{|}}{\overset{\overset{O}{\|}}{P}}-\text{OR} \rightleftharpoons \underset{\text{RO}}{\overset{\text{RO}}{\diagdown}}\ddot{P}-\text{OH}$$

This implies that these compounds have P–H type acidity and are therefore considerably less acidic than the corresponding P–OH type acids. It was established by means of ^{31}P NMR spectroscopy that in dimethoxy ethane, the acidity of dibutyl hydrogen phosphonate is close to that of ethanol, and that these P–H acids are stronger than the corresponding N–H acids [3]. The pK_a values of a series of H-phosphonate diesters have been calculated by the so-called premetallization method [4], according to the following equation:

$$\text{RO}-\underset{\underset{H}{|}}{\overset{\overset{O}{\|}}{P}}-\text{OR} + \text{A}^-\text{K}^+ \longrightarrow \text{RO}-\underset{\underset{H}{|}}{\overset{\overset{O}{\|}}{P}}-\text{O}^-\text{K}^+ + \text{AR}$$

where *AR is an indicator type CH acid.*

Table 3.1

pK_a values of phosphonic acid, phosphorus acid, and some of their esters [4[a],5[b],6[c]]

Acid	pK_{a1}	pK_{a2}	pK_{a3}
$(HO)_2P(O)H$	1.42 [5]	6.7 [5]	
	1.5 [6]	6.79 [6]	
$(C_2H_5O)(HO)P(O)H$	0.81 [5]		
	0.90 [6]		
$(CH_3O)_2P(O)H$	19.9 [4]		
$(C_2H_5O)_2P(O)H$	13.0 [6]		
	20.8 [5]		
$(C_4H_9O)_2P(O)H$	20.8 [4]		
$(C_5H_{11}O)_2P(O)H$	21.0		
$(C_6H_{13}O)_2P(O)H$	20.9		
$(HO)_3P$	7.4 [6]	11.9 [6]	16.4 [6]
$(C_2H_5O)(HO)_2P$	6.7 [6]	11.3 [6]	
$(C_2H_5O)_2(HO)P$	6.1 [6]		

[a]pK_a values determined by the premetallization method; [b]pK_a values determined by potentiometric titration; [c]pK_a values calculated from thermodynamic data for aqueous solution at 25 °C;

These results are summarized in Table 3.1, together with some other calculated [5] and experimental [6] pK_a values for phosphonic acid and its esters, as well as for phosphorus acid and its esters. The data in Table 3.1 provide a direct comparison between the strengths of the P–H and P–OH types of acids and indicate the significant difference in their acidities. The substitution of one OH group in the molecules of both phosphonic and phosphorus acids with an ethoxy group leads to an increase in the acidity of the remaining OH groups.

The pK_a value of 13.0 for diethyl phosphonate in Table 3.1 has been calculated from thermodynamic data based on the following scheme [6]:

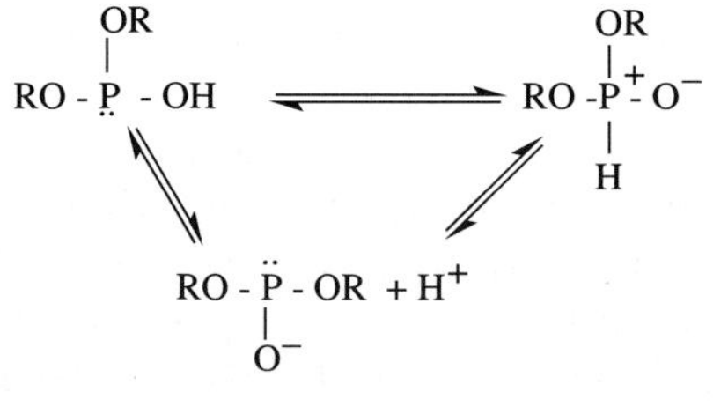

$$\text{RO-P(OR)-OH} \rightleftharpoons \text{RO-P}^+(\text{OR})(\text{O}^-)\text{H}$$
$$\searrow \qquad \swarrow$$
$$\text{RO-}\ddot{\text{P}}(\text{OR})(\text{O}^-) + \text{H}^+$$

This calculated pK_a value for diethyl phosphonate is with 7.8 pK_a units lower than that obtained by the premetallization method [4]. Both results, however, indicate that a diethyl phosphonate is a very weak acid, so that the equilibrium outlined below is almost completely shifted toward the neutral phosphonate form:

$$(C_2H_5O)_2P(O)H \longleftarrow [(C_2H_5O)_2PO^-] + H^+$$

3.2 DISPROPORTIONATION

Disproportionation is characteristic for the asymmetric dialkyl H-phosphonates [7].

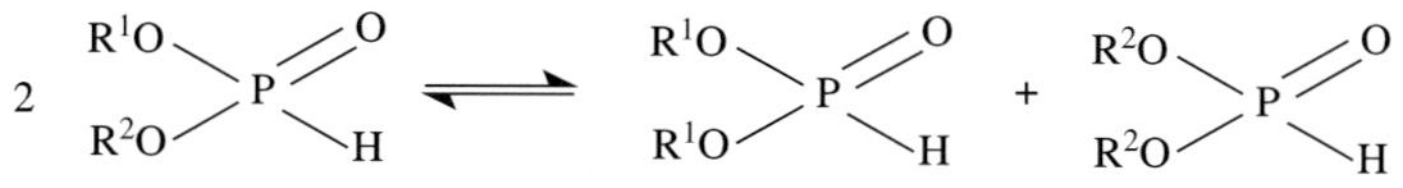

$$2 \; (R^1O)(R^2O)P(=O)H \; \rightleftharpoons \; (R^1O)_2P(=O)H \; + \; (R^2O)_2P(=O)H$$

The process has been shown to be reversible [8,9]. The equilibrium is established readily at room temperature. This disproportionation reaction can be used as a method for the synthesis of optically active asymmetric or symmetric dialkyl phosphonates.

3.3 REDUCTION

Dialkyl H-phosphonates are reduced by organomagnesium compounds to dialkyl phosphinous acids [10–13].

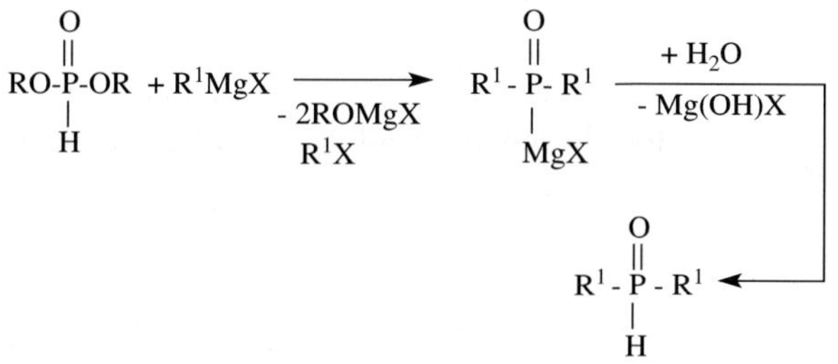

$$\text{RO-P(=O)(H)-OR} + R^1MgX \xrightarrow[R^1X]{-2ROMgX} R^1\text{-P(=O)(MgX)-}R^1 \xrightarrow[-Mg(OH)X]{+H_2O} R^1\text{-P(=O)(H)-}R^1$$

3.4 HYDROLYSIS

Dialkyl H-phosphonates are hydrolytically unstable compounds. The hydrolysis of H-phosphonate diesters and phosphate esters is believed to proceed via pentacoordinated intermediates and transition structures that are formed by nucleophilic attack of the tetracoordinated phosphorus atom [14–16]. These intermediates undergo further pseudorotation [17] and elimination of alcohol to form the final products of the nucleophilic substitution.

It has been established that the hydrolysis of dialkyl H-phosphonates is general-base and general-acid catalyzed [18]. The hydrolysis of phosphonic acid diesters is facilitated significantly under basic conditions. The mechanism of the base-catalyzed hydrolysis of dimethyl H-phosphonate has been proposed to take place according to the following scheme [19]:

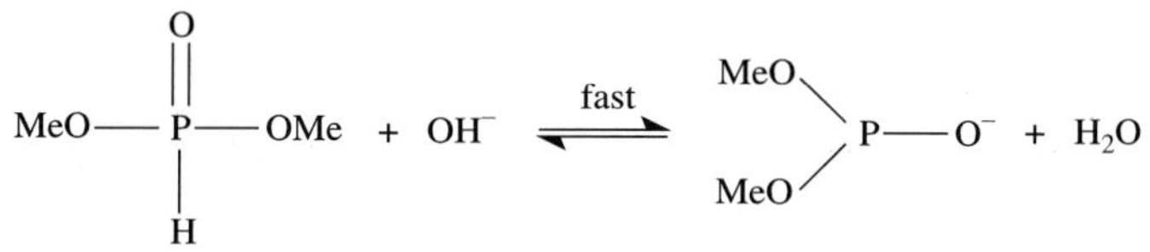

$$\text{MeO-P(=O)(H)-OMe} + OH^- \; \underset{}{\overset{fast}{\rightleftharpoons}} \; (MeO)_2P\text{-}O^- + H_2O$$

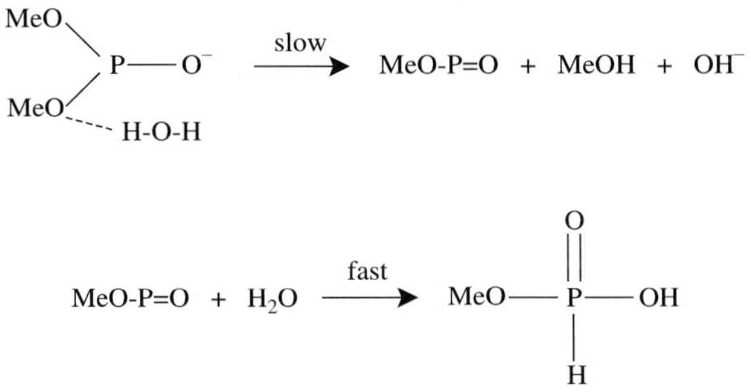

$$\text{MeO}\diagdown\!\!\!\!\!\!\diagup\text{P—O}^- \xrightarrow{\text{slow}} \text{MeO-P=O} + \text{MeOH} + \text{OH}^-$$
$$\text{MeO}\diagup \quad \cdots \text{H-O-H}$$

$$\text{MeO-P=O} + \text{H}_2\text{O} \xrightarrow{\text{fast}} \text{MeO—P(=O)(H)—OH}$$

The second methoxy group hydrolyzes in the same way to yield H-phosphonic acid as a final product. The rate of the alkaline hydrolysis of dialkyl H-phosphonate depends on the substituent type at the phosphorus atom [18] (Table 3.2).

In this study, the comparatively higher hydrolytic stability of diisopropyl H-phosphonate is attributed to the higher steric constant of the isopropyl group in comparison to *n*-propyl and *n*-butyl groups. These data imply the following sequel of increasing hydrolytic stability with respect to the type of alkoxy group (18):

$$CH_3O < C_2H_5O < C_3H_7O < C_4H_9O < i\text{-}C_3H_7O$$

Table 3.3 shows the values for some Taft constants of the substituents σ^*, σ^p, σ^p_I, and σ^p_R for the alkaline hydrolysis of dialkyl phosphonates.

By contrast, the mechanism of neutral and acid hydrolysis of phosphonate and phosphate esters has been suggested by Westheimer to involve initial attack of a water molecule on the phosphorus tetrahedron to form a neutral pentacoordinated intermediate with a zwitterionic structure [14].

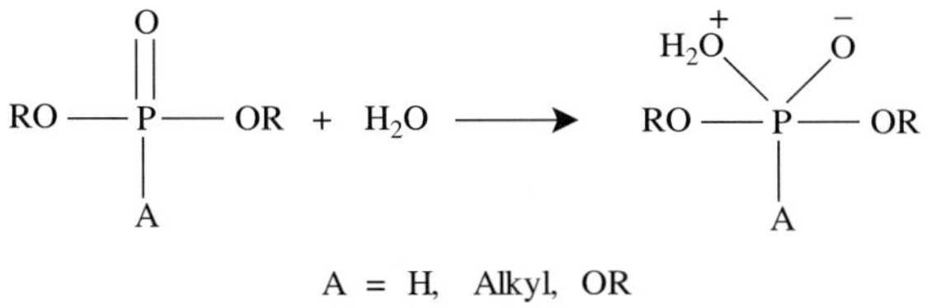

$$\text{RO—P(=O)(A)—OR} + \text{H}_2\text{O} \longrightarrow \text{RO—P(}^+\text{OH}_2)(\text{O}^-)(A)\text{—OR}$$

A = H, Alkyl, OR

Table 3.2

The rate of the alkaline hydrolysis of H-phosphonate diesters

H-Phosphonate diesters	Alkaline hydrolysis, mol/sec
$(CH_3O)_2P(O)H$	16×10^3
$(C_2H_5O)_2P(O)H$	8×10^3
$(C_3H_7O)_2P(O)H$	7.6×10^3
$(i\text{-}C_3H_7O)_2P(O)H$	1.9×10^3
$(C_4H_9O)_2P(O)H$	7.0×10^3

3.4 Hydrolysis

Table 3.3

Taft constants σ^*, σ^p, σ^p_I, and σ^p_R of the substituent R for the alkaline hydrolysis of dialkyl H-phosphonates with general formula $(RO)_2P(O)H$ [20]

R	σ^*	σ^p	σ^p_I	σ^p_R
H	1.31	−0.39	−1.65	−2.04
CH_3	1.73	−0.12	2.48	−2.61
C_2H_5	1.64	−0.21	2.3	−2.52
C_3H_7	1.57	−0.32	2.16	−2.48
$i\text{-}C_3H_7$	1.51	−0.29	2.04	−2.34
C_4H_9	1.55	−0.41	2.13	−2.54
C_6H_5	1.91	−0.06	2.83	−2.89

This intermediate undergoes further protonation and/or pseudorotation [17] to result finally in the loss of an alcohol group from the apical position. On the other hand, thermodynamic calculations by Guthrie of pK_a values for phosphate esters in aqueous solution predict that water addition to these compounds more likely occurs via a concerted cyclic proton-transfer mechanism to neutral adducts [21].

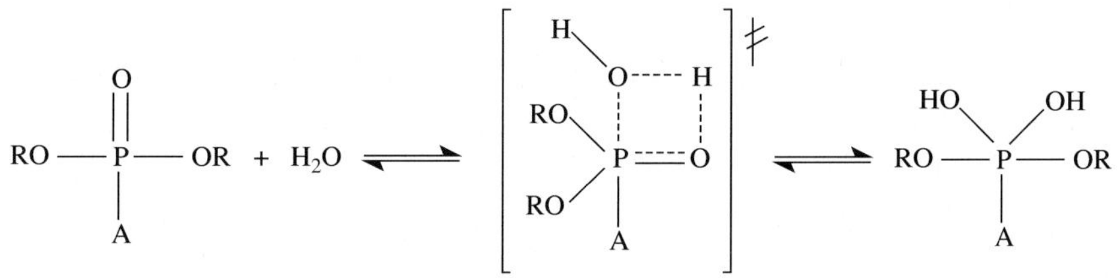

A = H, Alkyl, OR

A comparative *ab initio* study [22] on these two alternative reaction pathways concludes that a zwitterionic structure of the type suggested by Westheimer is not a stationary point on the explored potential energy surfaces for the systems $H_2O + (HO)_2P(O)H$ and $H_2O + (HO)_3P(O)$. Two types of critical points were found for these model systems. The first type corresponds to transition structures for the concerted addition of water to these phosphoryl compounds (Figure 3.1).

The second type of critical points corresponds to an energy minimum representing cyclic hydrogen-bonded complexes:

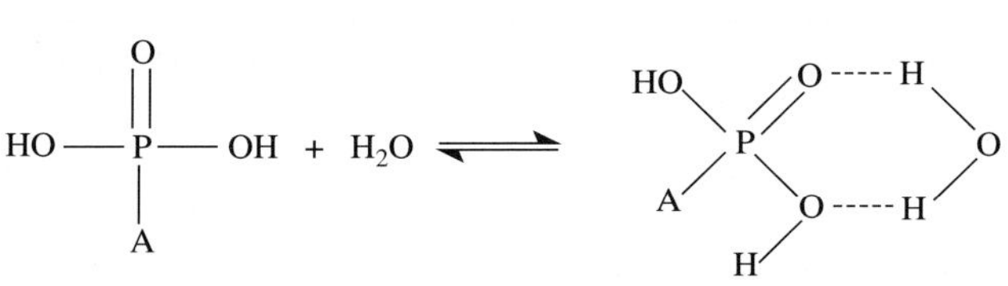

A = H, OH

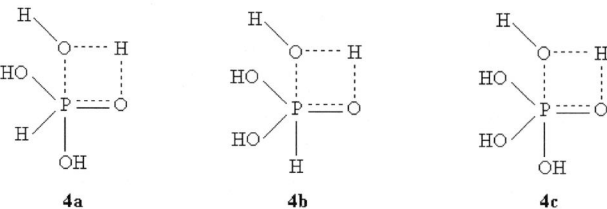

Figure 3.1 Four-centered transition structures of concerted water addition to phosphonic (**4a** and **4b**) and phosphoric acids (**4c**) have been calculated with the HF/6–31G* basis set.

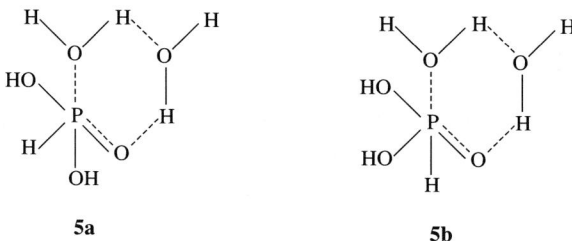

Figure 3.2 Six-centered transition structures of concerted water addition to phosphonic acid **5a** and **5b** have also been calculated with the HF/6–311G** basis set.

The energy barriers calculated at the MP4(SDTQ)/6–31G*//HF/6–31G* + ZPE level for the system H_2O + $(HO)_2P(O)H$ are 19.0 and 29.4 kcal/mol through transition structures **4a** and **4b**, respectively. An energy barrier of 26.7 kcal/mol is obtained for the system H_2O + $(HO)_3P(O)$. The transition structures **4a** and **4b** closely resemble those found for the non-catalyzed water addition to carbonyl compounds [23] referred to by Jencks as an 'enforced concerted addition' pathway [24]. This mechanism also includes four-centered planar transition structures, implying concerted electron and proton transfer from the nucleophile. In analogy with the catalyzed carbonyl addition [23b], catalysis for the phosphoryl addition by an ancillary water molecule also results in significant decrease of the calculated energy barriers. Lowering of the energy barrier on going from four-centered to six-centered transition structures has also been reported for water addition to the metaphosphate anion [25]. The six-centered analogues of the transition structures **5a** and **5b** are shown in Figure 3.2 [22]. The barriers for water addition to phosphonic acid computed on MP2/6–311G**// HF/6–311G** + ZPE level are 42.6 and 46.8 kcal/mol through **5a** and **5b**, respectively [25]. Surprisingly, these barriers are for **5a**, 23.8 kcal/mol and for **5b**, 25.2 kcal/mol higher in energy than the activation energies calculated with the same MP2/6–311G**// HF/6–311G** + ZPE basis through the analogous four-centered transition structures of the type shown in Figure 3.1 (**4a** and **4b**). This unexpected result may be due to the fact that the calculated species are neutral and not charged as in [25]. This allows for the adoption of a tetragonal pyramidal-type arrangement of ligands around the phosphorus atom in **5a** and **5b**.

The theoretical study [22] lends no support to the existence of zwitterionic species resulting from water addition to phosphonic and phosphoric acids in the gas phase. Instead, these reactions are calculated to proceed through enforcedly concerted transition

3.6 Substitution Reactions at the Phosphorus Atom

structures as suggested by Guthrie [21]. This result is also in good agreement with the experimental finding that the hydrolysis of ethylene phosphate is accompanied by rapid oxygen exchange in the unreacted phosphate [26]. Although the possibility of a stepwise mechanism involving the intermediacy of a solvated zwitterionic intermediate cannot be completely ruled out in aqueous solution, these gas-phase calculations [22] together with the experimental data-based results of Guthrie [21] give a strong indication that the concerted addition of water to phosphoryl compounds is probably the major reaction channel under the conditions of their acid or neutral hydrolysis.

3.5 ACIDOLYSIS

The acidolysis of dimethyl H-phosphonate to phosphonic acid was studied recently by ^{1}H- NMR and ^{31}P{H} NMR spectroscopy [27]. It was established that this is a two-stage process with the intermediate formation of monomethyl H-phosphonate.

$$CH_3O-\underset{H}{\overset{\overset{O}{\|}}{P}}-OCH_3 \xrightarrow[-CH_3COOCH_3]{+CH_3COOH} CH_3O-\underset{H}{\overset{\overset{O}{\|}}{P}}-OH \xrightarrow[-CH_3COOCH_3]{+CH_3COOH} HO-\underset{H}{\overset{\overset{O}{\|}}{P}}-OH$$

This two-stage equilibrium can be successfully shifted toward the formation of H-phosphonic acid by removing the formed methyl acetate by distillation.

3.6 SUBSTITUTION REACTIONS AT THE PHOSPHORUS ATOM

The strongly polar character of the phosphoryl group of the H-phosphonates is responsible to a great extent for the reactivity of this class of organophosphorus compounds.

3.6.1 Transesterification reaction

The charge distribution in the molecule of dimethyl H-phosphonate obtained with Milliken population analysis clearly indicates that the electron density at the phosphorus atom is the lowest, that is, the phosphorus atom is an electrophilic center. H-phosphonate diesters with nucleophiles undergo several characteristic reactions and the most typical one is the transesterification reaction, which presents an interaction between H-phosphonate diesters and hydroxyl-containing compounds.

$$RO-\underset{H}{\overset{\overset{O}{\|}}{P}}-OR + R^1OH \underset{-ROH}{\rightleftarrows} RO-\underset{H}{\overset{\overset{O}{\|}}{P}}-OR^1 \underset{-ROH}{\rightleftarrows} R^1O-\underset{H}{\overset{\overset{O}{\|}}{P}}-OR^1$$

The transesterification is carried out at elevated temperatures, usually in the range between 95 and 180 °C, depending on the type of H-phosphonate diesters, and proceeds both in the absence and in the presence of a catalyst (an acid or a base). The commonly used basic catalysts are alkali metals, alkali alkoxides, and tertiary amines, whereas H_3PO_4 and CH_3COOH are employed in the acid catalysis.

Reaction mechanism of noncatalyzed transesterification

It has been assumed that in the noncatalyzed transesterification of dialkyl H-phosphonates, these compounds react with their four-coordinated phosphonate form [28,29]. AM1 semiempirical calculations [30] of the model transesterification of dimethyl H-phosphonate with methanol indicate that in the first stage of this reaction, dimethyl H-phosphonate and the nucleophile form a pentacoordinated intermediate **II** via a four-centered cyclic transition structure **I**. The trigonal-bipyramidal intermediate **II** undergoes pseudorotation [17c,31] and via a new transition structure **III** of the same type as **I**, forms the monotransesterificated product. The rate of transesterification depends both on the type of substituents at phosphorus and the nucleophilicity of the corresponding alcohol.

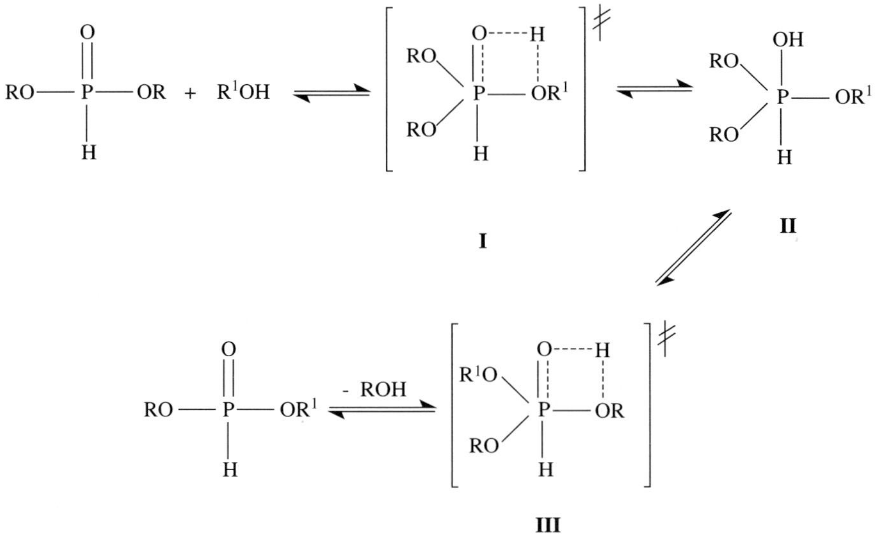

The transesterification rate decreases in the following order depending on the type of the alkoxy substituents at phosphorus [28]:

$$CH_3O > C_2H_5O > C_3H_7O > i\text{-}C_3H_7O$$

This order of reactivity correlates well with the electronegativity (induction constants) of these alkoxy groups. Compared to nucleophilic substitution at a tetrahedral carbon atom,

3.6 Substitution Reactions at the Phosphorus Atom

steric factors have significantly lower effects on the reactivity, since the phosphorus atom has a considerably larger atomic radius (1.28 Å) than carbon (0.91 Å). As a result, the distances between phosphorus and its substituents are larger than the comparable distances at a carbon atom, thus providing more space for the incoming nucleophile [20]. This explains the fact that for the nucleophilic substitution at the phosphorus atom where pentacoordinated intermediates are quite common, the corresponding five-coordinated species in the S_N2 substitution at a carbon atom are only transition structures.

Diphenyl H-phosphonate is reactive enough to undergo fast and quantitative reaction with various nucleophiles at room temperatures [32–34]. Different authors assume that this compound reacts by its tricoordinated tautomeric form.

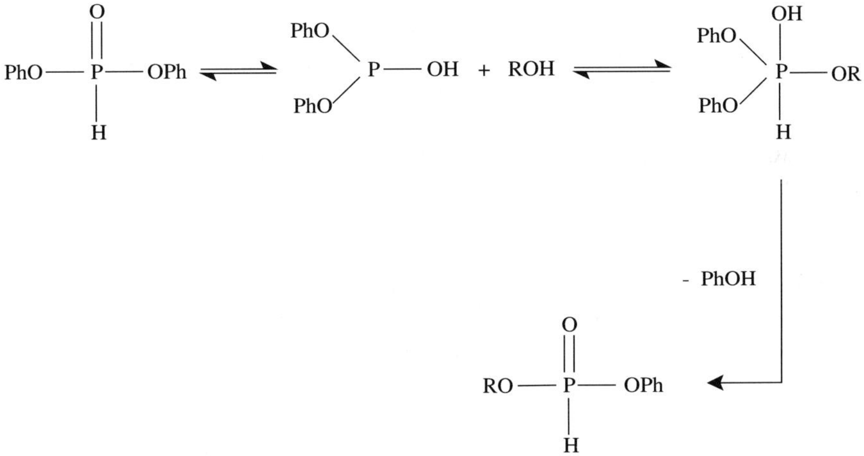

Nucleophilic substitutions at the phosphorus atom of the H-phosphonate diesters depend on the degree of occupation of its d-orbitals. The conjunction existing between the oxygen atom and the aromatic nucleus decreases the degree of p_π–d_π interaction between the oxygen and phosphorus; thus the degree of occupation of d-orbitals of the phosphorus atom in diphenyl H-phosphonate is lower than that in H-phosphonate dialkyl esters.

Reaction mechanism of catalyzed transesterification

It is suggested that when the transesterification reaction is carried out in the presence of a basic catalyst, dialkyl H-phosphonates react via their tricoordinated phosphite tautomeric form [32,35]. The first step of the reaction is tautomerization of phosphonate form into phosphite.

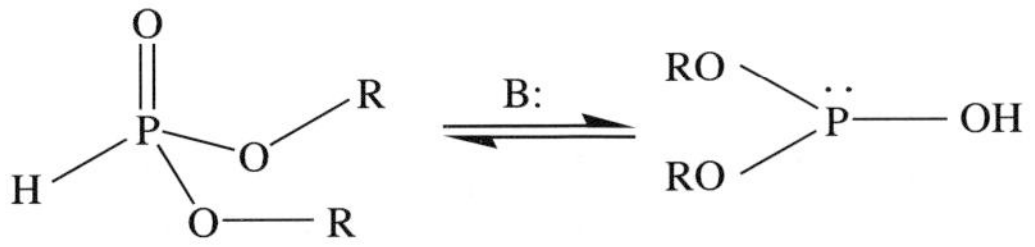

In the second step, the tricoordinated phosphorus atom is attacked by the nucleophile—the hydroxyl group of the alcohol.

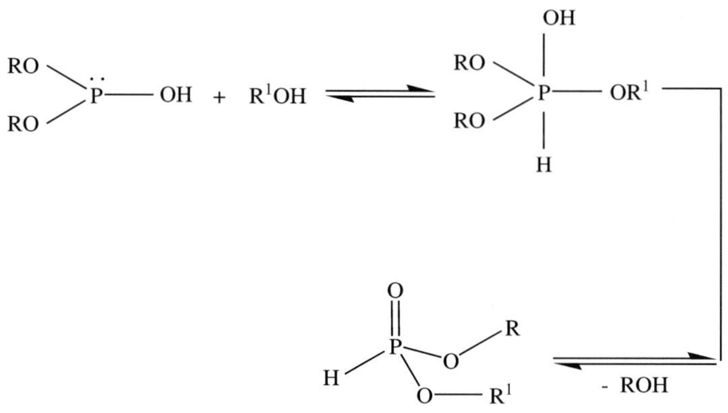

$$\begin{array}{c} RO \\ RO \end{array}\!\!\!\!\ddot{P}\!\!-\!\!OH + R^1OH \rightleftarrows \begin{array}{c} RO \\ RO \end{array}\!\!\!\!\overset{OH}{\underset{H}{P}}\!\!-\!\!OR^1 \longrightarrow$$

$$\underset{H}{\overset{O}{\underset{O-R^1}{\overset{\|}{P}}}}\!\!-\!\!O\!\!-\!\!R \quad \xrightarrow{-\ ROH}$$

More recent studies of the interaction between dialkyl H-phosphonates and amines indicate that the amine catalysis has a more complex character than initially assumed.

3.6.2 Side reactions

The transesterification of dialkyl H-phosphonates with hydroxyl-containing compounds is accompanied by side reactions, resulting in the formation of ether compounds and H-phosphonic acid monoalkyl esters [28,36,37]. The formation of these side products is due to the nucleophilic attack of the oxygen atom of the alcohol on the α-carbon atom of the alkoxy group, the second electrophilic center in the molecule of dialkyl H-phosphonates.

$$R\text{-}CH_2\text{-}O\text{-}\underset{H}{\overset{O}{\overset{\|}{P}}}\text{-}O\text{-}CH_2\text{-}R + R^1OH \rightleftarrows \left[R\text{-}CH_2\text{-}O\text{-}\underset{H}{\overset{O}{\overset{\|}{P}}}\text{-}O\text{-}CH_2\text{-}R \quad R^1OH \right]^{\neq}$$

$$RCH_2OR^1 + R\text{-}CH_2\text{-}O\text{-}\underset{H}{\overset{O}{\overset{\|}{P}}}\text{-}OH \longleftarrow R\text{-}CH_2\overset{+}{\text{-}O\text{-}R^1} + R\text{-}CH_2\text{-}O\text{-}\underset{H}{\overset{O}{\overset{\|}{P}}}\text{-}O^-$$

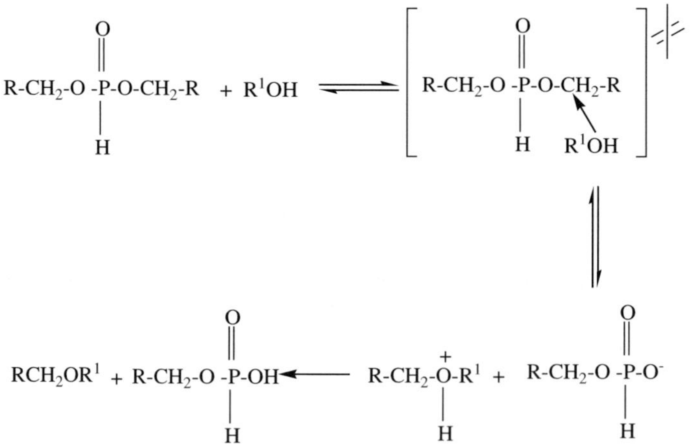

The content of the ether compounds depends on the reaction conditions and usually increases with increasing temperature. The low content of the ether compounds is determined by the fact that in the presence of two electrophilic centers, the phosphorus and the α-carbon atom

3.6 Substitution Reactions at the Phosphorus Atom

of the alkoxy groups, alcohols, which are regarded as 'hard' bases [38], show a tendency for selective attack at the phosphorus atom, which is a 'hard' acid due to its large positive charge. This renders the transesterification as the basic reaction and the reaction of formation of ether compounds as a side process.

Ether compounds are also formed as a result of an intramolecular dehydration of diols, which are catalyzed by the H-phosphonate diesters [39].

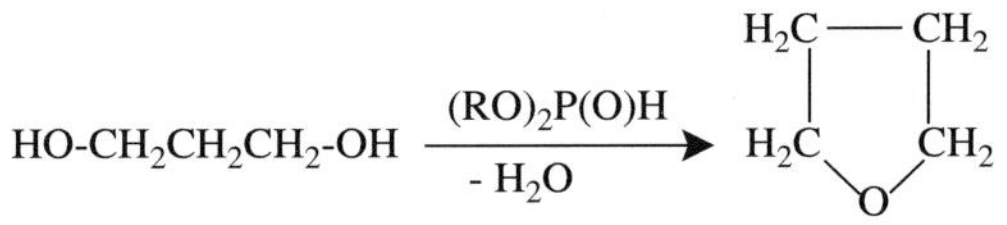

$$HO\text{-}CH_2CH_2CH_2\text{-}OH \xrightarrow[-H_2O]{(RO)_2P(O)H} \begin{array}{c} H_2C\text{---}CH_2 \\ | \quad\quad | \\ H_2C \quad CH_2 \\ \diagdown O \diagup \end{array}$$

Another possibility for the formation of cyclic ethers during the transesterification of H-phosphonate diesters with 1,4-butanediol or 1,5-pentadiol is the intramolecular cyclization of the monotransesterificated product. This cyclization occurs by the attack of the oxygen atom of the remaining hydroxyl group from the diol on the α-carbon atom of the same substitutent [28].

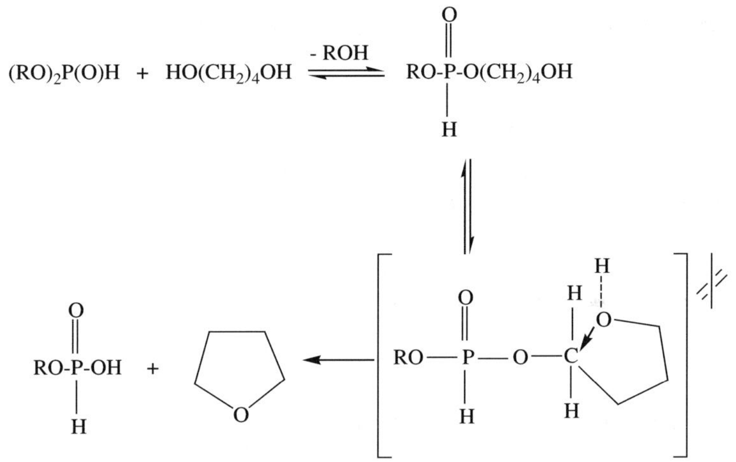

$$(RO)_2P(O)H + HO(CH_2)_4OH \xrightleftharpoons[]{-ROH} RO\text{-}P(O)(H)\text{-}O(CH_2)_4OH$$

It should be pointed out that the side reactions during the transesterification of diphenyl H-phosphonate are strongly depressed. In diphenyl H-phosphonate, there is only one electrophilic center—the phosphorus atom. It is shown that reaction mixtures resulting from transesterification of diphenyl H-phosphonate with alcohols in pyridine were contaminated with triphenyl phosphite and phenyl H-phosphonate, products of the disproportionation of diphenyl H-phosphonate [34].

3.6.3 Synthetic applications

Synthesis of H-phosphonates

Transesterification of H-phosphonate diesters is a simple synthetic route for the preparation of a variety of different esters of H-phosphonic acid (Scheme 3.1).

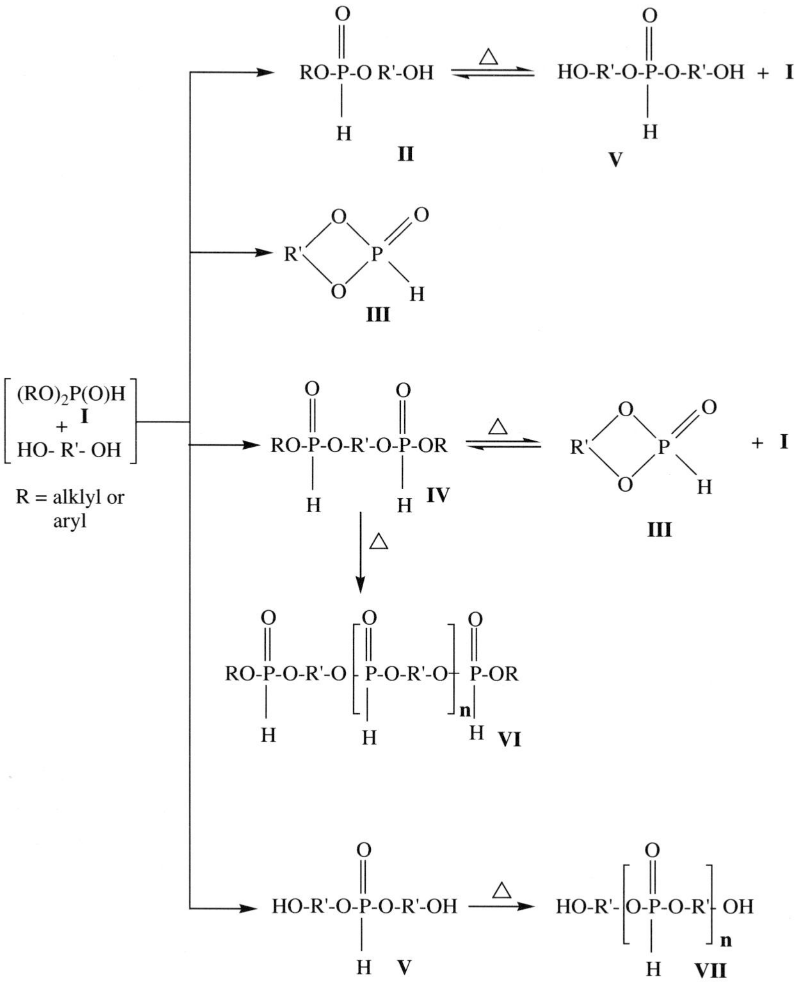

Scheme 3.1 General products of the transesterification of H-phosphonate diesters with diols.

Depending on the molar ratio between H-phosphonate diesters and hydroxyl-containing compounds and the reaction conditions, the following products can be obtained: monotransesterificated product **II** [40,41]; cyclic phosphonates, 1,3,2-dioxaphospholanes, or 1,3,2-dioxaphosphorinanes **III** [42,43]; dialkyl alkylene diphosphonates **IV** [39,44]; and bis(ω–hydroxyalkyl) phosphonates **V** [45]. Ethylene glycol [32,45], diethylene glycol [45,46], triethylene glycol [45–47], pentamethylene glycol [46], hexamethylene glycol [45,46,47–49], resorcinol [50], hydroquinone [48,51], pentaerytritol [49,52], glycerol [53], and diphenylpropanol [54] have been used as hydroxyl-containing compounds. The monotransesterified product **II** and dialkyl alkylene diphosphonates **IV** undergo disproportionation when heated to give bis(ω-hydroxyalkyl) phosphonates and cyclic phosphonate **III** and **I**, respectively [55]. Heating the products **IV** and **V** furnished poly(alkylene H-phosphonate)s **VI** and **VII**.

3.6 Substitution Reactions at the Phosphorus Atom

The transesterification of H-phosphonate diesters with 1,3 propanediol is accompanied however by the formation of 2-hydro-2-oxo-1,3,2-dioxaphosphorinane in 75 to 85% yield [28]. This compound results from the nucleophilic attack of the terminal hydroxyl group of the monotransesterificated product at the phosphorus atom.

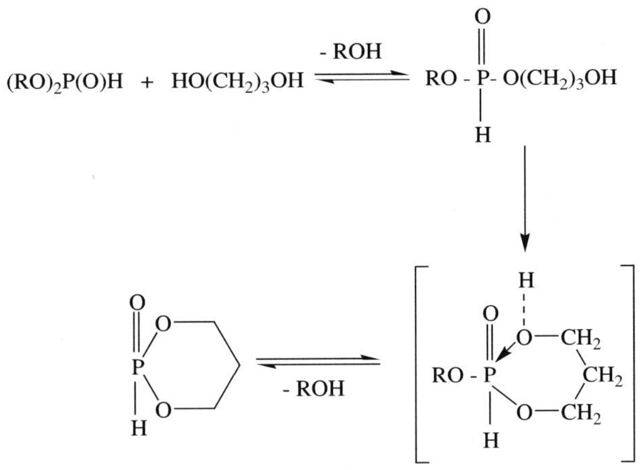

$$(RO)_2P(O)H + HO(CH_2)_3OH \xrightleftharpoons[]{-ROH} RO-\overset{\overset{O}{\|}}{\underset{H}{P}}-O(CH_2)_3OH$$

It was found that the transesterification of dimethyl H-phosphonate with 1,2-propanediol yields 4-methyl-2-hydro-2-oxo-1,3,2-dioxaphospholane [56]. Obviously, the first stage of the reaction furnished methyl-2-hydroxypropyl H-phosphonate. Subsequent intramolecular transesterification of the methyl-2-hydroxypropyl phosphonate yielded 4-methyl-2-hydro-2-oxo-1,3,2-dioxaphospholane. The specific reactivity of these esters of H-phosphonic acid is determined by the presence of a β–hydroxyl group. The role of the β–hydroxyl group may be regarded as an intramolecular catalysis. The reactivity enhancement of β–hydroxylethyl esters of H-phosphonic acid probably can be explained through hydrogen bonding, which favors the intramolecular transesterification reaction.

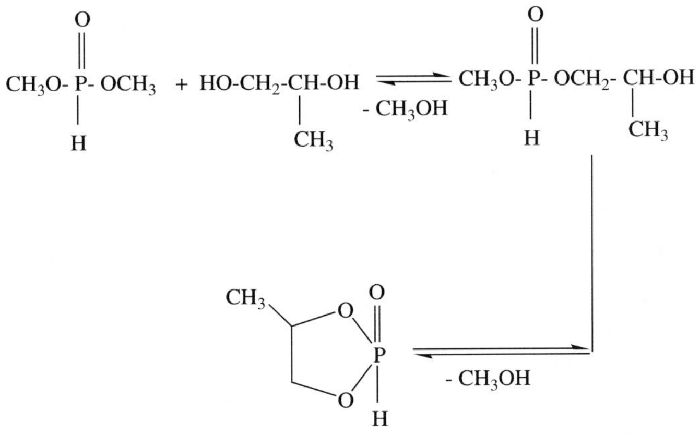

I

In the ^{31}P{H} NMR spectrum of 4-methyl-2-hydro-2-oxo-1,3,2-dioxaphospholane measured immediately after distillation, there are two signals at δ=23.92 ppm and 23.10 ppm in the ratio 1:1 [56]. After 6 h, in the ^{31}P{H} NMR spectrum, two new signals appear at 8.50 ppm and 7.38 ppm. The ratio between the signals at 23.90 ppm and 23.11 ppm is 1:1, and at 8.50 and 7.38 is 1:1 too. The presence of the signals at 8.50 ppm and 7.38 ppm in the ^{31}P{H} NMR spectrum can be explained by the existence of two tautomeric forms **I** and **II** of the 4-methyl-2-hydro-2-oxo-1,3,2-dioxaphospholane with P=O and P–OH bonds. Ovchinnikov *et al.* [57] are the first to accept that a new type of

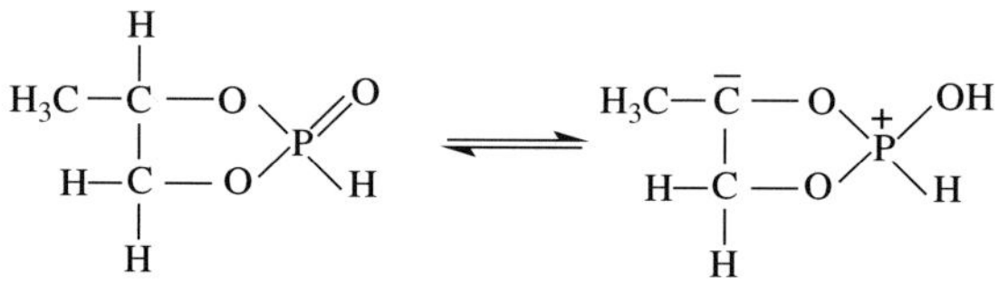

tautomerization exists at the cyclic H-phosphonate, connected with the migration of a proton to the phosphorus atom. The chemical shift of the phosphorus nucleus in form **I** is at 23.90 ppm, and for form **II** is at 8.50 ppm.

It is known that cyclic H-phosphonates may have the P–H atom in *cis* or *trans* to the ring substituents [58].

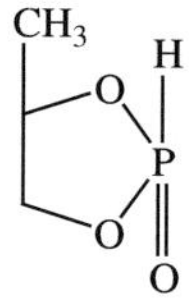

^{1}H NMR
cis: δ, ppm (CH$_3$) 1.38(d, ^{3}J(H,H) = 6.3 Hz)
      δ (PH) 7.25 ppm (d, ^{1}J(P,H) = 715 Hz)
^{31}P {H}NMR, δ = 23.11 ppm

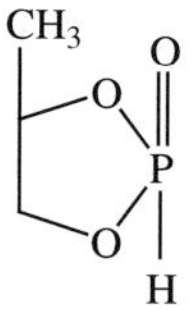

^{1}H NMR
trans: δ, ppm: (CH$_3$) 1.46 ppm (d, ^{3}J(H,H) = 6.3 Hz)
       (PH) 7.28 ppm (d, ^{1}J(P,H) = 717.5Hz)
^{31}P {H}NMR, δ = 23.90 ppm

The data from the ^{1}H NMR and ^{31}P{H} NMR spectra can be assigned as follows: the chemical shift of the phosphorus nucleus in the *cis* form of **I** was 23.11 ppm, and for the *trans* form **I** was 23.90 ppm. The difference in δp is 0.79 ppm.

The signals at 8.50 and 7.38 ppm can be assigned to the phosphorus nucleus of the *cis* form of **II**, and at 8.50 ppm for the phosphorus nucleus of the *trans* form of **II**.

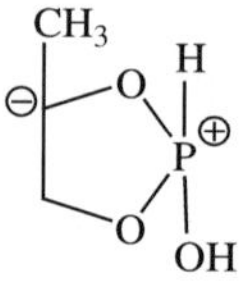

^{1}H NMR
cis : (CH$_3$) 1.08 (d, ^{3}J(H,H) = 6.3 Hz)
      (PH) 6.68 (d, ^{1}J(P,H) = 705.6 Hz)
^{31}P{H}NMR, δ = 7.38 ppm

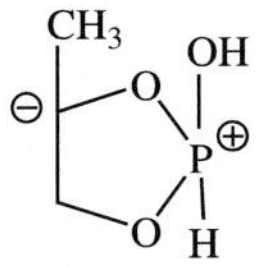

3.6 Substitution Reactions at the Phosphorus Atom

^{1}H NMR
trans : (CH$_3$) 1.14 (d, ^{3}J(H,H) = 7.0 Hz)
(PH) 6.80 (d, ^{1}J(P,H) = 706.8 Hz)
^{31}P{H}NMR, δ = 8.50 ppm

In the ^{31}P{H} NMR spectrum of the 2-hydro-2-oxo-1,3,2-dioxaphospholane obtained via transesterification of dimethyl H-phosphonate with 1,2-ethylene glycol, there is only one signal at 24.75 ppm. In this compound, mobile protons are not present in the ring; the form with P–OH bond is not observed [56].

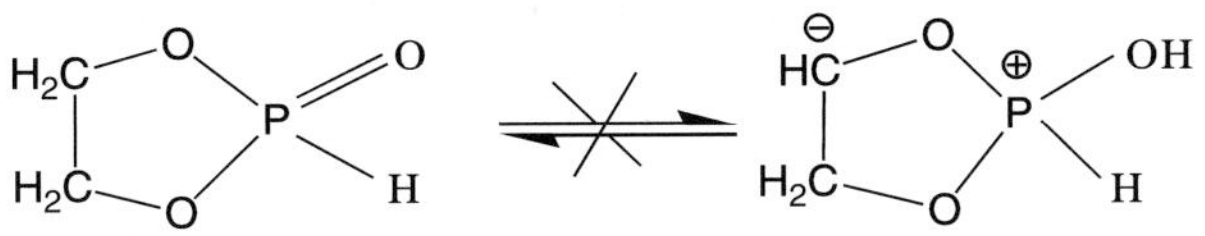

Diphenyl H-phosphonate reacts with aliphatic saturated mercaptans and unsaturated mercaptans with the formation of thio and dithio H-phosphonates [59].

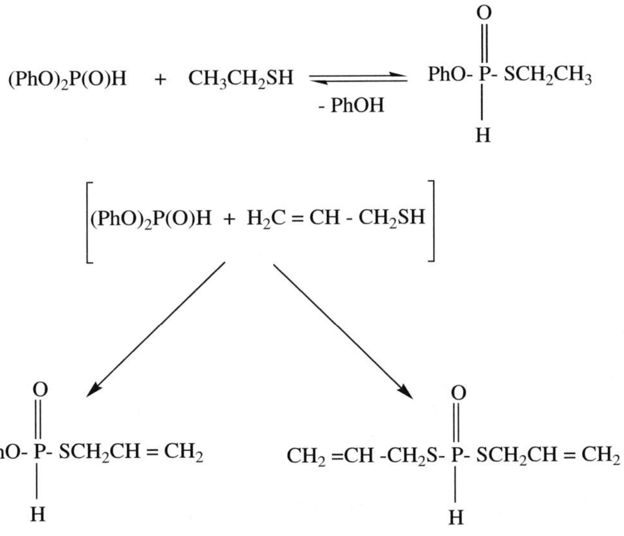

$$(PhO)_2P(O)H + CH_3CH_2SH \rightleftharpoons PhO-P(O)(H)-SCH_2CH_3 + PhOH$$

$$[(PhO)_2P(O)H + H_2C=CH-CH_2SH]$$

$$PhO-P(O)(H)-SCH_2CH=CH_2 \quad CH_2=CH-CH_2S-P(O)(H)-SCH_2CH=CH_2$$

H-phosphonate diesters having double or triple bonds in their alkoxy chains can be synthesized by transesterification of the lower homologues with unsaturated alcohols [60–62].

$$(RO)_2P(O)H + 2\ R'CH=CH-CH_2OH \longrightarrow (R'CH=CHCH_2O)_2P(O)H + 2\ ROH$$

Such H-phosphonate diesters are of interest as comonomers at the synthesis of phosphorus-containing polymers via polymerization.

Synthesis of polymers

Phosphorus-containing polyurethanes are generally synthesized by reacting phosphorus-containing diols with diisocyanates [63–66]. One of the readily available methods for incorporating phosphonate units in polyurethanes without the use of isocyanates is the transesterification of H-phosphonic diesters with hydroxycarbamates [67]. In the case of diphenyl H-phosphonate, when the transesterification proceeds at low temperatures (95°C), phosphorus-containing polyurethane with molecular weight 3267 Da was obtained. Thus-designed phosphorus-containing polyurerthanes have in the repeating unit the highly reactive P–H group, which participates in various reactions that can be used for the immobilization of bioactive substances onto the polymer (Scheme 3.2).

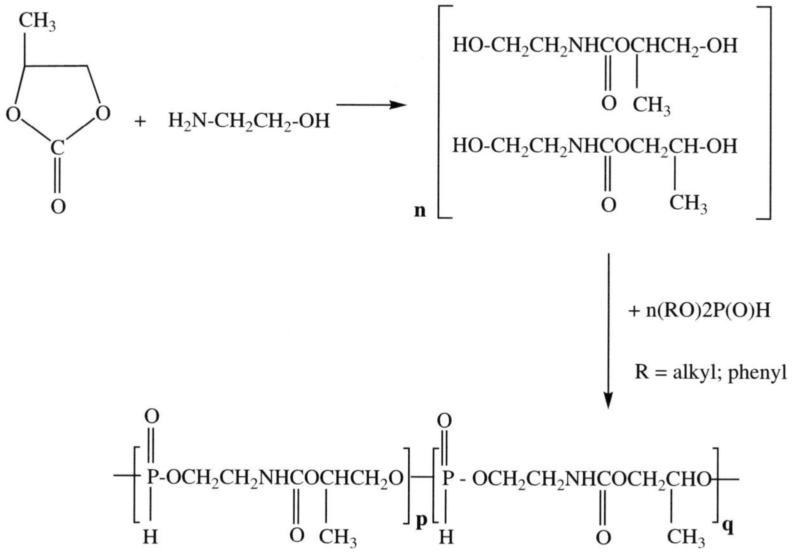

Scheme 3.2 Synthesis of phosphorus-containing polyurethanes via transesterification of diesters of H-phosphonic acids with urethane diols.

Phosphorus-containing poly(propylene ether carbonate)s were obtained reacting dimethyl H-phosphonate with poly(propylene ether carbonate)diols [68] (Scheme 3.3).

The structures of the products were established by ^{1}H and ^{13}C{H} NMR spectroscopy. Phosphorus-containing poly(propylene ether carbonate)s are of interest as intermediates, being capable of participating in various polymer-analogous reactions due to the presence of P–OCH$_3$ and P–H groups. The P–H group allows the introduction of new functional groups such as acid (P–OH), hydroxyl, amino, alkoxy, amido, and oxirane, imparting new properties. By choosing the proper derivatives and controlling the extent of substitutions, variations in hydrophilicity/hydrophobicity, physical properties, and biodegradation of the products can be achieved.

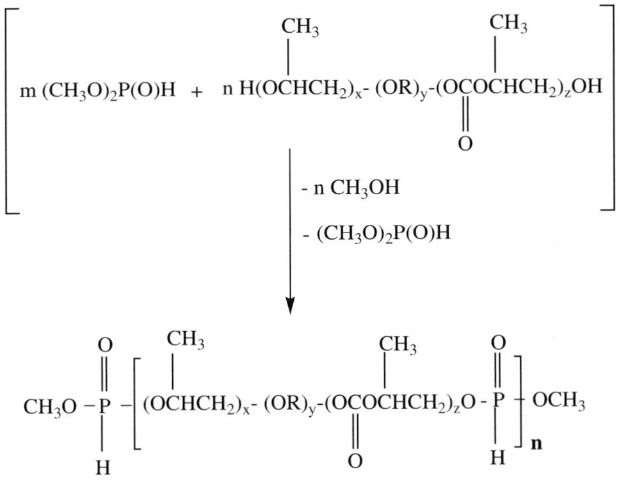

3.7 Characteristic Reactions of P–H Group

$$\begin{bmatrix} m\,(CH_3O)_2P(O)H \;+\; n\,H(OCHCH_2)_x\text{-}(OR)_y\text{-}(OCOCHCH_2)_zOH \\ \qquad\qquad\qquad\qquad CH_3 \qquad\qquad\qquad\qquad CH_3 \end{bmatrix}$$

with $-OCOCHCH_2-$ bearing CH_3 and C=O.

$$\downarrow \begin{array}{c} -n\,CH_3OH \\ -(CH_3O)_2P(O)H \end{array}$$

$$CH_3O-\underset{H}{\overset{O}{\overset{\|}{P}}}-\left[(OCHCH_2)_x\text{-}(OR)_y\text{-}(OCOCHCH_2)_zO-\underset{H}{\overset{O}{\overset{\|}{P}}}-OCH_3\right]_n$$

(with CH_3 substituents and $C=O$ group)

Scheme 3.3 Synthesis of phosphorus-containing poly(propylene ether carbonate)s via transesterification of dimethyl H-phosphonate with poly(propylene ether carbonate)diol.

3.7 CHARACTERISTIC REACTIONS OF P–H GROUP

The P–H group of diesters of H-phosphonic acid is highly reactive and participates in a variety of chemical reactions such as oxidation, addition to C=C double bond, C=N (Schiff bases), ketones, isocyanates, and condensation reactions.

3.7.1 Oxidation of P–H group

Owing to the presence of the P–H group, diesters of H-phosphonic acid can be oxidized by molecular oxygen to the corresponding phosphates [69].

$$(RO)_2P(O)H + 1/2\,O_2 \longrightarrow (RO)_2P(O)OH$$

The oxidation of dialkyl H-phosphonates with chlorine leads to the formation of the corresponding chlorophosphates [69].

$$(RO)_2P(O)H + Cl_2 \longrightarrow (RO)_2P(O)Cl + HCl$$

Oxidation of dialkyl H-phosphonate with nitrogen dioxide [70] or with nitrosyl chloride in pyridine [71] leads to the formation of tetraalkyl pyrophosphate.

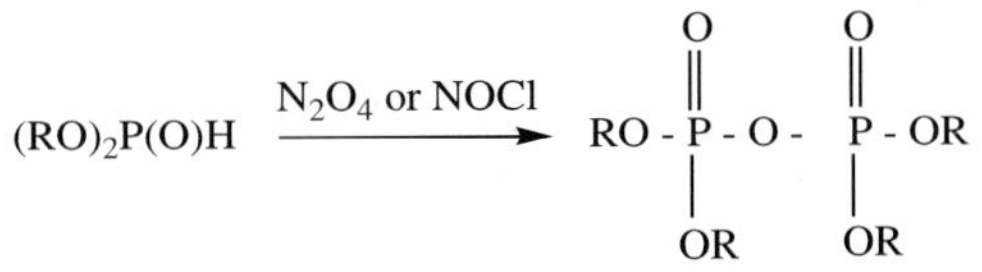

$$(RO)_2P(O)H \xrightarrow{N_2O_4 \text{ or NOCl}} RO-\underset{OR}{\overset{O}{\overset{\|}{P}}}-O-\underset{OR}{\overset{O}{\overset{\|}{P}}}-OR$$

Reaction of dialkyl H-phosphonates and sulfuric dichloride yields *bis*-(dialkoxyphosphonyl) disulfides [72].

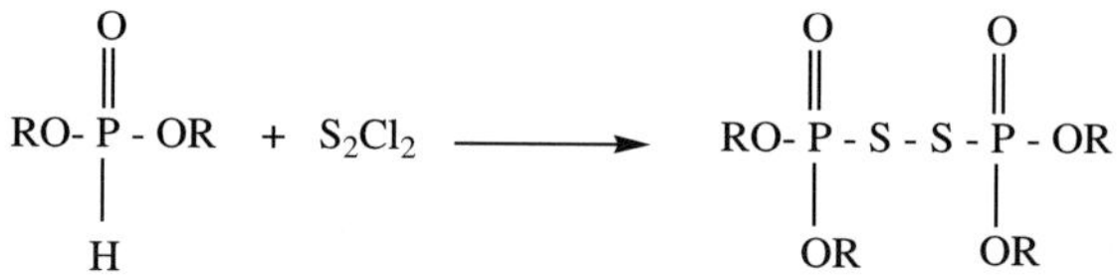

$$RO-\underset{H}{\underset{|}{P}}(=O)-OR + S_2Cl_2 \longrightarrow RO-\underset{OR}{\underset{|}{P}}(=O)-S-S-\underset{OR}{\underset{|}{P}}(=O)-OR$$

Diethyl H-phosphonate reacts with ethyl hypochloride to give triethyl phosphate [73].

$$(C_2H_5O)_2P(O)H + C_2H_5OCl \longrightarrow (C_2H_5O)_3P(O) + HCl$$

Diethyl H-phosphonate is oxidized by permanganate to the corresponding phosphates [74]. The process is pH-dependent and becomes more rapid at higher pH. Monoethyl H-phosphonate is also formed as a product of the subsequent oxidation of the initially obtained diethyl phosphate.

$$2(C_2H_5O)_2P(O)H + 2Mn(VII) \longrightarrow (C_2H_5O)_2P(O)OH$$
$$\downarrow$$
$$2Mn(IV) + CH_3CHO + (C_2H_5O)P(O)(OH)_2$$

Treatment of dialkyl H-phosphonates with diazonium salts leads to the formation of relatively stable dialkyl azophosphonates. Reduction of these compounds yields dialkyl hydrazin-phosphonates [75].

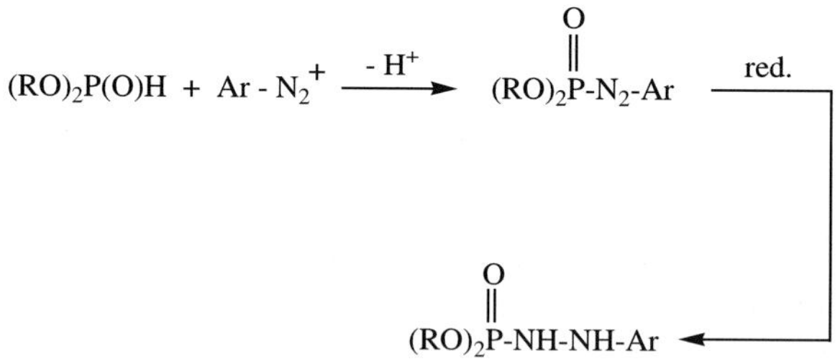

$$(RO)_2P(O)H + Ar-N_2^+ \xrightarrow{-H^+} (RO)_2P(O)-N_2-Ar \xrightarrow{red.}$$
$$(RO)_2P(O)-NH-NH-Ar$$

Trifluoromethanesulfonic acid catalyzes the reaction of phenyl diazomethane with dimethyl H-phosphonate. Diazoacetate and diethyl H-phosphonate react less effectively under similar conditions [76].

$$(RO)_2P(O)H + R^1R^2CN_2 \xrightarrow{CF_3SO_3H} (RO)_2P(O)CHR^1R^2$$

(a) $R^1 = R^2 = Ph$; $R = Me$
(b) $R^1 = H$; $R^2 = COOEt$; $R = Et$

3.7 Characteristic Reactions of P-H Group

The kinetics studies indicate that the ionization of dimethyl H-phosphonate by the base occurs before the attack of diazonium salt on the phosphonate anion [77]. According to this mechanism, the dimethylphosphonate anion or its O-protonated conjugate

$$(CH_3O)_2P(O)H + B \rightleftharpoons (CH_3O)_2PO^- + BH$$

$$(CH_3O)_2PO^- + ArN_2^+ \longrightarrow Ar-N=N-P(O)(OCH_3)_2$$

acid, dimethyl phosphite, is implicated as a reactive intermediate. Experimental results support the anion as a more acceptable intermediate. Most of the isolated arylazophosphonates probably have syn stereochemistry.

Diesters of H-phosphonic acid react with sulfur to give dialkyl thiophosphates [78].

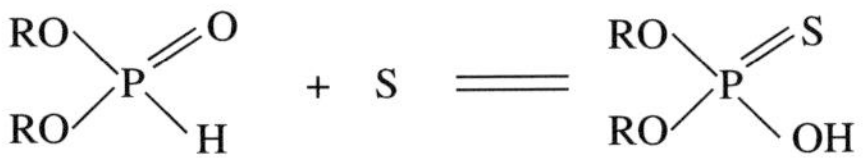

$$\begin{array}{c} RO \\ RO \end{array} P \begin{array}{c} O \\ H \end{array} + S \rightleftharpoons \begin{array}{c} RO \\ RO \end{array} P \begin{array}{c} S \\ OH \end{array}$$

Dialkylchlorophosphites can be obtained reacting dialkyl H-phosphonates with phosphorus pentachloride or thionylchloride [79].

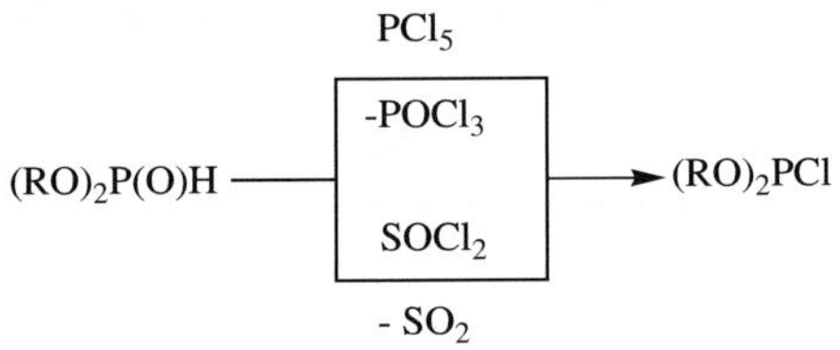

$$(RO)_2P(O)H \xrightarrow[\substack{-POCl_3 \\ SOCl_2 \\ -SO_2}]{PCl_5} (RO)_2PCl$$

Atherton–Todd reaction

The Atherton–Todd reaction is an interaction of dialkyl phosphonates with chlorocarbons in the presence of a base (see Appendix) [80]. This is a route for the oxidation of dialkyl H-phosphonates to the highly reactive dialkyl chlorophosphates, which are usually not isolated, but rather used *in situ* under mild conditions [81].

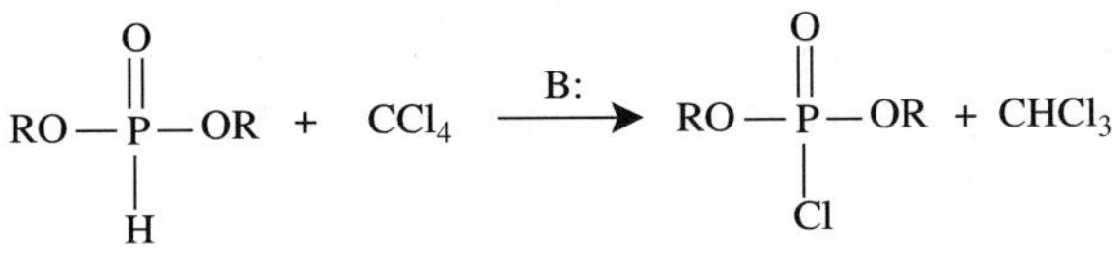

$$RO-\overset{\overset{O}{\|}}{\underset{H}{P}}-OR + CCl_4 \xrightarrow{B:} RO-\overset{\overset{O}{\|}}{\underset{Cl}{P}}-OR + CHCl_3$$

This reaction requires basic activation or basic catalysis to generate reactive trivalent phosphorus intermediates. The latter interact further with the corresponding substrates as strong nucleophiles. The formation of these reactive intermediates is often described in terms of simple phosphite salt formation in basic media.

$$(RO)_2 P(O)H + B: \longrightarrow (RO)_2 P\text{-}O^- + [HB]^+$$

Reaction mechanism. The commonly proposed mechanism of the reaction [81–85] is based primarily on the early kinetic investigations by Steinberg [81]. The initial step of this mechanism involves deprotonation of dialkyl H-phosphonate $(RO)_2P(O)H$ by a base B to give the dialkyl phosphite anion, $(RO)_2PO^-$. This anion then reacts as a nucleophile toward carbon tetrachloride, resulting in the sequence of reactions shown on the next scheme. In this scheme, step **1** leads to the formation of the active species $(RO)_2PO^-$, while steps **2** and **3** in the scheme represent the cycle of reactions that lead to the formation of the products.

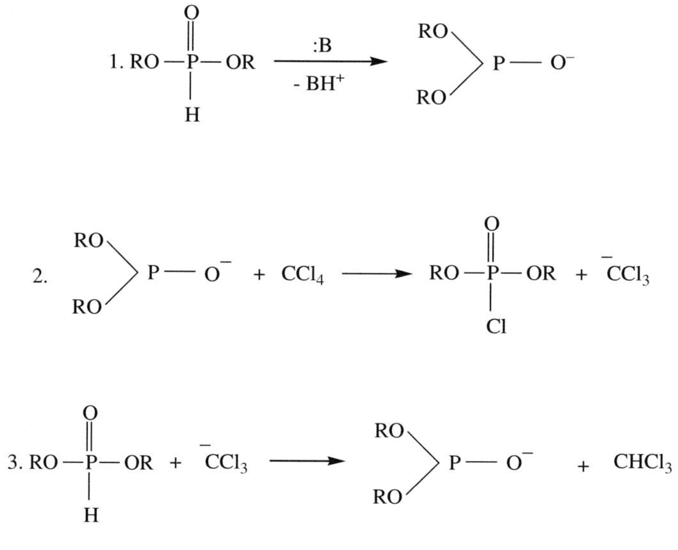

$$1.\quad RO-\underset{H}{\overset{\overset{O}{\|}}{P}}-OR \quad \xrightarrow[-BH^+]{:B} \quad \underset{RO}{\overset{RO}{>}}P-O^-$$

$$2.\quad \underset{RO}{\overset{RO}{>}}P-O^- + CCl_4 \quad \longrightarrow \quad RO-\underset{Cl}{\overset{\overset{O}{\|}}{P}}-OR + {}^-CCl_3$$

$$3.\quad RO-\underset{H}{\overset{\overset{O}{\|}}{P}}-OR + {}^-CCl_3 \quad \longrightarrow \quad \underset{RO}{\overset{RO}{>}}P-O^- + CHCl_3$$

Within this reaction cycle, dialkyl H-phosphonate and carbon tetrachloride react to form dialkyl chlorophosphate and chloroform as the final products of the Atherton–Todd reaction. A simple equilibrium shift (without salt formation) toward the dialkyl phosphite tautomer in step **1** [82] or formation of pentacoordinated phosphorane intermediates in step **2** [84,85] has also been discussed. It has been shown that the rate of this reaction depends on the strength of the applied base [86]. Amines are the most commonly used bases under Atherton–Todd conditions [80–85]. The validity of the deprotonation step in the above mechanism in the case of the base being an amine is, however, questionable, since it has been established that amines are alkylated and not protonated at the nitrogen by dialkyl phosphonates. It has been shown, however, [87] that in the case of basic activation with amines, which are the most commonly used bases, the phosphite intermediates are formed according to a different and more complex pathway (see Section 3.8.1). This pathway includes alkylation of the amine and formation of a monoalkyl H-phosphonate salt.

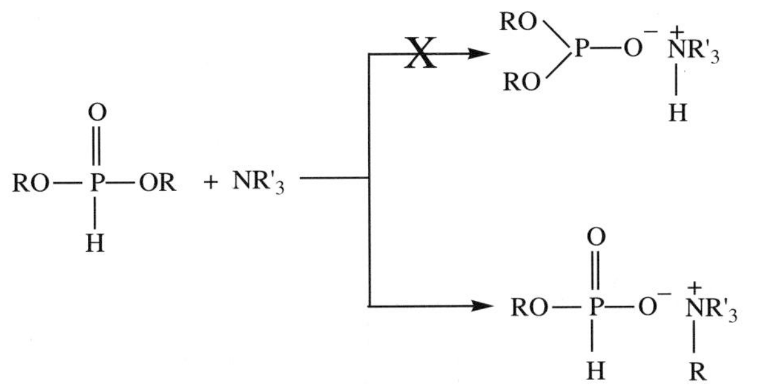

3.7 Characteristic Reactions of P-H Group

$$\text{RO-P(=O)(OR)-H} + NR'_3 \longrightarrow \begin{cases} X \rightarrow \text{RO}_2P(-O^-)(-O^+NR'_3H) \\ \\ \rightarrow \text{RO-P(=O)(H)-O}^- \; {}^+NR'_3R \end{cases}$$

This monoalkyl H-phosphonate anion is the actual base, which then deprotonates the dialkyl H-phosphonate. It has been established that alkylammonium or metal salts, containing monoalkyl H-phosphonate anions, promote the Atherton–Todd reaction [87]. Taking into account the observation of Kong and Engel that the Atherton–Todd reaction does not take place in the absence of a base, these results indicate that the monoalkyl H-phosphonate anion is playing a key role as an intermediate in the amine-promoted Atherton–Todd reaction. A recent theoretical *ab-initio* study [88] of the mechanism of the Atherton–Todd reaction favors the following reaction scheme, which is based on energetic evaluation of alternative reaction pathways:

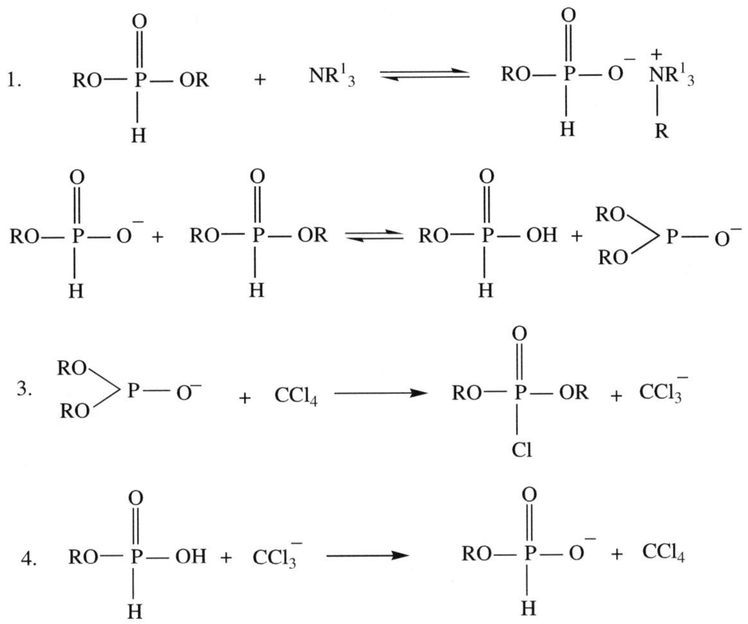

1. $\text{RO-P(=O)(H)-OR} + NR^1_3 \rightleftharpoons \text{RO-P(=O)(H)-O}^- \; {}^+NR^1_3R$

2. $\text{RO-P(=O)(H)-O}^- + \text{RO-P(=O)(H)-OR} \rightleftharpoons \text{RO-P(=O)(H)-OH} + (\text{RO})_2P-O^-$

3. $(\text{RO})_2P-O^- + CCl_4 \longrightarrow \text{RO-P(=O)(Cl)-OR} + CCl_3^-$

4. $\text{RO-P(=O)(H)-OH} + CCl_3^- \longrightarrow \text{RO-P(=O)(H)-O}^- + CCl_4$

In this new scheme, the initial step is the reaction of the amine with dialkyl H-phosphonate to form the corresponding monoalkyl phosphonium salt. The subsequent three steps form the catalytic cycle that leads to product formation. The anion of the monoalkyl H-phosphonate, acting as a base, first deprotonates the dialkyl H-phosphonate to generate the reactive dialkyl phosphite anion. In the subsequent step, which is the same as in the originally proposed mechanism of that reaction, dialkyl chlorophosphate is formed along with the trichloromethanide anion. The catalytic cycle is completed with the reaction of this trichloromethanide anion with monomethyl H-phosphonate to form chloroform and the monoalkyl H-phosphonate anion. The latter anion can then react with a further molecule of dialkyl H-phosphonate in a new catalytic cycle. Some results of this computational study [88], considering the alkylation step **1** leading to the formation of monoalkyl H-phosphonate anion in the above mechanism, are discussed in Section 3.8.1. The key result of these calculations is that the computations support the suggestion that the catalytically active trivalent phosphorus (III) species will be present in those reactions that involve both the diesters of H-phosphonic acid and the monomethyl H-phosphonate anion. This conclusion is based on the following energy profile, calculated for the model deprotonation of phosphonic acid by its anion using MP2/6–31+G*//HF/6–31+G* basis (Figure 3.3).

These results indicate that the energy gained in the formation of the early ion–dipole complex is sufficient to overcome the transition structure energy barrier. Since the energy of the transition structure is 8.6 kcal/mol lower than that of the separated reactants, a significant population of the late ion–dipole complex may be expected. These two ion–dipole complexes should be present on both sides of the transition structure along the reaction coordinate in the above scheme, but they do not exist for the simplified models. Preliminary AM1 studies for the proton-transfer reaction of the fully methylated species reveal, however, two weakly bonded ion–dipole complexes. The early complex has an O–H distance of 2.11 Å, and the late complex has a P–H distance of 1.90 Å. The computational study of the Atherton–Todd reaction (Table 3.4) is another indication supporting the

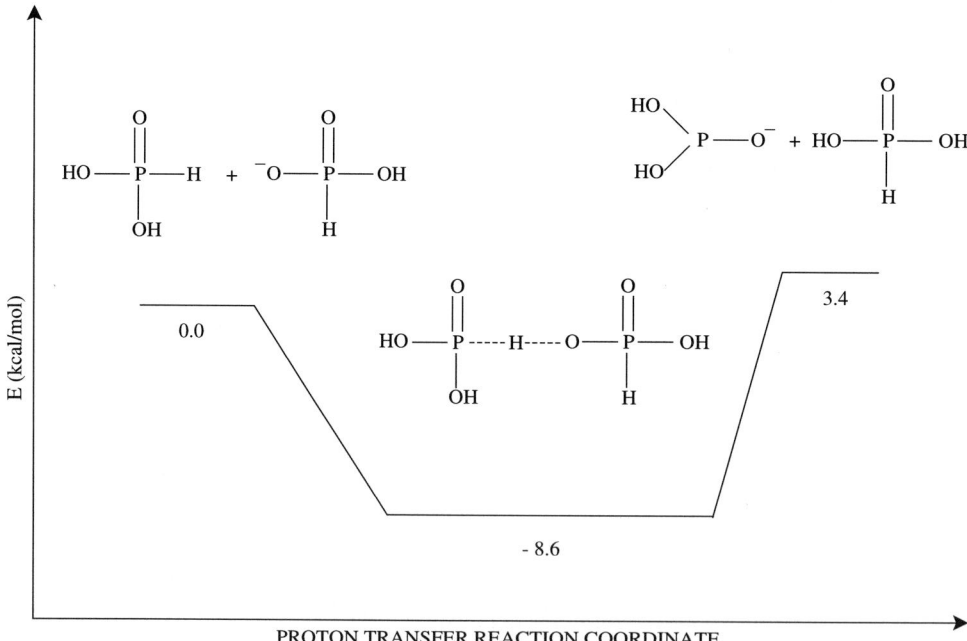

Figure 3.3 MP2/6–31+G*//HF/6–31+G* energy profile of the deprotonation of phosphonic acid by its anion.

3.7 Characteristic Reactions of P-H Group

Table 3.4

Calculated relative proton affinities of some of the bases used in the computational study of the Atherton–Todd reaction [86]

Base B[a]	$\delta\delta E$[b]		
	SCF	MP2	MP2 + ZPE
NH_3	−150.0	−201.1	−209.7
$[HO-P(O)(H)O]^-$	−26.9	−22.6	−21.0
$[CH_3O-P(O)(H)O]^-$	−23.8	−19.7	−18.8
$[HO-P(O)OH]^-$	−5.8	−4.7	−3.4
$[CH_3O-P(O)OH]^-$	−3.1	−2.4	−1.8
$[(CH_3O)_2PO]^-$	0.0	0.0	0.0

[a]Geometries optimized with HF/6–31+G* basis set.
[b]Energies in kcal/mol for the reaction $BH + [(CH_3O)_2PO]^- \rightarrow B + (CH_3O)_2P(O)H$.

assumption that the true base in the basic activation of the P–H bond in dialkyl H-phosphonates in the presence of amines is the corresponding monoalkyl H-phosphonate anion. Table 3.4 shows the calculated relative proton affinities of some of the bases used in the study. The above data indicate that the monomethyl H-phosphonate anion is a much stronger base than ammonia in this process, although it is still a weaker base than the anion of dimethyl phosphite. The rate of the reaction appears to be strongly dependent on the strength of the base. Triethylamine was only slightly more effective as a catalyst than tributylamine or triamylamine, but was 1000 times more effective than pyridine [81].

When the Atherton–Todd reaction is carried out in the presence of carboxylic acids, the formation of phosphate products is observed [89].

$$(RO)_2P(O)H + R^1C(O)OH \xrightarrow{Et_3N/CCl_4} (RO)_2P(O)OC(O)R^1$$

Amino acids were phosphorylated successfully using Atherton–Todd reaction conditions [89,90].

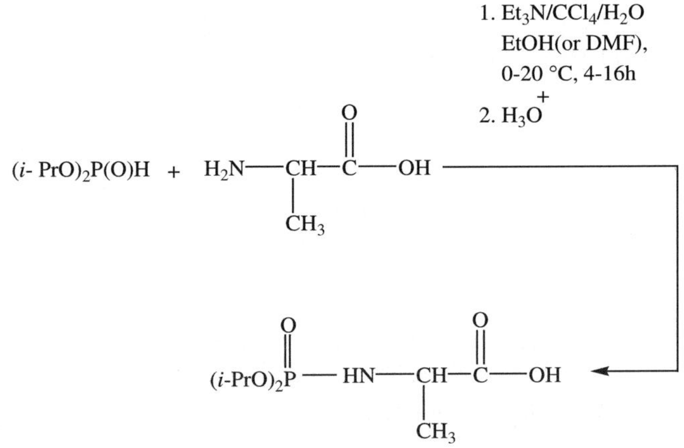

It was shown that the phosphorylation of amines by the Atherton–Todd reaction could be carried out conveniently and easily under phase-transfer conditions in the presence of catalytic amounts (about 5 mol %) of triethylbenzylammonium chloride (TEBA) [91].

$$(RO)_2P(O)H + CCl_4 + R^1R^2NH$$
$$\downarrow Et_3(PhCH_2)NCl$$
$$CH_2Cl_2 / 50\% \text{ NaOH}$$
$$(RO)_2P(O)NR^1R^2 + CHCl_3$$

This method allows for preparation of phosphoramides in high purity. Di-*tert*-butyl chlorophosphate was also obtained under phase-transfer conditions using TEBA [92].

$$(t\text{-BuO})_2P(O)H + CCl_4 \xrightarrow[CH_2Cl_2 / 20\% \text{ NaOH}]{Et_3(PhCH_2)NCl} (t\text{-BuO})_2P(O)Cl$$

Phosphorylation of alcohols [93] and phenols [94] using *n*-tetrabutylammonium bromide as a phase-transfer catalyst has also been reported. The mixed trialkyl phosphates obtained using this procedure are also of high purity. Hydrazine has also been successfully phosphorylated under phase-transfer conditions yielding the corresponding phosphoramide hydrazides [95].

$$(RO)_2P(O)H + CCl_4 + H_2N\text{-}NH_2 \xrightarrow[CH_2Cl_2 / K_2CO_3]{Et_3(PhCH_2)NCl} (RO)_2P(O)\text{-}NH\text{-}NH_2$$

It was shown that phosphoramides could also be obtained by the Atherton–Todd reaction between anilides and dialkyl H-phosphonates under phase-transfer conditions [96].

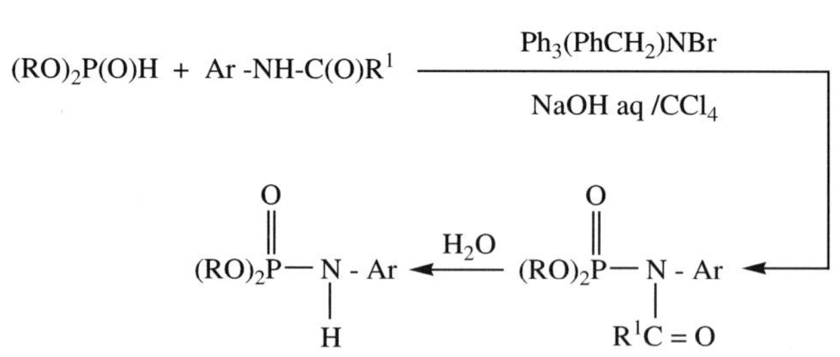

$$(RO)_2P(O)H + Ar\text{-}NH\text{-}C(O)R^1 \xrightarrow[NaOH \text{ aq} /CCl_4]{Ph_3(PhCH_2)NBr}$$

$$(RO)_2\overset{O}{\underset{}{P}}\text{-}\underset{H}{N}\text{-}Ar \xleftarrow{H_2O} (RO)_2\overset{O}{\underset{}{P}}\text{-}\underset{R^1C=O}{N}\text{-}Ar$$

Formanilides ($R^1 = H$) and chloracetanilides ($R^1 = CH_2Cl$) are used in these preparations. The formation of an Atherton–Todd product in this reaction is probably due to the high N–H acidity of formanilide and chloracetanilide.

3.7 Characteristic Reactions of P-H Group

The reaction of diethyl H-phosphonate with aziridines is reported to proceed without ring-opening under Atherton–Todd conditions [97].

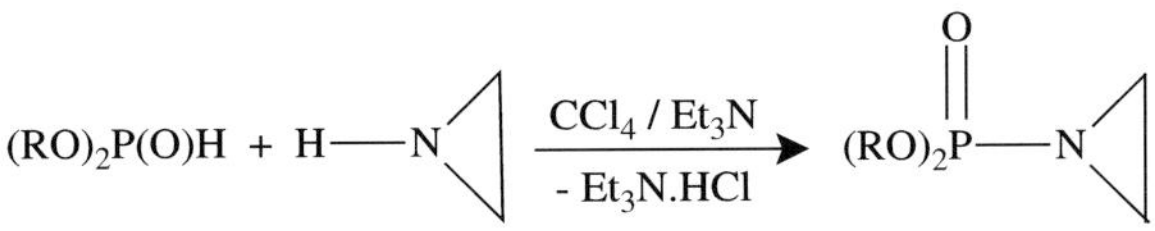

$$(RO)_2P(O)H + H\text{—}N\triangleleft \xrightarrow[-Et_3N \cdot HCl]{CCl_4 / Et_3N} (RO)_2\overset{\overset{O}{\|}}{P}\text{—}N\triangleleft$$

It is established that under Atherton–Todd conditions, dimethyl H-phosphonate reacts with Schiff bases containing a cyclopropyl fragment, yielding the corresponding amidophosphates [98].

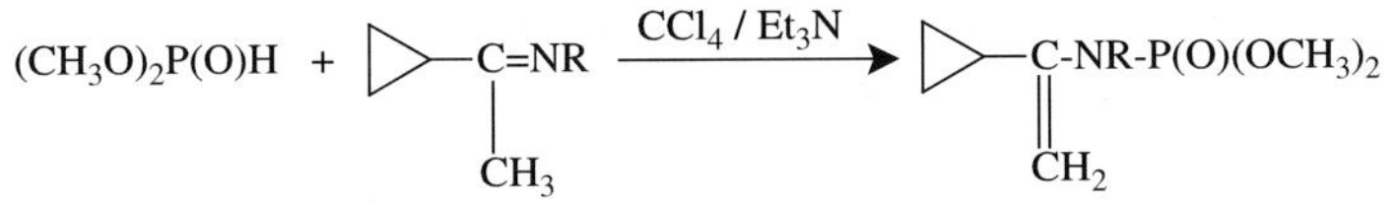

$$(CH_3O)_2P(O)H + \triangleright\!\!\!-\underset{CH_3}{C}=NR \xrightarrow{CCl_4 / Et_3N} \triangleright\!\!\!-\underset{CH_2}{\overset{\|}{C}}\text{-NR-P(O)(OCH_3)_2}$$

R = CH$_2$CH$_2$OCH=CH$_2$; CH$_2$CH$_2$OCH$_2$CH$_3$

Dialkyl chlorophosphates, except via the Atherton–Todd reaction, are also formed in the reaction between dialkyl H-phosphonates and copper dichloride in tetrachloromethane [99].

$$(RO)_2P(O)H + CuCl_2 \xrightarrow{CCl_4} (RO)_2P(O)Cl + CuCl + HCl$$

Dialkyl chlorophosphates are formed by the interaction of dialkyl H-phosphonates and trichloracetic acid in the presence of triethylamine [100].

$$(RO)_2P(O)H + CCl_3COOH \xrightarrow{Et_3N} (RO)_2P(O)Cl + HCCl_2COOH \cdot Et_3N.$$

When the Atherton–Todd reaction of dialkyl H-phosphonates is carried out in the presence of ethylene oxide, the final products are dialkyl 2-chloroethyl phosphates [101, 102].

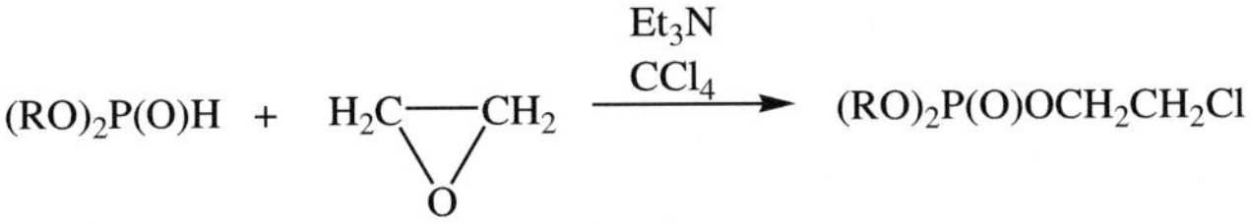

$$(RO)_2P(O)H + H_2C\underset{O}{\overset{}{\text{—}}}CH_2 \xrightarrow[CCl_4]{Et_3N} (RO)_2P(O)OCH_2CH_2Cl$$

The so-formed dialkyl chlorophosphates are reported to undergo reduction with sodium borohydride, yielding back dialkyl H-phosphonates (see Appendix) [103].

$$(RO)_2P(O)Cl + NaBH_4 \xrightarrow[\text{Dioxane}]{\Delta} (RO)_2P(O)H$$

Synthetic application. The synthetic applications of the Atherton–Todd reaction include preparation of dialkyl amidophosphates and di- or trialkyl phosphates by *in situ* condensation of the intermediately formed dialkyl chlorophosphates with N–H or O–H containing substrates. Thus, when the reaction takes place in an excess of the basic compounds (primary or secondary amines), dialkyl phosphoramides are formed as final products [81].

$$(RO)_2P(O)H + HNR^1R^2 \xrightarrow{Et_3N/CCl_4} (RO)_2P(O)NR^1R^2$$

Trialkyl phosphates are obtained when the Atherton–Todd reaction is carried out in the presence of aliphatic alcohols [104].

$$(RO)_2P(O)H + R^1OH \xrightarrow{Et_3N/CCl_4} (RO)_2P(O)OR^1$$

It was shown that when reacted with ethanolamine in chloroform solution, phosphofluoridates give exclusively O-phosphorylation, whereas phosphochlorides give mainly N-phosphorylation (95%) [105,106].

$$(C_2H_5O)_2P(O)H + HOCH_2CH_2NH_2 \xrightarrow[Et_3N]{CCl_4} (C_2H_5O)_2\overset{O}{\underset{\|}{P}}-NHCH_2CH_2OH \quad \delta p = 9.1 \text{ ppm}$$

$$\downarrow \text{heating}$$

$$(C_2H_5O)_3P(O)$$
$$\delta p = 1.0 \text{ ppm}$$

The phosphoramide undergoes a facile isomerization on heating to give the corresponding phosphate. Amidophosphate obtained by reacting diethyl H-phosphonate in the presence of chloroform and triethylamine does not undergo such isomerization.

The reaction of dialkyl chlorophosphates with 1,3-aminopropanol mainly furnished derivatives of the N-phosphorylation-γ-hydroxypropylamide of dialkylphosphoric acid (δ = 9.0 ppm for R = Et; δ = 11.0 ppm for *i*-Pr) [105]. The final product is stable and does not undergo isomerization resulting in the formation of corresponding phosphates, as in the case of the β-hydroxyethylamides of dialkylphosphates. This result indicates that the regioselectivity of this substitution (N–H versus O—H) correlates with the acidity of the corresponding protons.

N-phosphorylated proteins and amino acids play important roles in the regulation of enzyme activity and protein biosynthesis [107,108]. The Atherton–Todd reaction is used for the convenient synthesis of *N*-phosphoryldipeptide acids. The dialkyl phosphonate group has been successfully used for direct phosphorylation and for protection of the amine group [109–112].

3.7 Characteristic Reactions of P-H Group

$$(RO)_2P(O)H + H_3\overset{+}{N}-\underset{H}{\overset{R^1}{C}}-\underset{O}{\overset{H}{C}}-N-\underset{R^2}{\overset{H}{C}}-C(O)O^-$$

1. $Et_3N / CCl_4 / H_2O$
2. H_3O^+

↓

$$(RO)_2\overset{O}{\overset{\|}{P}}-\underset{H}{N}-\underset{H}{\overset{R^1}{C}}-\underset{O}{\overset{H}{C}}-N-\underset{R^2}{\overset{H}{C}}-COOH$$

Thus, the unprotected dipeptide acids are directly phosphorylated in high yields (78–94%) by dialkyl H-phosphonate in the mixed solvent system including CCl_4, water, ethanol, and THF.

It was established that the activities of many enzymes are regulated through phosphorylation–dephosphorylation at a serine residue [107,108,113]. It was shown that the N-phosphorylated serine revealed very unexpected reactivity [114]. Serine was

$$(R^1O)_2P(O)H + H_2N-\underset{\underset{OH}{|}}{CH}-COOH \xrightarrow[CCl_4]{Et_3N} (RO)_2\overset{O}{\overset{\|}{P}}-NH-\underset{\underset{OH}{|}}{CH}-COOH$$

1

step 1 ↓ R^2OH

$$\underset{R^2O}{\overset{R^2O}{>}}\overset{O}{\overset{\|}{P}}-O-\underset{R}{\overset{|}{CH}}-\underset{NH_2}{\overset{|}{CH}}-COOH \xleftarrow[\text{step 2}]{R^2OH} \underset{R^2O}{\overset{R^1O}{>}}\overset{O}{\overset{\|}{P}}-O-\underset{R}{\overset{|}{CH}}-\underset{NH_2}{\overset{|}{CH}}-COOH$$

3 **2**

phophorylated using Atherton–Todd reaction conditions. When the N-phosphorylated serine is treated with alcohol, several reactions proceed. Product **2** derives from the phosphoric ester exchanges of the phosphoramide **1** with the transfer reaction of the phosphoryl group from amino to the hydroxyl. Product **3** results from the phosphoric ester exchanges of the phosphoryl triesters **2**. The experimental results revealed that the hydroxyl and carboxylic groups in molecule **1** are necessary for the N→O migration (step **1**) and for the exchange reaction in steps **1** and **2**. It seems that the reaction intermediates in step **1** and step **2** might involve the participation of the three groups amino, hydroxyl, and carboxyl in a molecule, and the attack of an alcohol at these intermediates will result in the formation of compounds **2** and **3**. It is worthwhile to mention that the N→O migration of the phosphoryl group seems irreversible. No signals in the region of 0–20 ppm, which are typical for phosphor-amides, were observed in the ^{31}P{H} NMR spectra after **1** was heated at 40 °C in alcohol for 48 h.

Dialkyl H-phosphonates are used not only for the preparation of N-protected amino acids, but also as active coupling agents in the peptide synthesis [89,115,116]. Among a wide variety of methods for the synthesis of peptides, the procedure via mixed carboxylic–phosphoric anhydride-type intermediate has so far attracted attention because this compound plays an important role in the biosynthesis of proteins and peptides [117]. Zhao *et al.* [89] offer the following reaction scheme for the synthesis of peptides. Phenylalanine and tryptophane are used as amino acids.

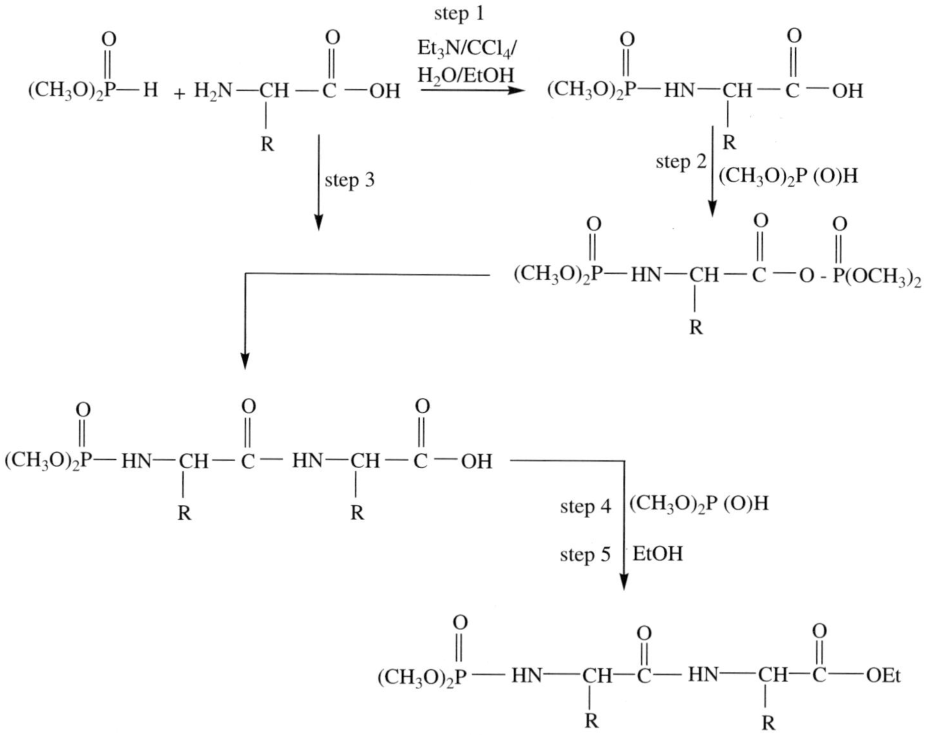

$$(CH_3O)_2P(O)H + H_2N-CH(R)-C(O)-OH \xrightarrow[H_2O/EtOH]{Et_3N/CCl_4/ \text{ step 1}} (CH_3O)_2P(O)-HN-CH(R)-C(O)-OH$$

$$\xrightarrow[\text{step 2}]{(CH_3O)_2P(O)H} (CH_3O)_2P(O)-HN-CH(R)-C(O)-O-P(OCH_3)_2$$

$$\xrightarrow{\text{step 3}} (CH_3O)_2P(O)-HN-CH(R)-C(O)-HN-CH(R)-C(O)-OH$$

$$\xrightarrow[\text{step 5 EtOH}]{\text{step 4 } (CH_3O)_2P(O)H} (CH_3O)_2P(O)-HN-CH(R)-C(O)-HN-CH(R)-C(O)-OEt$$

(+)–Pseudoephedrine reacts via the classical Atherton–Todd reaction with dimethyl H-phosphonate to give the corresponding amidophosphate, which undergoes stereoselective heterocyclization to give 2-methoxy-3,4-dimethyl-5-phenyl-2-oxo-1,3,2-oxazaphospholane **1** in 82% yield based on the amidophosphate [118].

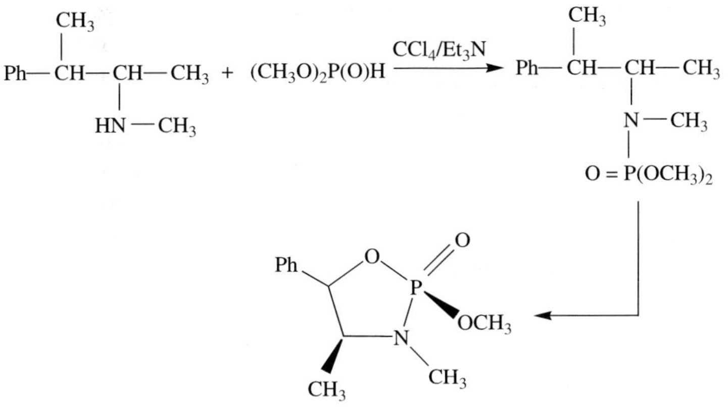

3.7 Characteristic Reactions of P-H Group

$$\text{Ph—CH(CH}_3\text{)—CH(NHCH}_3\text{)—CH}_3 + (CH_3O)_2P(O)H \xrightarrow{CCl_4/Et_3N} \text{Ph—CH(CH}_3\text{)—CH(N(CH}_3\text{)P(O)(OCH}_3\text{)}_2\text{)—CH}_3$$

1

The Atherton–Todd reaction is used for the phosphorylation of poly(*N*-vinylpyrrolidone-*co*-vinylamine) [119].

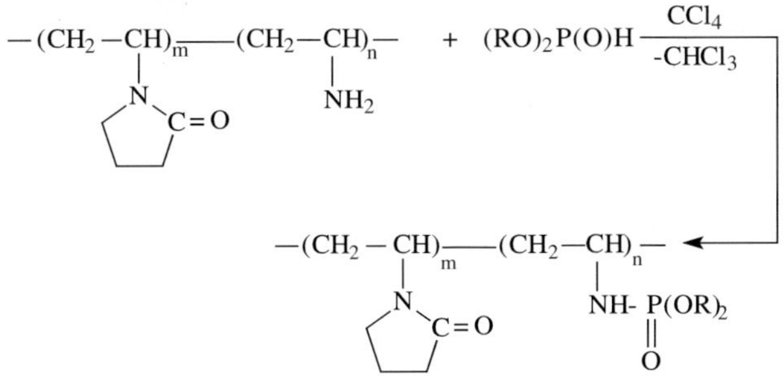

$$-(CH_2-CH)_m-(CH_2-CH)_n- + (RO)_2P(O)H \xrightarrow[-CHCl_3]{CCl_4} -(CH_2-CH)_m-(CH_2-CH)_n-$$

with pyrrolidone ring on m-unit, NH$_2$ on n-unit → NH-P(O)(OR)$_2$ on n-unit

R = CH$_3$; Cl-CH$_2$CH$_2$

The Atherton–Todd reaction has been recently used to convert linear poly(alkylene H-phosphonate)s into the corresponding poly(alkylene phosphate)s [120].

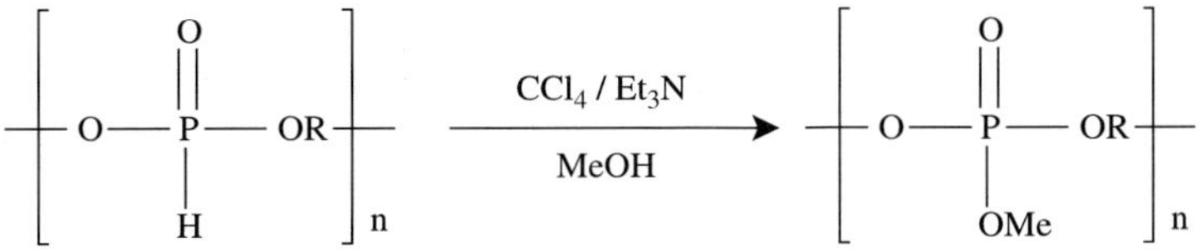

$$\left[-O-P(O)(H)(OR)- \right]_n \xrightarrow[\text{MeOH}]{CCl_4 / Et_3N} \left[-O-P(O)(OMe)(OR)- \right]_n$$

Aliphatic chlorocarbons and chlorofluorocarbons have found widespread commercial use as refrigerants, industrial solvents, and blow agents for the preparation of rigid polyurethane foams. Chlorofluorocarbons in the stratosphere are environmentally harmful because the photolytic cleavage of their carbon–chlorine bonds results in depletion of the ozone layer [121]. Chlorofluorohydrocarbons have been proposed as refrigerant substitutes for chlorofluorocarbons [122]. The latter molecules are considered environmentally less harmful because they contain a carbon–hydrogen bond that can be cleaved by chemical reactions at lower altitudes, thereby preventing the compounds from reaching the stratosphere. The Atherton–Todd reaction of halocarbons with dialkyl H-phosphonates is one potentially useful reaction that can be used to carry out such transformation under homogeneous conditions [123].

$$(CH_3O)_2 P(O)H + CCl_3F + CyNH_2 \longrightarrow (CH_3O)_2P(O)NHCy + CCl_2FH + [CyNH_3]Cl$$

The formation of dichlorofluoromethane is proved by ^{19}F NMR spectroscopy: $\delta = -80.8$ ppm, d, $^2J(HF) = 53.4$ Hz.

The experimentally found order of reactivity to dialkyl phosphonates follows the sequence $CCl_4 > CFCl_3 > CF_2Cl_2$, and no reaction has been reported yet at a carbon atom that also contains a carbon–hydrogen bond.

Direct azidation, cyanation, and thiocyanation of diesters of H-phosphonic acid were accomplished readily with sodium pseudohalides in acetonitrile under mild modified Atherton–Todd conditions [124].

[reaction scheme: dialkyl H-phosphonate + NaX, CCl$_4$, Et$_3$N^{cat}, CH$_3$CN → (RO)$_2$P(O)X]
R = C$_2$H$_5$; i-C$_4$H$_9$; C$_6$H$_5$
X = N$_3$; CN; NCS

Pseudohalophosphates $(RO)_2P(O)X$ (where $X = N_3$, CN, NCS) are versatile reagents in organic synthesis [125]. The Atherton–Todd reaction as a method for direct azidation, cyanation, and thiocyanation of diesters of H-phosphonic acid has a few advantages compared to the other methods for the preparation of phosphorazidates [126], phosphorcyanidates [127], phosphor(isothiocyanatidate)s [128]: (i) the fact that phosphorochloridates are the *in situ* intermediates, (ii) the feasibility of the pseudohalides of phosphorochloridates by alkali pseudohalogens under definite conditions, (iii) higher convenience of H-phosphonates than phosphorochloridates in preparation and availability as starting materials. More importantly, the methodology of direct pseudohalogenation of H-phosphonates may also be applied to other phosphorylating reactions of those compounds with the general functionality of P(O)H.

It was found that dialkyl H-phosphonates react with trialkyl orthoformates at room temperature (70% complete after one week) to give the corresponding alkyldialkyl phosphonates [129].

$$(RO)_2P(O)H + HC(OR)_3 \longrightarrow (RO)_2CH-P(O)(OR)_2 + ROH$$

The evaluation of amine or alcohol or thiol occurs by a mechanism involving deprotonation of the P–H group from the corresponding nucleophile—NR$_2$, SR, or OR, and formation of an active nucleophile—dialkyl phosphonate anion.

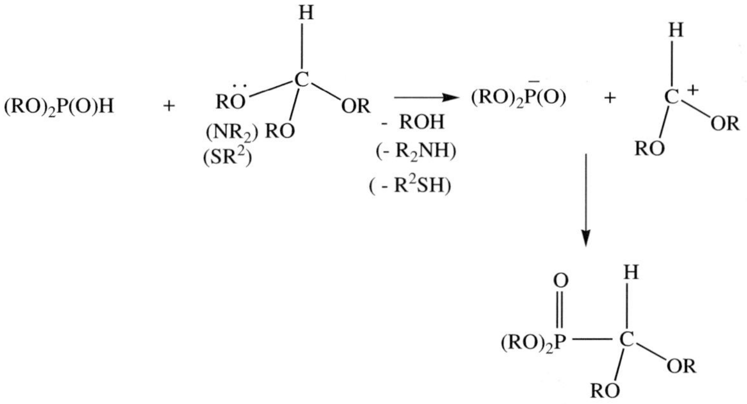

3.7 Characteristic Reactions of P-H Group

$$(RO)_2P(O)H + RO-CH(OR)(NR_2)(SR^2) \xrightarrow{-ROH, (-R_2NH), (-R^2SH)} (RO)_2\bar{P}(O) + {}^{+}CH(OR)(OR) \cdot RO$$

$$\downarrow$$

$$(RO)_2P(O)-CH(H)(OR)(OR)$$

The above-mentioned reactions are used for the P–H group protection. The P–H group can be protected by the dialkoxymethyl group [130], which can be removed from the phosphorus atom by acid hydrolysis [131].

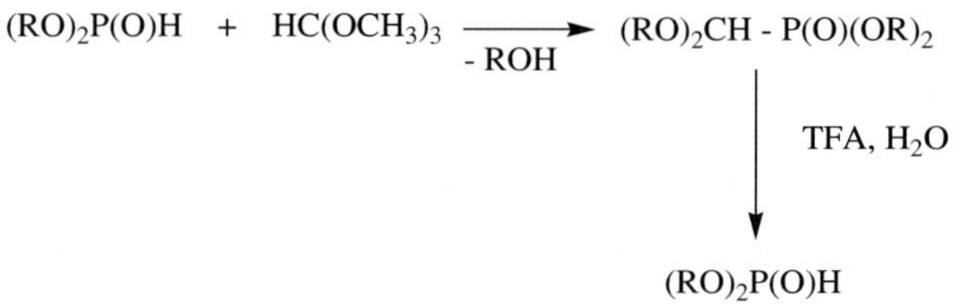

$$(RO)_2P(O)H + HC(OCH_3)_3 \xrightarrow{-ROH} (RO)_2CH-P(O)(OR)_2$$

$$\xrightarrow{TFA, H_2O} (RO)_2P(O)H$$

It is generally recognized that phosphorus–carbon (P–C) bonds of organophosphorus compounds are quite stable and are not cleaved easily under the usual conditions [131]. In contrast, those of dialkyl (dialkoxymethyl)phosphonates are labile.

Dialkyl H-phosphonates react with 4,5-dihalogen-dioxolon-2 yielding dialkylphosphonyl-vinylencarbonate [132].

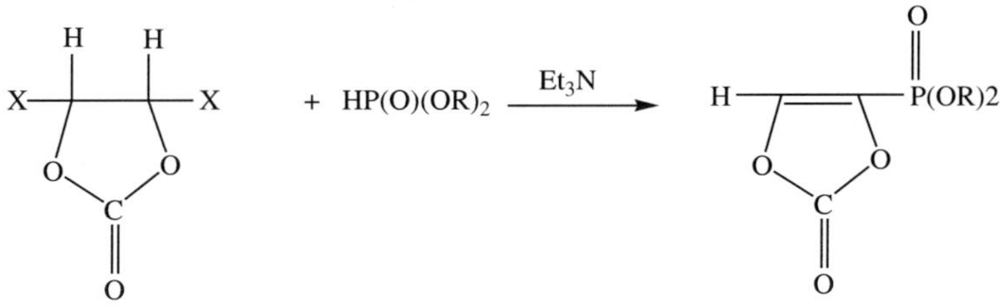

[Reaction scheme: 4,5-dihalo-1,3-dioxol-2-one + HP(O)(OR)$_2$ $\xrightarrow{Et_3N}$ dialkylphosphonyl vinylene carbonate]

X = Cl; Br; J

3.7.2 Addition reactions

Dialkyl alkyl phosphonates (1) are organophosphorus compounds bearing a P–C bond. They have received significant attention as a result of their ability to mimic biological phosphates (2), which play an integral role in the biochemical processes of all living systems [133]. Replacing a P–O linkage by a P–C bond obviously leads to differences in character, especially in pK_a, hydrogen-bonding, and metal-binding capacity.

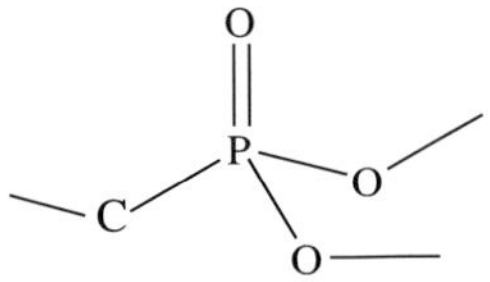

1 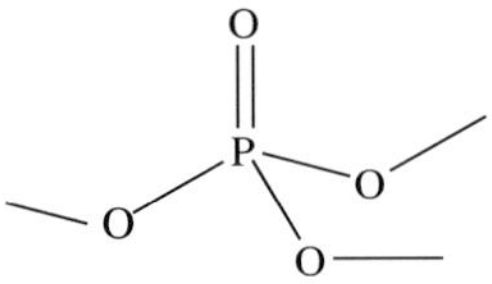   2

The identification of a naturally occurring phosphonate was by Horiguchi and Kandatsu [134], who identified 2-aminoethylphosphonic acid in an amino acid extract from a hydrolysate of rumen protozoal lipid. Subsequently, it has been shown that phosphonates occur naturally in a variety of organisms. Consequently, there is a growing need to develop new and efficient synthetic methods for preparation of organophosphorus compounds bearing a P–C bond.

Phosphonates are more reactive than the corresponding phosphate esters because the carbon atom has no unpaired electrons for a $p\pi$–$d\pi$ contribution to the C–P bond [135]. This makes the phosphorus atom of phosphonates more electrophilic than the phosphorus atom of the corresponding phosphate ester. The esters of phosphonic acid bearing a P–C bond are thus more susceptible to alkaline hydrolysis than the phosphate esters. The P–C bond is usually stable to hydrolytic cleavage. α-Hydroxyalkylphosphonates are an exception. In the presence of an alkali, they undergo rearrangement to give phosphate with cleavage of the P–C bond.

Addition to unsaturated compounds (Pudovik reaction)

H-phosphonate diesters react with electrophilic unsaturated compounds under basic conditions with the formation of dialkyl alkylphosphonic acids [136–139].

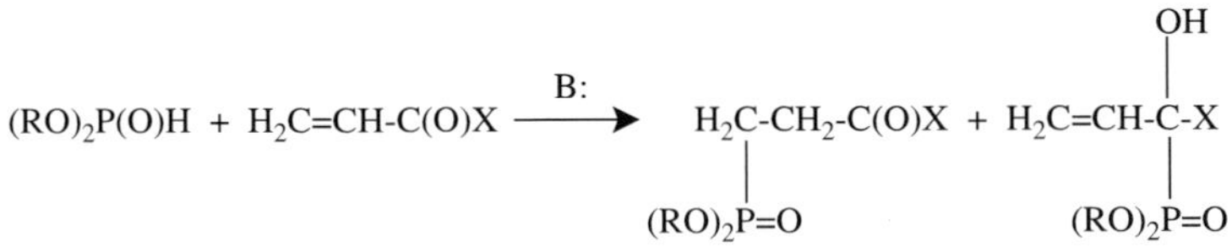

$$(RO)_2P(O)H + H_2C=CH-C(O)X \xrightarrow{B:} \underset{\underset{(RO)_2P=O}{|}}{H_2C-CH_2-C(O)X} + \underset{\underset{(RO)_2P=O}{|}}{H_2C=CH-\underset{\underset{}{|}}{\overset{\overset{OH}{|}}{C}}-X}$$

The addition of H-phosphonate diesters to activated double bonds is known as the Pudovik reaction (see Appendix).

Kinetic studies of the alcoholate-promoted Pudovik addition of dialkyl H-phosphonates to α-enones reveal that depending on the type of the substituents at the enone system, the following interactions take place [140,141]:

3.7 Characteristic Reactions of P-H Group

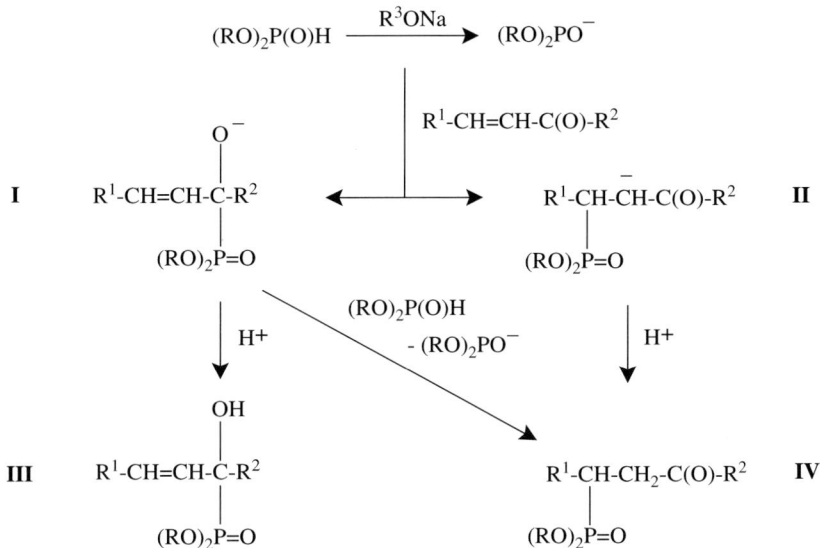

The first step of the reaction in the above scheme is the formation of the active nucleophile—the anion of dialkyl phosphite, obtained *in situ* by deprotonation of dialkyl H-phosphonate with the alcoholate anion. In the overwhelming majority of cases, the nucleophilic attack of the so-formed phosphorus (III) anion is directed toward the carbonyl group of the α-enone. Depending on the nature of the substituent R^2, the reaction may either cease at this stage with the formation of the α-hydroxyphosphonate **III** upon protonation, or proceed further with subsequent addition of dialkyl phosphite to the C=C double bond of the intermediate **I**. This addition is accompanied by simultaneous elimination of dialkyl phosphonate anion and formation of the corresponding γ-ketophosphonate **IV**, which is the thermodynamically stable product of this reaction. It has been established that only strong electronic or steric interactions are capable of blocking the carbonyl group of the enone in the second stage of this reaction, and lead directly to an attack of the phosphite anion at the C=C double bond to form the intermediate **II**. The addition of dialkyl H-phosphonates to butyl cinnamate, particularly, follows this reaction pathway [142].

Experiments have ascertained that after addition of small quantities of the base, a rapid reaction takes place but it does not result in a higher degree of completion. The reaction continues after addition of new quantities of the base. These facts have lead to the assumption that the reaction takes place as a chain process [143] and that the alkoxide anion is an initiator and not a catalyst.

The addition of dialkyl H-phosphonates to acetylene takes place in the presence of complex catalysts. Depending on the molar ratio between the reactants, two different products can be isolated [144].

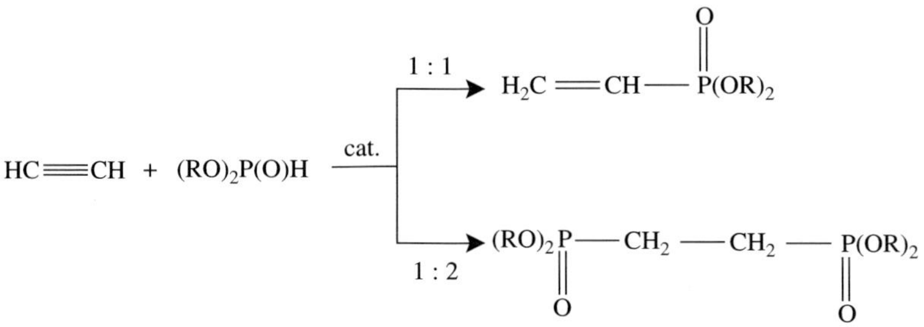

$$HC{\equiv}CH + (RO)_2P(O)H \xrightarrow{cat.} \begin{array}{l} \xrightarrow{1:1} H_2C{=}CH{-}P(O)(OR)_2 \\ \\ \xrightarrow{1:2} (RO)_2P(O){-}CH_2{-}CH_2{-}P(O)(OR)_2 \end{array}$$

Phosphorus-containing alkylene carbonates are obtained by the addition of H-phosphonate diesters to vinyl carbonate [145].

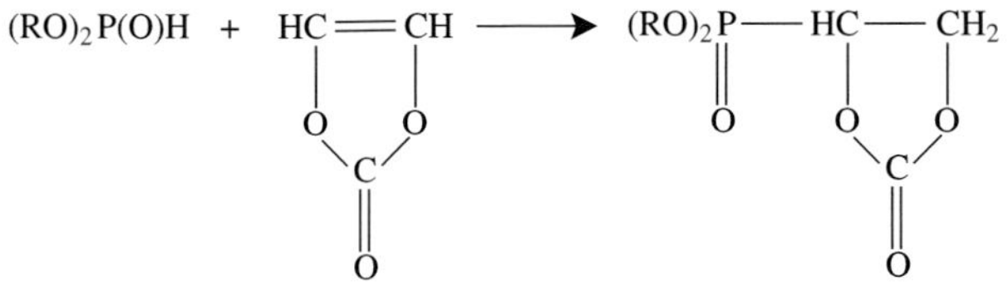

$$(RO)_2P(O)H + \underset{\underset{O}{\overset{\|}{C}}}{\overset{HC=CH}{\underset{O\quad\;\;O}{\diagdown\;\;\diagup}}} \longrightarrow (RO)_2P(O){-}\underset{\underset{O}{\overset{\|}{C}}}{\overset{HC{-}CH_2}{\underset{O\quad\;\;O}{\diagdown\;\;\diagup}}}$$

Diethyl H-phosphonate has been added to acrylonitrile in the presence of RONa [146].

$$(C_2H_5O)_2P(O)H + H_2C=CHCN \longrightarrow (C_2H_5O)_2P(O)CH_2CH_2CN$$

Ammonium or metal salts of monoethyl ester of H-phosphonic acid react with acrylonitrile to furnish the same product [147].

Cyclic phosphonates easily undergo Markovnikov's addition to enamines of the type **I**, yielding adducts with α-aminophosphonate structure **II** [148,149].

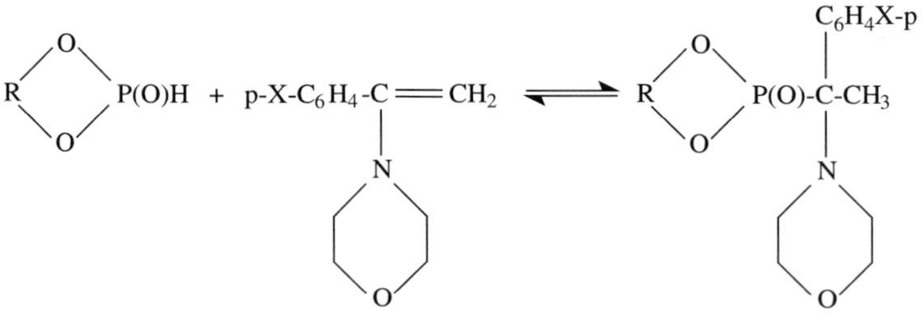

$$R{\diagup}\underset{O}{\overset{O}{\diagdown}}P(O)H + p{-}X{-}C_6H_4{-}\underset{\underset{\text{morpholine}}{N}}{\overset{}{C}}{=}CH_2 \rightleftharpoons R{\diagup}\underset{O}{\overset{O}{\diagdown}}P(O){-}\underset{\underset{\text{morpholine}}{N}}{\overset{C_6H_4X\text{-}p}{C}}{-}CH_3$$

I **II**

X = H; Me; Br

The Pudovik addition of diethyl H-phosphonate to nitrosubstituted alkenes of the type $Ph_2C{=}CRNO_2$ in DMSO or excess diethyl H-phosphonate yields the aziridine **I** (R = H, Me, Ph), formed presumably via the azirine **II** [150, 151].

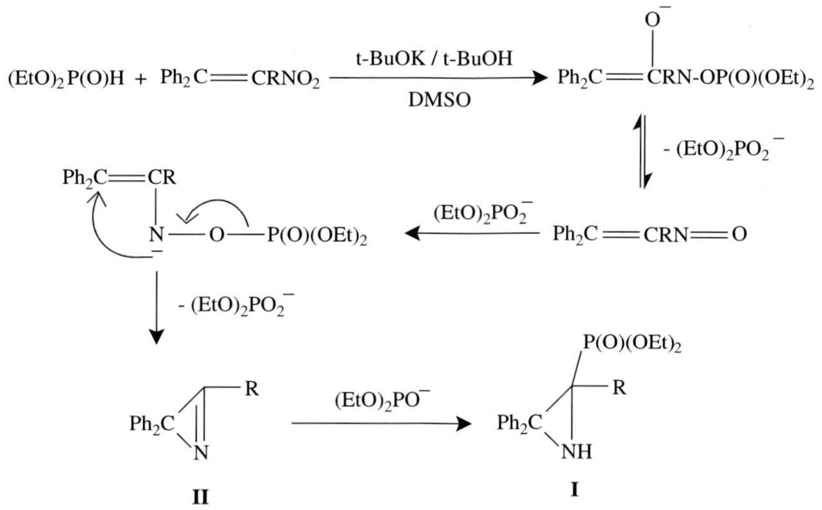

3.7 Characteristic Reactions of P-H Group

$$(EtO)_2P(O)H + Ph_2C=CRNO_2 \xrightarrow[DMSO]{t\text{-BuOK}/t\text{-BuOH}} Ph_2C=CRN\text{-}OP(O)(OEt)_2$$

[Reaction scheme showing the mechanism forming intermediates **II** (azirine) and **I** via loss of $(EtO)_2PO_2^-$ and attack by $(EtO)_2PO^-$]

Under similar conditions, the thiosubstituted alkene (R = SCMe$_3$) yielded only the intermediate azirine **II**. When the above addition reaction was carried out in the presence of potassium *tert*-butoxide, addition of the phosphorus nucleophile to the benzylidene carbon atom takes place [152].

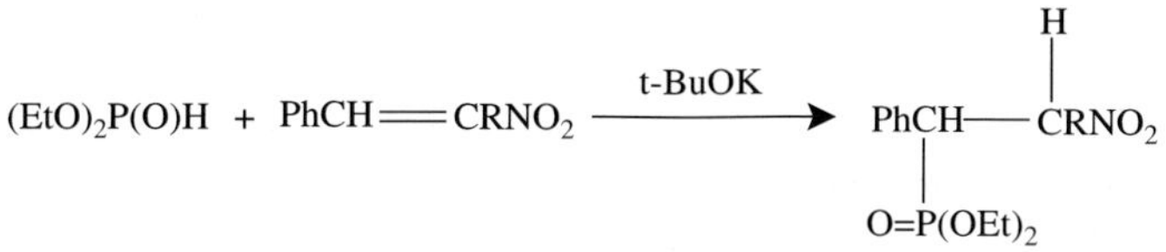

$$(EtO)_2P(O)H + PhCH=CRNO_2 \xrightarrow{t\text{-BuOK}} \underset{\underset{O=P(OEt)_2}{|}}{PhCH}-\underset{\underset{}{|}}{\overset{\overset{H}{|}}{CRNO_2}}$$

Dimethyl H-phosphonate adds to the C=C bond of the ethyl ester of 4,4,4-trichloro-2-cyano-2-butenoic acid [153].

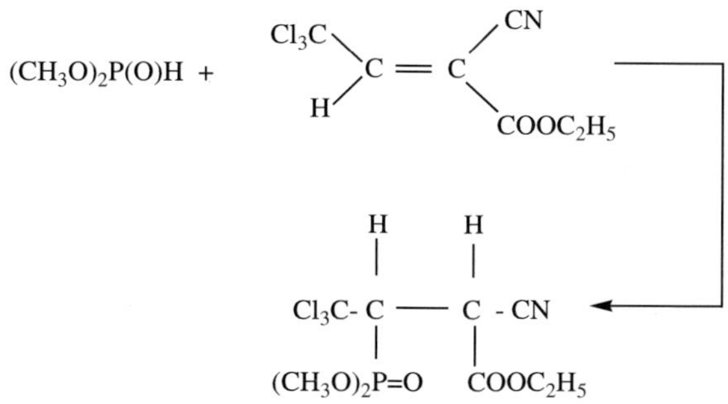

$$(CH_3O)_2P(O)H + \underset{H}{\overset{Cl_3C}{>}}C=C\underset{COOC_2H_5}{\overset{CN}{<}}$$

$$\longrightarrow Cl_3C\text{-}\underset{\underset{(CH_3O)_2P=O}{|}}{\overset{\overset{H}{|}}{C}}-\underset{\underset{COOC_2H_5}{|}}{\overset{\overset{H}{|}}{C}}\text{-}CN$$

This reaction takes place quantitatively at a low temperature. The C=C double bond of this substrate is activated by the presence of the three electron-withdrawing substituents, which emphasizes its strong electrophilic character. This is the reason why this addition reaction takes place in the absence of a catalyst.

The dimethyl H-phosphonate adds to the activated double bond of acrylamide to give dimethylphosphonopropionamide [153].

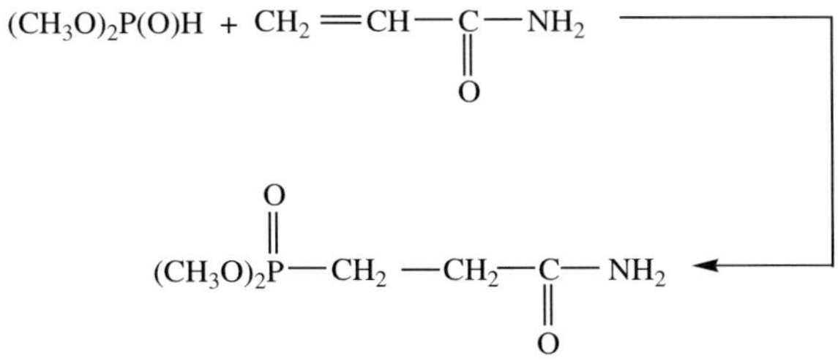

$$(CH_3O)_2P(O)H + CH_2=CH-C(O)-NH_2 \longrightarrow (CH_3O)_2P(O)-CH_2-CH_2-C(O)-NH_2$$

The hydroxymethyl derivative of dimethylphosphonopropionamide is made by reaction with formaldehyde under acidic conditions [154].

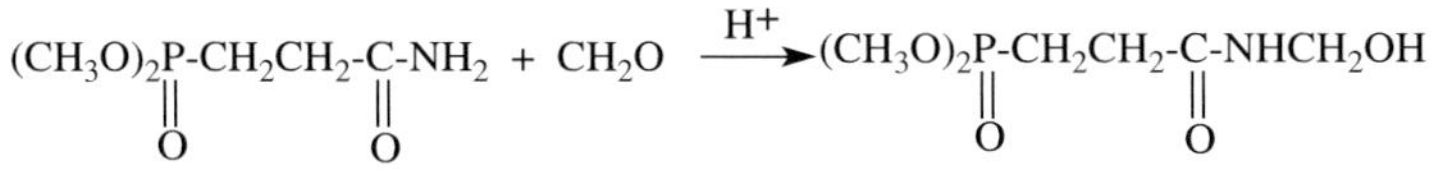

$$(CH_3O)_2P(O)-CH_2CH_2-C(O)-NH_2 + CH_2O \xrightarrow{H^+} (CH_3O)_2P(O)-CH_2CH_2-C(O)-NHCH_2OH$$

Dimethylphosphonopropionamide has been developed by the Ciba Giegy Company and is used as a flame retardant.

Acyclic dialkyl phosphonates do not react under Pudovik's conditions with unsaturated compounds whose double bond is not activated by the presence of an electron-withdrawing substituent. In contrast, their cyclic analogues easily add to styrene, anisole, and vinyl butyl ether in the absence of a catalyst, which is explained in terms of their increased acidity [146,153–157].

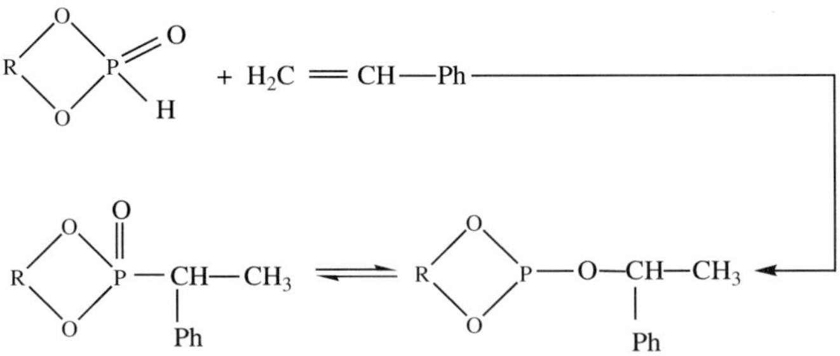

II   I

R = C$_6$H$_4$; CH(CH$_3$)CH(CH)$_3$

According to ^{31}P NMR data, pyrocatechol H-phosphonic acid (R = C$_4$H$_6$) forms with unsaturated compounds addition products of the type **I** having phosphite character with phosphorus chemical shifts in the range 126–130 ppm, typical for such compounds. The

3.7 Characteristic Reactions of P-H Group

formation of product **I** can be explained by the tautomerization of pyrocatechol H-phosphonic acid.

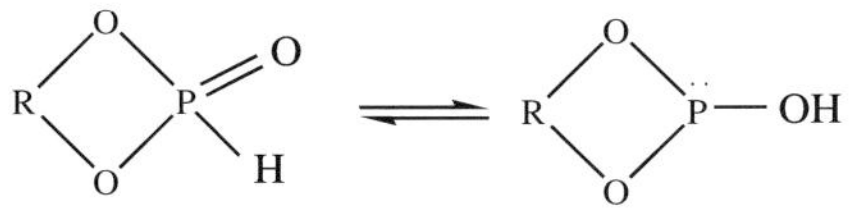

The formation of product **II** has been attributed to phosphite–phosphonate rearrangement from the corresponding phosphorus (III) intermediates of the type **I** [157]. Its analogue 2,3-butylene phosphonic acid (R=CH(CH$_3$)CH(CH$_3$)) forms only phosphonate-type adducts of the type **II** under these conditions.

α–Hydroxy aklanephosphonates (Abramov reaction)

α-Hydroxy alkanephosphonic esters are compounds of significant biological and medicinal applications, for example, in the inhibition of renin [158], HIV protease [159], inositol monophosphatase [160], EPSP synthatase [161], biophosphate mimics [133a,162], transition-state models [163], antibiotic [164], antiviral [165], and antitumor activity [166].

The addition of dialkyl H-phosphonates to aldehydes and ketones [167], known as Abramov reaction (phospho-aldol reaction), is a synthetic route for the preparation of α-hydroxy alkyldialkylphosphonates. This reaction is usually carried out in the presence of alcoholate anions [168] or other basic compounds (see Appendix) [169–174].

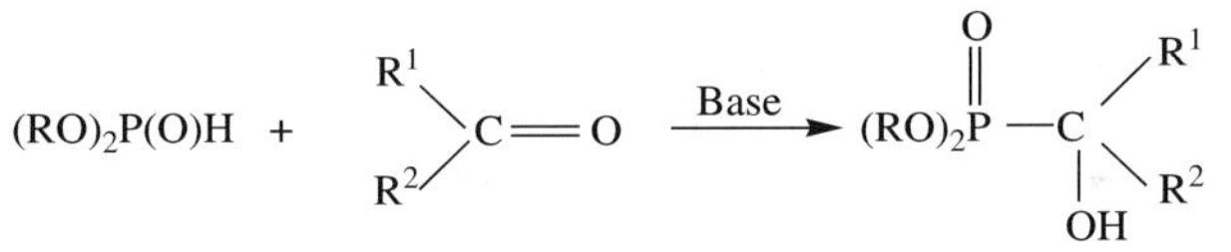

$$(RO)_2P(O)H + \underset{R^2}{\overset{R^1}{>}}C=O \xrightarrow{Base} (RO)_2\overset{O}{\overset{\|}{P}}-\underset{OH}{\overset{R^1}{\underset{|}{C}}}R^2$$

The reaction is exothermic and significant temperature increase is observed upon addition of small quantities of the base to the reaction mixture. α-Functionality introduces chirality at the α-carbon atom and consequently, a stereoisomeric influence upon biological properties.

The following mechanism for this reaction has been originally proposed by Abramov [168]:

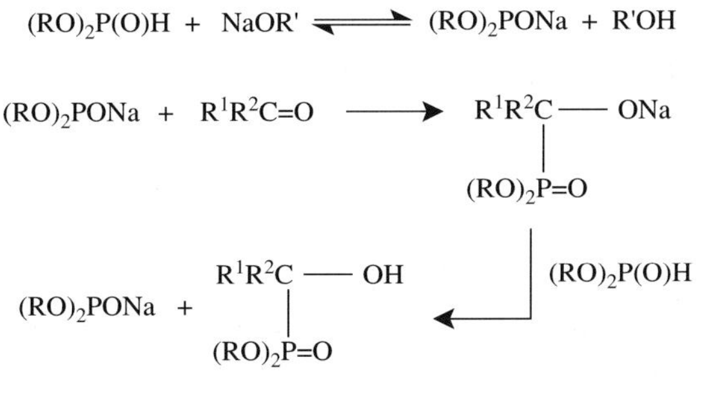

$$(RO)_2P(O)H + NaOR' \rightleftharpoons (RO)_2PONa + R'OH$$

$$(RO)_2PONa + R^1R^2C=O \longrightarrow R^1R^2C-ONa$$
$$\qquad\qquad\qquad\qquad\qquad\qquad |$$
$$\qquad\qquad\qquad\qquad\qquad\qquad (RO)_2P=O$$

$$(RO)_2PONa + R^1R^2C-OH \xleftarrow{(RO)_2P(O)H}$$
$$\qquad\qquad\qquad\qquad |$$
$$\qquad\qquad\qquad\qquad (RO)_2P=O$$

The possible role of the monoalkyl phosphonate anion [(RO)PH(O)O]⁻ as a proton acceptor from dialkyl H-phosphonate has also been discussed in the mechanism proposed by M. Kabachnik and E. Tsvetkov [175] for addition of dialkyl H-phosphonates to carbonyl compounds.

$$\underset{R^2}{\overset{R^1}{>}}C=O \rightleftharpoons (RO)_2\overset{O}{\overset{\|}{P}}-\overset{O^-}{\underset{|}{C}}R^1R^2 + \underset{HO}{\overset{RO}{>}}\overset{O}{\overset{\|}{P}}-H \longrightarrow$$

$$\left[\underset{RO}{\overset{RO}{>}}\overset{O}{\overset{\|}{P}}-H \cdots \underset{-O}{\overset{RO}{>}}\overset{O}{\overset{\|}{P}}-H\right] + (RO)_2\overset{O}{\overset{\|}{P}}-\underset{|}{\overset{OH}{C}}R^1R^2$$

Dimethyl phosphonate halohydrin is formed by the base-catalyzed reaction of dimethyl H-phosphonate and chloroacetone [176].

$$(CH_3O)_2P(O)H + CH_3-\overset{O}{\overset{\|}{C}}-CH_2Cl \xrightarrow{(C_2H_5)_3N} \underset{CH_3O}{\overset{CH_3O}{>}}\overset{O}{\overset{\|}{P}}-\underset{\underset{CH_3}{|}}{\overset{\overset{OH}{|}}{C}}-CH_2Cl$$

Halohydrin phosphonates are intermediates in some synthetic sequences [172].

It has been found that dialkyl H-phosphonates react with chloral in the absence of a catalyst [172,176–180].

$$(RO)_2P(O)H + Cl_3CCHO \longrightarrow (RO)_2\overset{O}{\overset{\|}{P}}-\underset{\underset{OH}{|}}{\overset{\overset{H}{|}}{C}}-CCl_3$$

In the absence of a solvent and excess of chloral, the reaction rate for this reaction may be expressed by a third-order equation–second order with respect to dimethyl H-phosphonate and first order with respect to the chloral [177]. In dioxane solution and excess dimethyl H-phosphonate, the dependence of the reaction rate on the chloral concentration is the same. In addition to the chloro-containing α-hydroxyalkyl phosphonate, which is the main product under the above conditions, a side product formed as a result of dehydrochlorination of the main product, followed by phosphonate–phosphate rearrangement, has been also isolated. The presence of a base such as triethylamine, alkali metal alkoxides and hydroxides, or sodium carbonate accelerates the dehydrochlorination process. An example of this side reaction is the transformation of dialkyl-2,2,2-trichloro-1-hydroxyethyl phosphonates into dialkyl-2,2-dichlorovinyl phosphates in the presence of sodium hydroxide [181,182].

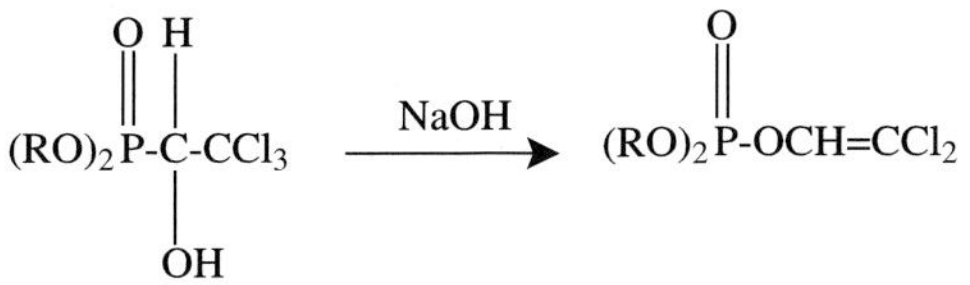

3.7 Characteristic Reactions of P-H Group

$$(RO)_2\overset{O}{\underset{\parallel}{P}}-\underset{\underset{OH}{|}}{\overset{\overset{H}{|}}{C}}-CCl_3 \xrightarrow{NaOH} (RO)_2\overset{O}{\underset{\parallel}{P}}-OCH=CCl_2$$

The synthesis of α-hydroxymethylene phosphonates involves reaction of dialkyl H-phosphonates with paraformaldehyde in the presence of a base [183].

$$(RO)_2P(O)H + CH_2O \xrightarrow{Et_3N} (RO)_2P(O)CH_2OH$$

$R = n\text{-}C_6H_{13}\ ;\ n\text{-}C_4H_9$

α-Hydroxymethylene phosphonates in 90–92% yield were obtained.

It has been established that instead of paraformaldehyde, a water solution of formaldehyde can also be applied when the reaction is carried out using a phase transfer catalyst [184].

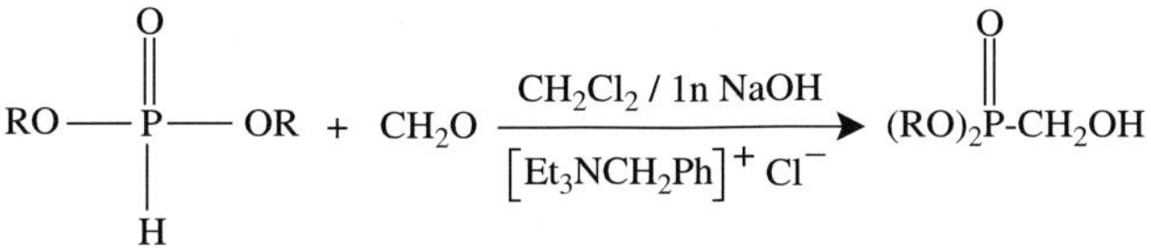

$$RO-\underset{\underset{H}{|}}{\overset{\overset{O}{\parallel}}{P}}-OR + CH_2O \xrightarrow[\text{[Et}_3\text{NCH}_2\text{Ph]}^+\text{Cl}^-]{CH_2Cl_2 / 1n\ NaOH} (RO)_2\overset{O}{\underset{\parallel}{P}}\text{-}CH_2OH$$

Dialkyl alkylene diphosphonates obtained by transesterification of dialkyl H-phosphonates with diols in the excess of H-phosphonate diesters are bifunctional monomers possessing two reactive P–H groups. The addition of ketones or aldehydes to dialkyl alkylene diphosphonates resulted in the formation of the corresponding bis(α-hydroxyalkylphosphonate)s [185].

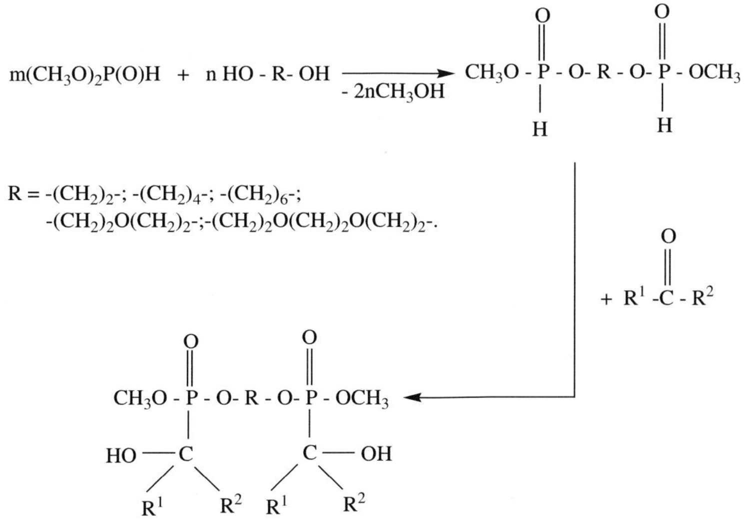

$$m(CH_3O)_2P(O)H + n\ HO\text{-}R\text{-}OH \xrightarrow{-2nCH_3OH} CH_3O\text{-}\underset{\underset{H}{|}}{\overset{\overset{O}{\parallel}}{P}}\text{-}O\text{-}R\text{-}O\text{-}\underset{\underset{H}{|}}{\overset{\overset{O}{\parallel}}{P}}\text{-}OCH_3$$

$R = -(CH_2)_2\text{-};\ -(CH_2)_4\text{-};\ -(CH_2)_6\text{-};$
$-(CH_2)_2O(CH_2)_2\text{-};-(CH_2)_2O(CH_2)_2O(CH_2)_2\text{-}.$

$+\ R^1\text{-}\overset{O}{\underset{\parallel}{C}}\text{-}R^2$

$$CH_3O\text{-}\overset{O}{\underset{\parallel}{P}}\text{-}O\text{-}R\text{-}O\text{-}\overset{O}{\underset{\parallel}{P}}\text{-}OCH_3$$
$$HO-\underset{\underset{R^1\ \ R^2}{\diagup\diagdown}}{C}\quad\quad\underset{\underset{R^1\ \ R^2}{\diagup\diagdown}}{C}-OH$$

$R^1 = R^2 = CH_3;\ R^1 = R^2 = H;\ R^1 = C_6H_5; R^2 = H$
$R^1 = CCl_3; R^2 = H;\ R^1 = C_{10}H_7; R^2 = H$

Bis(α–hydroxyalkylphosphonate)s are used in the polyurethanes synthesis.

The addition of dialkyl H-phosphonates to carbonyl compounds proceeds exothermically with the introduction of the initial quantities of the base in the reaction mixture. The reaction does not, however, go to completion and can be restarted with additional quantities of the base. These methods often are disadvantageous. The yields are not always good and mixtures of products are sometimes obtained. In a strongly alkaline medium, α-hydroxyalkanephosphonic esters are cleaved to regenerate the starting carbonyl compound [170]. The reaction of dialkyl H-phosphonates with α-haloketones in basic media gives epoxyalkanephosphonates or α-alkenephosphonates [186–188].

The above-mentioned disadvantages can be avoided when dialkyl H-phosphonates react with aldehydes or ketones (including α-haloketones) in the presence of potassium or cesium fluoride [189-191].

$$RO-\underset{H}{\underset{|}{\overset{O}{\overset{\|}{P}}}}-OR + \underset{R^1}{\overset{R^2}{>}}C=O \xrightarrow{KF \text{ or } CsF} R^1-\underset{OH}{\underset{|}{\overset{R^2}{\overset{|}{C}}}}-\underset{OR}{\overset{O}{\overset{\|}{P}}}\diagdown OR$$

This reaction provides an easy and rapid access to α-hydroxyalkanephosphonic esters in almost quantitative yield. Side reactions are not observed. Conducting the reaction in the presence of solvents (dimethylformamide or tetrahydrofuran) leads to lower yields of α-hydroxyalkanephosphonate, or when α-haloketones are used as starting materials, results in the formation of α-hydroxyalkanephosphonates and epoxyalkanephosphonates. These observations suggest that the reaction takes place on the surface of the alkali metal fluoride, the reaction rate depending on the quantity of fluoride salt used.

This method is very efficient for aldehydes and activated ketones. Recent results obtained in this field showed that the simple heterogeneuos mixture of finely powdered γ–alumina and potassium fluoride can promote the condensation of nonactivated ketones with dialkyl H-phosphonates at room temperature [192–194] to give α-hydroxyalkanephosphonates in good to quantitative yield [194].

Surface-mediated solid phase reactions are of growing interest [195] because of their ease of set-up and work, mild reaction conditions, rate of the reaction, selectivity, high yields, lack of solvent, and the low cost of the reactions compared to their homogeneous counterparts. Kaboudin et al. have described a method for synthesis of diethyl 1-hydroxyarylmethylphosphonates from aldehydes and diethyl H-phosphonate on a magnesia surface without a solvent [196] (see Appendix). The important features of this

$$Ar-\overset{O}{\overset{\|}{C}}H + H\overset{O}{\overset{\|}{P}}(OC_2H_5)_2 \xrightarrow{MgO} Ar-\underset{OH}{\underset{|}{C}H}-\overset{O}{\overset{\|}{P}}(OC_2H_5)_2$$

3.7 Characteristic Reactions of P-H Group

methodology are (i) the addition of external base is usually not required, (ii) side reactions are not observed; mostly high yields were obtained.

There is a growing need to develop new and efficient synthetic routes to & macmillan-functionalized phosphonic structures that permit control of stereochemistry. Various metal complexes have been found to act as effective catalysts for the reaction, including titanium [197] and lanthanide [198] systems. A key objective in phosphotransfer chemistry is the development of metal-based catalysts for the phosphor–aldol (PA) reaction. These complexes must be (i) simple to prepare, (ii) inexpensive, (iii) tunable, (iv) compatible with air and water, (v) reusable, (vi) stereoselective. There are a number of PA catalysts that satisfy the above criteria [199]. It was shown that the chiral salcyan complexes of aluminum are effective air- and water-tolerant enantioselective catalysts for hydrophosphonylation of carbonyls [200]. The initial step of the

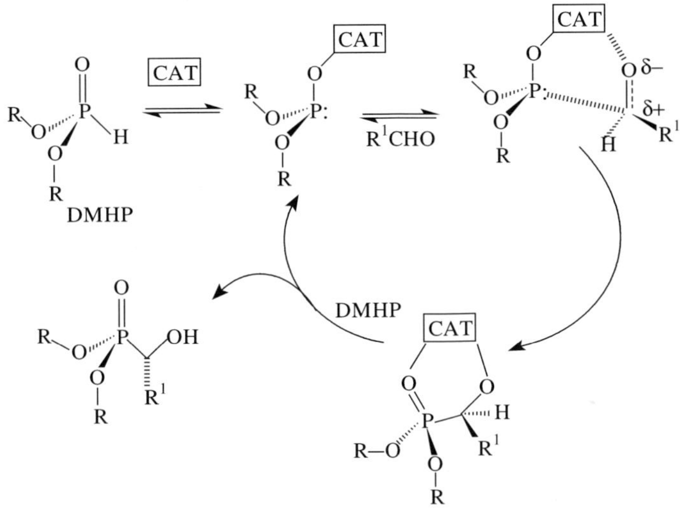

catalytic asymmetric phospho-aldol reaction involves deprotonation of H-phosphonate by a suitable polarized $M^{\delta+}-C^{\delta-}$ (CAT) M = Al bond to afford an intermediate metallo-phosphite. It should be pointed out that this intermediate has not been proven. Reaction of the metallo-phosphite complex with aldehyde is the important step in which a new stereocenter is created, P–C bond formation and the proton transfer from H-phosphonate occur to afford decomplexed α-hydroxyphosphonate ester and generate metallo-phosphite intermediate.

The alternative reaction pathway for the phospho-aldol reaction involves deprotonation of H-phosphonate by polarized $M^{\delta+} - C^{\delta-}$ (CAT) bond to afford an intermediate metallo-phosphonate, not metallo-phosphite. In this metallo-phosphonate intermediate, the phosphorus atom plays the role of a nucleophile.

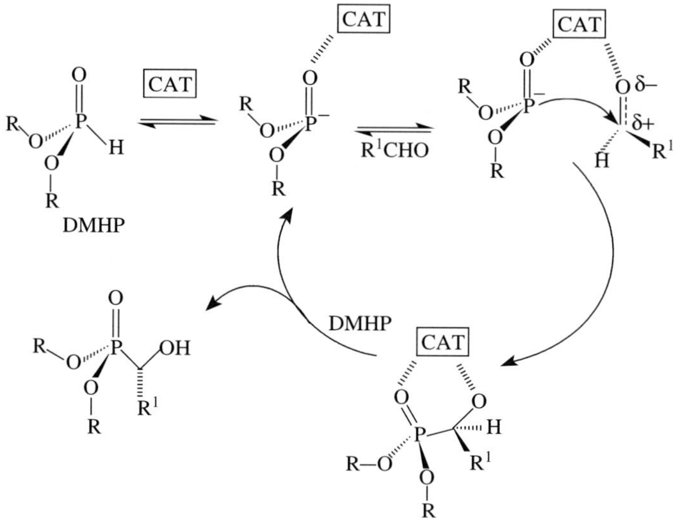

The role of the catalyst is to encourage both the phosphonate and carbonyl to come together in the crucial stereo-determining P–C bond formation step.

Heterobimetallic multifunctional catalysts such as LaLi$_3$tris (binaphtooxide) are believed to activate both nucleophiles and electrophiles [201–203]. For the hydrophosphonylation For the hydrophosphonylation of comparatively unreactive aldehydes, the activated phosphonates can react only with aldehydes that are precoordinated to lanthanum (route A). In the case of reactive aldehydes, the Li-activated phosphonate may be able to participate in a competing reaction with unactivated aldehyde (route B). If such aldehydes are added in one portion, the ee (enantiomeric excess) of the reaction product will be reduced. Slow addition of aldehyde facilitates aldehyde activation.

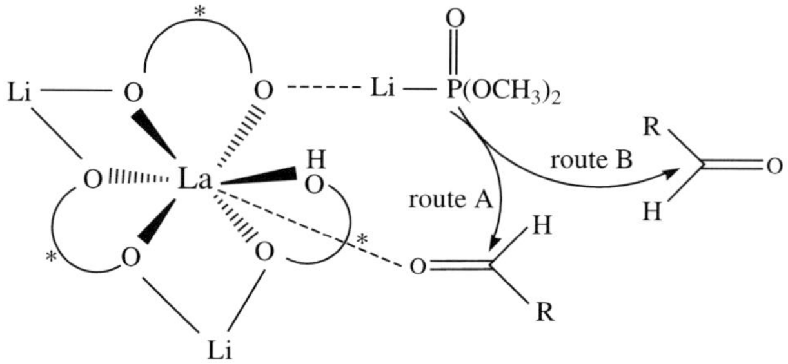

Enantioselectivities have been determined by ^{31}P{H} NMR spectroscopy using 1-(1-naphtyl)ethylamine [204] or quinine as a chiral solvating agent [205], a technique and reagent already used with considerable success for α-hydroxyphosphonate systems [206]. Absolute configurations are envisaged to be all R on the basis that the optical rotation $[\alpha]^D$

3.7 Characteristic Reactions of P-H Group

of $(MeO)_2P(O)CH(OH)C_6H_4Me-4$ (39% enantiomeric excess) was found to be $+17.1°$ ($c = 1$;$CHCl_3$).

The striking feature of the dialkyl α-hydroxyalkylphosphonates is their anomalous behavior in the cryoscopic determination of the molecular weight in benzene. The observed values indicate that the compounds are associated to a dimeric structure due to the formation of strongly hydrogen-bonded structures involving intermolecular reaction between the hydroxyl group and the strongly polar phosphoryl group (P=O) [186].

β- and γ-Hydroxyphosphonate. A facile synthesis of β- and γ-hydroxyphosphonate esters by $BF_3.OEt_2$-catalyzed regiospecific ring-opening of a series of epoxides by dialkyl phosphonate has been recently reported [207].

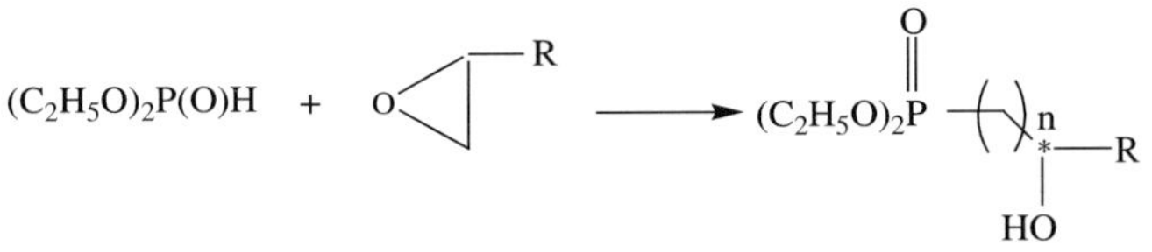

$$(C_2H_5O)_2P(O)H + \underset{}{\triangle}\!\!-R \longrightarrow (C_2H_5O)_2\overset{O}{\underset{}{P}}\!\!-\!(\ )_n\!-R$$
with HO group on the chiral carbon.

Peptidic α-hydroxyphosphonates. α-Hydroxyl esters of type **1** were good renin inhibitors [208,209]. Replacement of the ester moiety in **1** by its bioisostere, a phosphonate group, results in a novel potent human renin inhibitor **2**.

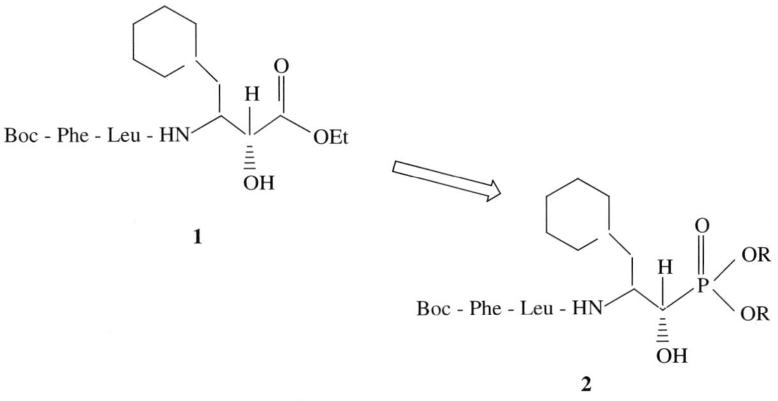

Boc - Phe - Leu - HN—CH(OH)—C(=O)—OEt

1

Boc - Phe - Leu - HN—CH(OH)—P(=O)(OR)₂

2

A simple approach for preparing peptidic α-hydroxyphosphonates involves coupling an appropriate amino acid or peptidic fragment with a β-amino-α-hydroxyphosphonate [158a]. α-Hydroxyphosphonates **3** (mixture of diastereomers *syn* and *anti* in ratio 12:1.0) obtained by reaction of the chiral aldehyde derived from L-phenylalanine with dimethyl H-phosphonate in the presence of potassium fluoride as a catalyst in dimethylformamide as solvent [158a] was utilized as an intermediate for the preparation of α-hydroxyphosphonate **4** [158a] or α-hydroxyphosphinyls [158b].

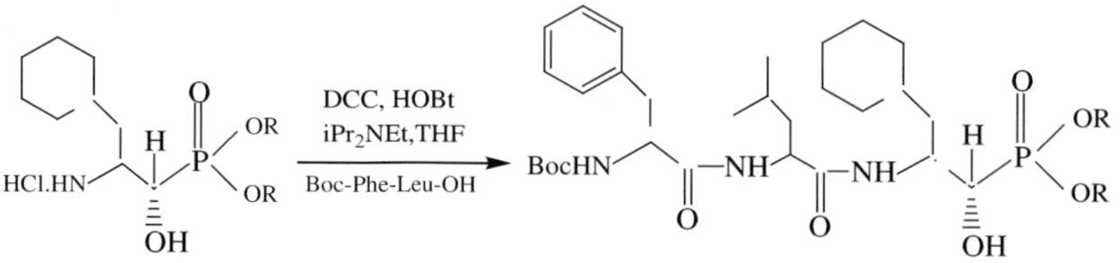

The peptidic α-hydroxyl phosphonates present a promising and structurally unique class of transition-state analogue inhibitors of proteolytic enzymes and make the first disclosure of their application in preparing inhibitors of human renin.

Nucleoside α-Hydroxyphosphonates.   In recent years, there has been a tremendous resurgence of interest in synthesis of modified nucleosides, primarily because of their potential antiviral activity. Three modified nucleosides, 3'-azidothymidine [210], dideoxyinosine [211], and dideoxycytidine [212], are the only drugs of recognized therapeutic value in the treatment of AIDS. Their toxic side effects [213], which limited the application of these compounds, are the driving force to develop new nucleoside derivatives with the potential to be more selective anti-HIV agents.

Reaction of the 2'-keto nucleoside [214] with diethyl H-phosphonate at −78 °C in the presence of lithium bis(trimethylsily)amide gave the desired 2'-hydroxyl-2'-phosphononucleosides as a single diastereomer in 80% yield [215] (see Appendix).

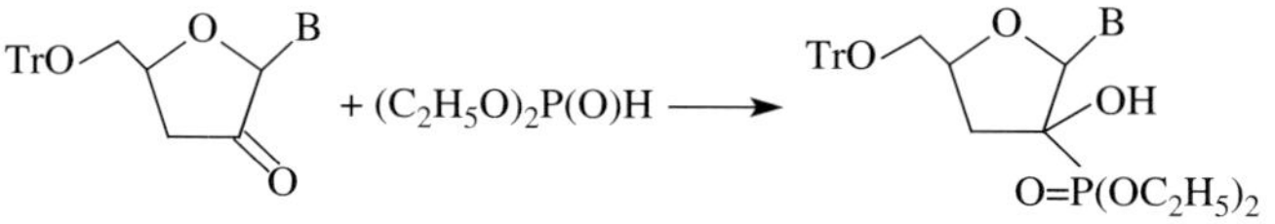

B = uridine; adenosine

At a higher temperature, a mixture of diastereomers was obtained. The protected adenosine (B = adenosine) derivative also undergoes a smooth reaction, affording the analogous phosphonate in 82% yield.

Reaction of diethyl H-phosphonate with the 3'-keto derivatives of uridine [216], adenosine [217], and 2'-deoxythymidine [218] furnished in high yield the desired 3'-hydroxy-3'-phosphononucleosides.

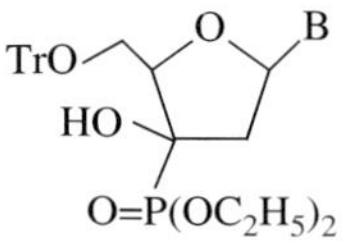

B = uridine; adenosine; thymidine

Despite the reported instability of these 3'-ketones, Wiemer et al. [215] have found that there is no significant elimination of the purine or pyrimidine systems under the mildly basic reaction conditions. The stereochemistry of 5'-trytyl-3'-β-hydroxyl-3-α–(diethyl)phosphonothymidine was determined by single-crystal diffraction analysis. The X-ray analysis clearly indicates that a phosphonate anion has added on to the carbonyl group from the sterically less hindered α-face of the nucleoside.

3.7 Characteristic Reactions of P-H Group

Nucleoside α-hydroxyl phosphonates present a new type of nucleoside derivatives. These compounds should be of interest for several reasons. The hydroxyl phosphonate group itself offers a variety of possibilities for transformation to other functionality. The corresponding phosphonic acids could be viewed as new analogues isomeric to nucleoside 3'-phosphates.

Such compounds may have some potential to interfere with viral replication and to be more selective anti-HIV agents.

The nucleoside 5'-hydroxylphosphonate was prepared by Abramov nucleophilic addition of H-phosphonate diesters to nucleoside 5'-aldehydes (see Appendix) [219].

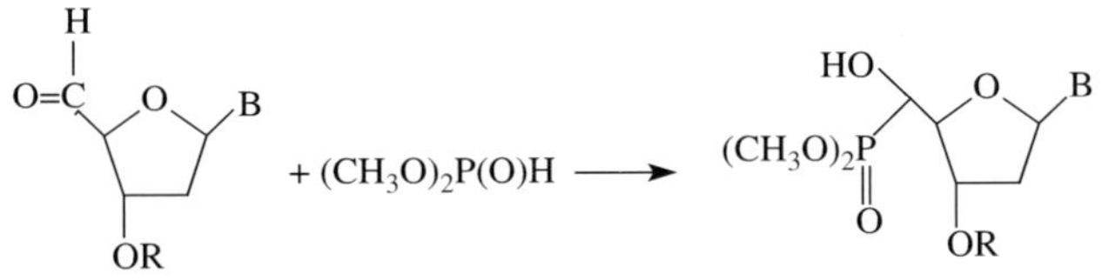

$R = Si^tC_4H_9(C_6H_5)_2$

It was found that CH_2Cl_2 provided better yield of nucleoside 5'-hydroxylphosphonate than THF. On the other hand, increasing the amount of triethylamine (from 1 to 5 equivalents) did not influence the ratio of epimers. It was found that little change in the epimeric ratios was found when dialkyl hydrogen phosphonates with bulkier ester groups (dimethyl to diethyl to diisopropyl) were used in the presence of triethylamine.

Chemical transformations of α-hydroxyphosphonates

Phosphonate–phosphate rearrangement. Dialkyl alkylphosphonates are, in general, more reactive than the corresponding phosphate esters because the carbon has no unpaired electrons to contribute to allow a pπ–dπ contribution to the P–C bond. This makes the phosphorus atom of phosphonates more electrophilic than the phosphorus atom of the corresponding phosphate ester. The P–C bond is usually stable to hydrolitic procedures. However, α-hydroxyalkylphosphonates, in the presence of alkali, rearrange to give phosphate with cleavage of the P–C bond [220].

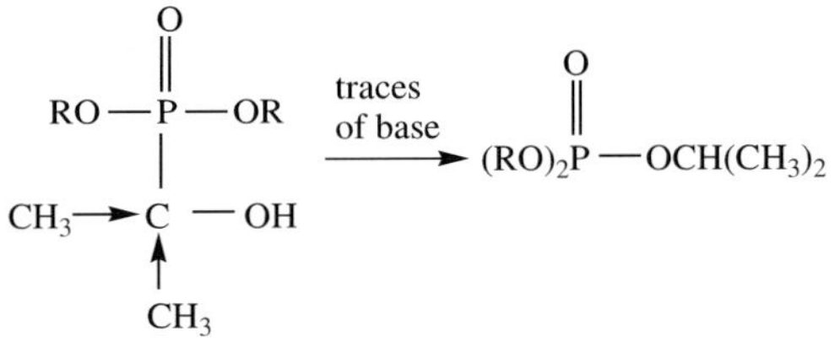

It was shown that α-hydroxyalkylphosphonate derived from dimethyl H-phosphonate and 2-nitrobenzaldehyde undergoes such rearrangement when traces of triethylamine are present [220d].

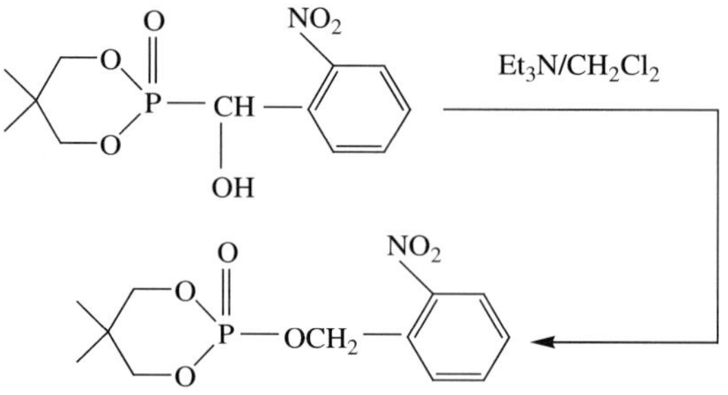

Depending on the type of the substituents at the carbon atom of the P–C fragment, the α-hydroxyalkylphosphonates undergo a phosphonate–phosphate rearrangement [220] or decompose upon heating to the starting materials. The presence of electron-donor substituents at the P–C carbon atom facilitates the phosphonate–phosphate rearrangement. On the other hand, electron-withdrawing substituents at the P–C carbon of the α-hydroxyalkyl phosphonates cause additional polarization of the P–C bond, and its thermal cleavage occurs with a higher rate toward the formation of the starting compounds.

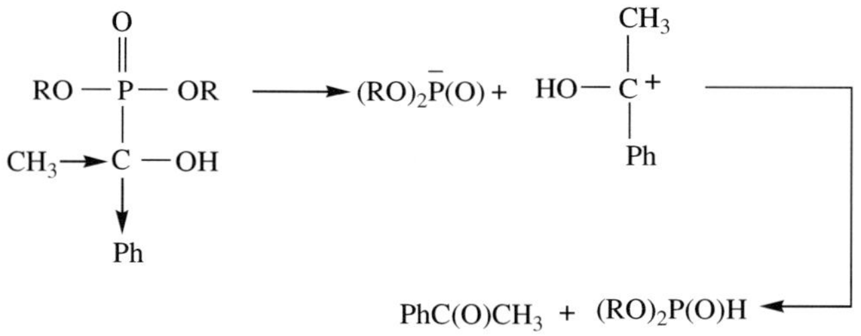

This assumption is supported by the fact that when a phenyl group is introduced at the P–C carbon atom, the thermal stability of the corresponding α-hydroxyalkylphosphonates decreases [221].

Hydrolysis. The saturated α-hydroxyalkylphosphonates are readily hydrolyzed in 18% aqueous hydrochloric acid to the corresponding α-hydroxyalkylphosphonic acids (route a) [186]. In the presence of a strongy alkali, the α-hydroxyalkylphosphonates are not hydrolyzed, but quantitatively regenerate the starting aldehyde (route b).

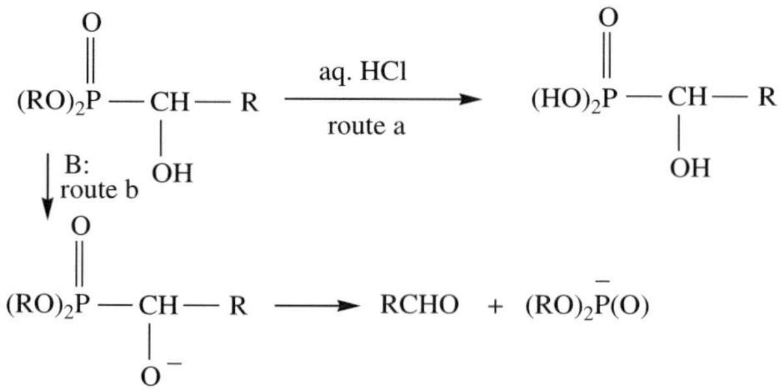

3.7 Characteristic Reactions of P-H Group

Oxidation: Synthesis of Dialkyl Acylphosphonates. Phosphonate esters have found a wide range of application in the areas of industrial, agricultural, and medicinal chemistry owing to their physical properties as well as their utility as synthetic intermediates. Phosphonates are interesting complements to phosphates in terms of biological activity [133a,222]. Within this class of compounds there exists an important subdivision, the α-ketophosphonates. The Michael–Arbuzov reaction is a general method for the preparation of α-ketophosphonates from acyl chlorides and trialkyl phosphites [223]. Oxidation of α-hydroxyphosphonates is another method for the preparation of α-ketophosphonates [163,224]. A new method for the preparation of acyl phosphonates by oxidation of α-hydroxyphosphonates on the solid surface is described [196b]. It is found that alumina (neutral)-supported CrO_3 under solvent-free conditions is capable of producing high yields of acyl phosphonates from α-hydroxyphosphonates under mild reaction conditions (see Appendix).

α-Ketophosphonates have been prepared by oxidation of α-hydroxyphosphonates by zinc dichromate trihydrate ($ZnCr_2O_7.H_2O$) under solvent-free conditions [225]. The limitation of this method is the toxicity of zinc dichromate trihydrate. This disadvantage is avoided using the environmentally friendly and inexpensive potassium permanganate ($KnMnO_4$) for the oxidation of α-hydroxyphosphonates with or without solvent [226].

In order to have convenient product isolation and a better reactivity, a neutral alumina-supported potassium permanganate (NASPP) is used. Solvent-free reactions have received more attention compared to their homogeneous counterparts, due to economical and environmental demands and simplicity in processes [227].

Dialkyl acylphosphonates possessing C(O)–P bonds are known to be labile, even toward moisture in air, and decompose into carboxylic acids and dialkyl phosphonates [228–233].

Kluger *et al.* [234] have shown that the hydrolysis (neutral aqueous solution) of dimethyl acetylphosphonate (DAP) results in cleavage of the carbonyl-to-phosphonyl bond with attack of water occurring at the carbonyl carbon atom. ^{1}H NMR spectroscopy reveals that DAP reacts with water initially to form a hydrated carbonyl derivative **2**.

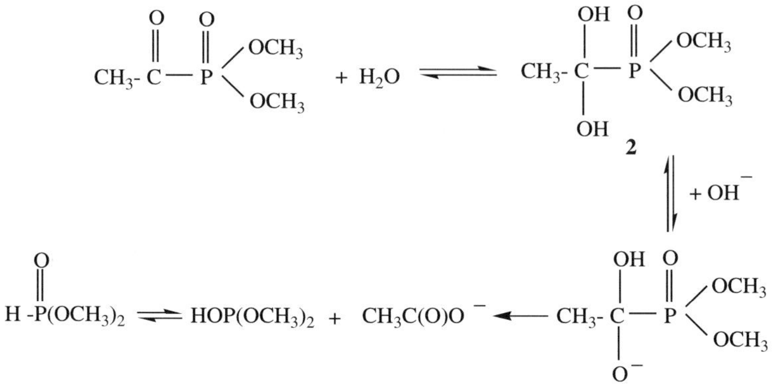

$$CH_3-C(O)-P(O)(OCH_3)_2 + H_2O \rightleftharpoons CH_3-C(OH)_2-P(O)(OCH_3)_2 \quad \mathbf{2}$$

$$\mathbf{2} + OH^- \downarrow\uparrow$$

$$H-P(O)(OCH_3)_2 \rightleftharpoons HOP(OCH_3)_2 + CH_3C(O)O^- \leftarrow CH_3-C(OH)(O^-)-P(O)(OCH_3)_2$$

Acylation of functional groups such as hydroxyl and amino groups is one of the important and fundamental reactions in organic synthesis. The facile nucleophilic cleavage of the P–C bond is a characteristic feature of acylphosphonates, indicating their potential usefulness as acylating agents.

Stoichiometric reactions of diethyl benzoylphosphonate with a variety of amines gave amides **1** as the main products along with diethyl H-phosphonate **2** and α-(phosphoryloxy)benzyl phosphonates **3** [235]. The use of hindered diethyl benzoylphosphonate resulted in high yield of amides in the case of primary amines and poorer yields of amides in the case of the reaction with a hindered amine such as diethylamine.

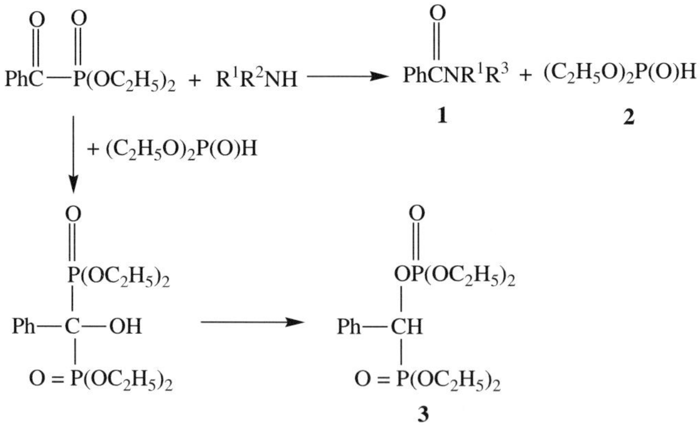

$$PhC(O)-P(O)(OC_2H_5)_2 + R^1R^2NH \longrightarrow PhCNR^1R^3 \; (\mathbf{1}) + (C_2H_5O)_2P(O)H \; (\mathbf{2})$$

$$\downarrow + (C_2H_5O)_2P(O)H$$

$$Ph-C(OH)(P(O)(OC_2H_5)_2)-O=P(OC_2H_5)_2 \longrightarrow Ph-CH(OP(O)(OC_2H_5)_2)-O=P(OC_2H_5)_2 \quad \mathbf{3}$$

In nonpolar solvents such as *n*-hexane and cyclohexane, benzoylation is remarkably rapid while in methylene chloride, it is much slower than in other solvents used. Addition of triethylamine and 4-(dimethylamino)pyridine slightly accelerates the benzoylation, but yields of amides are similar to those in the absence of the catalysts. α-(Phosphoryloxy)benzyl phosphonate **3** may be formed by the successive reaction of diethyl benzoylphosphonate with diethyl H-phosphonate, which accumulates during the formation of amide, followed by a phosphonates–phosphate rearrangement [220,221].

3.7 Characteristic Reactions of P-H Group

Dialkyl acylphosphonates are used for acylation of alcohols. Kabachnik [236] reported that dimethyl benzoylphosphonate formed crystalline 1:1 addition products with aliphatic alcohols, which can be converted in the presence of dry hydrogen chloride into carboxylic esters and dimethyl H-phosphonate in moderate yields. Sakurai reported that dialkyl acylphosphonates underwent slow acylation with a large excess amount of alcohols to give esters [237]. It was shown that triethylamine catalyzed acylation of alcohols with dialkyl acylphosphonates [238]. It was found that use of 1,5-diazabicyclo[5,4,0]undec-5-ene (DBU) increased the rate of the acylation of alcohols with diethyl benzoylphosphonate dramatically [239]. When one equivalent of DBU was used in CH_2Cl_2, the acylation was completed within 10 min and the corresponding ester **1** was obtained in 99% yield. The use of 0.1 equivalent of DBU gave **1** in 95% yield. Under this condition, the formation of **3** was suppressed remarkably and only a trace of **3** was formed.

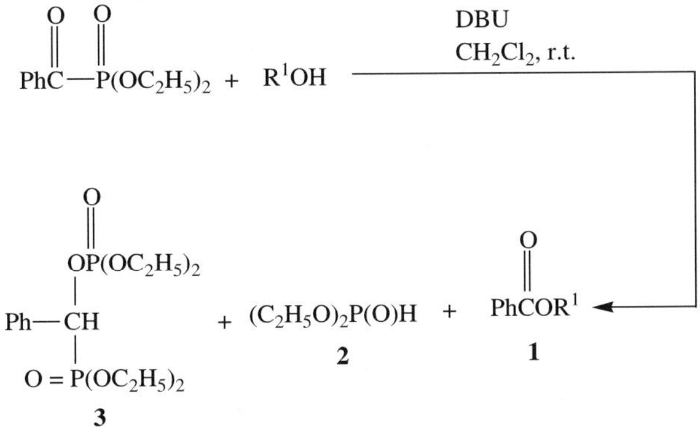

$$PhC(O)-P(O)(OC_2H_5)_2 + R^1OH \xrightarrow{DBU, CH_2Cl_2, r.t.}$$

$$Ph-CH(OP(O)(OC_2H_5)_2)(O=P(OC_2H_5)_2) \quad \textbf{3} \quad + \quad (C_2H_5O)_2P(O)H \quad \textbf{2} \quad + \quad PhCOR^1 \quad \textbf{1}$$

Halogenation. α-Hydroxyphosphonates have been converted to α-chloro- or α-bromophosphonates in excellent yield by simply treating them with thionyl chloride or bromide [220]. This reaction usually proceeds only in moderate yield, except when diphenyl 1-hydroxy-1-(aryl)methylphosphonates are used as starting materials [220b,d]. These results prompted the investigation of the possibility of direct hydroxyl–chloride exchange.

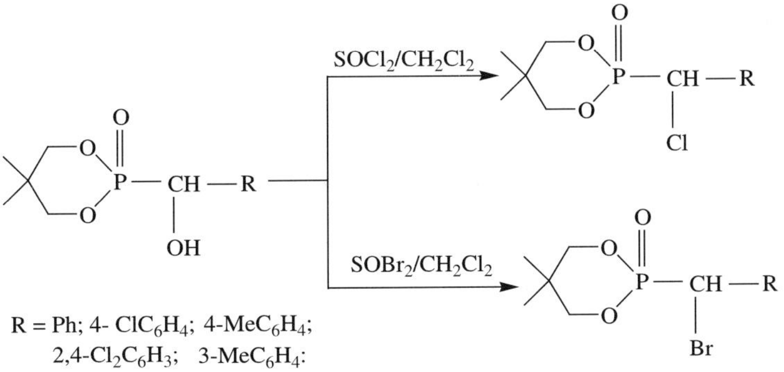

R = Ph; 4-ClC_6H_4; 4-MeC_6H_4; 2,4-$Cl_2C_6H_3$; 3-MeC_6H_4:

The reaction of alcohols with carbon tetrachloride in the presence of triphenylphosphine is known to provide, via oxyphosphonium intermediates, the appropriate chlorides [240].

It was shown that easily accessible diethyl 1-hydroxyalkylphosphonates **1** can be converted into diethyl 1-chloroalkylphosphonates **2** utilizing triphenylphosphine/carbon tetrachloride, affording high yield and purity under neutral conditions (see Appendix) [241].

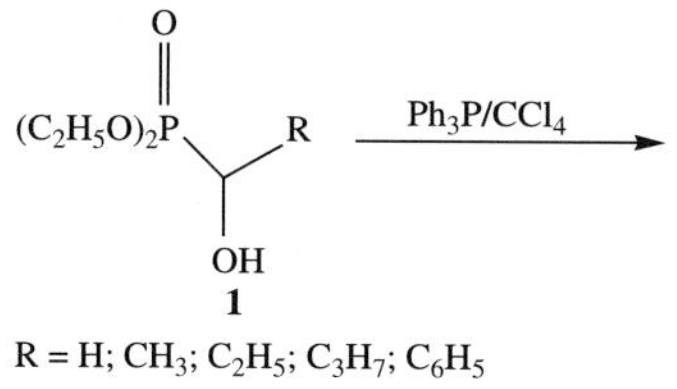

$$(C_2H_5O)_2P(O)CH(OH)R \xrightarrow{Ph_3P/CCl_4} (C_2H_5O)_2P(O)CH(Cl)R$$

$$\mathbf{1} \qquad\qquad \mathbf{2}$$

R = H; CH$_3$; C$_2$H$_5$; C$_3$H$_7$; C$_6$H$_5$

The reaction proceeds smoothly in boiling carbon tetrachloride, affording after vacuum distillation analytically pure products in high yield. The best results are obtained when a 50% molar excess of triphenylphosphine is used.

Diethyl 1-chloroalkylphosphonates **2** are key reagents for the preparation of different organic compounds. The diethyl 1-chloromethylphosphonate reacts readily with butyllitium to give diethyl α-lithiochloromethylphosphonate, which on treatment with tetrachloromethane gives the stable diethyl 1,1-dichloromethylphosphonate. The latter is suitable for the preparation of 1,1-dichloroalkenes [242 a,b].

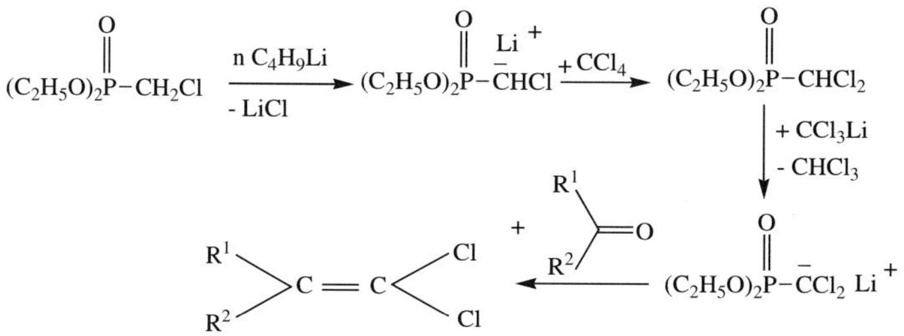

$$(C_2H_5O)_2P(O)-CH_2Cl \xrightarrow[-LiCl]{n\ C_4H_9Li} (C_2H_5O)_2P(O)-\overset{-}{C}HCl\ Li^+ \xrightarrow{+CCl_4} (C_2H_5O)_2P(O)-CHCl_2$$

$$\xrightarrow{+CCl_3Li,\ -CHCl_3} (C_2H_5O)_2P(O)-\overset{-}{C}Cl_2\ Li^+$$

$$\xrightarrow{+\ R^2\!-\!C(R^1)\!=\!O} \underset{R^2}{\overset{R^1}{>}}C=C\underset{Cl}{\overset{Cl}{<}}$$

The diethyl 1-chloromethylphosphonate is used for the preparation of diethyl carboxychloromethylphosphonate **1** in good yield (87–91%) [242 c].

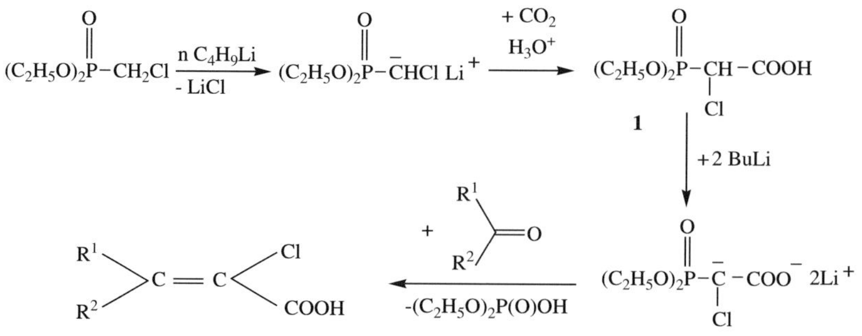

$$(C_2H_5O)_2P(O)-CH_2Cl \xrightarrow[-LiCl]{n\ C_4H_9Li} (C_2H_5O)_2P(O)-\overset{-}{C}HCl\ Li^+ \xrightarrow[H_3O^+]{+CO_2} (C_2H_5O)_2P(O)-CH(Cl)-COOH$$

$$\mathbf{1} \xrightarrow{+2\ BuLi} (C_2H_5O)_2P(O)-\overset{Cl}{\underset{|}{C}}-COO^-\ 2Li^+$$

$$\xrightarrow{+\ R^2\!-\!C(R^1)\!=\!O,\ -(C_2H_5O)_2P(O)OH} \underset{R^2}{\overset{R^1}{>}}C=C\underset{COOH}{\overset{Cl}{<}}$$

3.7 Characteristic Reactions of P-H Group

Treatment of **1** with two equivalents of *n*BuLi in a mixture of hexane –THF at –65 °C followed by the addition of the carbonyl compounds gave after hydrolysis the corresponding α-chloro-α,β-unsaturated acids [242c]. Diethyl 1-chloroalkylphosphonates are used for the preparation of diethyl 1,2-epoxyalkylphosphonates [243–246], and 1-(methylthio)methylphosphonate [247], as well as derivatives of alkyidenediphosphonates and vinylphosphonates [248].

Reduction. Treatment of the 2-[hydroxyl(4-chlorobenzen)methyl]-5,5-dimethyl-1,3,2-dioxaphosphinane-2-one with phosphorus triiodide resulted in the formation of the unexpected 2-(4-chlorobenzyl)-5,5-dimethyl-1,3,2- dioxaphosphinane-2-one [220f].

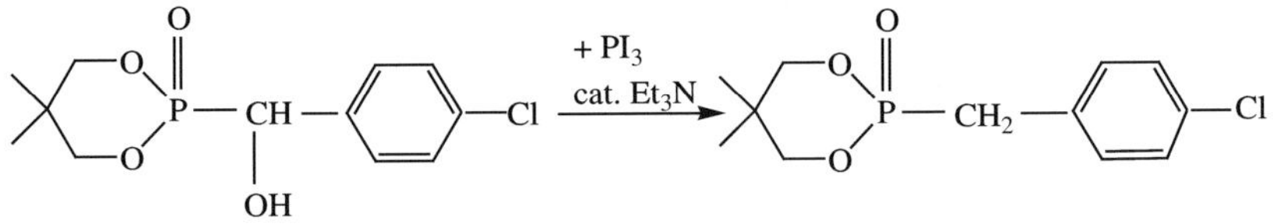

Dehydration. α-Hydroxyphosphonates **1** undergo dehydration in the presence of acetic anhydride and triethylamine to give vinylphosphates [220d].

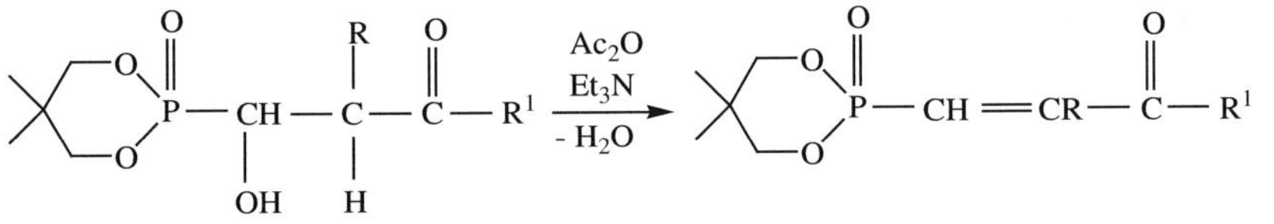

$R = R^1 = Ph$

Addition to isocyanates—carbamoyl phosphonates

The increased reactivity of the isocyanate group toward nucleophilic reagents [249] is also the driving force in their reaction with H-phosphonate diesters and results in the formation of carbamoyl phosphonates [250–252].

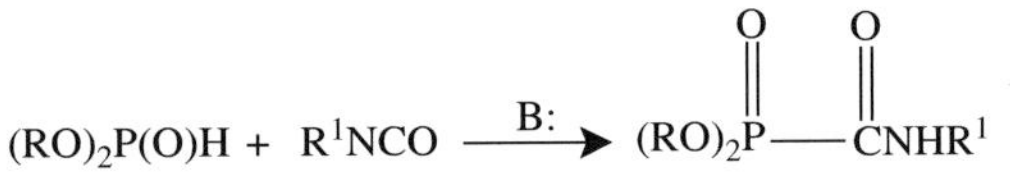

$$(RO)_2P(O)H + R^1NCO \xrightarrow{B:} (RO)_2P(O)-C(O)NHR^1$$

The reaction usually occurs in the presence of a base. As in the case of most of the preceding reactions in this chapter, the base activation of dialkyl phosphonates has been considered to take place as a simple deprotonation by the base to furnish the active phosphorus (III) intermediate anion. This is probably correct for base catalysis with alkali metals or their alkoxides, and for those cases, the following reaction mechanism has been proposed [253–261]:

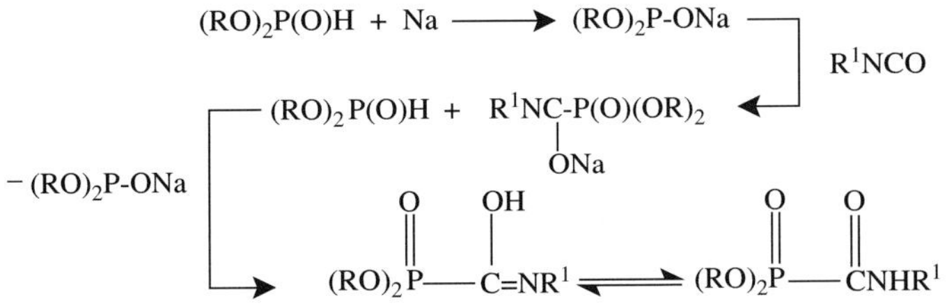

However, when amines are employed for base catalysis, one has to take into consideration the fact that in this case, the actual base is produced after alkylation of the amine as it was previously shown (see Section 3.8.1). When the reaction of dialkyl phosphonates with isocyanates is carried out in the absence of a catalyst, it is accompanied by carbon dioxide evaluation and results in the formation of three types of reaction products **I**, **II**, and **III** [262].

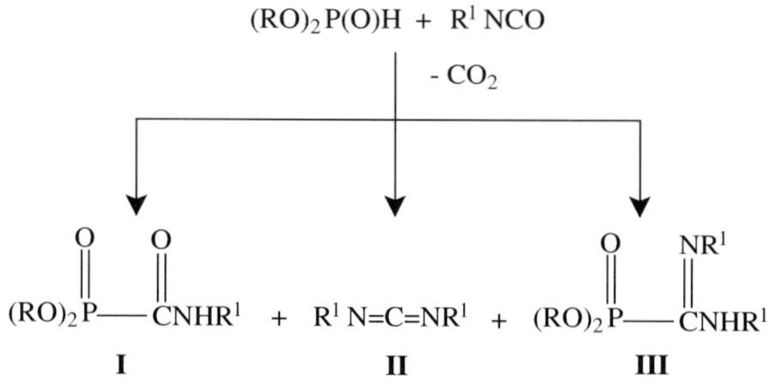

$$(RO)_2P(O)H + R^1NCO$$

$$\downarrow -CO_2$$

$$(RO)_2P(O)-C(O)NHR^1 \quad + \quad R^1N=C=NR^1 \quad + \quad (RO)_2P(O)-C(NR^1)NHR^1$$
$$\textbf{I} \qquad\qquad\qquad \textbf{II} \qquad\qquad\qquad \textbf{III}$$

The amount of carbon dioxide depends on the concentration of the reactants and increases with the increase of the isocyanate concentration. The type of the alkoxy substituents of the dialkyl phosphonates is also a determining factor in that process, and depending on that, the amount of the evolved carbon dioxide decreases in the following order [263]:

$$OCH_3 > OC_2H_5 > OC_3H_7 > OC_4H_9$$

Various mechanisms of the above reaction have been suggested [263–265]. The following scheme is based on data obtained by kinetic and IR spectroscopic studies of the noncatalyzed

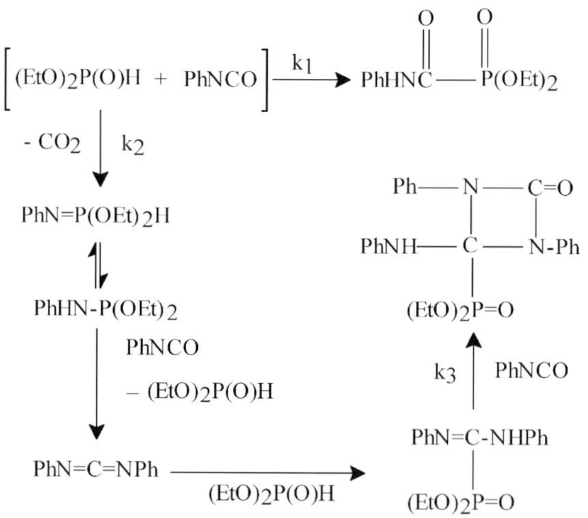

$$[(EtO)_2P(O)H + PhNCO] \xrightarrow{k_1} PhHNC(O)-P(O)(OEt)_2$$

$$-CO_2 \downarrow k_2$$

$$PhN=P(OEt)_2H$$

$$\updownarrow$$

$$PhHN-P(OEt)_2$$

$$\downarrow PhNCO$$
$$-(EtO)_2P(O)H$$

$$PhN=C=NPh \xrightarrow{(EtO)_2P(O)H}$$

$$\begin{array}{c} Ph-N-C=O \\ | \quad\quad | \\ PhNH-C-N-Ph \\ | \\ (EtO)_2P=O \end{array}$$

$$k_3 \Big\updownarrow PhNCO$$

$$\begin{array}{c} PhN=C-NHPh \\ | \\ (EtO)_2P=O \end{array}$$

reaction of diethyl phosphonate and phenylisocyanate [266]. According to this suggestion, the formation of carbon dioxide results from the interaction of the phosphoryl oxygen atom

3.7 Characteristic Reactions of P-H Group

and isocyanate group. At the first stage of the reaction, the oxygen atom of the phosphoryl group attacks the electrophilic carbon atom of the isocyanate group.

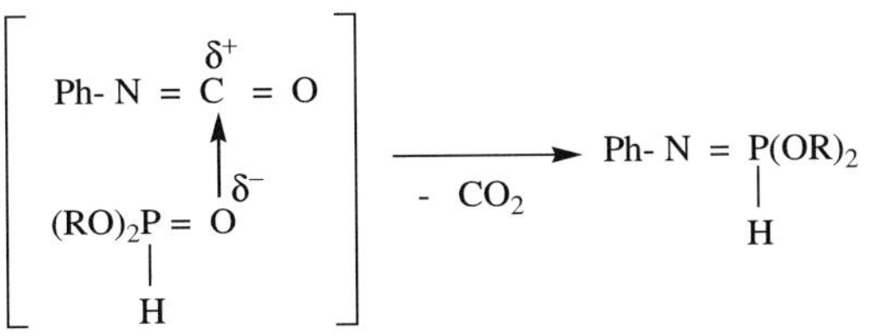

$$\begin{bmatrix} Ph-N = \overset{\delta^+}{C} = O \\ \uparrow \\ (RO)_2P = \overset{\delta^-}{O} \\ | \\ H \end{bmatrix} \xrightarrow{-CO_2} \begin{matrix} Ph-N = P(OR)_2 \\ | \\ H \end{matrix}$$

This initial suggestion has been confirmed by the reaction between phenyl isocyanate and diethyl thiophosphate $(EtO)_2P(S)OH$ where COS has been produced [266].

Carbamoyl phosphonate derivatives of methacrylic acid and α-methyl styrene were synthesized by reacting dimethyl and diethyl H-phosphonate with the corresponding

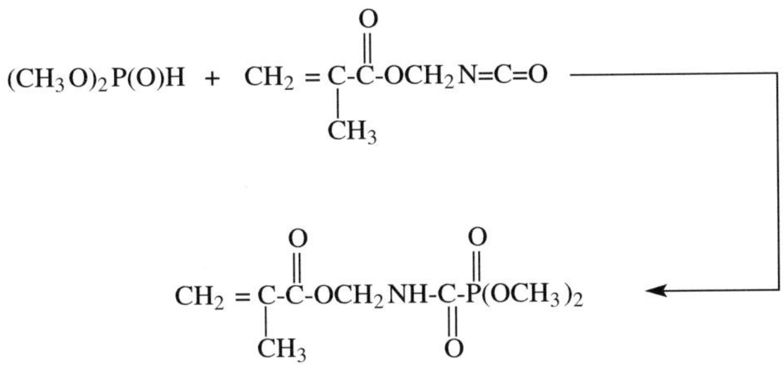

$$(CH_3O)_2P(O)H + CH_2 = \underset{CH_3}{\overset{O}{\underset{|}{C}}} - \overset{O}{\underset{\parallel}{C}} - OCH_2N=C=O \longrightarrow$$

$$CH_2 = \underset{CH_3}{\overset{O}{\underset{|}{C}}} - \overset{O}{\underset{\parallel}{C}} - OCH_2NH - \overset{O}{\underset{\parallel}{C}} - \overset{}{\underset{\parallel}{P}}(OCH_3)_2$$

isocyanate precursors [267].

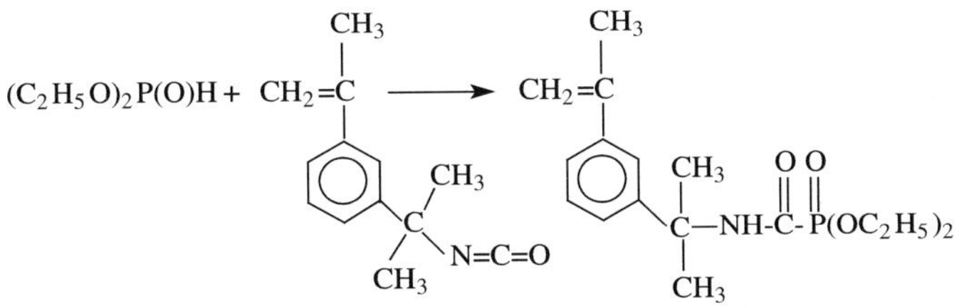

$$(C_2H_5O)_2P(O)H + CH_2=\overset{CH_3}{\underset{|}{C}} - \text{Ph} - \overset{CH_3}{\underset{|}{C}} - N=C=O \longrightarrow CH_2=\overset{CH_3}{\underset{|}{C}} - \text{Ph} - \overset{CH_3}{\underset{|}{C}} - NH - \overset{O}{\underset{\parallel}{C}} - \overset{O}{\underset{\parallel}{P}}(OC_2H_5)_2$$

These carbamoyl phosphonate monomers are very interesting as copolymers. When copolymerized with styrene and acrylates, they yielded soluble materials with better than 50 mol% incorporation of the phosphonate monomer.

Homolytic addition of dialkyl H-phosphonates to unsaturated hydrocarbons

The homolytic addition of dialkyl H-phosphonates to unsaturated hydrocarbons having C=C, or C=N double bonds or to compounds having the functionality Ar-Y-Y-Ar represents another useful synthetic route for the phosphorylation of compounds. The reaction is initiated by the formation of a phosphonic radical from dialkyl H-phosphonate during photolysis, irradiation (sonochemical conditions), or most frequently, in the presence of a radical initiator [157, 268].

$$(RO)_2P(O)H \xrightarrow{h\nu} (RO)_2\dot{P}(O) + \dot{H}$$

$$(RO)_2\dot{P}(O) + R^1CH=CH_2 \longrightarrow R^1\dot{C}H-CH_2-P(OR)_2(O)$$

$$\uparrow \qquad\qquad\qquad\qquad \downarrow +(RO)_2P(O)H$$

$$(RO)_2\dot{P}(O) + R^1CH_2-CH_2-P(OR)(O) \longleftarrow$$

Catalytic quantities of acetic or oxalic acids accelerate the addition of dialkyl phosphonates to olefins under the above conditions [269].

It has been suggested that free radicals more easily abstract the hydrogen from the P–H group than a hydrogen atom from the alkoxy groups of the phosphonate [270]. Depending on the type of dialkyl H-phosphonate, their reactivity to the phenyl radical [C_6H_5]· is nearly 1.5 to 4 times higher than that toward the carbon tetrachloride radical [CCl_4]· [271]. In general, the higher homologues of dialkyl H-phosphonates exhibit increased reactivity in the above process. The irradiation (sonochemical conditions) of the reaction mixture of dialkyl H-phosphonate and imine results in enhanced addition of the P–H group to the C=N group [272].

Arylselenophosphates are useful intermediates in organic chemistry [273] and biochemistry [274]. Due to its importance, a variety of procedures have been developed to prepare arylselenophosphates [272,275]. It was shown that arylselenophosphates can be obtained in mild condition and high yield by reacting dialkyl H-phosphonates with diphenyl diselenide in the presence of AIBN [276].

$$(RO)_2P(O)H + Ar\text{-}Se\text{-}Se\text{-}Ar \xrightarrow[C_6H_6,\ 60\text{-}65\ °C,\ 24\ h]{AIBN\ (50\%)} Ar\text{-}Se\text{-}P(O)(OR)_2$$

The initial step of the free radical reaction is the AIBN or thermal-induced generation of phosphonyl radical. Then abstraction of the arylseleno group from diaryl diselenide by the phosphonyl radical affords the product and an equivalent of the arylselenyl radicals, respectively. The arylselenyl radical might combine with the phosphonyl radical to form another molecule of the product.

$$(RO)_2P(O)H \xrightarrow{AIBN\ or\ Heat} (RO)_2\dot{P}(O)$$

$$(RO)_2\dot{P}(O) + Ar\text{-}Se\text{-}Se\text{-}Ar \longrightarrow Ar\text{-}Se\text{-}P(O)(OR)_2 + Ar\dot{S}e$$

3.8 Reactions with the Participation of the α-Carbon Atom

$$(RO)_2\overset{\bullet}{P}(O) + Ar\overset{\bullet}{Se} \longrightarrow Ar\text{-}Se\text{-}P(O)(OR)_2$$

Recently, Hubert *et al.* [272] have reported that the phosphonyl radical can be thermal-induced. Huang *et al.* confirmed this observation [276].

Dialkyl H-phosphonate radicals, generated via UV photolysis, add to C_{60} fullerene to form adducts of the type RC_{60} [277].

$$(RO)_2P(O)H \xrightarrow[C_{60}]{UV} [(RO)_2P(O).C_{60}]^{\bullet}$$

Dimethyl H-phosphonate adds to the double bond of alkenyl cyclopropanes in the presence of benzoil peroxide with the formation of the corresponding alkyl dimethyl phosphonates [278].

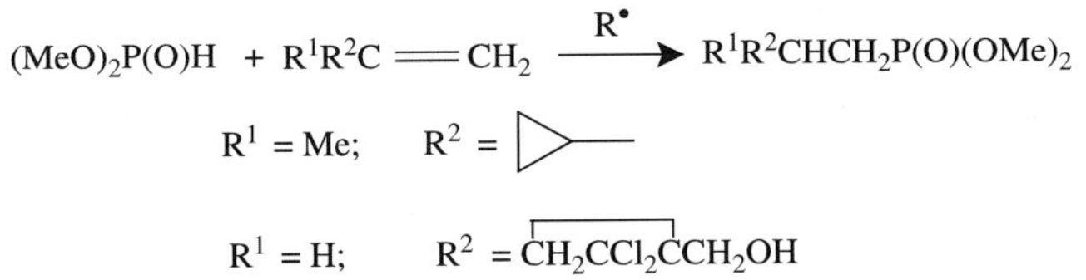

$$(MeO)_2P(O)H + R^1R^2C=CH_2 \xrightarrow{R^{\bullet}} R^1R^2CHCH_2P(O)(OMe)_2$$

$R^1 = Me; \quad R^2 = \triangleright\!\!\!-$

$R^1 = H; \quad R^2 = \overline{CH_2CCl_2CCH_2OH}$

3.8 REACTIONS WITH THE PARTICIPATION OF THE α-CARBON ATOM

The α-carbon atom of the alkoxy group is another potential site of nucleophilic attack in dialkyl H-phosphonates [279–281]. It is a considerably weaker electrophilic center than the phosphorus atom. Nevertheless, there are a number of nucleophiles that regioselectively attack the α-carbon atom of the alkoxy group in dialkyl H-phosphonates instead of at the more electropositive phosphorus. It has been established that the rate of nucleophilic substitution at the α-carbon atom in dialkyl H-phosphonates decreases for a given nucleophile in the following order, depending on the type of the alkoxy substituents [279]:

$$OCH_3 > OCH_2CH_3 > OCH_2CH_2CH_3 > OCH(CH_3)_2$$

The α-carbon atom in the molecule of dialkyl H-phosphonates is the preferred site of a nucleophilic attack in the reactions between dialkyl H-phosphonates and amines, aminoalcohols, some metal and amine salts, hydrogen halides, and chlorosilanes.

3.8.1 Reaction of dialkyl H-phosphonates with amines

The only products formed in the reaction between dialkyl H-phosphonates and amines are alkylammonium salts of the monoalkyl ester of H-phosphonic acid **I** resulting from alkylation of the corresponding amine [279–285].

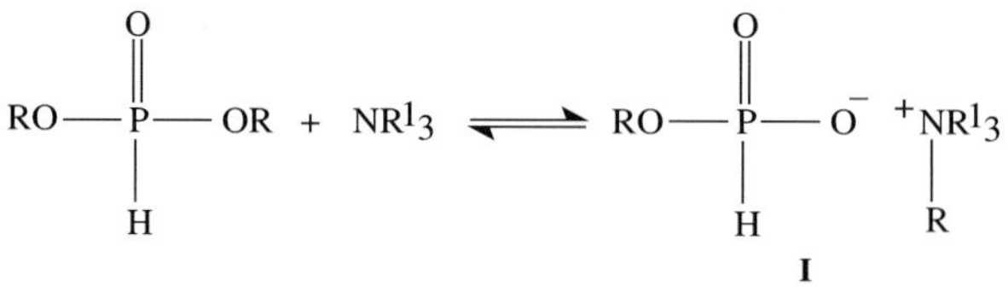

$$RO-\underset{\underset{H}{|}}{\overset{\overset{O}{\|}}{P}}-OR + NR^1{}_3 \rightleftharpoons RO-\underset{\underset{H}{|}}{\overset{\overset{O}{\|}}{P}}-O^- \ {}^+NR^1{}_3 \underset{R}{|}$$

I

The nitrogen atom of amines attacks the α-carbon atom of the alkoxy group of dialkyl H-phosphonates to give the corresponding alkylammonium salts of the monoalkyl ester of phosphonic acid. These results are in contrast to the often-expressed opinion—especially in the cases of amine-promoted reactions of dialkyl phosphonates—that amines deprotonate dialkyl H-phosphonates to form dialkyl phosphite anions **II** (Scheme 3.4.).

In the recent *ab initio* study [286], a comparison has been made between deprotonation and alkylation pathways for the model system dimethyl H-phosphonate–ammonia.

The calculated δG^{298} for the alkylation reaction at an MP2/6–31+G*//HF/6–31+G* + ZPE level is predicted to be 100.5 kcal/mol lower than that of the deprotonation process. Both reactions are predicted, however, to be thermodynamically unfavorable in the gas phase, since these calculations do not account for the stabilization effects of the solvation and hydrogen bonding in the liquid phase. In addition, the computational search of the deprotonation pathway with a 3–21*G basis set does not indicate the presence of a stationary point—an activated complex or a transition structure—along the N–H–P reaction coordinate. In contrast to this result, the alkylation pathway search resulted in locating the transition structure for this S_N2 substitution at the carbon atom, as well as locating the ion-pair adduct to be between the anion of monomethyl H-phosphonate and the methylammonium cation (Figures 3.4 and 3.5). The energy profile of this gas-phase reaction is presented in Figure 3.6. The important result of these calculations [286] is that the formation of the hydrogen-bonded contact ion pair is predicted to be an exothermic process with its total energy being 13.0 kcal/mol lower than that of the reactants. This result is also an indication of the importance of solvation effects for the stabilization of the phosphorus-containing alkylammonium salts obtained in the course of the alkylation reaction.

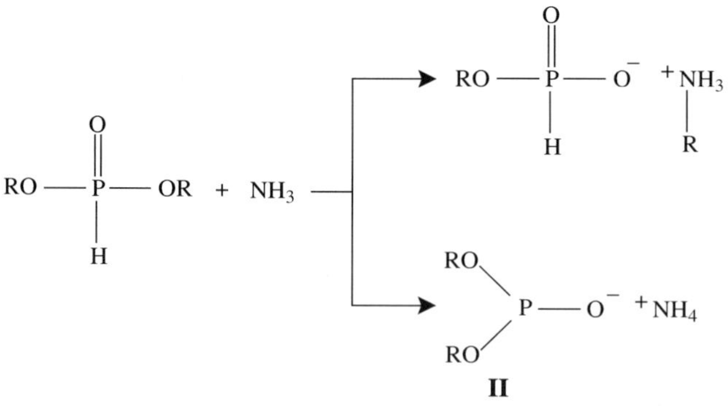

Scheme 3.4 Deprotonation and alkylation pathways for the model system dimethyl H-phosphonate–ammonia.

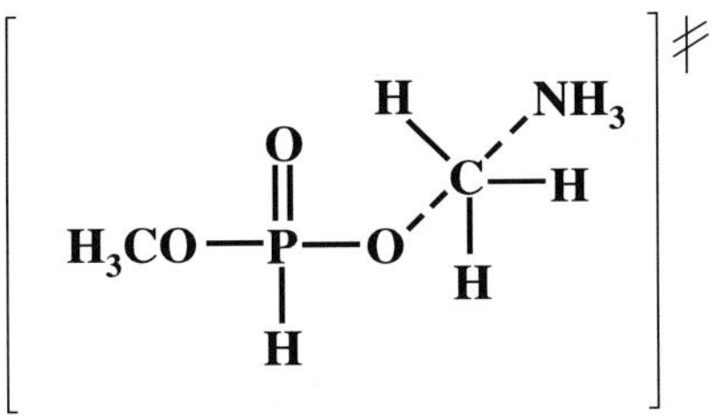

3.8 Reactions with the Participation of the α-Carbon Atom

Figure 3.4 S$_N$2 transition structure for the alkylation of ammonia with dimethyl H-phosphonate obtained with the HF/6–31+G* basis set.

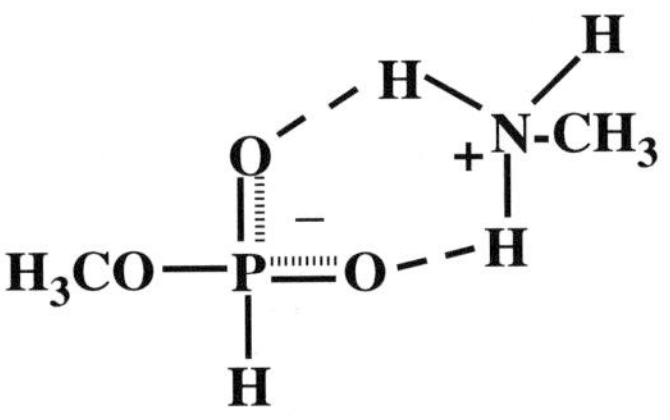

Figure 3.5 Ion pair adduct between the monoalkyl H-phosphonate anion and the methylammonium cation obtained with the HF/6–31+G* basis set.

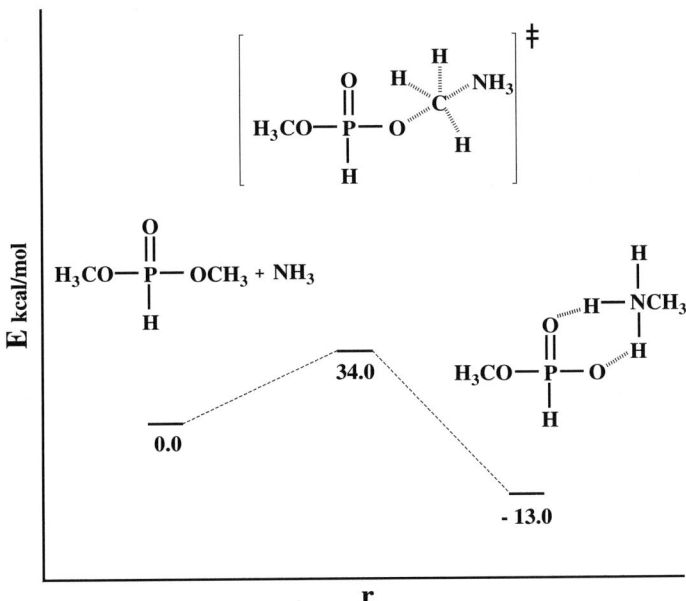

Figure 3.6 MP2/6–31+G*//HF/6–31+G* + ZPE gas-phase energy profile of the alkylation reaction between dimethyl H-phosphonate and ammonia.

The formation of alkylammonium salts of monoalkyl H-phosphonates **I** is characterized by specific changes of the ^{1}H, ^{31}P, and IR spectra as compared to the starting materials. These changes reflect predominantly the ionic character of the products and in particular, the delocalization of the negative charge in the monoalkyl H-phosphonate anion. Compared to the starting dialkyl H-phosphonates, the resonances for the phosphorus atom in the ^{31}P NMR spectra of the products are shifted upfield by 3 to 8 ppm, which indicates increased electronic shielding. The ^1J(PH) coupling constants decrease by 100–150 Hz. For example, ^1J(PH) is reduced from 710 Hz in dimethyl H-phosphonate to 578 Hz in the alkylammonium salt of phosphonic acid monomethyl ester. In the ^{31}P NMR spectrum of the same salt, the phosphorus signal appears at 4.55 ppm, indicating a phosphonate type of structure [285]. This signal is shifted 6.52 ppm upfield from that of dimethyl H-phosphonate. In the IR spectrum of the alkylammonium salts, the v(P–H) and v(P=O) stretching frequencies are each lowered by approximately 100 cm^{-1} as a result of the charge delocalization in the anion. The ^{31}P, ^{1}H NMR, and IR spectroscopic data provide direct evidence for the presence of a P–H bond in the structure of the final products of the reaction between amines and dialkyl H-phosphonates [282–285]. The alkylation reaction can be carried out with or without a solvent. The most suitable solvents are alcohols [281]. It is impossible to obtain diammonium salts of phosphonic acid [287]. A possible reason for this phenomenon is the presence of a delocalized negative charge that hinders nucleophilic substitution at the second α-carbon atom of the monoalkyl H-phosphonate anion.

Reaction of bis-(2-chloroethyl) H-phosphonate with triethylamine produces a cyclic ester of β-chloroethyl phosphorous acid [288]. The rate of this reaction is very low. It has been proposed that in the presence of triethylamine, bis-(2-chloroethyl) H-phosphonate isomerizes into bis-(2-chloroethyl) phosphite. The latter forms the corresponding 2-chloroethyl-1,3,2-dioxaphospholane, accompanied by hydrogen chloride elimination.

Another possibility that has been considered in connection with this reaction is the deprotonation of bis-(2-chloroethyl) H-phosphonate by triethylamine, followed by an attack of the oxygen atom at the β-carbon atom of the 2-chloroethoxy group.

3.8 Reactions with the Participation of the α-Carbon Atom

Both mechanisms are questionable, however, since it has been shown that a direct P–H deprotonation step of dialkyl H-phosphonates by amines is rather unlikely.

Aminoalcohols are bifunctional compounds having two functional groups—hydroxyl and amino, can react with dialkyl H-phosphonates attacking the phosphorus atom and/or the α-carbon atom of the alkoxy groups.

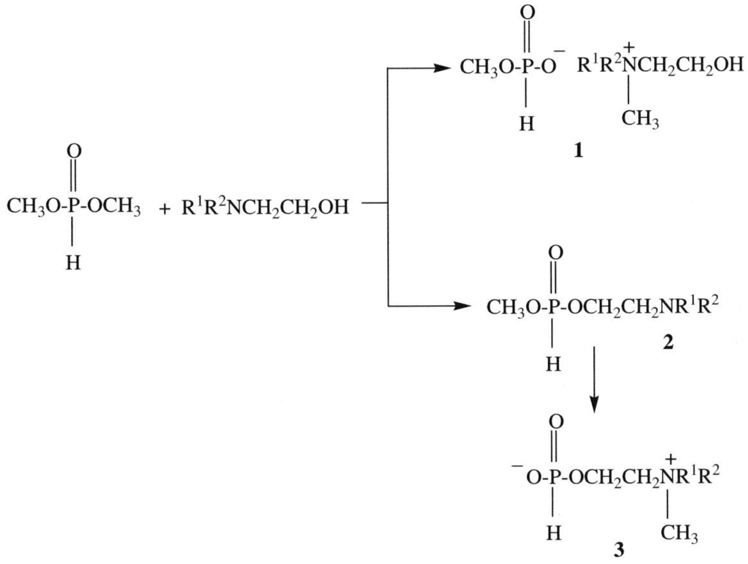

$$CH_3O\text{-}P(\text{=}O)(H)\text{-}OCH_3 + R^1R^2NCH_2CH_2OH \longrightarrow$$

$$CH_3O\text{-}P(\text{=}O)(H)\text{-}O^-\ \ R^1R^2\overset{+}{N}(CH_3)CH_2CH_2OH \quad \mathbf{1}$$

$$CH_3O\text{-}P(\text{=}O)(H)\text{-}OCH_2CH_2NR^1R^2 \quad \mathbf{2}$$

$$\downarrow$$

$$^-O\text{-}P(\text{=}O)(H)\text{-}OCH_2CH_2\overset{+}{N}(CH_3)R^1R^2 \quad \mathbf{3}$$

The $^{31}P\{H\}$ NMR spectrum of the reaction mixture obtained after mixing dimethyl H-phosphonate with N,N-dimethylethanolamine at temperatures lower than 70 °C showed only one signal at 7.69 ppm, which is a doublet of quartets with $^1J(P,H) = 654.0$ Hz and $^3J(P,H) = 11.8$ Hz. These results revealed that there proceeds alkylation reaction—product **1** is formed [290,293]. When the reaction is carried out at temperatures higher than 70 °C, in the $^{31}P\{H\}$ NMR spectrum, there are two signals at 7.69 ppm, a doublet of quartets with $^1J(P,H) = 654.0$ Hz and $^3J(P,H) = 11.8$ Hz and at 5.05 ppm, a doublet of triplets with $^1J(P,H) = 632.8$ Hz and $^3J(P,H) = 8.9$ Hz, showing that both reactions—alkylation and transesterification—take place [291,294]. Salt **1** is formed as a result of an alkylation reaction whereas salt **3** results from an intramolecular alkylation of the monotransesterificated product **2**. The latter compound is formed by the transesterification of dimethyl H-phosphonate with aminoalcohol. The low degree of completion of the transesterification of dimethyl phosphonate with N,N-dimethylethanolamine (the ratio between the integral intensities of the signals at 7.69 ppm and 5.05 ppm is 7:3) implies that the alkylation reaction takes place at low temperature and at a higher rate than the transesterification. Under the above reaction conditions (130 °C), the monoalkylated product participates in the transesterification reaction with low rate. Presumably, the delocalized negative charge in the monoalkyl H-phosphonate anion hinders further nucleophilic attack at the phosphorus atom.

3.8.2 Reaction of diaryl H-phosphonates with amines

The reactivity of diaryl H-phosphonates is significantly different from that of dialkyl H-phosphonates. The ^{31}P{H} NMR spectrum [289] of diphenyl H-phosphonate dissolved in pyridine shows signals at $\delta = 1.19$ ppm, the major resonance, and a weak signal at 128.3 ppm. In the same spectrum, after 24 h, some changes occurred: the signal at $\delta = 1.19$ ppm strongly decreases (by 30%) and the signal at 128.3 ppm increases by 30%. There is a new signal at 1.09 ppm (40%). The signal at 128.3 ppm can be assigned to the phosphorus atom of the triphenyl phosphite **3**, and the one at 1.09 ppm to the phosphorus atom of the phenyl H-phosphonate **2**.

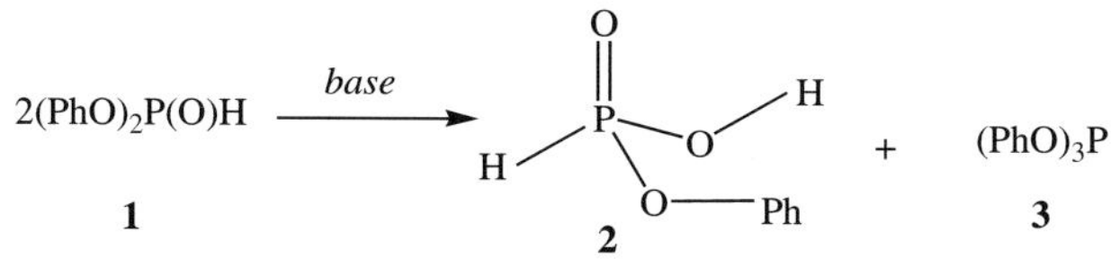

$$2(PhO)_2P(O)H \xrightarrow{base} \underset{\mathbf{2}}{\text{(H)(PhO)(HO)P=O}} + \underset{\mathbf{3}}{(PhO)_3P}$$

The disproportionation of diphenyl H-phosphonate to the corresponding triester and the H-phosphonate monoester was found to be practically an irreversible process. Both compounds were completely stable and did not undergo any reactions. Kers *et al.* [289] proposed a plausible mechanism for the disproportionation of diphenyl H-phosphonate to take into account the characteristic chemical features of diaryl H-phosphonates. The presence of electron-withdrawing substituents (two phenyl groups) makes the phosphorus more electrophilic in diphenyl H-phosphonate and enables abstraction of the proton from the P–H group even by a weak base (e.g.; pyridine).

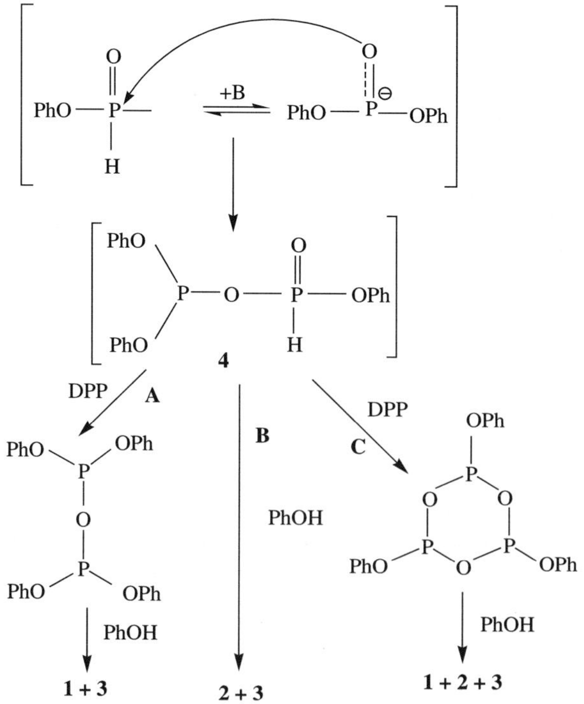

Probably, the reaction involves as an initial step a nucleophilic attack of the oxygen atom of the diphenyl phosphonate anion at the phosphorus atom of the diphenyl H-phosphonate. A resulting intermediate **4** may then react with phenol, affording the final products **2** and **3** (pathway **B**) or alternatively, be converted into another intermediates (pathway **B** and **C**).

Synthetic application

Alkylation of amino groups bonded to a polymer chain with H-phosphonates was studied in order to obtain physiologically active polymers. Polymeric physiologically active substances have numerous advantages over low-molecular ones, for example, lower toxicity, prolonged action, and higher selectivity. Poly(*N*-vinylpyrrolidone-*co*-vinylamine) was phosphorylated by dimethyl H-phosphonate [290,293].

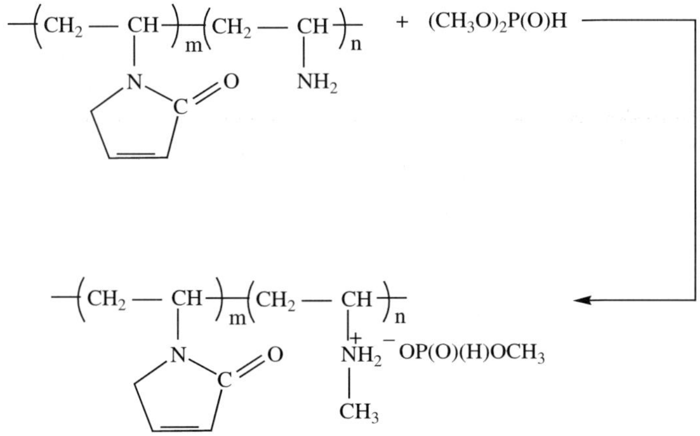

The signal in the ^{31}P NMR spectrum of the reaction product at 7.34 ppm, which appears as a doublet of quartets with $^1J(P,H) = 623.9$ Hz and $^3J(P,H) = 11.8$ Hz, confirms its structure. Study of the growth-regulating activity of this polymer showed that it exhibits growth-regulating (retardant and ageing delay) and herbicide activity [291,294]. The same polymer has a low cytotoxicity [292,295].

Dimethyl and diethyl H-phosphonates were used as alkylating agents of polyurethanes (see Section 3.2).

Chemical properties of alkylated products

Alkylammonium salts of monoalkyl H-phosphonic acid, having at least one ethyl group in the cation, undergo Hoffman's elimination [283].

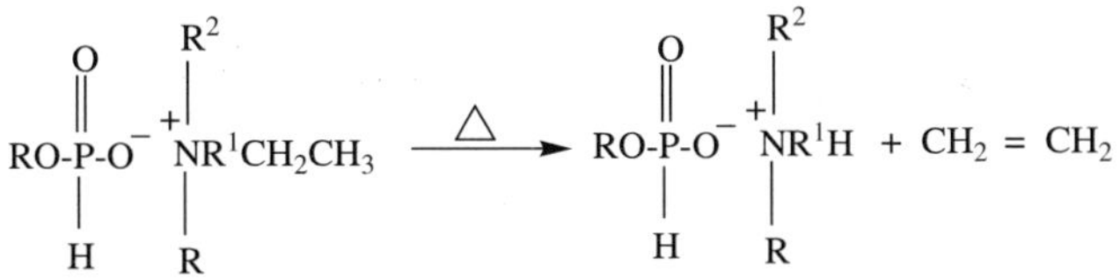

This result suggests that these salts exist as free ions in solution. In this case, the phosphonate anion ROP(O)HO⁻ facilitates the β-elimination, which leads to the formation of ethylene and the final phosphorus-containing product.

Methylalkylammonium salts, obtained by alkylation of primary amines with dimethyl H-phosphonate, are thermally unstable and easily yield the corresponding alkylammonium salts [292,295,296].

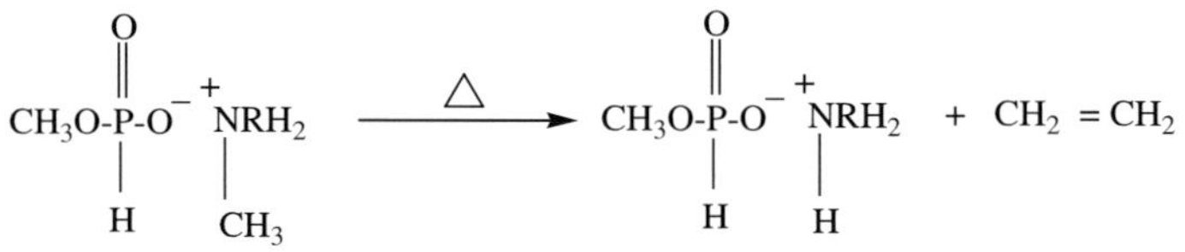

$$CH_3O\text{-}P(O)(H)\text{-}O^- \; {}^+NRH_2(CH_3) \xrightarrow{\Delta} CH_3O\text{-}P(O)(H)\text{-}O^- \; {}^+NRH_2(H) + CH_2=CH_2$$

This transformation takes place readily at room temperature and can be carried out quantitatively by stirring the reaction mixture at 50–70 °C under vacuum for several hours. Ethylene is formed as a result of recombination of the carbon that probably forms at the first step of the decomposition. On the other hand, it was found [292, 295, 296] that quaternary methylammonium salts obtained by alkylation of tertiary amines (N,N-dimethyl aniline or triethyl amine) with dimethyl H-phosphonate are quite stable with respect to the above conversion when heated up to 130 °C and 70 °C, respectively.

Quaternary alkylammonium salts of the monoethyl H-phosphonic acid, having at least one methyl group connected to the nitrogen atom undergo, the following type of rearrangement [284]:

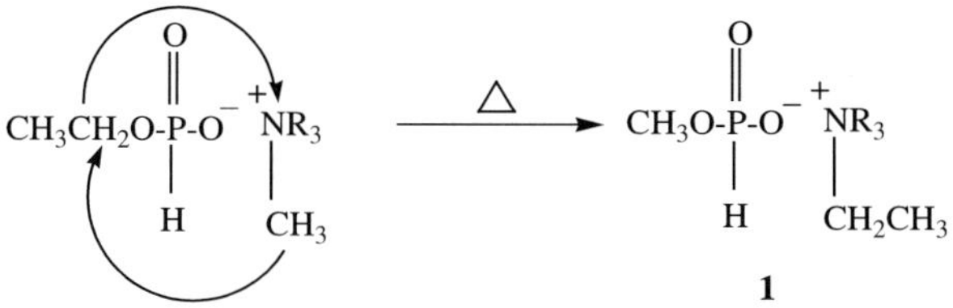

$$CH_3CH_2O\text{-}P(O)(H)\text{-}O^- \; {}^+NR_3(CH_3) \xrightarrow{\Delta} CH_3O\text{-}P(O)(H)\text{-}O^- \; {}^+NR_3(CH_2CH_3)$$
1

The ³¹P NMR spectrum of the reaction mixture obtained by heating diethyl H-phosphonate and dimethyl aniline shows signals for two types of phosphorus atoms. The signal at 4.25 ppm appears as a doublet of triplets and is characteristic of a phosphorus atom in the $CH_3CH_2OP(O)HO^-$ anion, whereas the second one at 6.15 ppm representing a doublet of quartets has been assigned to the phosphorus atom of the monomethyl H-phosphonate anion $CH_3OP(O)HO^-$. According to the ³¹P{H} NMR spectrum, the content of **1** is 24%.

Similar rearrangement occurs during the reaction of phenyl trichloroacetate and triethylamine [297].

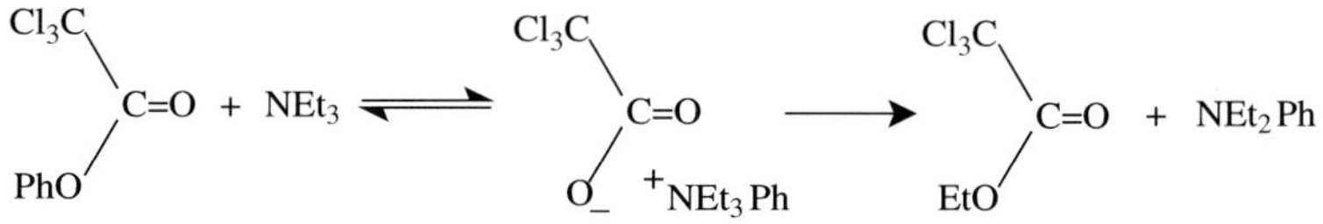

3.8 Reactions with the Participation of the α-Carbon Atom 85

$$Cl_3C\diagdown C=O + NEt_3 \rightleftharpoons Cl_3C\diagdown C=O \longrightarrow Cl_3C\diagdown C=O + NEt_2Ph$$
$$PhO\diagup \qquad O^-\ {}^+NEt_3Ph \qquad EtO\diagup$$

Treatment of phosphorus-containing alkylammonium salts, formed by alkylation of primary and secondary amines with bis-(β-chloroethyl) H-phosphonate, with aqueous sodium hydroxide leads to the formation of the corresponding aziridine salts of β-chloroethyl H-phosphonate [298].

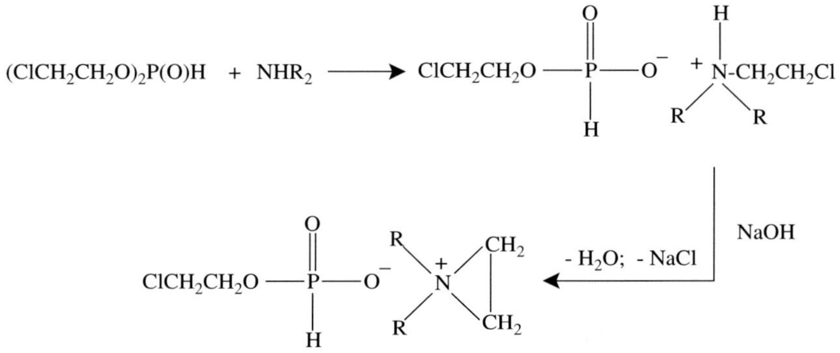

$$(ClCH_2CH_2O)_2P(O)H + NHR_2 \longrightarrow ClCH_2CH_2O-\underset{H}{\overset{\overset{O}{\|}}{P}}-O^-\ {}^+\underset{R}{\overset{H}{N}}-CH_2CH_2Cl$$

$$ClCH_2CH_2O-\underset{H}{\overset{\overset{O}{\|}}{P}}-O^-\ \underset{R}{\overset{+}{N}}\diagup\overset{CH_2}{\underset{CH_2}{|}} \xleftarrow[\text{- }H_2O;\text{ - NaCl}]{NaOH}$$

3.8.3 Reaction with amides and urethanes

It was shown for the first time that the reaction of dimethyl H-phosphonate with acetanilide represents an alkylation of the amido group [299].

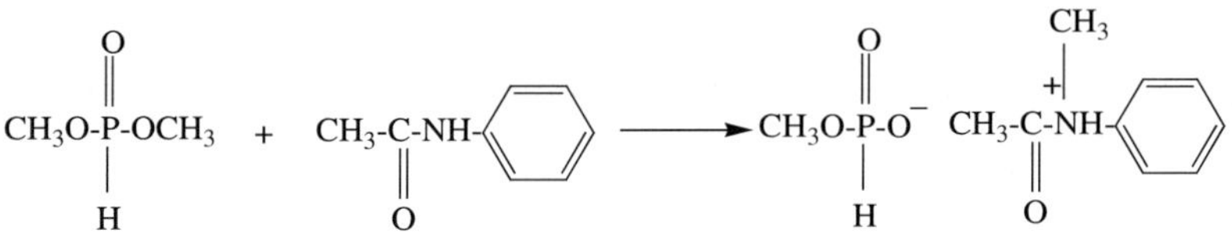

$$CH_3O-\underset{H}{\overset{\overset{O}{\|}}{P}}-OCH_3 + CH_3-\underset{O}{\overset{\|}{C}}-NH-\!\!\bigcirc \longrightarrow CH_3O-\underset{H}{\overset{\overset{O}{\|}}{P}}-O^-\ CH_3-\underset{O}{\overset{\|}{C}}-\overset{\overset{CH_3}{\overset{+|}{}}}{N}H-\!\!\bigcirc$$

The $^{31}P\{H\}$ NMR spectrum showed a signal at δ = 7.37 ppm, a doublet of quartets with $^3J(P,H) = 12.3$ Hz and $^1J(P,H) = 662.4$ Hz. The measured value of the $^1J(P,H)$ constant confirms the ionic structure of the final product. The observation of an alkylation of the amido group by the methoxy group of dimethyl H-phosphonate represents a new method for the phosphorylation of polyamide.

^{31}P NMR studies of the reaction mixture obtained after heating dimethyl H-phosphonate and urethane (ethyl carbamate) at 160 °C showed that this interaction includes an alkylation of the urethane group, an exchange reaction between the methoxy group of the dimethyl H-phosphonate and urethane group, and disproportionation of methylethyl H-phosphonate [299].

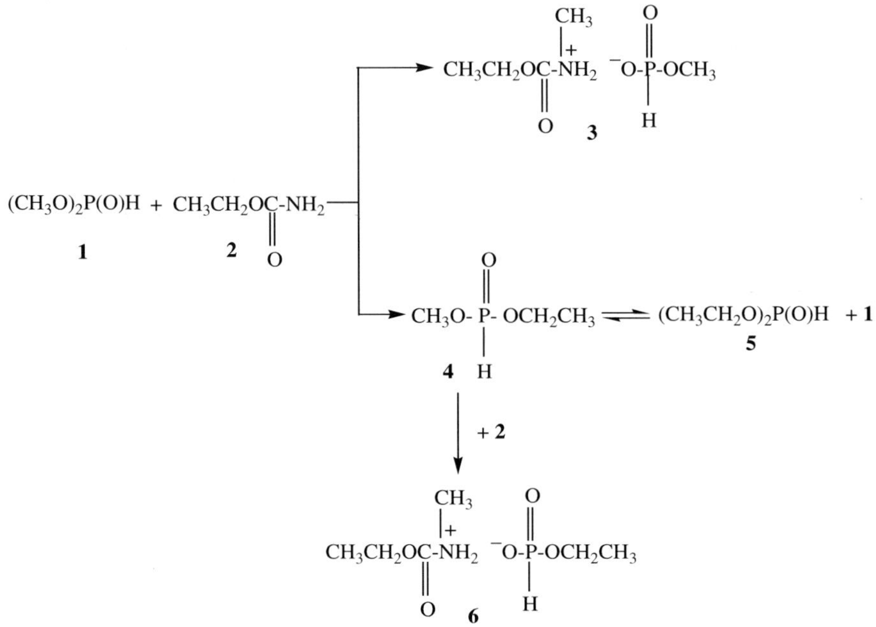

$(CH_3O)_2P(O)H + CH_3CH_2OC\underset{\underset{O}{\|}}{-}NH_2$
    1   **2**

$\longrightarrow CH_3CH_2O\overset{CH_3}{\underset{\underset{O}{\|}}{\overset{|+}{C}}}-NH_2 \quad ^-O-\overset{O}{\underset{H}{\overset{\|}{P}}}-OCH_3$
                                    3

$\longrightarrow CH_3O-\overset{O}{\underset{H}{\overset{\|}{P}}}-OCH_2CH_3 \rightleftharpoons (CH_3CH_2O)_2P(O)H + \mathbf{1}$
           4   **5**

↓ + **2**

$CH_3CH_2O\overset{CH_3}{\underset{\underset{O}{\|}}{\overset{|+}{C}}}-NH_2 \quad ^-O-\overset{O}{\underset{H}{\overset{\|}{P}}}-OCH_2CH_3$
                              6

The ^{31}P NMR spectrum of the reaction mixture revealed signals at 7.04 ppm, a doublet of quartets with $^3J(P,H) = 11.8$ Hz, and $^1J(P,H) = 640.9$ Hz, which is of the phosphorus atom of the product **3**. This product is formed as result of the alkylation of the urethane group by dimethyl H-phosphonate. The product **6**, with a doublet of triplets at 5.10 ppm with $^3J(P,H) = 9.1$ Hz and $^1J(P,H) = 644.5$ Hz, is obtained from the alkylation of urethane **2** by methylethyl H-phosphonate **4**. The alkylation of urethane group is unexpected, keeping in mind the low basicity of the nitrogen atom of this group.

The signal at 9.80 ppm with $^3J(P,H) = 11.8$ and 9.1 Hz, and $^1J(P,H) = 695.8$ Hz is due to the phosphorus atom of the product **4**, a result of the exchange reaction between the urethane and methoxy groups.

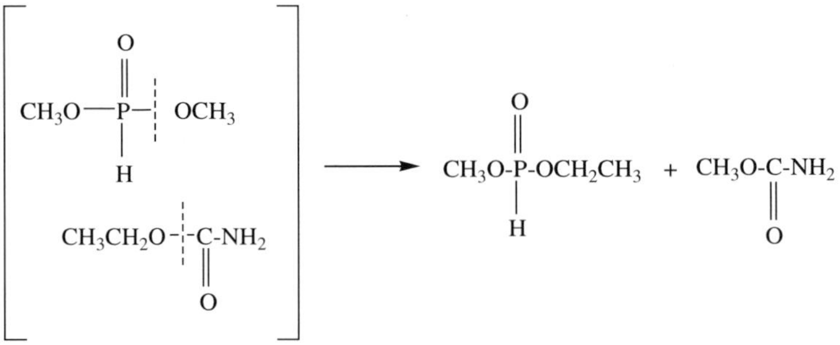

$$\left[\begin{array}{c} CH_3O-\overset{O}{\underset{H}{\overset{\|}{P}}}-OCH_3 \\ CH_3CH_2O-\underset{\underset{O}{\|}}{C}-NH_2 \end{array} \right] \longrightarrow CH_3O-\overset{O}{\underset{H}{\overset{\|}{P}}}-OCH_2CH_3 + CH_3O-\underset{\underset{O}{\|}}{C}-NH_2$$

These observations represent an important scientific result for polymer chemistry. The results obtained make the concept of an exchange reaction an alternative to simple

3.8 Reactions with the Participation of the α-Carbon Atom

degradation of polyurethane waste products and an alkylation reaction an alternative to phosphorylation of polyurethanes.

The disproportionation of methylethyl H-phosphonate **4** leads to the formation of diethyl H-phosphonate **5** (δ = 8.3 ppm, quintets with $^3J(P,H)$ = 9.5 Hz and $^1J(P,H)$ =693.6 Hz).

3.8.4 Reaction with hydrogen halides

Hydrogen halides react with dialkyl H-phosphonates to form monoesters of H-phosphonic acid and alkyl halides as a result of nucleophilic attack of the halide anion at the α-carbon atom of the alkoxy group [300].

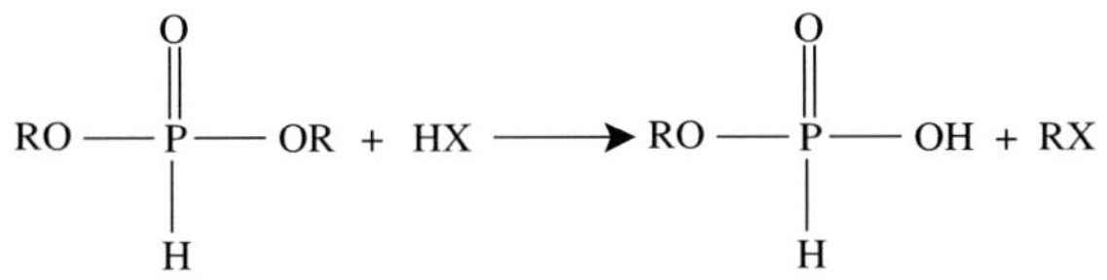

$$\text{RO}-\underset{\underset{H}{|}}{\overset{\overset{O}{\|}}{P}}-\text{OR} + \text{HX} \longrightarrow \text{RO}-\underset{\underset{H}{|}}{\overset{\overset{O}{\|}}{P}}-\text{OH} + \text{RX}$$

Kinetic investigations have shown that dealkylation of diisopropyl H-phosphonate with hydrogen chloride is a second-order reaction with respect to hydrogen chloride [301]. The suggested reaction mechanism involves the initial formation of an activated complex between dialkyl H-phosphonate and hydrogen chloride, which further undergoes dealkylation by nucleophilic attack of a second hydrogen halide molecule.

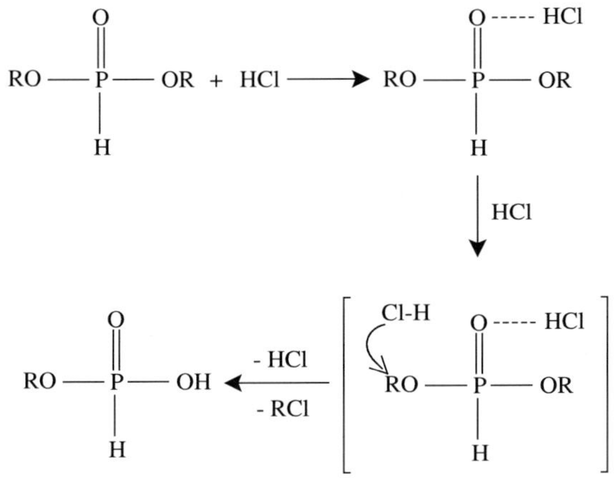

The dealkylation of the second alkoxy group proceeds in general at a considerably lower rate. The ratio of the rate constants for the dealkylation of the first and second alkyl groups is $k_1/k_2 > 9$ [301]. The reactivity of the different hydrogen halides toward dialkyl H-phosphonates decreases in the following order:

$$HI > HBr > HCl$$

3.8.5 Reaction with metal salts

The reaction between dialkyl H-phosphonates and Ce (III) or La (III) chlorides furnishes the corresponding metal salts of the monoalkyl esters of H-phosphonic acid [302].

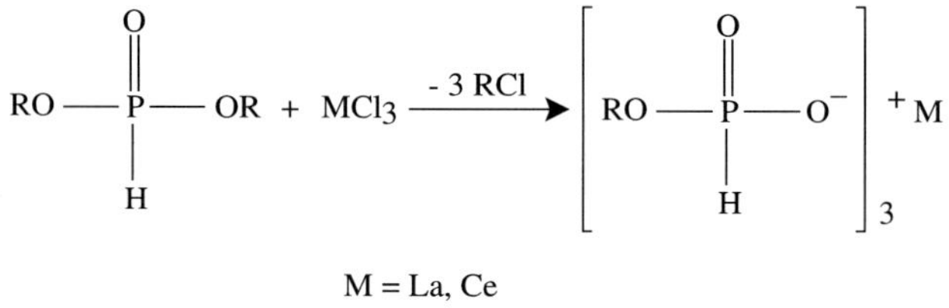

$$RO-\overset{\overset{O}{\|}}{\underset{H}{P}}-OR + MCl_3 \xrightarrow{-3\,RCl} \left[RO-\overset{\overset{O}{\|}}{\underset{H}{P}}-O^-\right]_3 {}^+M$$

M = La, Ce

It has been proposed that initially dissociation of the salts in dialkyl H-phosphonate takes place [303], most probably connected with complex formation at room temperature [304]. The reason for this assumption is the observed lowering of the vibrational frequency corresponding to the P = O stretch in the IR spectrum of the reaction mixture of ca. 25 cm^{-1} with respect to the starting dialkyl H-phosphonates. Such complex formation activates the α-carbon atoms toward nucleophilic substitution, which is the second step of this reaction [304].

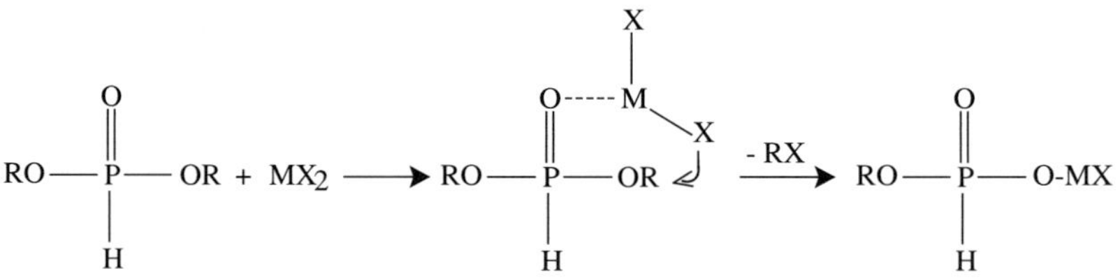

$$RO-\overset{\overset{O}{\|}}{\underset{H}{P}}-OR + MX_2 \longrightarrow RO-\overset{\overset{O-----M\overset{\overset{X}{|}}{\diagdown X}}{\|}}{\underset{H}{P}}-OR \xrightarrow{-RX} RO-\overset{\overset{O}{\|}}{\underset{H}{P}}-O-MX$$

Dimethyl H-phosphonate reacts with calcium dichloride at 70 °C, with manganese acetate at 112 °C, and with calcium nitrate at 115 °C. Kinetic investigations [304] have shown that the reaction takes place as a bimolecular substitution of the second order. There is no difference in the reaction rate of the substitution of both alkyl groups. In general, it decreases with increase in the length of the carbon chain in the alkoxy group.

Depending on the molar ratio between the reactants, the following dealkylation products have been synthesized [305]:

3.8 Reactions with the Participation of the α-Carbon Atom

$$nRO\text{-}P(\text{=}O)(H)\text{-}OR + mMX_2 \longrightarrow$$

- n:m = 1:1 → RO-P(=O)(H)-O-M-X **1**
- n:m = 2:1 → RO-P(=O)(H)-O-M-O-P(=O)(H)-OR **2**
- n:m = 1:1 → RO-P(=O)(H)-O-M-[O-P(=O)(H)-O-M]$_n$-X **3**

R = CH$_3$; C$_2$H$_5$; C$_4$H$_9$
M = Ca; Zn; Cd; Co; Cu; Mn
X = Cl; NO$_3$; CH$_3$COO

The structure of the phosphorus- and metal-containing salts **1** and **2** was proved by ^{1}H and ^{31}P NMR and IR spectroscopy. The phosphorus chemical shift for product **1** (R = C$_2$H$_5$ and M = Zn) appears as doublet of triplets at 4.84 ppm with $^1J(P,H) = 663.0$ Hz and $^3J(P,H) = 8.1$ Hz. The presence of a negative charge in the molecule of this salt determines the decrease of the phosphorus chemical shift as well as in the $^1J(P,H)$ coupling constant as compared to their values in the pure dialkyl H-phosphonates. The IR spectrum of the products **1** and **2** revealed absorption bands at 2410 cm^{-1} for the P–H group, 1185 cm^{-1} for the P=O group, 1080 cm^{-1} for the P–O–C group, and 575 cm^{-1} for the M–O. The strong shift in the absorption of the P=O group at 1185 cm^{-1} and the P–H group at 2410 cm^{-1} of the products **1** and **2**, compared to the shift of the P=O group (1258 cm^{-1}) and the P–H group (2440 cm^{-1}) for diethyl H-phosphonate, can be explained by the presence of the ionic bond in these salts. The structure of the oligomeric Zn^{2+} and Ca^{2+} salts **3**, obtained by dealkylation of diethyl H-phosphonate with the corresponding metal chlorides, has been confirmed by ^{31}P{H} NMR and ^{1}H NMR analysis. The ^{31}P{H} NMR spectrum of the Zn oligomer shows signals for two types of phosphorus atoms at δ = 0.38 ppm, triplet with $^3J(P,H) = 9.2$ Hz and $^1J(P,H) = 642.7$ Hz and at –0.88 ppm, doublet with $^1J(P,H) = 618.7$ Hz [305]. The signal at δ = 0.38 ppm is for the phosphorus atom at the end group and the one at δ = –0.88 ppm, for the phosphorus atom in the repeating unit. The ratio between the integral intensities of the phosphorus atom in the repeating unit and the phosphorus atom in the end group gave the degree of the polymerization 'n'. Products with molecular mass between 1000 and 1500 Da were synthesized. The oligomeric structure of the products is also supported by the ^{1}H NMR data. In the ^{1}H NMR spectrum of the Zn-containing product **3**, there are signals at δ = 6.73 ppm and at 6.61 ppm, for the P–H protons in the end group and in the repeating unit, respectively. The formation of oligomeric products in the reaction between dialkyl

H-phosphonates and metal salts of the type MX_2 implies that nucleophilic substitution at both α-carbon atoms takes place. This is a substantial difference from the alkylation reaction with amines where dialkyl H-phosphonates react as monofunctional compounds, that is, the reaction terminates at the stage of the formation of the monoanion. This difference in the reactivity of the salts **I** and **II** shown below is obviously determined by the type of cation, since they are both salts of monoalkyl H-phosphonate.

$$\underset{\textbf{I}}{\text{RO}-\overset{\overset{\text{O}}{\|}}{\underset{\text{H}}{\text{P}}}-\text{O}^-\ {}^+\text{MX}} \qquad \underset{\textbf{II}}{\text{RO}-\overset{\overset{\text{O}}{\|}}{\underset{\text{H}}{\text{P}}}-\text{O}^-\ {}^+\underset{\text{R}}{\text{NR}^1{}_3}}$$

The ^{13}C NMR chemical shifts are very sensitive with respect to changes in the electron density at a given carbon atom and provide useful information about differences in the structures of the salts **I** and **II**. Thus, the signal of the α-carbon atom in the sodium salt of monoethyl H-phosphonate (type **I**) is shifted 2.3 ppm downfield (Table 3.5) with respect to the same signal in diethyl H-phosphonate, which indicates a decrease in its electron density. In contrast, the signal of the α-carbon atom in the triethyl methylammonium salt of monomethyl H-phosphonate (type **II**) is shifted 2.1 ppm upfield with respect to dimethyl H-phosphonate as a result of an increased electron density, due to the negative charge delocalization.

The increased reactivity of the monoalkyl H-phosphonate salt of the type **I** toward nucleophilic substitution at the second α-carbon has been explained in terms of complex formation between the phosphonate anion and the metal through the oxygen donors [306].

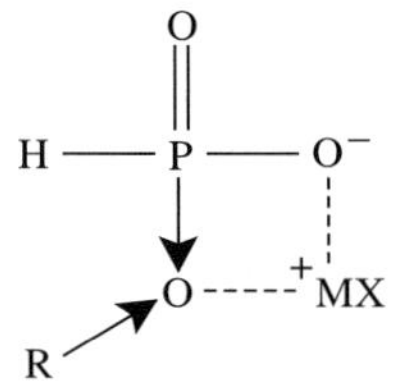

$$\text{H}-\overset{\overset{\text{O}}{\|}}{\underset{\underset{\text{R}}{\overset{\downarrow}{\text{O}}-----{}^+\text{MX}}}{\text{P}}}-\text{O}^-$$

Table 3.5

^{31}C{^{1}H} NMR data for some salts of H-phosphonic acid monoalkyl esters with general formula [RO–PHOO]⁻ A⁺. Chemical shift for the α-carbon atoms of the corresponding dialkyl H-phosphonates are given in parentheses

R	A	δ (ppm)
Et	Na	18.72 (CH_3); 63.18 {60.78} (CH_2)
Me	MeN(Et)$_3$	6.57 (CH_3); 45.48 (CH_2); 48.98 {51.08} (POCH$_3$); 54.61 (NCH$_3$)

3.8 Reactions with the Participation of the α-Carbon Atom

This complex formation results in a reduction of the electron density at the α-carbon atom of the alkoxy group by the so-called 'push–pull' mechanism, since the metal cation is effectively withdrawing electron density from the phosphonate ligand. In contrast, the alkylammonium cation is considered noncomplexing, which results in the α-carbon atom in the salts of the type **II** having increased electron density, due to the delocalization of the negative charge in the anion. These changes of the electron density at the &macmillan-carbon atom in the salts **I** and **II**, indicated by the $^{13}C\{^1H\}$ NMR data in Table 3.5, correlate well with the observed reactivity toward dealkylation of the second alkoxy group in that increased electron density will hinder nucleophilic attack at that carbon center, and vice versa.

It has been established that salts of monoalkyl H-phosphonic acids add more easily to double bonds than do the corresponding acids [306]. The addition takes place in the presence of different catalysts, and depending on their type, the reaction mechanism is either ionic or radical. Addition of alkali and ammonium salts of monoalkyl H-phosphonates to unsaturated compounds or ketones results in the formation of alkyl monoalkyl phosphonates.

$A = NH_4^+;\ Ca^{2+}: X = \text{electron-withdrawing substituent}$

Monoalkyl H-phosphonates exhibit a different reactivity toward addition to double bonds depending on the type of the cation. The metal salts, which add more easily, have been shown to exist as contact-ion pairs [287]. In this case, coordination of the corresponding metal cation to the phosphoryl oxygen is suggested to activate the P–H bond. Similar complex formation is not possible for ammonium salts, which exist as free ions in polar solvents.

The phosphorus- and metal-containing monomers and oligomers are used as modifiers of poly(ethylene terephthalate) and as catalysts of the polycondensation of bis(2-hydroxyethyl terephthalate).

3.8.6 Reaction with ammonium halides

It was established that some ammonium halides dealkylate dialkyl H-phosphonates with the formation of the ammonium salts of the corresponding monoalkyl ester of H-phosphonic acid [307].

Kinetic studies of the reaction between dialkyl H-phosphonates and hydrochlorides of diethylamine and pyridine indicate that this reaction is a second-order bimolecular nucleophilic substitution [308]. The reaction rate depends on the type of alkyl substituents of the H-phosphonate diester as well as on the basicity of the amine. The low reaction rate observed for dibutyl H-phosphonate and the absence of reaction with diisopropyl H-phosphonate is not solely associated with the inductive effect of the alkyl substituents on the α-carbon atom but is also due to steric effects, which is especially apparent in the case of diisopropyl H-phosphonate.

3.8.7 Reaction with chlorosilanes

Dealkylation of dialkyl H-phosphonates with chlorosilanes takes place with formation of a P–O–Si linkage according to the following scheme [309]:

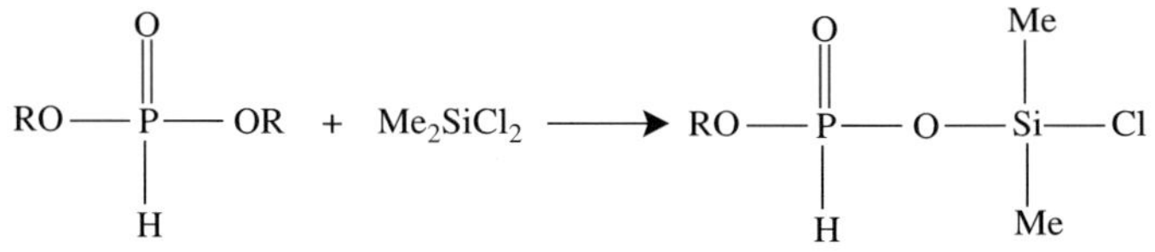

$$\text{RO-P(O)(H)-OR} + \text{Me}_2\text{SiCl}_2 \longrightarrow \text{RO-P(O)(H)-O-Si(Me)_2-Cl}$$

The reaction between dimethyl H-phosphonate and dimethyl chlorosilane proceeds at room temperature. In the ^{31}P{H} NMR spectrum of the reaction product, there are signals for three types of phosphorus atoms—at 11.78 ppm, a doublet of septets with $^3J(P,H) = 12.0$ Hz and $^1J(P,H) = 698.2$ Hz, which can be assigned to the phosphorus atom in dimethyl H-phosphonate; at 11.40 ppm, a doublet of quartets with $^3J(P,H) = 11.7$ Hz and $^1J(P,H) = 690$ Hz, which can be assigned to the phosphorus atom in the end groups; and at 8.55 ppm,

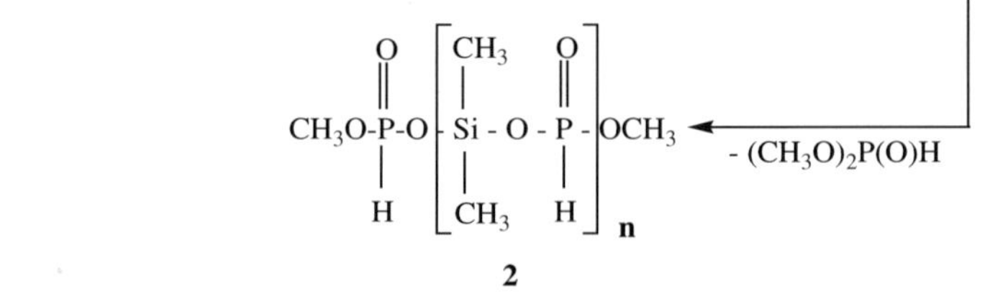

$$2n\,\text{CH}_3\text{O-P(O)(H)-OCH}_3 + \text{Cl-Si(CH}_3)_2\text{-Cl} \xrightarrow{-2n\,\text{CH}_3\text{Cl}} \text{CH}_3\text{O-P(O)(H)-O-Si(CH}_3)_2\text{-O-P(O)(H)-OCH}_3$$
1

$$\text{CH}_3\text{O-P(O)(H)-O-[Si(CH}_3)_2\text{-O-P(O)(H)-OCH}_3]_n \xleftarrow{-(\text{CH}_3\text{O})_2\text{P(O)H}}$$
2

a doublet $^1J(P,H) = 696.4$ Hz, which can be assigned to the phosphorus atom with O–Si groups on both sides. The presence of such a phosphorus atom and the presence of dimethyl H-phosphonate is unexpected since the material balance (amount of CH_3Cl) of the reaction indicates that the dimethyl H-phosphonate has reacted completely. It has been suggested that the silicon-containing diphosphonate **1** is initially formed as a result of dealkylation of dimethyl H-phosphonate. This intermediate undergoes further condensations with evaluation of dimethyl H-phosphonate, yielding a phosphorus- and silicon-containing oligomer with terminal phosphonate groups **2**. The experimental data revealed that in a molar ratio 2:1, oligomeric product **2** of the structure given below is obtained [309]. When the above reaction is carried out in a molar ratio 1:1, oligomeric products of the type **1** are formed [310].

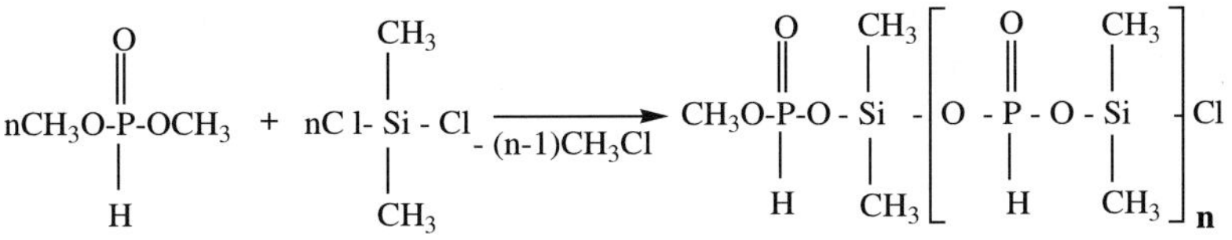

$$nCH_3O\text{-}P(=O)(H)\text{-}OCH_3 + nCl\text{-}Si(CH_3)_2\text{-}Cl \xrightarrow{-(n-1)CH_3Cl} CH_3O\text{-}P(=O)(H)\text{-}O\text{-}Si(CH_3)_2\text{-}[O\text{-}P(=O)(H)\text{-}O\text{-}Si(CH_3)_2]_n\text{-}Cl$$

1

The $^{31}P\{H\}$ NMR spectrum of the reaction product showed signals for two types of phosphorus atoms—at 11.40 ppm, a doublet of quartets with $^3J(P,H) = 11.7$ Hz and $^1J(P,H) = 690.0$ Hz and at 8.55 ppm, a doublet $^1J(P,H) = 696.4$ Hz. These phosphorus- and silicon-containing monomers and oligomers are interesting products as modifiers of polymers.

REFERENCES

1. M. Grayson, C. F. Fareley, C. A. Strenli, *Tetrahedron*, **1967**, 23, 1065.
2. G. Doak, L. Freedman, *Chem. Rev.*, **1961**, 61, 31.
3. K. Moedritzer, *Inorg. Nucl. Chem.*, **1967**, 22, 19.
4. E. Tsvetkov, M. Terehova, E. Petrov, R. Maleviannaia, S. Measets, A. Shatenshtein, M. Kabachnik, *Zh. Obshch. Khim.*, **1977**, 47, 1981.
5. J. Guthrie, *Can. J. Chem.*, **1979**, 57, 235.
6. A. Razumov, S. D. Hen, *Zh. Obshch. Khim.*, **1956**, 26, 2233.
7. G. Imaev, V. Masslennikov, V. Gorina, O. Krasheninikova, *Zh. Obshch. Khim.*, **1965**, 35, 75.
8. Ger. Pat. 1,059,425 (**1959**); C. A., 56, 11446d (**1962**); A. DeRose, W. Gerrard, E. Moonly, *Chem. Ind.*, **1961**, 36, 1449.
9. K. Troev, G. Borisov, *Izv. Otd. Khim Nauki, Bulg. Acad. Sci.*, **1984**, 17, 482.
10. R. Williams, L. Hamilton, *J. Am. Chem. Soc.*, **1952**, 74, 5418.
11. R. Williams, L. Hamilton, *J. Am. Chem. Soc.*, **1955**, 77, 3411.
12. R. Miller, J. Bradley, L. Hamilton, *J. Am. Chem. Soc.*, **1956**, 78, 5299.
13. U. S. Pat. 3,591,501 (**1971**).
14. F. H. Westheimer, *Acc. Chem. Res.*, **1968**, 1, 70.
15. R. F. Hudson, *Ibid.*, **1972**, 5, 204.
16. R. R. Holmes, *Pentacoordinated Phosphorus, ACS Monograph Series 175 & 176*; American Chemical Society; Washington D., **1980**, Vol. 1 & 2.

17. (a) R. S. Berry, *J. Chem. Phys.*, **1960**, 32, 933; (b) K. Mislow, *Acc. Chem. Res.*, **1970**, 3, 321; (c) I. Ugi, D. Marquarding, H. Kulasacek, P. Gillespie, F. Ramirez, *Acc. Chem. Res.*, **1971**, 4, 288.
18. P. Nylen, *Svensk. Kem. Tid.*, **1937**, 49, 29.
19. F. H. Westheimer, S. Huang, F. Covitz, *J. Am. Chem. Soc.*, **1988**, 110, 181.
20. T. Mastrukova, M. Kabachnik, *Uspehi Khimii*, **1969**, 38, 1751.
21. P. Guthrie, *J. Am. Chem. Soc.*, **1977**, 99, 3991.
22. E. M. Georgiev, J. Kaneti, K. Troev, D. M. Roundhill, unpublished results.
23. (a) I. H. Williams, G. M. Maggiora, R. L. Showen, *J. Am. Chem. Soc.*, **1980**, 102, 7831; (b) I. H. Williams, D. Spangler, D. A. Femec, G. M. Maggiora, R. L. Showen, *J. Am. Chem. Soc.*, **1983**, 105, 31; (c) I. H. Williams, D. Spangler, G. M. Maggiora, R. L. Showen, *J. Am. Chem. Soc.*, **1985**, 107, 7717; (d) R. R. Fraser, F. Kong, M. Stanculescu, Y. D. Wu, K. N. Houk, *J. Org. Chem.*, **1993**, 58, 4431.
24. (a) W. P. Jencks, *Acc. Chem. Res.*, **1976**, 9, 425; (b) W. P. Jencks, *Ibid.*, **1980**, 13, 161.
25. (a) B. Ma, Y. Xie, M. Shen, H. F. Schaefer III, *J. Am. Chem. Soc.*, **1993**, 115, 1943; (b) Y. D. Wu, K. N. Houk, *J. Am. Chem. Soc.*, **1993**, 115, 11997.
26. P. C. Haake, F. H. Westheimer, *J. Am. Chem. Soc.*, **1961**, 83, 1102.
27. M. A. Kaslina, L. V. Krinitskaya, A. V. Kessenikh, T. V. Balaghova, Yu. P. Reshetnikov, *Zh. Obshch. Khim.*, **1992**, 62, 1531.
28. W. Vogt, S. Balasubramanian, *Macromol. Chem.*, **1973**, 163, 111.
29. V. Vassileva, G. Lihina, *Izv. Tomsk. Politech. Inst.*, **1976**, 258, 83.
30. J. Kaneti, E. Georgiev, K. Troev, unpublished.
31. (a) R. S. Berry, *J. Chem. Phys.*, **1960**, 32, 933; (b) K. Mislow, *Acc. Chem. Res.*, **1970**, 3, 321.
32. K. Petrov, E. Nifantief, R. Goltsova, A. Tsegolev, B. Bushmin, *Zh. Obshch. Khim.*, **1962**, 32, 3723.
33. E. Walsh, *J. Am. Chem. Soc.*, **1959**, 81, 3023.
34. J. Jankowska, M. Sobkowski, J. Stawinski, A. Kraszewski, *Tetrahedron Lett.*, **1994**, 35, 3355.
35. V. Pessin, K. Haletskii, *Trudi Leningr. Chim. Pharm. Inst.*, **1958**, 5, 16.
36. V. Kodolov, S. Spasskii, *Trudi Akad. Nauk SSSR Uralskii Filial, Elementoorg. Soed.*, **1968**, 15, 15.
37. K. Troev, M. Stefanov, *Phosphorus and Sulfur*, **1984**, 19, 363.
38. R. Pearson, J. Sougstad, *J. Am. Chem. Soc.*, **1967**, 89, 1827.
39. USSR Pat. 199,144 (**1963**).
40. G. Borisov, K. Troev, *Izv. Otd. Khim. Nauki, Bulg. Acad. Sci.*, **1971**, 4, 369.
41. E. Nifantief, A. Zavalishina, I. Nosonovskii, I. Komlev, *Zh. Obshch. Khim.*, **1968**, 38, 2538.
42. A. Oswald, *Can. J. Chem.*, **1959**, 37, 1498.
43. M. Sobkowski, M. Wenska, A. Kraszewski, J. Stawinski, *Nucleosides, Nucleotides & Nucleic Acids,* **2000**, 19, 1487.
44. G. Borisov, K. Troev, *Izv. Otd. Khim. Nauki, Bulg. Acad. Sci.*, **1972**, 5, 175.
45. G. Borisov, K. Troev, *European Pol. J.*, **1972**, 9, 1077.
46. G. Borissov, Ts. Nedjalkova, *Izv. Otd. Khim. Nauki Bukg. Acad. Sci.*, **1971**, 4, 359.
47. Ger. Pat. 1,178,596 (**1964**).
48. Ger. Pat. 1,190,186 (**1965**); C. A., 62, 15765 (**1965**).
49. K. Petrov, E. Nifantief, R. Goltsova, *Vysokomol. Soed.*, **1964**, 6, 1545.
50. U.S. Pat. 2,963,451 (**1956**).
51. K. Petrov, E. Nifantief, R. Goltsova, *Vysokomol. Soed.*, **1962**, 4, 1291.
52. USSR Pat. 132,404 (**1960**); C. A. 55, 8933 (**1961**).
53. K. Petrov, E. Nifantief, R. Goltsova, *Geterocyclenye vysokomol, Soed., Nauka,* Moskow, **1960**, 170.

References

54. Brit. Pat. 679,834 (**1952**).
55. A. Zavalishina, S. Sorokina, E. Nifantief, *Zh. Obshch. Khim.*, **1968**, 38, 2271.
56. K. Troev, unpublished data, **2001**.
57. (a) V. V. Ovchinnikov, G. I. Galkin, E. G. Jarkova, R. A. Cherkasov, A. N. Pudovik, *Dokl. Akad. Nauk SSSR*, **1977**, 235, 118; (b) V. V. Ovchinnikov, G. I. Galkin, E. G. Jarkova, L. E. Markova, R. A. Cherkasov, A. N. Pudovik, *Zh. Obshch. Khim.*, **1978**, 48, 2224.
58. M. Mikolajczyk, *Chem. Comm.*, **1969**, 1221.
59. U.S. Pat. 3,147,298 (**1960**); C. A., 64, 80866, (**1964**).
60. Brit. Pat. 965785 (**1959**); C. A., 63, 495c, (**1963**).
61. Fr. Pat. 1,162,199 (**1959**); C. A., 55, 1733d, (**1961**).
62. Ger. Pat. 2,002,678 (**1970**); C. A., 75, 206090, (**1971**).
63. G. Borisov, K. Troev, *Eur. Polym. J.*, **1973**, 9, 1077.
64. K. Troev, E. Tashev, G. Borisov, *Eur. Polym. J.*, **1982**, 18, 223.
65. K. Troev, E. Tashev, G. Borisov, *Acta Polym.*, **1985**, 36, 531.
66. E. Tashev, St. Shenkov, K. Troev, G. Borisov, L. Zabski, Z. Edlinski, *Eur. Polym. J.*, **1988**, 24, 1101.
67. K. Troev, R. Tsevi, T. Bourova, S. Kobayashi, H. Uayama, D. M. Roundhill, *J. Polym. Sci., Part A: Polym. Chem.*, **1996**, 34, 621.
68. K. Troev, K. Kossev, D. Tortopov, *Eur. Polym. J.*, **1994**, 30, 757.
69. E. Nifantief, *Khimia Gidrophosphorilnih Soedinenii*, Mir, Moskow, **1983**, 173pp.
70. K. Petrov, E. Nifantief, T. Lisenko, *Zh. Obshch. Khim.*, **1961**, 31, 1709.
71. J. Michalski, A. Zwierzak, *Roczn. Chem.*, **1961**, 35, 619.
72. B. Haskin, N. Tolmaheva, V. Promenenkov, *Zh. Obshch. Khim.*, **1980**, 50, 2455.
73. K. Petrov, G. Sokol'skii, *Zh. Obshch. Khim.*, **1956**, 26, 3337.
74. E. Zahoyi-Budo, L. I. Simandi, *Inorg. Chem. Acta*, **1993**, 205, 207.
75. F. Snekfull, H. Hanbrick, *Angew. Chem.*, **1958**, 70, 238.
76. A. M. Polozov, A. Kh. Mustafin, *Zh. Obshch. Khim.*, **1992**, 62, 1429.
77. E. S. Lewis, E. C. Nieh, *J. Org. Chem.*, **1973**, 38, 4402.
78. M. I. Kabachnik, E. I. Golubeva, *Dokl. Akad. Nauk SSSR*, **1955**, 105, 1258.
79. M. M. Kabachnik, E. V. Snjatkova, Z. C. Novikova, U. F. Lutsenko, *Zh. Obshch. Khim.*, **1980**, 50, 227.
80. A. Atherton, A. Todd, *J. Chem. Soc.*, **1947**, 674; A. Atherton, H. Openshaw, A.Todd, *J. Chem. Soc.*, **1945**, 660.
81. G. Steinberg, *J. Org. Chem.*, **1950**, 15, 637.
82. E. Nifantief, *Khimia gidrophosphorilnih soedinenii*, Mir, Moskow, **1983**, 121.
83. E. Nifantief, V. Blagovestenski, A. Sokurenko, L. Skliarskii, *Zh. Obshch. Khim.*, **1973**, 45, 108.
84. L. Riesel, J. Steinbach, E. Herrmann, *Z. Anorg. Allgem. Chem.*, **1983**, 502, 21.
85. A. Kong, R. Engel, *Bull. Chem. Soc. Jpn.*, **1985**, 58, 3671.
86. E. Nifantief, A. Kruchkova, *Dokl. Akad. Nauk SSSR*, **1972**, 203, 841.
87. K. Troev, E. M. G. Kirilov, D. M. Roundhill, *Bull. Chem. Soc. Jpn.*, **1990,** 63, 1284.
88. E. M. Georgiev, J. Kaneti, K. Troev, D. M. Roundhill, *J. Am. Chem. Soc.*, **1993**, 115, 10678.
89. G. J. Ji, C. B. Xue, J. N. Zeng, L. P. Li, W. G. Chai, Y. F. Zhao, *J. Org. Chem.*, **1988**, 444.
90. Y. F. Zhao, S. K. Xi, A. T. Song, G. J. Ji, *J. Org. Chem.*, **1984**, 49, 4549.
91. A. Zwierzak, *Synthesis*, **1975**, 507.
92. T. Gajda, A. Zwierzak, *Synthesis*, **1976**, 243.
93. A. Zwierzak, *Synthesis*, **1976**, 305.
94. B. S. Barta and Purnanand, *Indian J. Chem.*, **1991**, 30B, 57.
95. A. Zwierzak, A. Sulewska, *Synthesis*, **1976**, 835.
96. L. Lukanov, R. Venkov, N. Mollov, *Synthesis*, **1985**, 971.

97. E. Nifantief, A. Zavalishina, *Zh. Obshch. Khim.*, **1967**, 37, 1854.
98. B. Minbaev, S. Mataeva, B. Shostakovskii, B. Abiurov, *Zh. Obshch. Khim.*, **1989**, 59, 2638.
99. T. Smith, *J. Chem. Soc.*, **1962**, 1122.
100. TB. Abramov, S. Ilina, *Zh. Obshch. Khim.*, **1967**, 39, 2234.
101. Fr. Pat. 1413874 (**1965**).
102. Fr. Pat. 86531 (**1966**).
103. R. Engel, S. Chakraborty, *Synth. Commun.*, **1988**, 18, 665.
104. A. Brel, O. Bogdanova, A. Rahimov, *Zh. Obshch. Khim.*, **1979**, 49, 715.
105. M. Pudovik, A. Pudovik, *Zh. Obshch. Khim.*, **1976**, 46, 21.
106. R. Greenhalgh, M. A. Weinberger, *Can. J. Chem.*, **1967**, 45, 495.
107. *Recently Discovered Systems of Enzyme Regulation by Reversible Phosphorylation* (ed. P. Cohen), Elsevier, NY, **1980**.
108. O. M. Rosen, E. C. Krebs, *Protein Phosphorylation*, Cold Spring Harbor, NY, **1981**.
109. Y. F. Zhao, S. K. Xi, Y. F. Tiau, A. T. Song, *Tetrahedron Lett.*, **1983**, 24, 1617.
110. Y. F. Zhao, S. K. Xi, A. T. Song, G. J. Ji, *Phosphorus & Sulfur*, **1983**, 18, 155.
111. X. B. Ma, Y. F. Zhao, *Phosphorus, Sulfur & Silicon*, **1991**, 61, 9.
112. X. B. Ma, Y. F. Zhao, *J. Org. Chem.*, **1989**, 54, 4005; *Synthesis*, **1992**, 759.
113. G. Taborsky, *Adv. Protein Chem.*, **1974**, 28, 1.
114. C. B. Xue, Y. W. Yin, Yu-Fen Zhao, *Tetr. Lett.*, **1988**, 29, 1145.
115. J. N. Zeng, C. B. Xue, Q. W. Chen, Y. F. Zhao, *Bioorg. Chem.*, **1989**, 17, 434.
116. Y. M. Li, D.Q. Zhang, H. W. Zhang, G. J. Ji, Y. F. Zhao, *Bioorg. Chem.*, **1992**, 20, 285.
117. J. H. Jones. In: *The Peptides* (eds. E. Gross, J. Meienhofer), Academic Press, New York, **1979**, 75.
118. O. A. Nurkenov, A. M. Gazaliev, I. V. Markova, S. N. Balitskii, *Zh. Obshch. Khim.*, **1998**, 68, 1401.
119. D. Trednafilova-Gercheva, K. Troev, M. Georgieva, V. Vassileva, *Angew. Makromol. Chem.*, **1992**, 199, 137.
120. R. Tzevi, G. Todorova, K. Kossev, K. Troev, D. M. Roundhill, *Makromol. Chem.*, **1993**, 194, 3261.
121. M. J. Moluna, F. S. Rowland, *Nature*, **1974**, 249, 810; F. S. Rowland, *Am. Sci.*, **1989**, 77, 36.
122. M. McFarland, *Environ. Sci. Technol.*, **1989**, 23, 1203.
123. E. M. Georgiev, D. M. Roundhill, K. Troev, *Inorg. Chem.*, **1992**, 31, 1965.
124. E. Shi, C. Pei, *Synthesis*, **2004**, 18, 2995.
125. *Encyclopedia of Reagents for Organic Synthesis* (ed. L. A. Paquette), Wiley, New York, **1995**, Vol. 1–8.
126. T. Shori, S. Yamada, *Org. Synth.*, **1984**, 62, 187.
127. (a) E. Shi, C. Pei, *Phosphorus, Sulfur Silicon Relat. Elements*, **2003**, 178, 1093; (b) *Synth. Commun.*, **2004**, 34, 1285.
128. (a) A. Łopusiski, L. Łuczak, J. Michalski, *Tetrahedron*, **1982**, 38, 679; (b) A. N. Pudovik, G. V. Romanov, T. Y. Stepanova, *Zh. Obshch. Khim.*, **1979**, 49, 1425.
129. (a) A. I. Razumov, V.V. Moskva, *Zh. Obshch. Khim.*, 1964, 34, 3125; **1965**, 35, 1595; (b) M. J. Gallagher, H. Honegger, *Aust. J. Chem.*, **1980**, 33, 287.
130. P. Alexsander, A. Holy, M. Masojidkova, *Collect. Czech. Chem. Commun.*, **1994**, 59, 1870.
131. (a) Jpn. Pat. 79,135,724 (**1978**); C. A. 92, 146905v (**1980**); (b) Jpn. Pat. 79,144,383 (**1979**); C. A. 93, 26547 (**1980**); (c) V. I. Krutikov, A. V. Erkin, P. A. Pautov, M. M. Zolotukhina, *Zh. Obshch. Khim.*, **2003**, 73, 205;
132. Ger. Pat. 1,815,946 (**1970**).
133. (a) R. L. Hildebrand, *The Role of Phosphonates in Living Systems*, CRC Press, Boca Raton, FL, **1983**; (b) R. Engel, *Chem. Rev.*, **1977**, 77, 349; (c) G. M. Blackburn, *Chem. Ind.*, **1981**,

134; (d) J. Krapcho, C. Turk, D. W. Cushman, J. R. Powel, J. M. DeForrest, E. R. Spitzmiller, D. S. Karanewsky, M. Duggan, G. Rovnak, J. Schwartz, S. Natarajan, J. D. Godfrey, D. E. Ryono, R. Neubeck, Jr. E. W. Petrillo, *J. Med. Chem.*, **1988**, 31, 1148.
134. M. Horiguchi, M. Kandatsu, *Nature*, **1959**, 184, 901.
135. M. Eto, *Organophosphorus Pesticides: Organic and Bioorganic Chemistry*, CRC Press, Boca Raton, FL, **1979**, 73, 151.
136. A. N. Pudovik, *Tesisi Dokl., Otd. Khim. Nauki, Dokl. AN SSSR*, **1950**, 73, 499.
137. G. Kosolapoff, *Organophosphorus Compounds*, Wiley, New York, **1950**, 124, 358.
138. A. N. Pudovik, B. Arbuzov, *Izv. Akad. Nauk SSSR, Otd. Khim Nauki*, **1949**, 522.
139. A. N. Pudovik, M. Frolova, *Zh. Obshch. Khim.*, **1952**, 22, 2052.
140. V. Galkin, A. B. Khabibulina, I. V. Bakhtiyarova, R. A. Cherkasov, A. N. Pudovik, *Zh., Obshch. Khim.*, **1988**, 58, 1002.
141. V. Galkin, K Kurdi, A. Khabibulina, R. Cherkasov, *Zh. Obshch. Khim.*, **1990**, 60, 92.
142. V. Galkin, A. B. Khabibulina, E. R. Gaifutdinova, I. V. Bakhtiyarova, R. A. Cherkasov, A. N. Pudovik, *Dokl. Akad. Nauk SSSR*, **1987**, 296, 107.
143. L. Nesterov, N. Krepisheva, N. Aleksandrov, *Zh. Obshch. Khim.*, **1984**, 54, 54.
144. U. S. Pat. 3,681,48 (**1972**).
145. Ger. Pat. 2,646,350 (**1978**).
146. A. N. Pudovik, T. M. Sudakova, *Zh. Obshch. Khim.*, **1966**, 36, 1113.
147. B. N. Laskorin, V. V. Jakshin, V. B. Bulgakova, *Zh. Obshch. Khim.*, **1976**, 46, 2477.
148. Yu. G. Safina, G. Sh. Malkova, R. A. Cherkasov, *Zh. Obshch. Khim.*, **1991**, 61, 620.
149. Yu. G. Safina, G. Sh. Malkova, R. A. Cherkasov, V. V. Ovchinnikov, *Zh. Obshch. Khim.*, **1990**, 60, 221.
150. Yu. G. Safina, G. Sh. Malkova, R. A. Cherkasov, *Zh. Obshch. Khim.*, **1992**, 62, 1557.
151. G. A. Russell, C. F. Yao, H. I. Tashtoush, J. F. Russell, D. F. Dedolph, *J. Org. Chem.*, **1991**, 56, 663.
152. G. A. Russell, C. F. Yao, *J. Org. Chem.*, **1992**, 57, 6508.
153. F. Bardone, M. Mladenova, M. Gaudemar, *Synthesis*, **1988**, 611.
154. V. Ovchinnikov, V, Galkin, *Izv. Akad. Nauk SSSR, Otd. Khim. Nauki*, **1977**, 2021.
155. V. Ovchinnikov, R. Cherkasov, *Izv. Akad. Nauk SSSR, Otd. Khim. Nauki*, **1978**, 689.
156. V. Ovchinnikov, V. Galkin, R. Cherkasov, A. Pudovik, *Dokl. Akad. Nauk SSSR*, **1979**, 249, 128.
157. V. Ovchinnikov, S. Cherezov, V, Galkin, R. Cherkasov, A. N. Pudovik, *Zh. Obshch. Khim.*, **1980**, 50, 2615.
158. (a) D. V. Patel, K. Rielly-Gauvin, D. E. Ryono, *Tetrahedron Lett.*, **1990**, 31, 5587; (b) **1990**, 31, 5591.
159. B. Stowasser, K. H. Budt, Qi. L. Jian, A. Peyman, D. Ruppert, *Tetrahedron Lett.*, **1992**, 33, 6625.
160. A. M. MacCleod, R. Baker, M. Hudson, K. James, M. B. Roe, M. Knowles, G. MacAllister, *Med. Chem. Res.*, **1992**, 2, 96.
161. J. A. Shiroski, M. J. Miller, D. S. Bacclino, d. g. Cleary, S. D. Corey, J. L. Font, K. J. Gruys, C. Y. Han, K. C. Lin, P. D. Pansegrau, J. E. Ream, D. Schnur, A. Shan, M. C. Walker, *Phosphorus, Sulfur, Silicon and Related Elements*, **1993**, 76, 115.
162. (a) T. Gaida, *Tetrahedron: Asymmetry*, **1994**, 5, 1965; (b) T. Khushi, K. J. O' Toole, J. T. Smith, *Tetrahedron Lett.*, **1993**, 34, 2375; (c) A. Heisler, C. Rabiller, R. Douillard, N. Goalou, G. Haegele, F. Levayer, *Tetrahedron: Asymmetry*, **1993**, 4, 959; (d) D. M. Pogatchnik, D. F. Wiemer, *Tetrahedron Lett.*, **1997**, 38, 3495.
163. M. S. Smyth, H. Ford Jr., T. R. Burke Jr., *Tetrahedron Lett.*, **1992**, 33, 4137.
164. F. Hammerschmidt, *Angew. Chem., Int. Ed. Engl.*, **1994**, 33, 341.
165. L. J. Jennings, M. Macchia, A. Parkin, *J. Chem. Soc., Perkin Trans.1*, **1992**, 2197.

166. G. Lavielle, P. Hautefaye, C. Schaffer, J. A. Boutin, C. A. Cudennec, A. Pierre, *J. Med. Chem.*, **1991**, 34, 1998.
167. B. Abramov, *Zh. Obshch. Khim.*, **1952**, 22, 647.
168. B. Abramov, *Dokl. Akad. Nauk SSSR*, **1950**, 73, 487.
169. U. S. Pat. 2,579,810 (**1951**).
170. M. A. Ruveda, S. A. de Licastro, *Tetrahedron*, **1972**, 28, 6013.
171. T. Agawa, T. Kubo, Y. Ohshiro, *Synthesis*, **1971**, 27.
172. B. Springs, P. Haake, *J. Org. Chem.*, **1976**, 41, 1165.
173. H. Timmler, L. Kurz, *Chem. Ber.*, **1971**, 104, 3740.
174. P. G. Baraldi, M. Guarnieri, F. Moroder, G. Pollini, D. Simoni, *Synthesis*, **1982**, 653,
175. M. Kabachnik, E.Tsvetkov, *Izv. Akad. Nauk SSSR, Otd. Khim. Nauki*, **1963**, 7, 1227.
176. B. Springs, P. Haake, *J. Org. Chem.*, **1977**, 42, 472.
177. W. Barthel, P. Giang, S. Hall, *J. Am. Chem. Soc.*, **1954**, 76, 4186.
178. I. Vorontsov, O. Vlasov, N. Mel'nikov, *Zh. Obshch. Khim.*, **1978**, 48, 2663.
179. I. Vorontsov, O. Vlasov, L. Batel'fer, N. Mel'nikov, *Zh. Obshch. Khim.*, **1977**, 47, 2698.
180. A. Vasil'ev, I. Vorontsov, O. Vlasov, N. Mel'nikov, *Dokl. Akad. Nauk SSSR*, **1979**, 248, 875
181. W. Barthel, B. Alexxander, P. Giang, S. Hall, *J. Am. Chem. Soc.*, **1955**, 77, 2424.
182. W. Lorenz, A. Henglein, G. Schroder, *J. Am. Chem. Soc.*, **1955**, 77, 2554.
183. K. A. Petrov, V. A.Chuzov, I. V. Pastuhova, *Zh. Obshch. Khim.*, **1976**, 46, 1413.
184. L. Uzlova, O. Eruzeva, E. Hurtsidse, A. Zdanov, *Zh. Obshch. Khim.*, **1988**, 58, 1160.
185. G. Borisov, K. Troev, *Commun Dept. Chem., BAS*, **1972**, 5, 175.
186. M. S. Kharash, R. A. Mosher, I. S. Bengeldorf, *J. Org. Chem.*, **1960**, 25, 1000.
187. B. A. Arbuzov, V. S. Vinogradova, N. A. Polezhaeva, *Dokl. Akad. Nauk SSSR*, **1956**, 111, 107; **1958**, 121, 641.
188. A. Meisters, J. M. Swan, *Aust. J. Chem.*, **1965**, 18, 168.
189. J. H. Clark, *Chem. Rev.*, **1980**, 80, 429.
190. (a) F. Texier-Boullet, A. Foucaud, *Tetrahedron Lett.*, **1980**, 2161; (b) F. Texier-Boullet, A. Foucaud, *Synthesis,* **1982,** 165; (c) F. Texier-Boullet, A. Foucaud, *Synthesis,* **1982,** 916; (d) F. Texier-Boullet, D. Villemin, M. Ricard, H. Moison, A. Foucaud, *Tetrahedron*, **1985**, 41, 1259.
191. J. Boyer, R. J. P. Corri, R. Perz, C. Reye, *Synthesis*, **1981**, 122.
192. J. Yamawaki, T. Kawate, T. Ando, T. Hanafusa, *Bull. Chem. Soc. Jpn.*, **1983**, 56, 1885.
193. D. Villemin, R. Racha, *Tetrahedron Lett.*, **1986**, 27, 1789.
194. F. Texier-Boullet, M. Lequitte, *Tetrahedron Lett.*, **1986**, 27, 3515.
195. (a) A. Fadel, R. Yefash, J. Salaun, *Synthesis*, **1987**, 37; (b) G. Rosini, R. Galarni, E. Marotta, R. Right, *J. Org. Chem.*, **1990**, 55, 781; (c) M. Kodomari, T. Sakamoto, S. Yoshitomi, *J. Chem. Soc.Chem. Commun.*, **1990**, 701; (d) P. J. Kropp, K. A. Daus, S. D. Crawford, M. W. Tubergen, K. D. Kepler, S. L. Craig, V. P. Wilson, *J. Am. Chem. Soc.*, **1990**, 112, 7433; (e) G. Hondrogiannis, R. M. Pagni, G. W. Kabalka, P. Anosike, R. Kurt, *Tetrahedron Lett.*, **1990**, 31, 5433; (f) H. K. Patney, *Tetrahedron Lett.*, **1991**, 32, 2259; (g) F. Pautel, M. Daudon, *Tetrahedron Lett.*, **1991**, 32, 1457.
196. (a) A. R. Sardarian, B. Kaboudin, *Synthetic Commun.*, **1997**, 27, 543; (b) B. Kaboudin, *Tetrahedron Lett.*, **2000**, 41, 3169.
197. M. D. Groaning, B. J. Rowe, C. D. Spilling, *Tetrahedron Lett.*, **1998**, 39, 5485.
198. (a) M. Shibasaki, H. Sasai, T. Arai, *Angew. Chem., Int. Ed. Engl.*, **1997**, 36, 1236; (b) T. Yokomatsu, T. Yamagishi, K. Matsumoto, S. Shibuya, *Tetrahedron*, **1996**, 52, 11725; (c) N. P. Rath, C. D. Spilling, *Tetrahedron Lett.*, **1994**, 35, 227.
199. (a) H. Wynberg, A. A. Smaardijk, *Tetrahedron Lett.*, **1983**, 24, 5899; (b) A. A. Smaardijk, S. Noorda, F. van Bolhuis, H. Wynberg, *Tetrahedron Lett.*, **1985**, 26, 493; (c) T. Yokomatsu, T. Yamagishi, S. Shibuya, *Tetrahedron*: *Asymmetry*, **1993**, 4, 1783; (d) 1779;(d) H. Sasai, S. Arai,

Y. Tahara, M. Shibasaki, *J. Org. Chem.*, **1995**, 60, 6656; (e) T. Yokomatsu, T. Yamagishi, S. Shibuya, *J. Chem. Soc., Perkin Trans. 1*, **1997**, 1527;
200. (a) J. P. Duxbury, A. Cawley, M. Thornton-Pett, L. Wantz, J. N. D. Warne, R. Greatrex, D. Brown, T. P. Kee, *Tetrahedron Lett.*, **1999**, 40, 4403–4406; (b) J. P. Duxbury, J. N. D. Warne, R. Mushtaq, C. Ward, M. Thornton-Pett, M. Jiang, R. Greatrex, T. P. Kee, *Organometallics*, **2000**, 19, 4445;(c) C. V. Ward, M. Jiang, T. P. Kee, *Tetrahedron Lett.*, **2000**, 41, 6181.
201. H. Sasai, T. Arai, Y. Satow, K. N. Houk, M. Shibasaki, *J. Am. Chem. Soc.*, **1995**, 117, 6194.
202. T. Arai, M. Bougauchi, H. Sasai, M. Shibasaki, *J. Org. Chem.*, **1996**, 61, 2926.
203. H. Sasai, M. Bougauchi, T. Arai, M. Shibasaki, *Tetrahedrton Lett.*, **1997**, 38, 2717.
204. Z. Glowacki, M. Hoffmann, J. Rachon, *Phosphorus, Sulfur, and Silicon*, **1993**, 82, 39.
205. D. Parker, *Chem. Rev.*, **1991**, 91, 1441.
206. D. E. Zymanczyk, M. Skwarczynski, B. Lejczak, P. Kafarski, *Tetrahedron: Asymmetry*, **1996**, 7, 1277.
207. S. Racha, Z. Li, H. El-Subbagh, E. Abushanab, *Tetrahedron. Lett*, **1992**, 33, 5491.
208. K. Iizuka, T. Kamijo, T. Kubota, K. Akahane, H. Umeyama, Y. Kiso, *J. Med. Chem.*, **1988**, 31, 704.
209. J. Boger, N. S. Lohr, E. H. Ulm, M. Poe, E. H. Blaine, G. M. Fanelli, T. Y. Lin, L. S. Payne, T. W. Schorn, B. I. Lamont, T. C. Vassil, I. I. Stabilito, D. F. Veber, D. H. Rich, A S. Bopari, *Nature*, **1983**, 303, 81.
210. H. Mitsuya, K. J. Weinhold, P. A. Furman, M. H. St. Clair, S. N. Lehrman, R. C. Gallo, D. Bolognesi, D. W. Barry, S. Broder, *Proc. Natl. Acad. Sci. U. S. A.*, **1985**, 82, 7096.
211. R. Yarchoan, H. Mitsuya, R. V. Thomas, J. M. Pluda, N. R. Hartman, C. F. Perno, K. S. Marczyk, J. P. Allain, D. G. Johns, S. Broder, *Science*, **1989**, 245, 412.
212. H. Mitsuya, S. Broder, *Proc. Natl. Acad. Sci. USA*, **1986**, 83, 1911.
213. (a) D. D. Richman, M. A. Fischl, M. H. Grieco, M. S. Gottlieb, P. A. Volberding, O. L. Laskin, J. M. Leedom, J. E. Groopman, N. D. Mildvan, M. S. Hirsch, G. G. Jackson, D. T. Durack, S. N. Lehrman, *N. Engl. J. Med.*, **1987**, 317, 192; (b) J. P. Sommadossi, R. Carlisle, *Antimicrob. Agents Chemother.*, **1987**, 31, 452.
214. S. Juntenen, J. Chattopadhyaya, *Acta Chem. Scand.*, **1985**, 39, 149.
215. W. L. McEldoon, K. Lee, D. F. Wiemer, *Tetrahedron Lett.*, **1993**, 34, 5843.
216. F. Hansske, D. Madej, M. J. Robins, *Tetrahedron*, **1984**, 40, 125.
217. V. Samano, M. J. Robins, *J. Org. Chem.*, **1991**, 56, 7108.
218. M. L. Froehlich, D. J. Swartling, R. E. Lind, A. W. Mott, D. E. Bergstrom, *Nucleosides Nucleotides*, **1989**, 8, 1529.
219. S. Kralikova, M. Budesinsky, M. Masojidkova, I. Rosenberg, *Tetrahedron Lett.*, **2000**, 41, 955.
220. (a) A. N. Pudovik, I. Konovalova, I. Aladzov, *Some Questions in Organic Chemistry, Khimia*, Khazan, **1969**, 37; (b) H. Zimmer, P. J. Bercz, O. J. Maltenieks, M. W. Moore, *J. Am. Chem. Soc.*, **1965**, 87, 2777; (c) A. N. Pudovik, M. G. Zimin, A. A. Sobanov, *Zh. Obhch. Khim.*, **1972**, 42, 2174; (d) G. Takeshi, Y. Hiroshi, O. Tsuyoshi, I. Hiromi, I. Saburo, *Nippon Kagaku Kaishi*, **1974**, 1093; C. A. 81, 91649 (**1974**); (e) A. N. Pudovik, I. V. Konovalova, *Synthesis*, **1979**, 81; (f) H. Wynberg, A. A. Smaardijk, *Tetrahedron Lett.*, **1983**, 24, 5899; (g) S. Kumaraswamy, R. S. Selvi, K. C. K. Swamy, *Synthesis*, **1997**, 207;(h) R. Gancarz, I. Gancarz, *Tetrahedron Lett.*, **1993**, 34, 145.
221. A. Pudovik, I. Konovalova, M. Ygfarov, E. Goldfarb, G. Rhomanov, *Zh. Obshch. Khim.*, **1973**, 43, 556.
222. (a) R. Engel, *Synthesis of Carbon–Phosphorus Bonds*, CRC Press, Boca Raton, FL, **1988**; (b) T. Hori, M. Horiguchi, A. Hayashi, *Biochemistry of Natural P–C Compounds*, Maruzen, Kyoto Branch Publishing Service, Kyoto, Japan, **1984**.
223. (a) K. D. Berlin, D. M. Hellwege, M. Nagabhushanam. *J. Org. Chem.*, **1965**, 1265; (b) K. D. Berlin, H. A. Taylor, *J. Am. Chem. Soc.*, **1964**, 86, 3862.

224. (a) P. G. Baraldi, M. Guarneri, F. Moroder, G. P. Polloni, D. Simoni, *Synthesis*, **1982**, 653; (b) A. Sardarian, B. Kaboudin, *Synth. Commun.*, **1997**, 27, 543; (c) Y. Liao, H. Shabany, D. S. Christopher, *Tetrahedron Lett.*, **1998**, 39, 8389.
225. H. Firouzabadi, N. Iranpoor, S. Sobhani, A. R. Sardarian, *Tetrahedron Lett.*, **2001**, 42, 4369.
226. H. Firouzabadi, N. Iranpoor, S. Sobhani, *Tetrahedron Lett.*, **2002**, 43, 477.
227. (a) K. Tanaka, F. Toda, *Chem. Rev.*, **2000**, 100, 1025; (b) A. Shaabani, D. G. Lee, *Tetrahedron Lett.*, **2001**, 42, 5833.
228. P. Alexsander, A. Holy, M. Masojidkova, *Collect. Czech. Chem. Commun.*, **1994**, 59, 1870.
229. (a) P. C. Crofts, *Q. Rev. Chem. Soc.*, **1963**, 12, 624; (b) L. D. Freedman, G. O. Doak, *Chem. Rev.*, **1957**, 57, 479.
230. B. Ackerman, T. A. Jordan, C. R. Eddy, D. Swern, *J. Am. Chem. Soc.*, **1956**, 78, 4444.
231. K. D. Berlin, H. A. Tsaylor, *J. Am. Chem. Soc.*, **1964**, 86, 3862.
232. R. Kluger, J. Chin, *J. Am. Chem. Soc.*, **1978**, 100, 7382.
233. K. S. Narayanan, K. D. Berlin, *J. Am. Chem. Soc.*, **1979**, 101, 109.
234. R. Kluger, D. C. Pike, J. Chin, *Can. J. Chem.*, **1978**, 56, 1792.
235. M. Sekine, M. Satoh, H. Yamagata, T. Hata, *J. Org. Chem.*, **1980**, 45, 4162.
236. M. I. Kabachnik, P. A. Rossiiskaya, *Izv. Akad. Nauk SSSR, Ser. Khim.*, **1945**, 597.
237. K. Terauchi, H. Sakurai, *Bull. Chem. Soc. Jpn.*, **1970**, 43, 883.
238. A. P. Pashinkin, T. K. Gazikov, A. N. Pudovik, *Zh. Obshch. Khim.*, **1970**, 40, 28.
239. M. Sekine, A. Kume, T. Hata, *Tetrahedron Lett.*, **1981**, 22, 3617.
240. (a) J. B. Lee, I. M. Downie, *Tetrahedron*, **1967**, 23, 359; (b) R. Appel, *Angew. Chem.*, **1975**, 87, 863; *Angew. Chem., Int. Ed. Engl.*, **1975**, 14, 801.
241. T. Gajda, *Synthesis*, **1990**, 717.
242. (a) P. Savignac, M. Dreux, P. Coutrot, *Tetrahedron Lett.*, **1975**, 609; (b) P. Savignac, J. Petrova, M. Dreux, P. Coutrot, *Synthesis*, **1975**, 535; (c) P. Savignac, M. Snoussi, P. Coutrot, *Synth. Commun.*, **1978**, 8, 19.
243. V. F. Martynov, V. E. Timofeev, *Zh. Obshch. Khim.*, **1962**, 32, 3383.
244. D. Redmore, *Chem. Rev.*, **1971**, 71, 326.
245. P. Perriot, J. Villieras, J. F. Normant, *Synthesis*, **1978**, 33.
246. P. Coutrot, P. Savignac, *Synthesis*, **1978**, 34.
247. W. Wagner, K. Courault, *Z. Chem.*, **1980**, 20, 337.
248. M. P. Teulade, P. Savignac, E. E. Aboujaoude, S. Lietege, N. Collignon, *J. Organomet. Chem.*, **1986**, 304, 283.
249. H. Saunders, R. Slocumbe, *Chem. Rev.*, **1948**, 43, 203.
250. T. Reetz, D. Chadwick, E. Hardy, S. Caufman, *J. Am. Chem. Soc.*, **1955**, 77, 3813.
251. B Arbuzov, N. Rizpolozenskii, *Izv. Akad. Nauk SSSR, Otd. Khim. Nauki*, **1952**, 847.
252. B. Arbuzov, N. Rizpolozenskii, *Izv. Akad. Nauk SSSR, Otd. Khim. Nauki*, **1954**, 631.
253. R. Fox, D. Venezky, *J. Am. Chem. Soc.*, **1956**, 78, 1661.
254. A. Pudovik, A. Kuznetsova, *Zh. Obshch. Khim.*, **1955**, 25, 1369.
255. A. Pudovik, I. Konovalova, R. Krivonosova, *Zh. Obshch. Khim.*, **1956**, 26, 3110.
256. E. Kuznetsov, M. Bahitov, *Trudi Kazanskii Khim. Techn. Inst.*, **1960**, 29, 105.
257. E. Kuznetsov, M. Bahitov, *Dokl. Akad. Nauk SSSR*, **1960**, 134, 830.
258. E. Kuznetsov, M. Bahitov, *Dokl. Akad. Nauk SSSR*, **1961**, 141, 1105.
259. E. Kuznetsov, M. Bahitov, *Zh. Obshch. Khim.*, **1962**, 31, 3025.
260. E. Kuznetzov, M. Bahitov, *Zh. Obshch. Khim.*, **1962**, 32, 278.
261. Kuznetsov, M. Bahitov, *Sintesi i Svoistva Monomerov*, Nauka, Moskow, **1964**.
262. G. Borisov, K. Troev, *Izv. Otd. Khim. Nauki, Bulg. Akad. Sci.*, **1972**, 5, 545.
263. M. Bahitov, Doctoral Thesis, Kazan University, Kazan, SSSR, **1963**.
264. F. Samigulin, I. Kafengauz, A. Kafengauz, E. Gefter, *Zh. Obshch. Khim.*, **1968**, 38, 8.

265. E. Nifantief, *Khimia Fosfororganicheskih Soedinenii*, Izd. Moskow University, Moskow, **1971**.
266. M. Bahitov, V. Zarkov, E. Kuznetsova, F. Kligman, *Zh. Obshch. Khim.*, **1981**, 51, 1035.
267. F. Ignatious, A. Sein, I. Cabasso, J. Smid, *J. Polym. Sci., Part A: Polym. Chem.*, **1992**, 31, 239.
268. Y. Levin, E. Vorkunova, *Homolit. Khim. Phosphora, Akad. Nauk SSSR*, **1978**, 318.
269. E. Nifantief, R. Magdeeva, H. Tsepentieva, *Zh. Obshch. Khim.*, **1980**, 50, 1744.
270. Y. Levin, E. Grutneva, B. Ivanov, *Zh. Obshch. Khim.*, **1974**, 44, 1443.
271. Y. Levin, E. Vorkunova, B. Ivanov, F. Halitov, O. Raevskii, *Zh. Obshch. Khim.*, **1974**, 44, 1301.
272. C. Hubert, A. Munoz, B. Garrigues, J. L. Luche, *J. Org. Chem.*, **1995**, 60, 1488.
273. (a) L. B. Han, N. Choi, M. Tanaka, *J. Am. Chem. Soc.*, **1996**, 118, 7000;(b) T. F. Herpin, J. S. Houlton, W. B. Motherwell, B. P. Roberts, J. M. Weibel, *J. Chem. Soc., Comm.* **1996**, 613; (c) T. F. Herpin, W. B. Motherwell, B. P. Roberts, S. Roland, J-M. Weibel, *Tetrahedron*, **1997**, 53, 15085.
274. R. S. Glass, W. P. Singh, W. Jung, Z. Veres, T. D. Scholz, T. Stadman, *Biochemistry*, **1993**, 32, 12555.
275. (a) D. J. Chen, Z. C. Chen, *Synth. Commun.*, **2001**, 31, 421; (b) D. J. Chen, Z. C. Chen, *J. Chem. Res.*, **2000**, 370; (c)V. A. Potapov, A. A. Tarkova, S. V. Amosova, A. I. Albanov, B. V. Petrov, *Russ. Chem. Bull.*, **1998**, 47, 2042.
276. Q. Xu, C-G. Liang, X. Huang, *Synth. Commun.*, **2003**, 33, 2777.
277. J. R. Morton, K. F. Preston, P. J. Krusic, E. Wasserman, *J. Chem. Soc., Perkin Trans. II*, **1991**, 1425.
278. V. V. Kormachev, Yu. N. Mitrasov, E. A. Anisimova, O. A. Kolyamshin, *Zh. Obshch. Khim.*, **1992**, 62, 1428.
279. U. S. Pat. 2,815,345 (**1957**); C. A., 52, 5485a (**1958**).
280. U. S. Pat. 2,842,111 (**1958**); C. A., 52, 11110f (**1958**).
281. N. Thoung, *Bull. Soc. Chem. France*, **1971**, 3, 928.
282. K. Troev, G. Borissov, *Phosphorus and Sulfur*, **1987**, 29, 129.
283. K. Troev, D. Roundhill, *Phosphorus and Sulfur*, **1988**, 37, 243.
284. K. Troev, D. Roundhill, *Phosphorus and Sulfur*, **1988**, 37, 247.
285. K. Troev, E. M. G. Kirilov, D. M. Roundhill, *Bull. Chem. Soc. Jpn.*, **1990**, 63, 1284.
286. E. M. Georgiev, J. Kaneti, K. Troev, D. M. Roundhill, *J. Am. Chem. Soc.*, **1993**, 115,10964.
287. K. Troev, D. M. Roundhill, *Phosphorus and Sulfur*, **1988**, 36, 189.
288. V. Pesin, A. Haletskii,, *Zh. Obshch. Khim.*, **1961**, 31, 2515.
289. A. Kers, I. Kers, J. Stawinski, *Tetrahedron*, **1996**, 52, 9931.
290. D.Trendafilova-Gercheva, V. Vassileva, M. Georgieva, K. Troev, E. F. Panarin, *Angew. Makromol. Chem.*, **1991**, 187, 135.
291. E. Karanov, D. Trendafilova-Gercheva, M. Georgieva, V. Alexieva, V. Vassileva, K. Troev, *Comptes Rend. Acad. Bulg. Sci.*, **1992**, 45, 53.
292. J. Mircheva, D. Trendafilova-Gercheva, M. Georgieva, K. Troev, V. Vassileva, D. Dabov, *Pharmacia*, **1995**, 43, 17.
293. E. Tashev, St. Shenkov, K. Troev, *Phosphorus and Sulfur*, **1991**, 56, 225.
294. K. Troev, E. Tashev, G. Borisov, *Acta Polymerica*, **1985**, 36, 531.
295. E. M. Georgiev, R. Tzevi, V. Vassileva, K. Troev, D. M. Roundhill, *Phosphorus, Sulfur and Silicon*, **1994**, 88, 139.
296. V. Vassileva, E. M. Georgiev, K. Troev, D. M. Roundhill, *Phosphorus, Sulfur& Silicon*, **1994**, 92, 101.
297. A. Pierce, M. Joullic, *J. Org. Chem.*, **1962**, 27, 3968.
298. D. Trendafilova-Gercheva, K. Troev, M. Georgieva, *Bull. Chem. Soc. Jpn.*, **1991**, 64, 2033.
299. K. Troev, *Heteroatom Chem.*, **2000**, 11, 205.
300. C. Campbell, D. Chadwick, *J. Am. Chem. Soc.*, **1955**, 77, 3379.

301. V. Cooke, W. Gerrard, *J. Chem. Soc.*, **1955**, 1978.
302. H. Tihomirova, N. Ostankekevich, *Zh. Obshch. Khim.*, **1970**, 40, 1412.
303. V. Orlovskii, B. Vovsi, A. Mishkevich, *Zh. Obshch. Khim.*, **1970**, 42, 1930.
304. K. Troev, G. Borisov, *Phosphorus and Sulfur*, **1983**, 17, 267.
305. K. Troev, G. Borisov, *Phosphorus and Sulfur*, **1983**, 17, 257.
306. B. Laskorin, V. Ykshin, V. Bulgakova, *Zh. Obshch. Khim.*, **1976**, 46, 2477 .
307. K. Troev, E. Tashev, G. Borisov, *Phosphorus& Sulfur*, **1981**, 11, 363.
308. K. Troev, E. Tashev, G. Borisov, *Phosphorus& Sulfur*, **1982**, 13, 359.
309. K. Troev, St. Shenkov, U. Delimarinova, *Phosphorus& Sulfur*, **1992**, 68, 107.
310. K.Troev, St. Shenkov, U. Delimarinova, *Eur. Pol. J.*, **1992**, 29, 613.

APPENDIX

General procedure for Atherton–Todd reaction (ref. J. Chem. Soc., **1945**, *660).*

A slight excess of the amine was added to a solution of dialkyl H-phosphonate in carbon tetrachloride at room temperature without external cooling, the mixture left overnight, and then washed with dilute acid and water. Evaporation of the carbon tetrachloride solution gave the crude product.

General procedure for the reduction of dialkylphosphorochloridate (ref. Syn. Commun., **1988**, *18, 665).*

A 100-mL three-necked flask equipped with reflux condenser, drying tube, compensating dropping funnel, and magnetic stirrer was charged with 30 mL of dry dioxane and sodium borohydride (1.66 g, 0.044 mol). The mixture was stirred rapidly to provide a highly dispersed state for the sodium borohydride. To this suspension was added in small portions the di-*n*-butyl phosphorochloridate (5.0 g, 0.022 mol). Upon completion of the addition, the reaction mixture was heated at reflux for 5 h. After cooling, water (10 mL) was cautiously added followed by sufficient 6 M hydrochloric acid to bring the pH to less than 2. Any remaining solid material was filtered and the filtrate was evaporated under reduced pressure. The semisolid residue was washed with 2 5 × 50 mL of methylene chloride and the washings were dried over magnesium sulfate, filtered, and evaporated under reduced pressure. Thus pure di-*n*-butyl H-phosphonate (3.62 g, 85.4%) was isolated.

General procedure for the phosphorylation of aminoalcohols using Atherton–Todd reaction (ref. Zh. Obshch. Khim., **1974**, *46, 21).*

7.5 g of propylamine and 10.1 g of triethylamine in 20 mL of benzene was added with stirring to 13.8 g of diethyl H-phosphonate in 60 mL of CCl_4. The reaction temperature was kept below 15–20 °C. After 12 h, the solvent was removed. Pure product γ-hydroxypropylamide of diethyl phosphonic acid is obtained by distillation under reduced pressure (b.p. 150 °C, 6 × 10^{-3} mmHg) in 56% yield.

General procedure for the phosphorylation of amino acids using Atherton–Todd reaction (*ref. J. Org. Chem.*, **1988**, 444).

A solution or suspension of the amino acids (10 mmol) in Et$_3$N (5 mL), H$_2$O (3 mL), and EtOH (2 mL) is cooled to 0 °C. A mixture of diisopropyl H-phosphonate (10 mmol) and CCl$_4$ (4 mL) is added dropwise and the mixture is stirred at 20 °C for 4–16 h. The reaction is quenched by acidifying the mixture to pH = 2 with dilute HCl. The mixture is then extracted with EtOAc (3 × 20 mL), the extract is dried (MgSO$_4$), the solvent is evaporated, and the oily residue is purified by crystallization from EtOAc/petroleum ether.

General procedure for the preparation of peptides using Atherton-Todd reaction (*ref. J. Org. Chem.*, **1988**, 444).

To a stirred solution of phenylalanine or tryptophan (10 mmol) in a mixture of H$_2$O (5 mL), EtOH (5 mL), and Et$_3$N (7 mL) at –10 °C, a solution of dimethyl H-phosphonate (3.3 g, 30 mmol) in CCl$_4$ (6 mL) is added. The mixture is stirred for 8 h, then diluted with H$_2$O (50 mL) and extracted with EtOAc (3 × 20 mL). The organic layer is washed with dilute HCl (3 × 20 mL) and H$_2$O (3 × 20 mL), dried (MgSO$_4$), and evaporated. The residue is purified by column chromatography.

General procedure in addition reactions (Pudovik reaction) (*ref. Zh. Obshch. Khim.*, **1965**, 35, 358).

The reactions were carried out with equimolecular amounts of the unsaturated compound and the dialkyl H-phosphonate. On addition of a saturated alcoholic solution of a sodium alkoxide (containing a lower alkyl), the temperature of the reaction mixture rose to 30–80 °C. To complete the reaction, the mixture was heated in a flask with reflux condenser for 0.5–7 h at 90–110 °C, after which it was vacuum-fractionated in a stream of dry carbon dioxide.

General procedure for Abramov reaction (*ref. Synthesis*, **1982**, 165).

Commercial potassium fluoride (dehydration not necessary) (50 mmol), or finely powdered commercial cesium fluoride (20 mmol) is stirred with dialkyl H-phosphonate (10 mmol) and the carbonyl compound (10 mmol) at room temperature for a period of 0.5 to 1 h. The reaction mixture is extracted with dichloromethane (50 mL) by shaking for 5 min. The solid fluoride salt is filtered off and the dichloromethane evaporated under reduced pressure. The residual solid or oily product is recrystallized from cyclohexane.

General procedure for the synthesis of nucleoside α-hydroxylalkylphosphonates (*ref. Tetrahedron Lett.*, **1993**, 34, 5843).

To a solution of diethyl H-phosphonate (1mmol) in anhydrous THF (1 mL) at –78 °C, lithium bis(trimethylsily)amide (2 equiv, 1.0 M) in THF under a nitrogen atmosphere was added dropwise via a syringe. After 10–15 min, a solution of keto nucleoside (1 mmol) in 6 mL THF was added, and the reaction mixture was allowed to warm to room temperature over 1.5 h. The reaction was quenched by slow addition of acetic acid in diethyl ether and the resulting mixture was filtered. After concentration *in vacuo*, the residue was purified by radial chromatography.

General procedure for the preparation of nucleoside 5′-hydroxyl phosphonates (ref. Tetrahedron Lett., **2000**, 41, 955).

Dimethylsulfoxide (0.213 mL, 3 mmol) was added with stirring to a solution of oxalyl chloride (0.131 mL, 1.5 mmol) in CH_2Cl_2 (3.5 mL) at –78 °C under an argon atmosphere. After 10 min, a solution of 3′macmillan-*O-tert*-butyldiphenylsilylthymidine (484 mg, 1 mmol) in CH_2Cl_2 (7 mL) was added dropwise and the reaction mixture was stirred for a further 30 min. Then the reaction was quenched by addition of triethylamine (0.7 mL, 5 mmol); the resulting suspension was stirred at low temperature for a further 5 min and after bringing to room temperature, dimethyl H-phosphonate (0.183 mL, 2 mmol) was added. The reaction mixture was either set aside overnight at room temperature or heated at 40 °C for several hours. The course of the reaction was checked by thin-layer chromatography on silica gel plates in chloroform:ethanol (9:1, v/v) and the product was detected both by UV monitoring and by spraying with 1% ethanolic solution of 4-(4-nitrobenzyl)pyridine (after a short heating and exposing to ammonia vapors, the product dimethyl ether afforded an intense blue spot). The reaction mixture was diluted with chloroform, extracted with water, and dried over anhydrous sodium sulfate. Chromatography of the crude product on a silica gel column in chloroform–ethanol mixture afforded 459 mg (78%) of the expected 3′-*O-tert*-butyldiphenylsilyl-5′-hydroxyl-5′-dimethylphosphonothymidine.

General procedure for the oxidation of α-hydroxyalkylphosphonates to α-ketophosphonates (ref. Tetrahedron Lett., **2000**, 41, 3169).

CrO_3 (30 mmol, 3 g, finely ground) and alumina (Al_2O_3, neutral, 5.75 g) are placed in a mortar and pestle by grinding them together until a fine, homogeneous, orange powder is obtained (5–10 min). The α-hydroxyalkylphosphonate (10 mmol) is added to this reagent. After 2.5–12 h of vigorous stirring, the reaction mixture is washed with CH_2Cl_2 (200 mL), dried (Na_2SO_4), and the solvent evaporated to give the crude product. Pure product is obtained by distillation under reduced pressure in 65–90% yields.

General procedure for the preparation of α-hydroxyalkylphosphonates (ref. Tetrahedron Lett., **2000**, 41, 3169).

MgO (2 g) was added to a stirred mixture of diethyl H-phosphonate (0.02 mol) and aldehyde (0.02 mol) at room temperature. After 2 h, the mixture was washed by dichloromethane (4 × 50 mL) and dried with $CaCl_2$; evaporation of the solvent gave the crude product. The product was crystallized from CH_2Cl_2: *n*-hexane (1:2) or distilled under reduced pressure.

General procedure for the oxidation of α-hydroxyalkylphosphonates to α-ketophosphonates with $KMnO_4$ under heterogeneous conditions (ref. Tetrahedron Lett., **2002**, 43, 477).

A solution of the α-hydroxyalkylphosphonate (5 mmol) in dry benzene (50 mL) was prepared. Powdered potassium permanganate (0.316–0.949 g, 2–6 mmol) was added and the resulting mixture was stirred at room temperature (3–12 h). The reaction mixture was then filtered and the solvent was evaporated to give the crude product. The pure product was obtained (82–98% yields) by bulb-to-bulb vacuum distillation.

General procedure for the chlorination diethyl 1-hydroxyalkylphosphonates to diethyl 1-chloroalkylphosphonates (ref. Synthesis, **1990**, 717).

A solution of diethyl 1-hydroxyalkylphosphonate (0.02 mol) and triphenylphosphine (7.86 g, 0.03 mol) in dry CCl_4 (35 mL) is refluxed for 8 h. Then, the mixture is evaporated under reduced pressure, and the semisolid residue is extracted with hexane (3 × 50 mL). The combined extracts are filtered, and the solvent is removed *in vacuo*. The oily residue is distilled under reduced pressure to afford the analytically pure diethyl 1-chloroalkylphosphonate.

General procedure for the preparation of bis(ω-hydroxyalkyl)H-phosphonates (ref. Eur. Polym. J., **1973**, 9, 1077).

Dimethyl H-phosphonate (11.7 g, 0.11 mol) and 1,6-hexandiol (50.2 g, 0.43 mol) were placed in a three-necked flask fitted with a capillary for nitrogen, a thermometer, and a condenser, and heated to 130–135 °C until the theoretical amount of methanol was liberated. The excess 1,6-hexandiol was removed by distillation under reduced pressure (45 °C, 2 × 10–2 mmHg). Yield was 28.2 g (94%).

General procedure for the preparation of alkylene bis(H-phosphonates) (ref. Commun. Dept. Chem., **1972**, V, 175).

Dimethyl H-phosphonate (15.0 g, 0.14 mol) and 1,6-hexandiol (5.36 g, 0.045 mol) were placed in a three-necked flask fitted with a capillary for nitrogen, a thermometer, and a condenser, and heated to 130–135 °C until the theoretical amount of methanol was liberated. The excess of dimethyl H-phosphonate was removed by extraction with diethyl ether. Yield was 10.0 g (81%).

– 4 –

Important Classes of Compounds

4.1 AMINOPHOSPHONATES AND AMINOPHOSPHONIC ACIDS

Aminophosphonic acids constitute an important class of biologically active compounds, and their synthesis has been a focus of considerable attention in synthetic organic chemistry as well as in medicinal chemistry. Although the phosphonic and carboxylic acid groups differ considerably with respect to shape, size, and acidity, α-aminophosphonic acids are consider to be structural analogues of the corresponding amino acids and transition-state mimics of peptide hydrolysis. α-Aminophosphonic acids occupy an important place among the various compounds containing a P–C bond and an amino group, because they are analogues of natural α-amino acids, the 'building blocks' of peptides and proteins. The so-called 'phosphorus analogues' of amino acids, in which the carboxylic acid group is replaced by a phosphonic acid group, $P(O)(OH)_2$, have attracted particular interest and have reached a position of eminence in fields of research directed to the discovery, understanding, and modification of physiological processes in living organisms.

The replacement of carbon by phosphorus has a number of important consequences: (a) an additional substituent group is present in the molecule (hydroxyl), (b) the central atom (phosphorus) has a tetrahedral configuration, whereas the carbonyl atom is planar, (c) there are significant differences in steric bulk and in acidity (pK_a). The tetrahedral configuration of phosphorus has important implications in the design of transition state–analogue enzyme inhibitors, which have a far-reaching potential in medicinal chemistry.

In 1959, Horiguchi and Kandatsu discovered 2-aminoethanephosphonic acid (AEP) in ciliated sheep rumen protozoa [1], the simplest natural aminophosphonate. This acid is present in some organisms in remarkably high amounts. Freshly laid eggs contain over 95% of total phosphorus in the phosphonate form, principally as AEP [2].

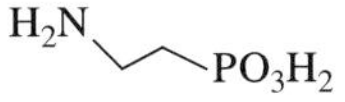

$$H_2N\text{-}CH_2\text{-}CH_2\text{-}PO_3H_2$$

The only naturally occurring aminophosphonic acid is (−)-1-amino-2-(4-hydroxyphenyl)ethylphosphonic acid.

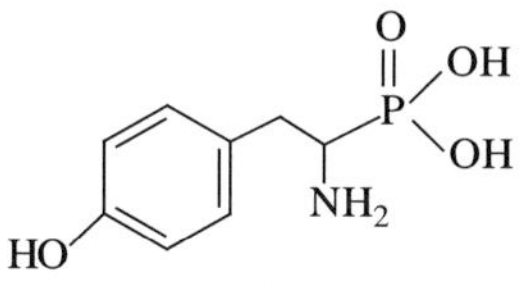

4.1.1 Methods for the preparation of aminophosphonates

Synthesis of aminophosphonic acids is an active area of research and many methods are now available. Several reviews have been devoted to the synthesis of aminoalkanephosphonic acids [3–7]. However, most of the described methods for the synthesis of α-aminophosphonic acids use carbonyl compounds such as aldehydes, ketones, or carboxylic acids as starting compounds. In this chapter, only the synthetic approaches in which the starting phosphorus compounds are H-phosphonate diesters will be discussed.

Kabachnik –Fields reaction

One of the very first methods for the preparation of α-aminophosphonic acids appears to be the one described by Kabachnik and Medved [8]. The Kabachnik–Fields reaction is still very useful, especially for the preparation of dialkyl 1-aminoalkanephosphonates. According to this method, α-aminophosphonates were obtained reacting ammonia, carbonyl compounds (aldehydes and ketones), and dialkyl H-phosphonate. A little later, Fields [9] presented a method of synthesis of 1-aminoalkylphosphonic acids by replacing ammonia with amine—reacting both (aldehydes and ketones) with ammonia, or amine and dialkyl H-phosphonate to give dialkyl esters of 1-aminoalkylphosphonic acid (see Appendix). Hydrolysis of the esters produced free aminoalkylphosphonic acids. Yields of aminophosphonates vary from 40 to 47%.

$$(RO)_2P(O)H + CH_2O + R^1NH_2 \longrightarrow (RO)_2P(O)\text{-}CH_2\text{-}NHR^1 + H_2O$$

This synthetic procedure is to some extent similar to the Mannich aminomethylation of aldehydes and ketones having a labile α-hydrogen atom with primary or secondary amines [10].

$$R^1R^2CH\text{-}CHO + CH_2O + HNR_2 \longrightarrow R_2N\text{-}CH_2\text{-}CR^1R^2\text{-}CHO + H_2O$$

Since Kabachnik and Medved found that α-hydroxyalkylphosphonates are present in the reaction mixture of ammonia, carbonyl compound, and dialkyl H-phosphonate, they accepted that the reaction proceeds via formation of α-hydroxyalkylphosphonate followed

4.1 Aminophosphonates and Aminophosphonic Acids

by replacement of the hydroxy group by the amino group. Petrov et al. [11], based on the experimental data, suggested that the synthesis of α-aminophosphonates includes the following reactions: (a) reaction between dialkyl H-phosphonate with formaldehyde to give the corresponding hydroxymethyl phosphonic acid, (b) reaction of amine with formaldehyde to give the corresponding hydroxyl compound. Thus, there are at least two possible reaction pathways [12a]. For the ternary system dialkyl H-phosphonate–benzaldehyde–aniline, the following mechanism, involving imine intermediate formation [12b], was

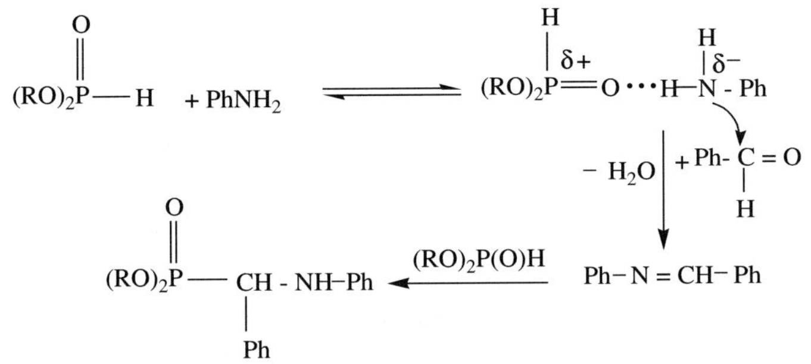

$$(RO)_2P(O)H + PhNH_2 \rightleftharpoons (RO)_2P=O \cdots H-N(H)-Ph \xrightarrow[-H_2O]{+Ph-CHO} Ph-N=CH-Ph \xrightarrow{(RO)_2P(O)H} (RO)_2P(O)-CH(Ph)-NH-Ph$$

accepted: Due to the low basicity and correspondingly high N–H acidity of aniline, the role of an electron-acceptor component in the resulting complex belongs to the dialkyl H-phosphonate. As a result, in the complex the partial negative charge is mostly localized on the nitrogen atom, which functions as a nucleophilic center in the subsequent reaction with the carbonyl compounds to give the corresponding imine.

For the system dialkyl H-phosphonate–benzaldehyde–cyclohexylamine, the 'hydroxyphosphonate' mechanism is operative, involving cyclohexylamine-catalyzed addition of dialkyl H-phosphonate to benzaldehyde (Abramov reaction), followed by substitution of the hydroxyl group by a cyclohexylamino group [12b]. In this case, a strongly basic

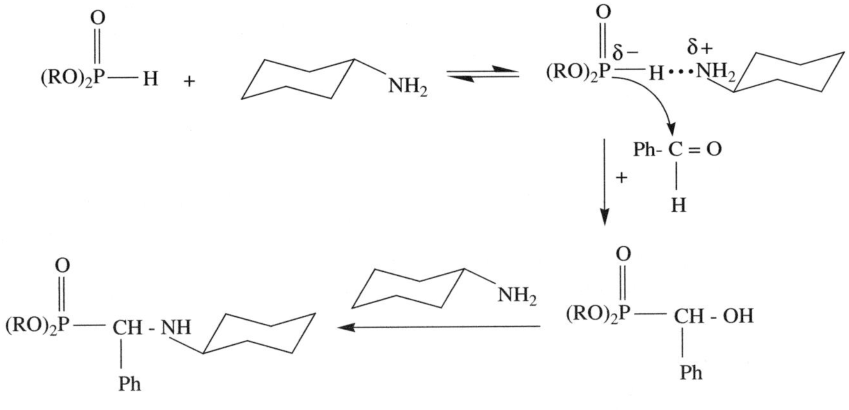

$$(RO)_2P(O)H + C_6H_{11}NH_2 \rightleftharpoons (RO)_2P(O)-H \cdots NH_2-C_6H_{11} \xrightarrow{+ Ph-CHO} (RO)_2P(O)-CH(Ph)-OH \xrightarrow{C_6H_{11}NH_2} (RO)_2P(O)-CH(Ph)-NH-C_6H_{11}$$

cyclohexylamine (low N–H acidity) plays the role of an electron-acceptor compound in the resulting complex with dialkyl H-phosphonate. As a result, in the complex, the partial negative charge is mostly localized on the phosphorus atom, which functions as a nucleophilic center in the subsequent reaction with the carbonyl compounds, to furnish α-hydroxyalkylphosphonate.

Experimental results revealed that the first stage of the Kabachnik–Fields reaction, namely, the formation of a dialkyl H-phosphonate–amine complex is of critical importance for further reaction direction. Depending on the acidity–basicity relationship between the dialkyl H-phosphonate and the amine, the complex can have a structure like **1** or **2** with mutually opposite polarization of the reagents, or a relatively symmetrical and readily polarizable structure like **3** [12c].

$$(RO)_2P(O)H + R^1NH_2 \underset{K_2}{\overset{K_1}{\rightleftarrows}} \begin{bmatrix} (RO)_2\overset{\delta+}{P}=O\cdots\overset{\delta-}{H-N}-R^1 \\ || \\ H\mathbf{1}H \\ \\ \overset{\delta-}{(RO)_2P}-\overset{\delta+}{H}\cdots NH_2R^1 \\ \|\mathbf{2} \\ O \end{bmatrix} \rightleftarrows \begin{matrix} (RO)_2P-H\cdots NHR^1 \\ \|| \\ O\cdots\cdots H \\ \mathbf{3} \end{matrix}$$

If the complexes are structurally similar to **1**, as in the case of dialkyl H-phosphonate and weakly basic amines, they further react with carbonyl compounds by route *a* with irreversible imine formation **4** ; the latter react with dialkyl H-phosphonate to give

$$R^2-\underset{H}{\overset{O}{C}} \underset{b}{\overset{a}{\longrightarrow}} \begin{bmatrix} \overset{1, k_1}{\longrightarrow} R^2-CH=N-R^1 \xrightarrow{(RO)_2P(O)H} \\ \mathbf{4} \\ \underset{k_{-2}}{\overset{2\,k_2}{\rightleftarrows}} (RO)_2\overset{O}{\overset{\|}{P}}-CH-R^2 \xrightarrow[-H_2O]{R^1NH_2,\,k_3} \\ \mathbf{5}| \\ OH \end{bmatrix} \begin{matrix} O \\ \| \\ (RO)_2P-CH-R^2 \\ \mathbf{6}| \\ NHR^1 \end{matrix}$$

aminophosphonate **6**. In this case, $k_1 > k_2$ and the Kabachnik–Fields reaction strictly follows the route *a*. If the structure of the complexes more resembles **2**, as in the case of dialkyl H-phosphonate and strongly basic amines, then $k_2 > k_1$, and route *b* leading to α-hydroxyphosphonate **5** becomes kinetically preferred. It was found that replacement of dialkyl H-phosphonates by phosphonites and phosphinites affects the rate and mechanism of the Kabachnik–Fields reaction [12d].

The reaction rate of the Kabachnik–Fields reaction by the first mechanism (route *a*) can be given by the following equation:

$$v(a) = K_1 k_1 [R^2CHO][R^1NH_2][(RO)_2P(O)H]$$

4.1 Aminophosphonates and Aminophosphonic Acids

where [R²CHO], [R¹NH₂], and [(RO)₂P(O)H] are the concentrations of the carbonyl, amine, and dialky H-phosphonate, respectively.

The reaction rate for the second mechanism (route b) has equation

$$v(b) = K_2 \frac{k_2}{k_{-2}} [R^2CHO][(RO)_2P(O)H]$$

From the condition $v(a) \rightleftharpoons v(b)$, it follows that for the two reaction pathways to be realized simultaneously, the following condition should be met:

$$K_1 k_1 \approx K_2 (k_2/k_{-2}) k_3$$

As $k_2/k_{-2} = K_e$, where K_e is the equilibrium constant of α-hydroxyphosphonate formation, the final expression takes the form

$$K_1 k_1 \approx K_2 K_e k_3$$

Analysis of this equation enables one to predict conditions under which the alternative Kabachnik–Fields reaction mechanisms will be realized in the 'pure' state. If one of the constants on the left-hand side of this equation tends to zero, the corresponding reaction pathway will immediately change to the alternative direction, and this latter will automatically turn out to be an exclusive reaction direction.

The α-hydroxyphosphonate formation depends on the type of the carbonyl compounds and the type of amine [13,14]. On replacing acetone with fluorenone, the yield of aminophopsphonates is 0. Replacing butylamine with aniline resulted in over 70% aminophosphonate formation, when fluorenone was used as the carbonyl compound. Obviously, aromatic amines as a weak base do not activate dialkyl H-phosphonates and the reaction furnished aminophosphonate. In contrast to aromatic amines, aliphatic amines activate dialkyl H-phosphonate and α-hydroxyphosphonate is formed. Reversibility of this reaction allows the formation of aminophosphonate unless there is a phosphonate–phosphate rearrangement, which is irreversible. The α-hydroxyphosphonate formation is reversed only by about 12–19%, which means that it is six to eight times slower than the phosphonate–phosphate rearrangement for all amines and all H-phosphonate esters used. For aromatic and especially diaromatic ketones and aliphatic amines, the formation of α-hydroxyphosphonate is so fast that it is impossible to obtain a desired aminophosphonate in a good yield by heating a mixture of three reagents, as is usually done. In this case, the imine must be prepared first, followed by addition of dialklyl H-phosphonate. Removing the water under reduced pressure results in 86–94% yield. On the basis of the data obtained, Gancarz offers the following reaction scheme for the Kabachnik–Fields reaction:

[Path A: RNH$_2$ + R^{1}R^{2}C=O + HP(O)(OC$_2$H$_5$)$_2$ → R^{1}R^{2}C=NR → R^{1}R^{2}C(NHR)(P(O)(OC$_2$H$_5$)$_2$)]

[Path B: R^{1}R^{2}C(OH)(P(O)(OC$_2$H$_5$)$_2$) ⇌ R^{1}R^{2}C(H)(O-P(O)(OC$_2$H$_5$)$_2$)]

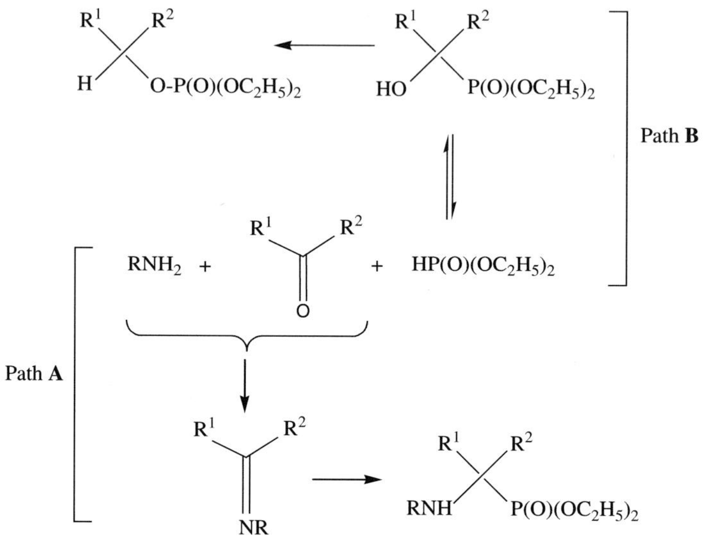

An alternative reaction mechanism suggests the formation of methylenediamine in the first stage of the Kabachnik–Fields reaction [8].

$$2\ RNH_2\ +\ CH_2O\ \xrightarrow{-H_2O}\ RNH\text{-}CH_2\text{-}NHR$$

It has been established that higher yields in the Kabachnik–Fields reaction can be achieved when secondary amines are employed. The yield of N-substituted aminomethyl phosphonates is considerably lower with primary amines because of the participation of the reaction product in further aminomethylation, leading to the substitution of the second N–H hydrogen atom with a methylphosphonate functionality [8].

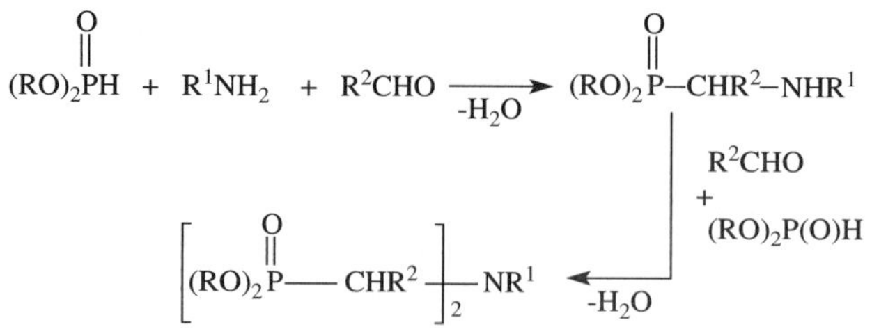

$$(RO)_2PH(O)\ +\ R^1NH_2\ +\ R^2CHO\ \xrightarrow{-H_2O}\ (RO)_2P(O)\text{-}CHR^2\text{-}NHR^1$$

$$\left[(RO)_2P(O)\text{—}CHR^2\right]_2 NR^1 \xleftarrow[-H_2O]{} \begin{array}{c} R^2CHO \\ + \\ (RO)_2P(O)H \end{array}$$

4.1 Aminophosphonates and Aminophosphonic Acids

A simple and general method for asymmetric synthesis of α-aminophosphonic acids has been introduced, based on the Kabachnik–Fields reaction of dimethyl H-phosphonate with 2-phenyl ethanolamine [15].

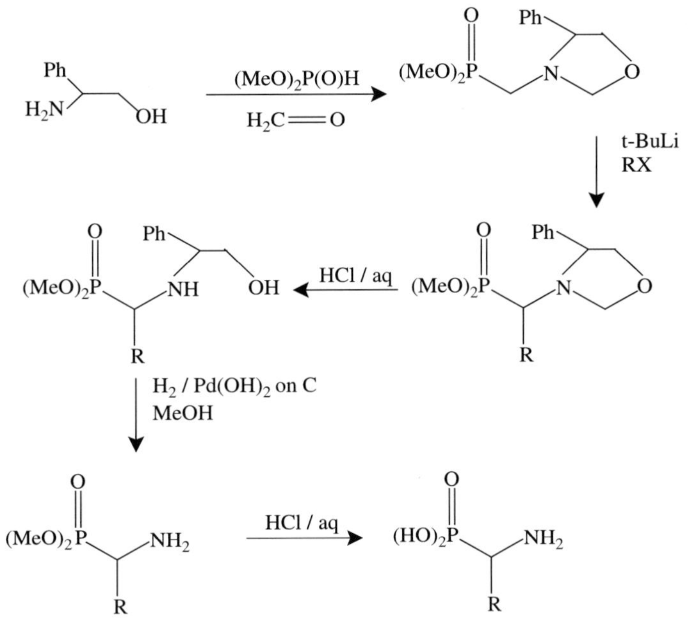

A novel route has been developed for the preparation of a series of diethyl 1-aminoarylmethylphosphonates. The route involves facile reaction among aromatic aldehydes, diethyl H-phosphonate, and hexamethyldisilazane on the surface of alumina [16].

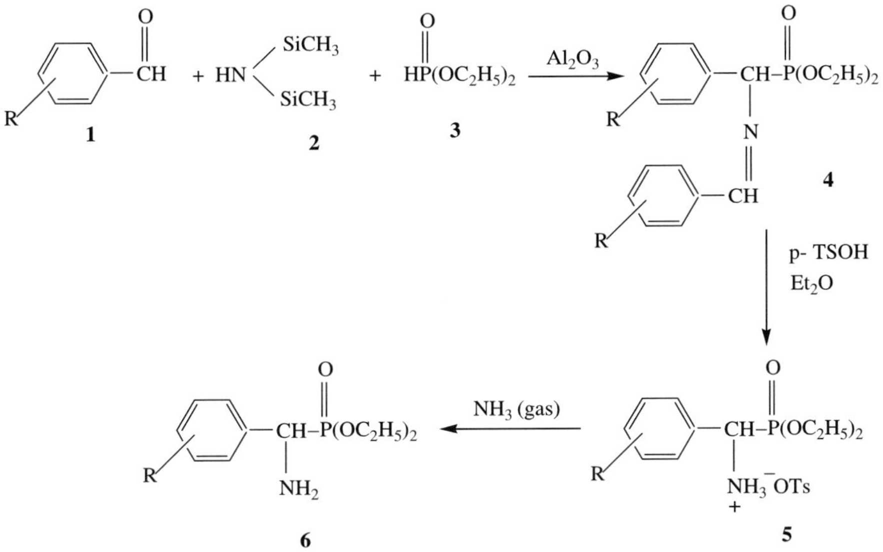

114   4. Important Classes of Compounds

Thus, the aromatic aldehyde **1** is treated with hexamethyldisilazane **2** in the presence of diethyl H-phosphonate **3** on the surface of acidic alumina. The products are benzylidene derivatives of 1-aminoarylmethylphosphonate **4**. The removal of the benzylidene group by treatment of **4** with *p*-toluenesulfonic acid monohydrate in ether and the subsequent neutralization of the ammonium salts **5** with ammonia resulted in 1-aminoarylmethylphosphonate **6**. A simple set-up, low consumption of solvents, fast reaction rates, mild reaction conditions, good yields, and selectivity of the reaction make this method very attractive.

An efficient one-pot synthesis of anilinobenzylphosphonates was achieved from the reaction of aniline, benzaldehyde, and dialkyl H-phosphonates in the presence of $BF_3 \cdot OEt_2$ as a catalyst [17].

$$PhNH_2 + C_6H_5CHO + (RO)_2P(O)H \xrightarrow{BF_3 \cdot OEt_2} \begin{array}{c} RO \\ \\ RO \end{array} \!\!\!\! P\text{-}CH\text{-}NH\text{-}Ph \;\; | \;\; C_6H_5$$

The advantage of this synthetic method is the use of $BF_3 \cdot OEt_2$ as a catalyst. The generation of imine and addition of dialkyl H-phosphonate were greatly accelerated by addition of $BF_3 \cdot OEt_2$ to give a correspondingly high yield of aminophosphonate within 3 h.

A simple and general one-pot method was developed to give α-aminophosphonates from aldehydes, amines, and dimethyl H-phosphonate in the presence of lithium perchlorate/diethylether catalyst (LPDE) [18]. Optically active α-aminophosphonates were synthesized using (R) – (+), (S) –(–) α-methylbenzylamine, (R) – (–), (S) – (+) 2-phenylglycinol. (R) – (+), (S) – (–) α-methylbenzylamine affords predominantly (S) – α-aminophosphonates. The configurations of the major and minor diastereomeric products were determined by ^{1}H NMR spectroscopy. For example, the ^{1}H NMR spectrum of the

crude (**1** +**2**) (R^1 = *cyclo*-hexyl) showed two double/doublets, one at 2.5 ppm (*J* (P,H) = 12.7 Hz and the other at 2.6 ppm (*J*(P,H) = 19.5 Hz) in ratio 79 : 21. Each double/doublet is derived from the proton attached to the carbon bonded to the phosphorus. On the basis of the literature

4.1 Aminophosphonates and Aminophosphonic Acids

experiment for compound **1** [19], the upfield (major) double/doublet is from the *S* chiral center of the aminophosphonate proton and the downfield (minor) double/doublet is from the *R* chiral center. (*R*) – (–) 2-phenylglycinol leads predominantly to (*R*) – α-aminophosphonates.

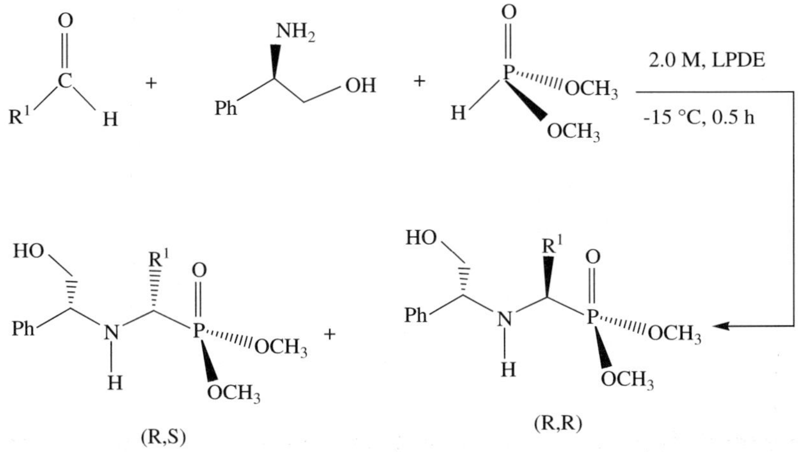

The experimental results showed that the lithium perchlorate/diethylether promotes the direct conversion of aldehydes into α-aminophosphonates. This method has a few noteworthy features: (a) excellent yields can be obtained for both aliphatic and aromatic aldehydes, (b) great operational simplicity.

Lanthanide triflates were found to be efficient catalysts in the one-pot reaction of aldehydes, amines, and diethyl H-phosphonate to afford α-aminophosphonates in good to excellent yield under mild conditions [20].

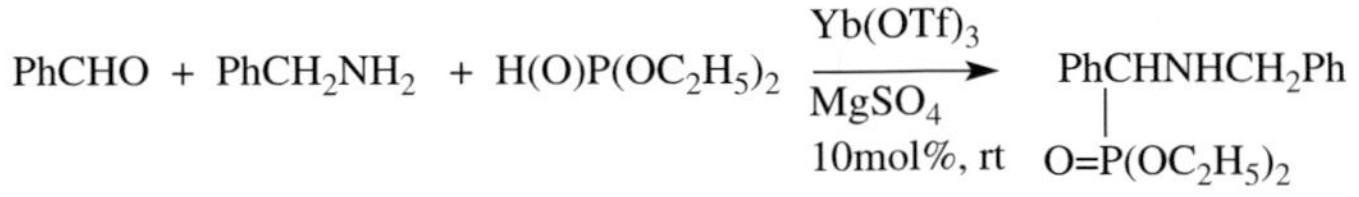

$$PhCHO + PhCH_2NH_2 + H(O)P(OC_2H_5)_2 \xrightarrow[\text{MgSO}_4]{\text{Yb(OTf)}_3} \underset{\underset{O=P(OC_2H_5)_2}{|}}{PhCHNHCH_2Ph}$$
$$10\text{mol\%, rt}$$

1-Amino-2-phenylethylphosphonic acid, the phosphonic analogue of phenylalanine, was first synthesized by Kosolappoff *et al.* by reacting 2-phenylacetaldehyde, ammonia, and diethyl H-phosphonate followed by hydrolysis with HCl [21].

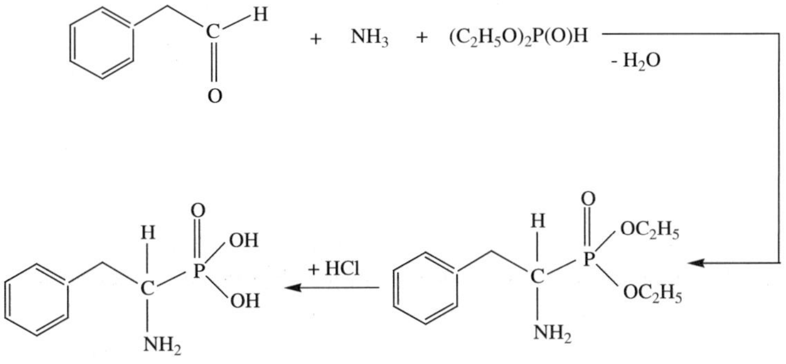

1-Aminoethylphosphonic acid (α-alanine analogue) was synthesized reacting acetaldehyde, diethyl H-phosphonate, and ammonia.

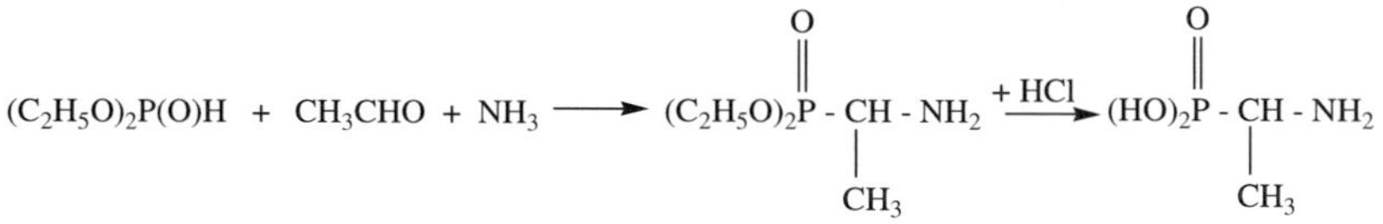

$$(C_2H_5O)_2P(O)H + CH_3CHO + NH_3 \longrightarrow (C_2H_5O)_2\overset{\overset{O}{\|}}{P}-\underset{\underset{CH_3}{|}}{CH}-NH_2 \xrightarrow{+HCl} (HO)_2\overset{\overset{O}{\|}}{P}-\underset{\underset{CH_3}{|}}{CH}-NH_2$$

The rather low yields of the aminophosphonic acids can be fairly substantially improved over those reported by Kabachnik and Medved if anhydrous ammonia and the aldehyde are premixed in alcohol solution. Using this procedure, a few amino-substituted phosphonic acids are prepared: α-aminobenzylphosphonic acid (benzaldehyde, dibutyl H-phosphonate, solution of dry ammonia in absolute ethyl alcohol); α-aminopropylphosphonic acid propionaldehyde, diethyl H-phosphonate, and a solution of dry ammonia in absolute ethyl alcohol); *p*-methoxybenzyl-α-aminophosphonic acid (*p*-hydroxybenzaldehyde, diethyl H-phosphonate, and solution of dry ammonia in absolute ethyl alcohol).

α-Aminophosphonates are prepared by processes in which primary amines having formula **1** are contacted with aldehydes with formula **2** under the conditions and time effective for forming imines having formula **3** [22,23].

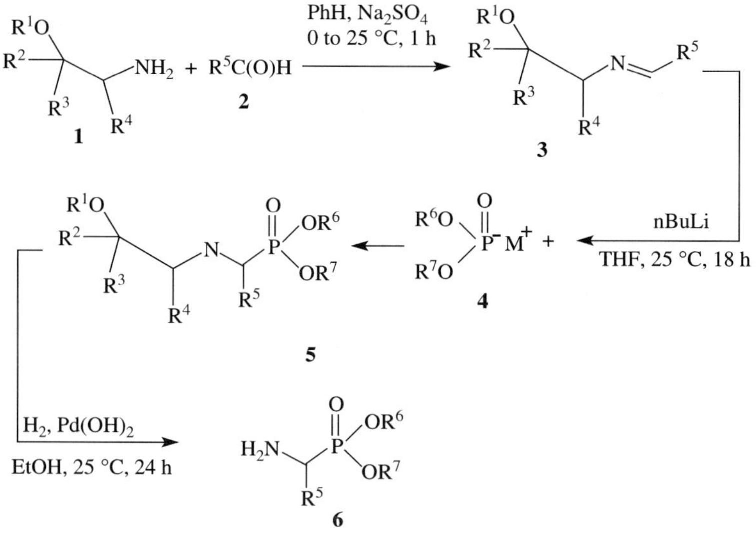

The imine compounds are then reacted with metal salts of dialkyl phosphonates with formula **4** to yield secondary aminophosphonates **5**. Reduction of the secondary aminophosphonates furnished the corresponding aminophosphonates having structure **6**.

4.1 Aminophosphonates and Aminophosphonic Acids

By this method, novel α-aminophosphonates of relatively high optical purity can be prepared.

The best method for synthesis of various pyridine aminophosphonates is a thermal addition of diethyl phosphonate to pyridine aldimines. Pyridine aldehydes react easily with aliphatic or aromatic amines in toluene solution to form corresponding aldimines (Schiff bases). The obtained imines were reacted *in situ* with diethyl H-phosphonate to give aminophosphonate diethyl esters [24].

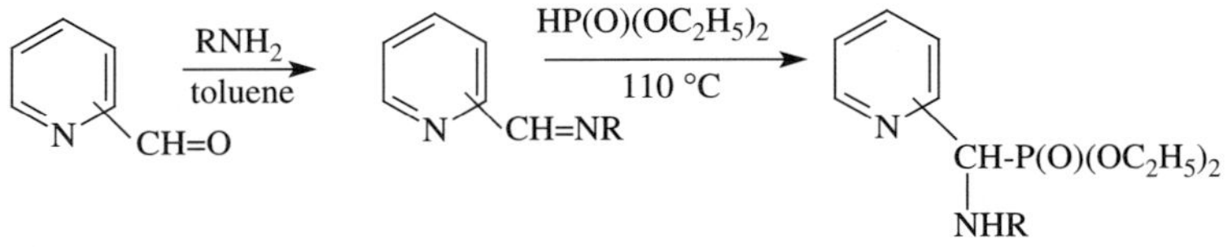

$$\text{Pyridine-CH=O} \xrightarrow[\text{toluene}]{RNH_2} \text{Pyridine-CH=NR} \xrightarrow[110\,°C]{HP(O)(OC_2H_5)_2} \text{Pyridine-CH}(NHR)\text{-P(O)(OC}_2H_5)_2$$

The first synthesis of an optically active α-aminophosphonic acid was reported by Gilmore and McBride [25]. Diastereoselective addition of diethyl H-phosphonate to imine, derived from enantiomerically pure α-methylbenzylamine, furnished α-aminobenzylphosphonic acid enantiomers.

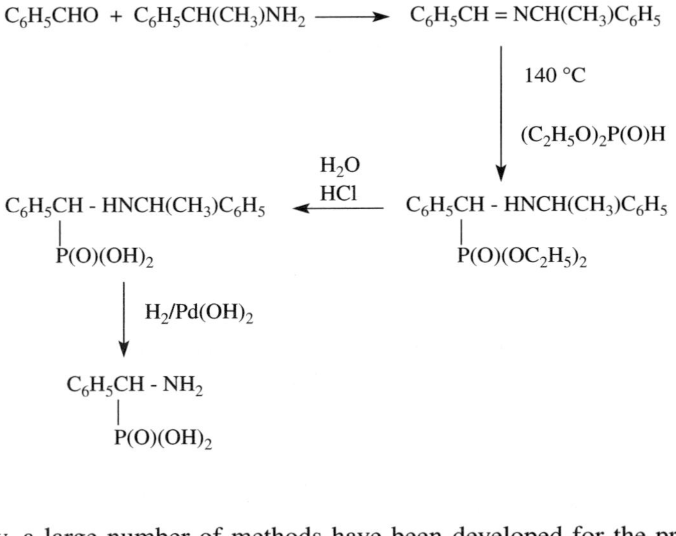

$$C_6H_5CHO + C_6H_5CH(CH_3)NH_2 \longrightarrow C_6H_5CH=NCH(CH_3)C_6H_5$$

$$\downarrow 140\,°C,\ (C_2H_5O)_2P(O)H$$

$$C_6H_5CH\text{-}HNCH(CH_3)C_6H_5 \;\;|\;\; P(O)(OC_2H_5)_2$$

$$\xrightarrow{H_2O,\ HCl} C_6H_5CH\text{-}HNCH(CH_3)C_6H_5 \;\;|\;\; P(O)(OH)_2$$

$$\downarrow H_2/Pd(OH)_2$$

$$C_6H_5CH\text{-}NH_2 \;\;|\;\; P(O)(OH)_2$$

Subsequently, a large number of methods have been developed for the preparation of optically active α-aminophosphonates: nucleophilic addition of lithium dibenzyl phosphonate to the spironitrone [26]; Lewis acid–catalyzed addition of diethyl H-phosphonate to N-galactosylimine [27]; addition of lithium diethyl phosphonate to chiral chelating imines [28].

Couty *et al.* have described an efficient asymmetric synthesis of enantiomerically pure azetidinic 2-phosphonic acids, starting from readily available β-amino alcohols [29].

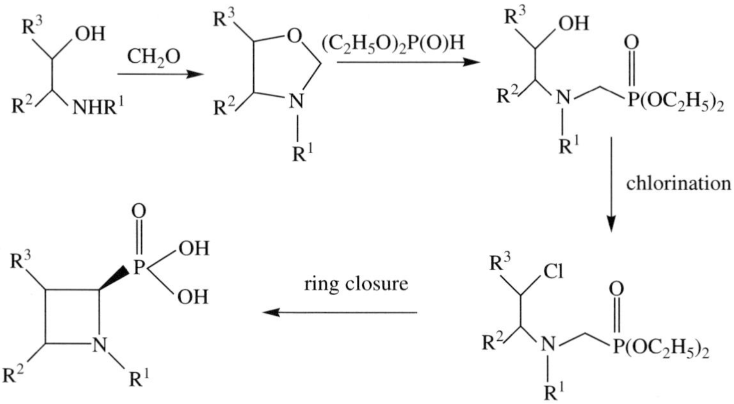

1-Amino-2-nitrobenzylphosphonates were prepared by an addition reaction of diethyl H-phosphonate to aldimines, obtained from *o*-nitrobenzaldehyde and amine (butylamine, benzylamine, or *p*-toluidine) in diluted toluene solution [30].

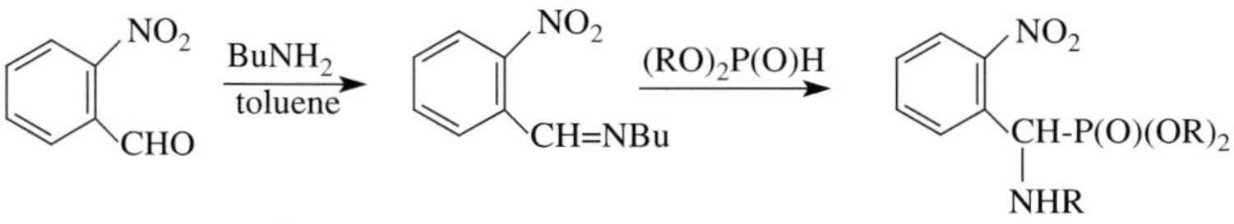

Ferrocenylaminomethylphosphonates were synthesized by condensation of readily available formylacetylferrocenes with dialkyl hydrogen phosphonates in the presence of primary or secondary amines [31].

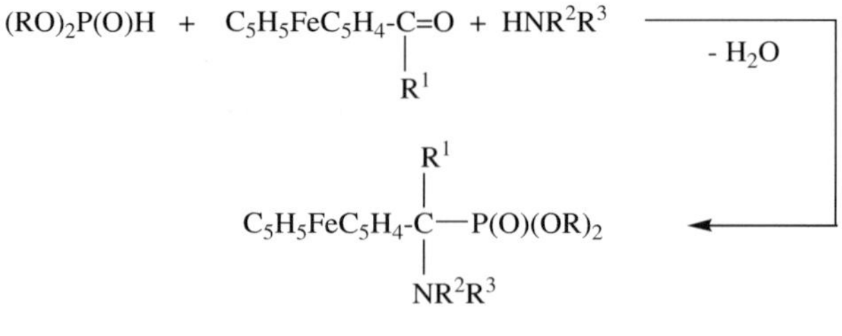

4.1 Aminophosphonates and Aminophosphonic Acids

Two aminophosphonic groups have been introduced in the 'upper rim' [32] of the calix[4]arenes, reacting amino-containing calix[4]arenes with diethyl H-phosphonate and a carbonyl compound.

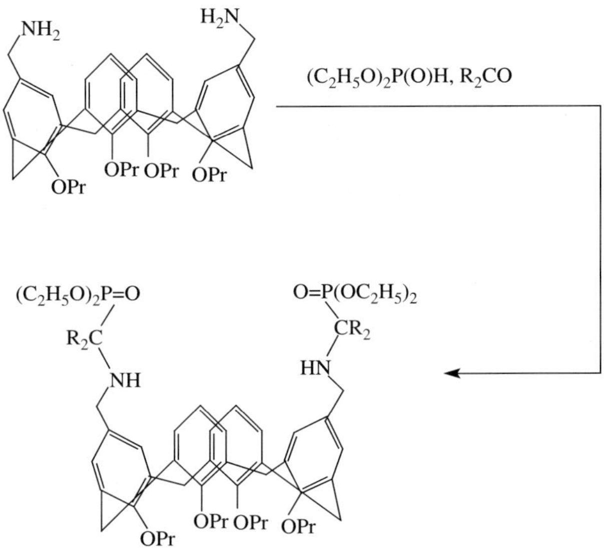

Two aminophosphonic groups have been introduced in the 'lower rim' [33], exploring the same reaction conditions.

The sterically hindered aminomethylphosphonates were synthesized reacting p-substituted benzaldehyde, benzyl amine, and dialkyl H-phosphonates [34].

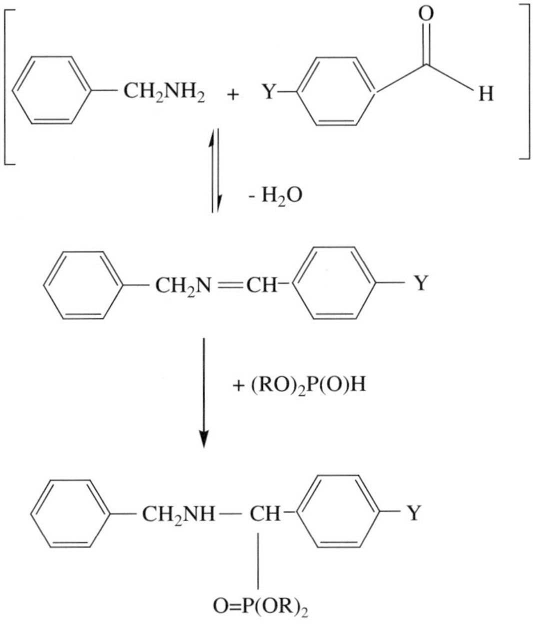

Since the reaction of the aliphatic amines with an aldehyde forming imine is reversible, the yields are comparatively low (60–70%). It is probably due to the incomplete removal of water from the reaction system. In order to increase the yield, water was thoroughly removed under reduced pressure, resulting in yields of 86–94%. The addition of dialkyl H-phosphonates to imine was carried out in the presence of hydrochloric acids as a catalyst. In the presence of this catalyst, the reaction time could be reduced. The aminophosphonates thus obtained showed activity against leukemia.

N-phosphonylation proceeds when there are a few electron-acceptor groups at the carbon atom of the C=N double bond [35]. Thus, 3-(O-methoxyphenylimino)indolin-2-one, obtained in 74% yield from isatin and *o*-anisidine, reacted with H-phosphonates in the presence of RONa to give 39–40% indolinones **1** and 21–26% indolinones **2**.

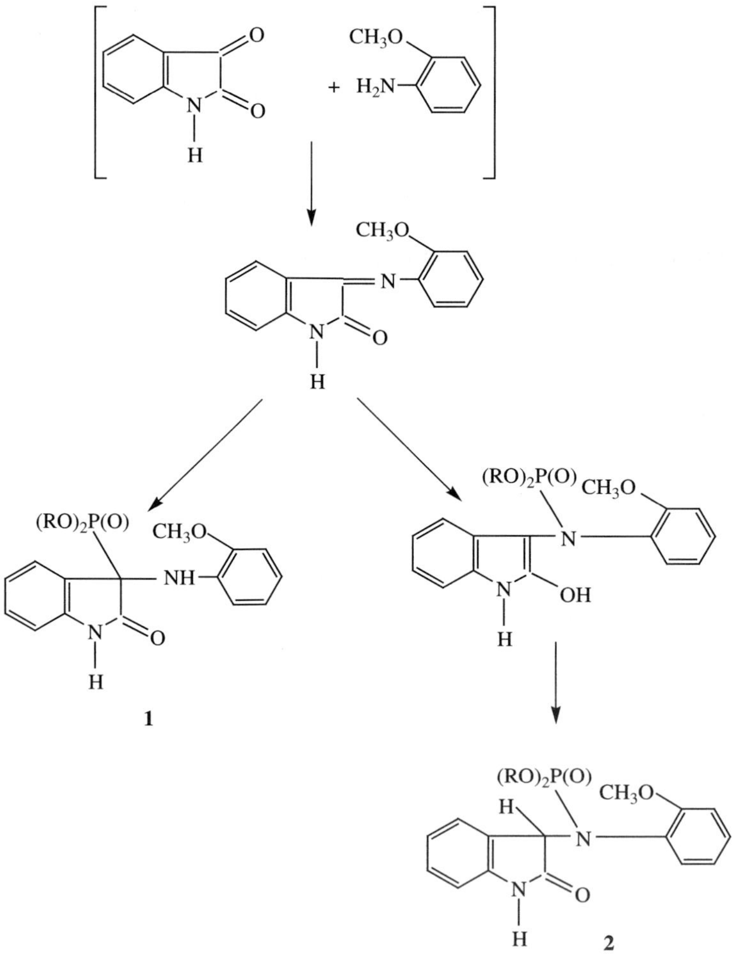

4.1 Aminophosphonates and Aminophosphonic Acids

Kabachnik–Fields reaction was used for the preparation of poly(*N*-vinyl- pyrrolidone-*co*-vinyl amine) bearing aminophosphonate groups [36].

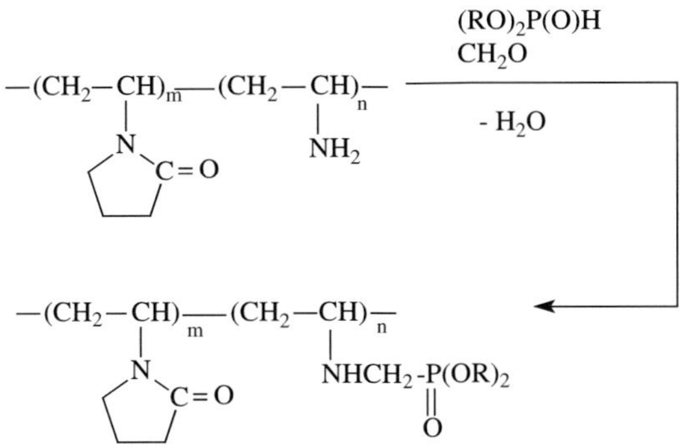

The structure of the poly(N-vinylpyrrolidone-co-vinyl amine) bearing aminophosphonate groups was proved by ^{1}H and ^{31}P NMR spectroscopy. This polymer is interesting because it has its own biological activity.

Another interesting application is the use of the Kabachnik–Fields reaction for the preparation of cyclic aminophosphonates–1,2-oxaphospholanes and 1,2-oxaphosphorinanes [37]. 3-Amino-3-methyl-2-ethoxy-2-oxo-1,2-oxaphosphorinane is obtained by aminophosphonylation of 5-hydroxy-2-pentanone.

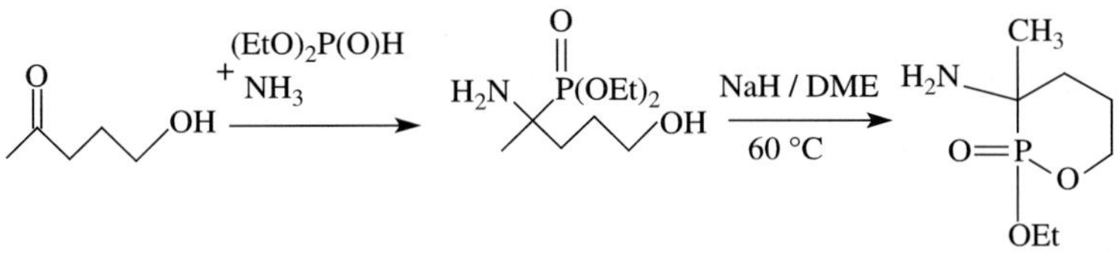

3-Amino-3-methyl-2-ethoxy-2-oxo-1,2-oxaphospholane can be prepared in a similar fashion by aminophosphonylation of 4-benzoyloxy-2-butanone, followed by reduction of the first intermediate over palladium catalyst and cyclization of the second intermediate using catalytic amounts of base.

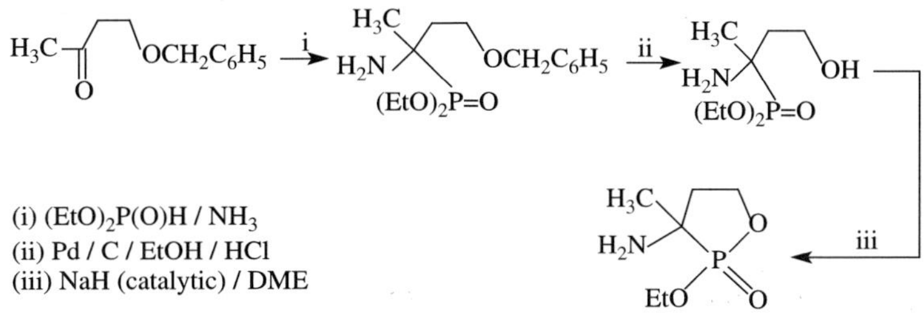

(i) (EtO)$_2$P(O)H / NH$_3$
(ii) Pd / C / EtOH / HCl
(iii) NaH (catalytic) / DME

Some dialkyl *N*-substituted α-amino-α-furylmethanephosphonates were synthesized via addition of dialkyl H-phosphonates to corresponding imines in mild conditions [38,39].

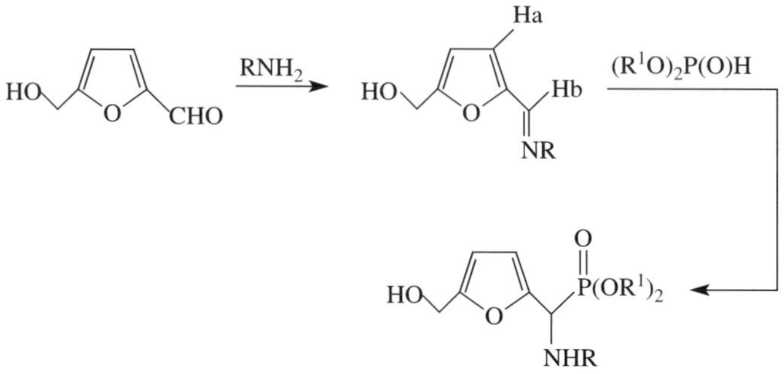

A highly efficient solvent-free and catalyst-free method for the synthesis of α-aminophosphonates is a microwave-assisted three-component Kabachnik–Fields reaction involving aldehyde, amine, and dimethyl H-phosphonate [40].

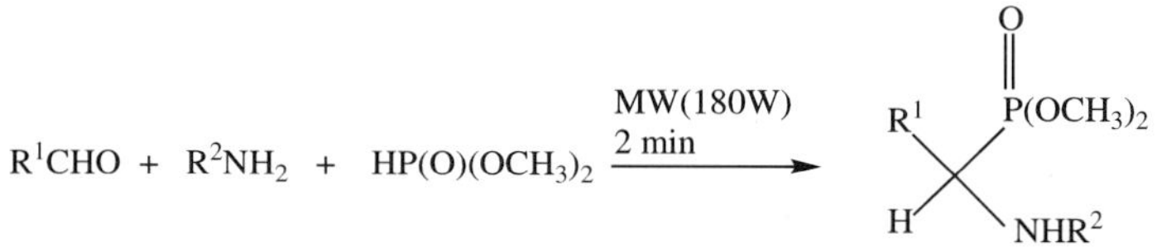

Microwave-promoted solvent-free reactions have received much attention due to their high efficiency, cost effectiveness, and environmentally friendly characteristics. The irradiation of the reaction mixture of benzaldehyde, aniline, and dimethyl H-phosphonate in a multi-mood microwave reactor at 180 W for only 2 min furnished **1** in a 98% yield (see Appendix). The yield of **1** using conventional heating is 80% for 2 h. Reactions of aliphatic aldehydes, and heterocyclic aldehydes such as 2-furylaldehyde and 2-thiophene aldehyde, and aliphatic amines and dimethyl H-phosphonate produced corresponding aminophosphonates in good to excellent yields.

α-Aminophosphonates bearing porphyrin nucleii were synthesized using microwave irradiation in excellent yield within a few minutes in the presence of CdI_2 as a catalyst [41].

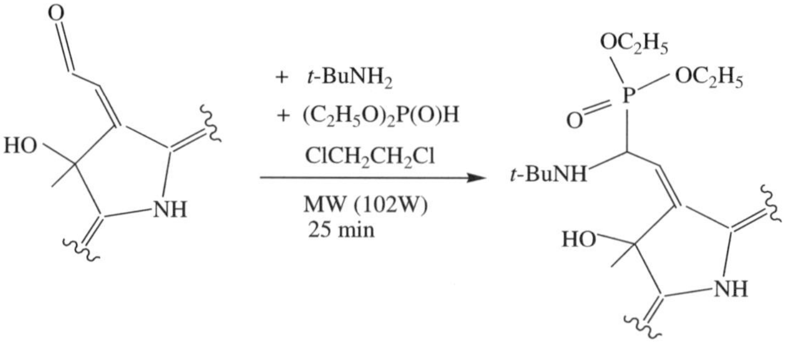

4.1 Aminophosphonates and Aminophosphonic Acids

α-Aminophosphonates bearing a C=C double bond are synthesized by the reaction of dialkyl H-phosphonates or thiophosphonates with N-benzylidene-2-(vinyloxy)ethylamine or N-(4-fluorobenzylidene)-2-(vinyloxy)ethylamine [42].

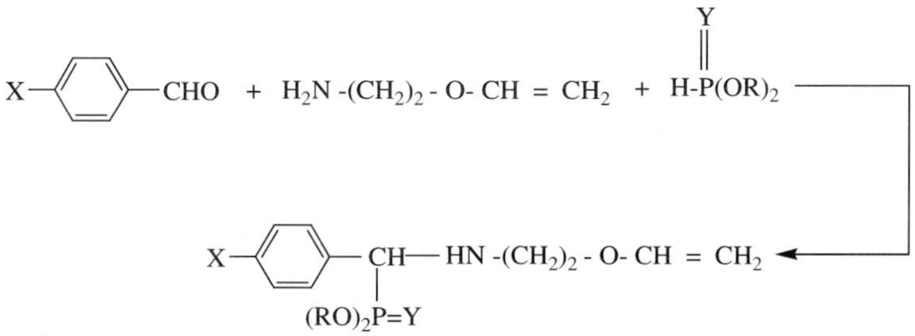

$$X\text{-}C_6H_4\text{-}CHO + H_2N\text{-}(CH_2)_2\text{-}O\text{-}CH=CH_2 + H\text{-}P(OR)_2(=Y) \longrightarrow$$

$$X\text{-}C_6H_4\text{-}CH(P(=Y)(OR)_2)\text{-}HN\text{-}(CH_2)_2\text{-}O\text{-}CH=CH_2$$

X = H; F.
Y = O; S.
R = CH$_3$O; C$_2$H$_5$; i-C$_3$H$_7$.

The reaction proceeds at room temperature in the absence of catalyst in 73–82% yield.

The Kabachnik–Fields reaction is used for phosphorylation of nitrogen-containing heterocycles. Phosphorylated heterocycles are promising plant growth regulators and stability inductors. These compounds combine biological activity of the heterocyclic system and organophosphorus moiety. Phosphorylated imidazoles, benzimidazoles, and benzotriazoles occupy a specific place among biologically active heterocycles.

Phosphinoylmethyl-substituted benzimidazoles are obtained according to the following scheme [43]. The substituent in positions 2, 5, and 6 of the benzimidazole ring affects the

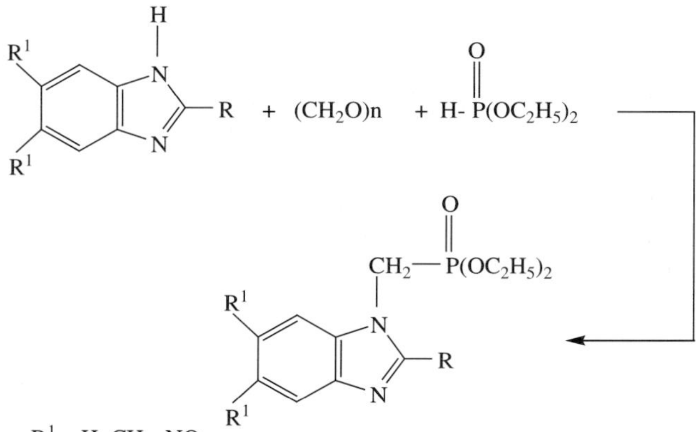

[benzimidazole(R^1, R^1)-2-R] + (CH$_2$O)$_n$ + H-P(O)(OC$_2$H$_5$)$_2$ ⟶ [N-CH$_2$-P(O)(OC$_2$H$_5$)$_2$ benzimidazole(R^1, R^1)-2-R]

R^1 = H; CH$_3$; NO$_2$.
R = H; CH$_3$; CH$_2$Ph; CH$_2$Cl; CH$_2$N(C$_2$H$_5$)$_2$.

reactivity of the heterocycle. Introduction of electron-donor substituents into the benzene and imidazole ring facilitates the reaction, whereas electron-acceptor substituents slow down the process [44].

Addition of dialkyl H-phosphonates to the Schiff base

The addition of H-phosphonic acid diesters neat or in solvent to the arylidenalkylamines [45] or to the Schiff base precursors (readily available by the condensation of primary with aldehydes) is one of the best synthetic routes to α-aminophosphonic acids [15,46–50].

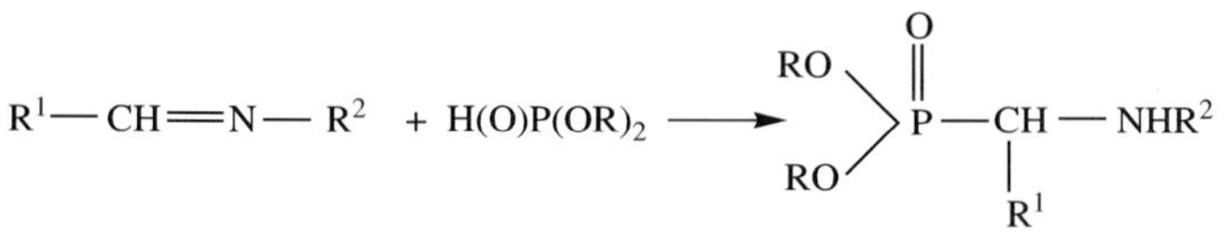

$$R^1-CH=N-R^2 + H(O)P(OR)_2 \longrightarrow \begin{matrix} RO \\ RO \end{matrix} \!\!\!\!> P(O)-CH(R^1)-NHR^2$$

α-Aminophosphonic acid dialkyl esters are obtained in good yield in the presence of a catalytic amount of NaH, which facilitates the addition reaction.

The addition of diethyl H-phosphonate to symmetrical diimines proceeds stereospecifically with the predominant formation of one of the two possible diastereomeric forms [50]. The addition of diethyl H-phosphonate to the diimines should generate, owing to the

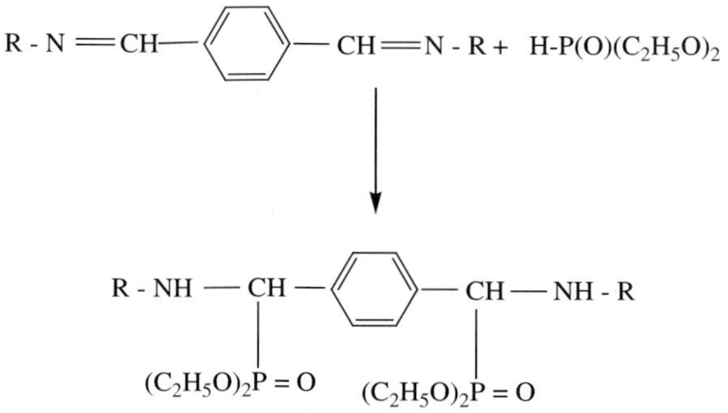

$$R-N=CH-C_6H_4-CH=N-R + H-P(O)(C_2H_5O)_2$$
$$\downarrow$$
$$R-NH-CH(P(O)(OC_2H_5)_2)-C_6H_4-CH(P(O)(OC_2H_5)_2)-NH-R$$

chiralities of the groups present in the diphosphonate molecule, two diastereomeric products (*meso* and racemic forms). The presence of only one signal for methylene (a sharp doublet with coupling constant in the range 20–24 Hz) and the ethoxy group indicates that only one of the two possible diastereomers is formed stereospecifically. ^{31}P{H} NMR spectroscopy confirmed the formation of only one diastereomer.

The reaction of Schiff bases with dialkyl H-phosphonates, similar to their reactions with all other compounds having labile hydrogen atoms, takes place by electrophilic addition. It is assumed that the reaction is initiated by protonation of the imine nitrogen atom, followed by attack of the nucleophile at the methine carbon atom. The reaction occurs both in the presence and the absence of a catalyst. The noncatalyzed addition of dialkyl phosphonates to Schiff bases is suggested to proceed according to the following scheme [51]:

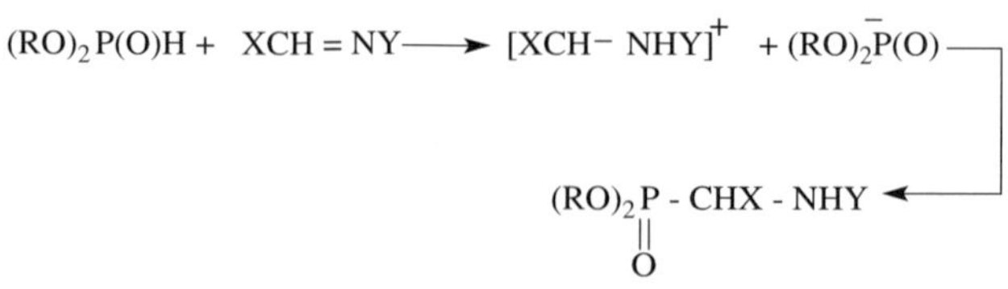

$$(RO)_2P(O)H + XCH=NY \longrightarrow [XCH-NHY]^+ + (RO)_2\bar{P}(O) \longrightarrow (RO)_2P(O)-CHX-NHY$$

4.1 Aminophosphonates and Aminophosphonic Acids

The first stage of this reaction includes protonation of the imine nitrogen atom by dialkyl H-phosphonate. The reaction occurs at a low rate, while the addition of the so-formed phosphite anion to the electrophilic methine carbon atom (the second step) is a rapid process. Alkaline metals, alkoxides [52], concentrated hydrochloric, or glacial acetic acid [53] are used as catalysts in this addition process. For the basic catalysis with alkaline metals and alkoxides, the same mechanism as in the case of the addition of dialkyl phosphonates to unsaturated compounds has been suggested [52]. When hydrochloric or glacial acetic acid are used as catalysts, their catalytic action involves protonation of the imine nitrogen atom [53].

$$XCH=NY + HCl \longrightarrow [XCH=NHY]^+ + Cl^-$$

A kinetic study of the reaction of parasubstituted *N*-benzylidene methylamines with dialkyl phosphonates reveals that below the isokinetic temperature β (where for a given azomethine the rates of addition for the different phosphonic acid diesters are equal), the following order of reactivity is observed, depending on the type of the alkoxy substituents at the phosphorus atom [51]:

$$OCH_3 > OC_2H_5 > OC_3H_7 > OC_4H_9$$

The observed decrease of the reaction rates for the higher homologues of dialkyl phosphonates has been explained in terms of their decreasing acidity with the increase of the length of the aliphatic chain [52]. On the other hand, it is known that the reactivity of the Schiff bases depends largely on their basicity. Therefore, substituents that increase the basic character of Schiff's bases will also increase the reaction rate.

The following examples illustrate the application of acidic catalysis for addition of dialkyl phosphonates to Schiff bases [53].

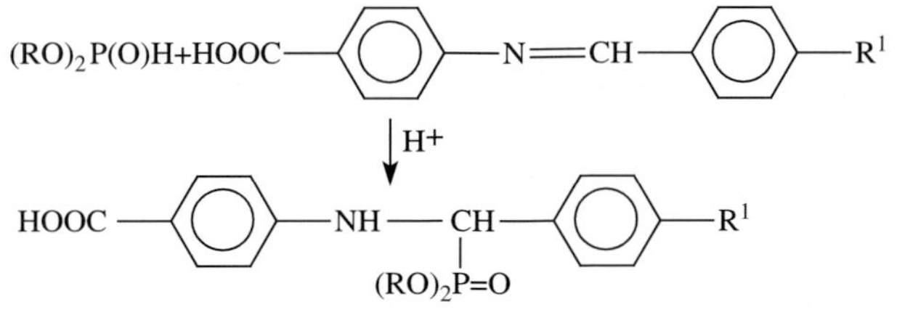

$$(RO)_2P(O)H + HOOC\text{-}C_6H_4\text{-}N=CH\text{-}C_6H_4\text{-}R^1 \xrightarrow{H^+} HOOC\text{-}C_6H_4\text{-}NH\text{-}CH(P(O)(OR)_2)\text{-}C_6H_4\text{-}R^1$$

The above reaction can be carried out in the presence of concentrated hydrochloric or glacial acetic acid. Addition of dialkyl phosphonates to other azomethines under acid catalysis has also been reported [45].

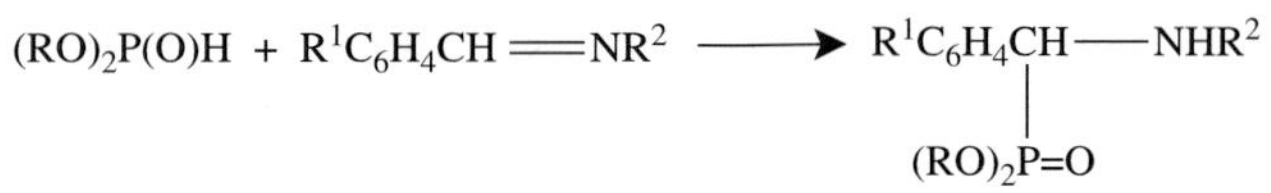

$$(RO)_2P(O)H + R^1C_6H_4CH=NR^2 \longrightarrow R^1C_6H_4CH(P(O)(OR)_2)\text{---}NHR^2$$

Aminophosphonic acids containing groups sensitive to hydrolysis and hydrogenation are obtained easily from aldimines by addition of di(*p*-methylbenzyl) H-phosphonate followed by selective removal of *p*-methylbenzyl group by solvolysis with formic acid [54].

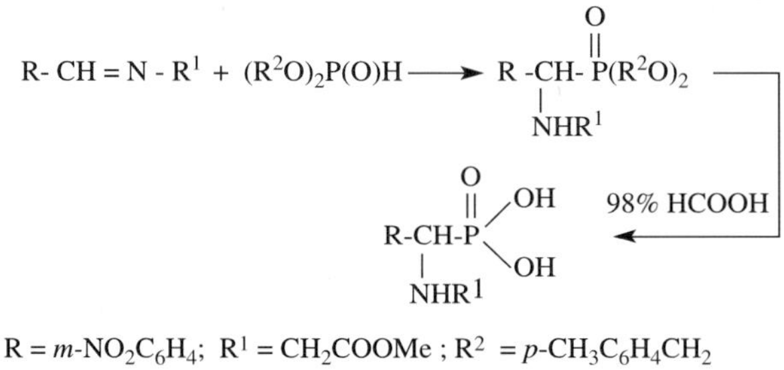

$$R-CH=N-R^1 + (R^2O)_2P(O)H \longrightarrow R-\underset{NHR^1}{\underset{|}{CH}}-\overset{O}{\overset{\|}{P}}(R^2O)_2$$

$$R-\underset{NHR^1}{\underset{|}{CH}}-\overset{O}{\overset{\|}{P}}\underset{OH}{\overset{OH}{\diagup}} \xleftarrow{98\% \text{ HCOOH}}$$

R = *m*-NO$_2$C$_6$H$_4$; R^1 = CH$_2$COOMe ; R^2 = *p*-CH$_3$C$_6$H$_4$CH$_2$

α-Aminophosphonic acids are obtained when imines are heated with H-phosphonic acid in the absenceof solvent [55a].

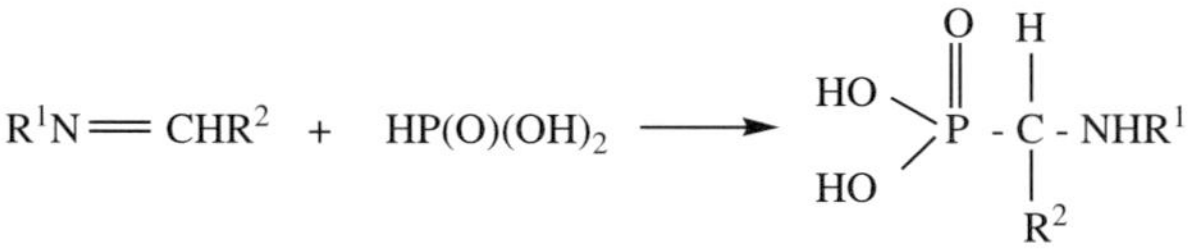

$$R^1N=CHR^2 + HP(O)(OH)_2 \longrightarrow \underset{HO}{\overset{HO}{\diagdown}}\overset{O}{\overset{\|}{P}}-\underset{R^2}{\underset{|}{\overset{H}{\overset{|}{C}}}}-NHR^1$$

N-benzylidenebenzylamine (R$_1$ = PhCH$_2$, R$_2$ = Ph) when heated with phosphorous acid gave an almost quantitative yield of *N*-benzyl-α-aminobenzylphosphonic acid (see Appendix). It was found that heating *N*-benzylidenebenzylamine with phosphorous acid in the presence of triethylamine dramatically altered the course of the reaction. Under these conditions, no aminophosphonic acid was formed, but instead, an efficient reduction reaction yielding dibenzylamine (>95%) took place.

The addition of diethyl H-phosphonate to 3,4-dihydroisoquinoline yielded 1,2,3,4,-tetrahydroiosquinoline-1-phosphonic acid [55b].

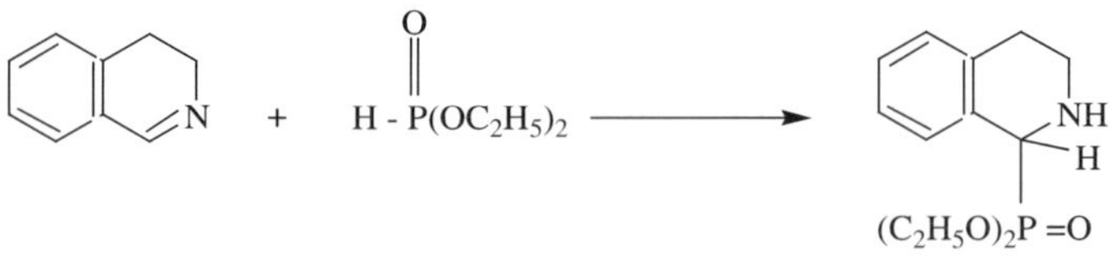

The pharmacological importance of aminophosphonates, and the biological activity of some furan derivatives [56], was the driving force to synthesize aminophosphonates by

4.1 Aminophosphonates and Aminophosphonic Acids

addition of diethyl H-phosphonate to *N,N'*-difurfuryllidene-*m*-phenylenediamine in the presence of base [57].

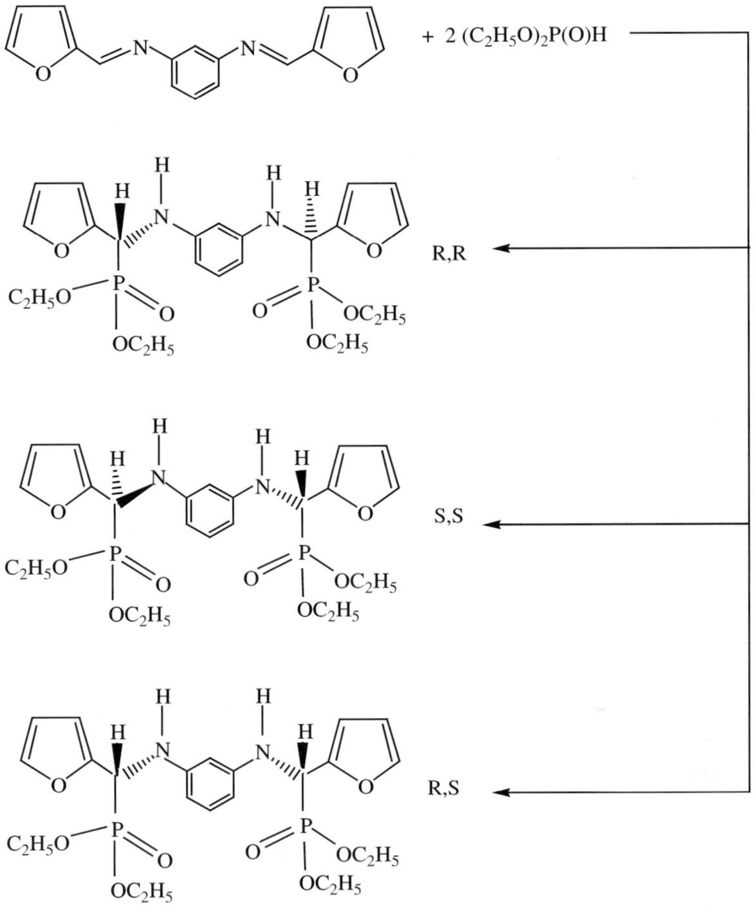

It is known that the addition of dialkyl or diaryl H-phosphonates to bis-imines should generate two diastereomeric products (*meso* and racemic forms) owing to the chirality of the groups preset in the bis(aminophosphonate) molecule [58]. The NMR studies revealed that the reaction product is a mixture of the two possible diastereomeric forms: *R, S* (*meso*) and the enantiomeric pair *R, R* and *S, S*.

Anthracene-derived bis(aminophosphonate), 1,3-bis[*N*-methyl(diethoxyphosphynyl)-1-(9-anthryl)diaminobenzene was synthesized through the addition of diethyl H-phosphonate to the bis-imine, *N,N'*-di(9-anthrylidene)-1,3-phenylenediamine **1** [59]. The analysis of the NMR spectra of reaction products revealed that the addition of diethyl H-phosphonate to

N,N'-di(9-anthrylidene)-1,3-phenylenediamine leads to one of diastereomeric forms—*meso* (R, S), or *racemes* (R, R and S, S).

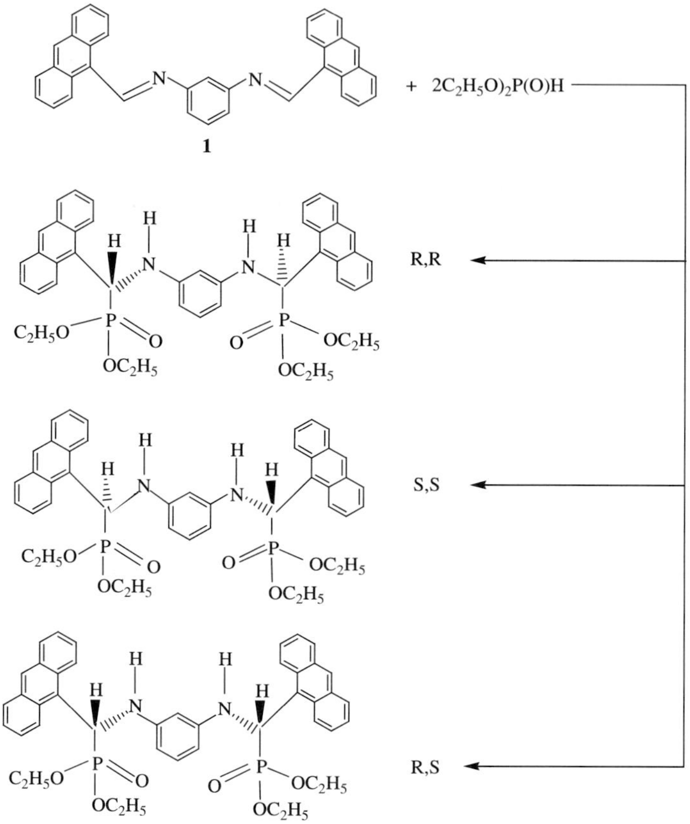

Anthracene-derived aminophosphonates are of particular interest due to the fact that the DNA-intercalating anthracene ring is the main pharmacophoric fragment of some cytostatic drugs [60]. Also, anthracene-derived aminophosphonates can find applications in analytical biochemistry [61] due to the fluorescent properties of the anthracene moiety [62].

Mannich-type reaction

The Mannich-type reaction of amines with formaldehyde and phosphorous acid is a very useful procedure for the preparation of aminomethylenephosphonic acid. Moedritzer and Irani [63] reacted 2-aminoethanol with phosphorous acid and aqueous formaldehyde in the presence of concentrated hydrochloric acid and isolated [(2-hydroxyethyl)imino]dimethylene diphosphonic acid.

$$HOCH_2CH_2NH_2 + 2CH_2O + 2 H_3PO_3 \xrightarrow[-H_2O]{H^+} HOCH_2CH_2N[CH_2\overset{O}{\overset{\|}{P}}(OH)_2]_2$$

4.1 Aminophosphonates and Aminophosphonic Acids

It was shown that very low pH is required for the formaldehyde/amine/H_3PO_3 reaction to be efficient.

Worms and Wollmann [64] reported that the reaction of 2-aminoethanol with phosphorous acid and aqueous formaldehyde in the presence of concentrated hydrochloric acid resulted in the formation of 4-(phosphonomethyl)-2-hydroxyl-2-oxo-1,4,2-oxazaphosphorinane.

$$HOCH_2CH_2NH_2 + 2CH_2O + 2\,H_3PO_3 \xrightarrow[-H_2O]{H^+}$$

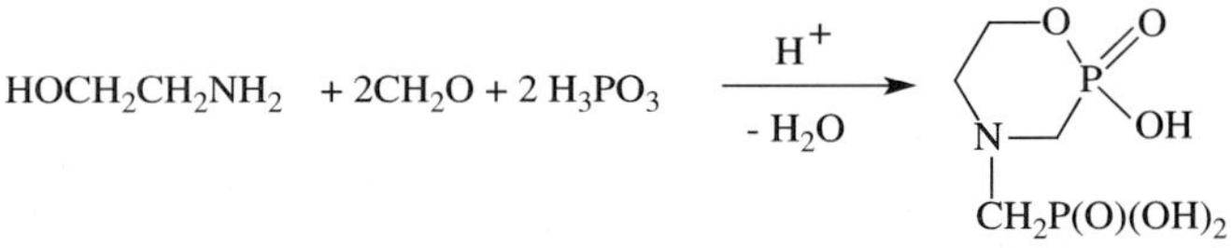

The phosphonomethylation of several aminoalcohols using the Moedritzer and Irani conditions was studied by Redmore et al. [65]. The structure of the reaction products are proved by ^{31}P, ^{13}C, ^{1}H NMR spectroscopy. It was established that the treatment of aminoalcohols with phosphorous acid and formaldehyde in the presence of concentrated hydrochloric acid gave mixtures of [(2-hydroxyalkyl)imino]dimethylene diphosphonic acid **1** and 4-(phosphonomethyl)-2-hydroxyl-2-oxo-1,4,2-oxazaphosphorinane **2** (see Appendix).

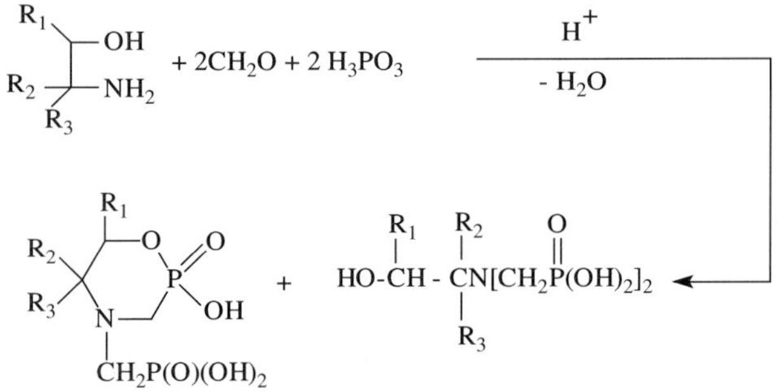

a $R_1 = R_2 = R_3 = H$
b $R_1 = R_2 = H, R_3 = C_2H_5$
c $R_1 = H, R_2 = R_3 = CH_3$
d $R_1 = CH_3, R_2 = R_3 = H$
e $R_1 = C_6H_5, R_2 = R_3 = H$

The crude product obtained exhibited signals at 4.99 ppm (d, $J = 1.8$ Hz), 8.12 (d, $J(P,P) = 1.8$ Hz), and 8.97 ppm in its ^{31}P NMR spectrum. The ^{31}P NMR spectrum of the solid product, isolated after addition of methanol to the crude product, contained signals at 4.90 ppm (d, $J = 2.4$ Hz) for the phosphorus atom in the heterocyclic ring (product **2**) and at 7.53 ppm (d, $J = 2.4$ Hz) for the phosphorus atom in the methylene phosphonic acid. After the removal of the solid product, the filtrate exhibited a major signal at 8.60 ppm that split into a triplet when coupled to protons. On the basis of this observation, the signal at 8.97 ppm is assigned to the phosphorus atom in **1**. From the $^{31}P\{H\}$NMR spectrum, the crude product appeared to be a 1:1 mixture of **1** and **2**.

Phosphonomethylation of 2-amino-2-methyl-1,3-propanediol gave the dimethylene diphosphonic acid of 5-amino-5-methyl-1,3-dioxane.

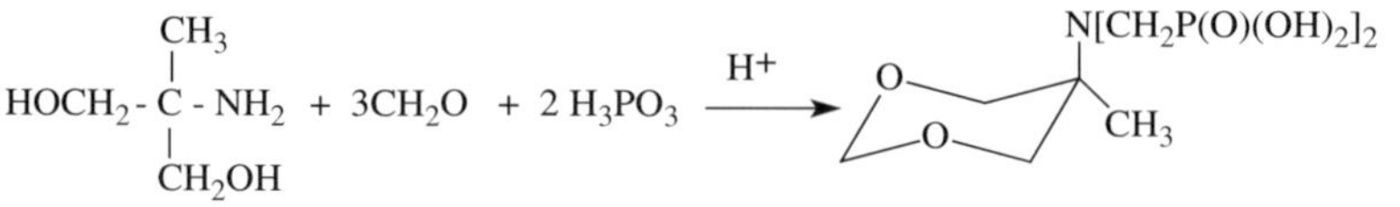

$$HOCH_2-\underset{\underset{CH_2OH}{|}}{\overset{\overset{CH_3}{|}}{C}}-NH_2 + 3CH_2O + 2H_3PO_3 \xrightarrow{H^+} \text{[1,3-dioxane with } N[CH_2P(O)(OH)_2]_2 \text{ and } CH_3\text{]}$$

The ^{31}P NMR spectrum of the white crystalline solid obtained when the crude product was dissolved in ethanol revealed a single ^{31}P signal at 8.30 ppm that split into a triplet when coupled to protons. In the $^{13}C\{H\}$ NMR spectrum, this solid exhibited a signal at 95.45 ppm that is characteristic of the C-2 of the 1,3-dioxane structure.

One of the limitations of the Mannich-type reaction for the preparation of aminomethylenephosphonic acids is that primary amines treated with 1 equivalent of formaldehyde and phosphorous acid yield a mixture of mono- and bis(methylenephosphonic) acids. A further limitation appears to be in the choice of carbonyl compounds.

Additional methods for preparation of aminophosphonates based on dialkyl H-phosphonates

Dialkyl H-phosphonates react with methylenediamines yielding amino-substituted methanephosphonic acids [66].

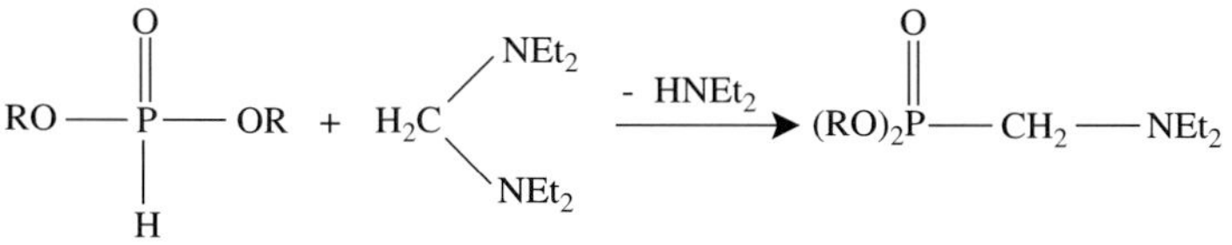

$$RO-\overset{\overset{O}{\|}}{\underset{\underset{H}{|}}{P}}-OR + H_2C\overset{NEt_2}{\underset{NEt_2}{\diagdown}} \xrightarrow{-HNEt_2} (RO)_2\overset{\overset{O}{\|}}{P}-CH_2-NEt_2$$

Physiologically active nitrogen-containing heterocycles have been obtained reacting heterocycles **1** with methylenediamines and H-phosphonate diesters [67]. In the first stage of the reaction, diamine reacts with dialkyl H-phosphonate to give dialkyl

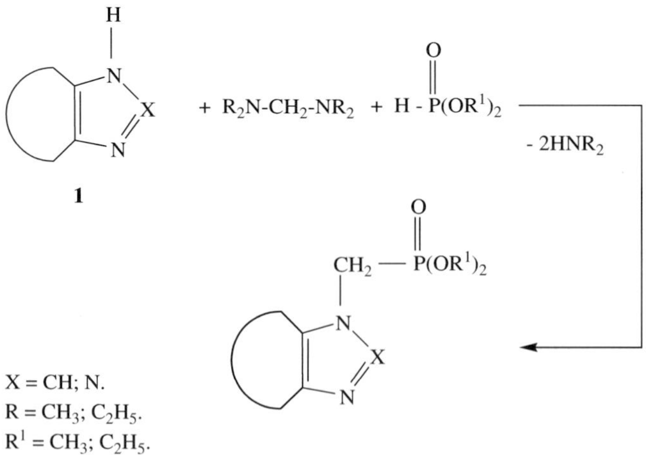

[heterocycle **1**] $+ R_2N-CH_2-NR_2 + H-\overset{\overset{O}{\|}}{P}(OR^1)_2 \xrightarrow{-2HNR_2}$ [heterocycle with $CH_2-\overset{\overset{O}{\|}}{P}(OR^1)_2$]

X = CH; N.
R = CH_3; C_2H_5.
R^1 = CH_3; C_2H_5.

4.1 Aminophosphonates and Aminophosphonic Acids

aminomethylphosphonate, which then reacts with azole **1**.

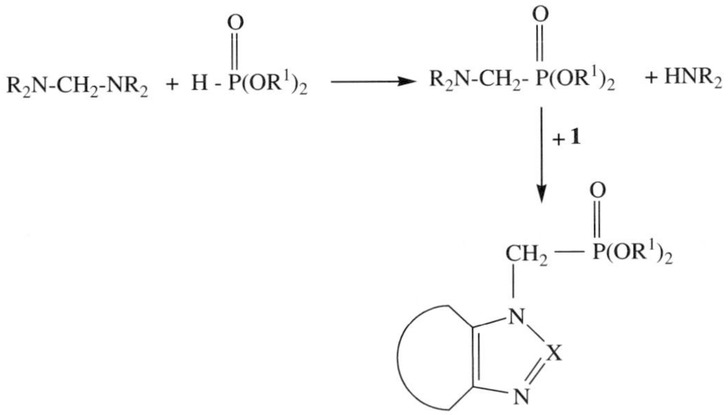

$$R_2N\text{-}CH_2\text{-}NR_2 + H\text{-}P(OR^1)_2 \longrightarrow R_2N\text{-}CH_2\text{-}P(O)(OR^1)_2 + HNR_2$$

$$\downarrow +\mathbf{1}$$

[azole-N-CH$_2$-P(O)(OR^1)$_2$]

It was shown that the interaction of dialkyl H-phosphonates with α,α-bis-2-pyridil-aminotoluene furnished pyridilaminobenzylphosphonate with 76–86% yield [68].

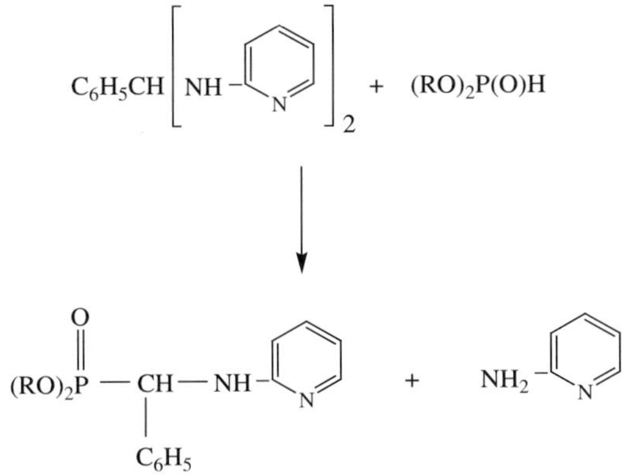

$$C_6H_5CH[NH\text{-}(2\text{-Py})]_2 + (RO)_2P(O)H \longrightarrow (RO)_2P(O)\text{-}CH(C_6H_5)\text{-}NH\text{-}(2\text{-Py}) + NH_2\text{-}(2\text{-Py})$$

Dialkyl H-phosphonates also react with methyl-(*N*-diethyl)aminomethyl ether [69], furnishing aminomethylphosphonates.

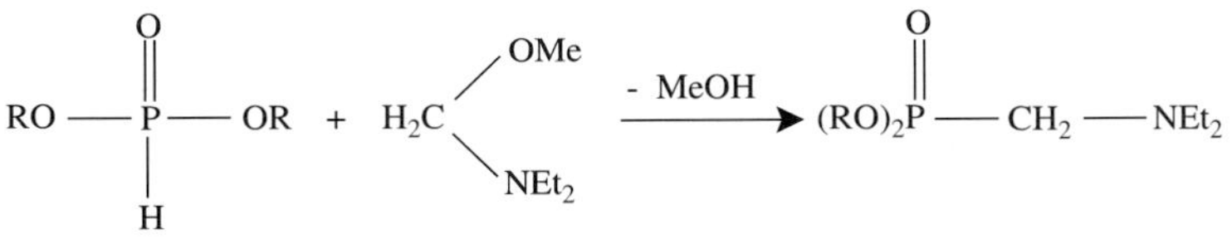

$$RO\text{-}P(O)(H)\text{-}OR + H_2C(OMe)(NEt_2) \xrightarrow{-\text{MeOH}} (RO)_2P(O)\text{-}CH_2\text{-}NEt_2$$

The above-mentioned reactions give nearly quantitative yields of aminophosphonates and avoid the formation of water entirely. The analogous reaction of dialkyl H-phosphonates with α-alkylamino dialkyl mercaptenes proceeds just to the stage of the formation of the corresponding S, N acetals [70].

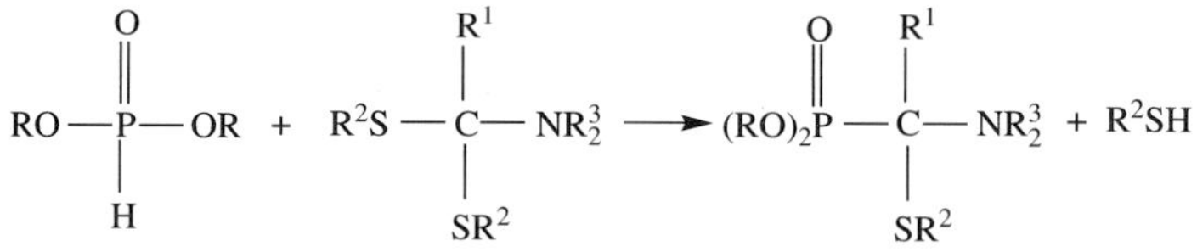

$$RO-\overset{\overset{O}{\|}}{P}-OR + R^2S-\underset{\underset{SR^2}{|}}{\overset{\overset{R^1}{|}}{C}}-NR_2^3 \longrightarrow (RO)_2\overset{\overset{O}{\|}}{P}-\underset{\underset{SR^2}{|}}{\overset{\overset{R^1}{|}}{C}}-NR_2^3 + R^2SH$$

It was found that N,N-bis-(2-hydroxyalkyl)-aminomethane phosphonic acid dialkyl esters, corresponding to compounds with formula **3**, can be obtained by reacting dialkyl H-phosphonates **1** with oxazolidines **2** in the presence of an acid ion exchanger [71].

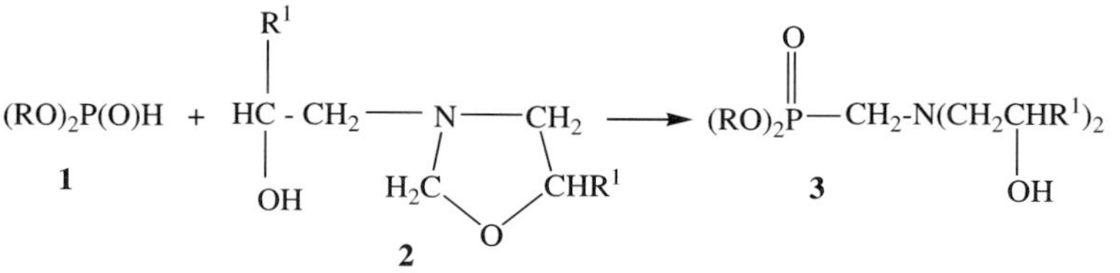

$$(RO)_2P(O)H + \underset{\mathbf{2}}{HC\text{-}CH_2-N-CH_2} \longrightarrow (RO)_2\overset{\overset{O}{\|}}{P}-CH_2\text{-}N(CH_2CHR^1)_2$$
(with OH and oxazolidine ring substituents; product **3** bears OH)

The characteristic groups of the ion exchanger may be sulfonic acid, phosphonic acid, or boric acid.

Aminoalkylbisphosphonic derivatives are also formed by the addition of dialkyl H-phosphonates to nitriles in the presence of hydrogen chloride [72].

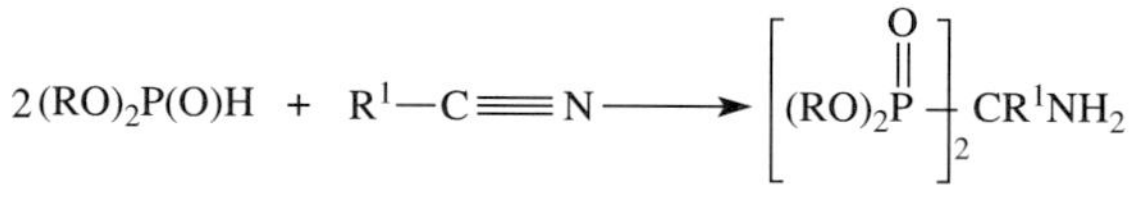

$$2(RO)_2P(O)H + R^1-C\equiv N \longrightarrow \left[(RO)_2\overset{\overset{O}{\|}}{P}\right]_2 CR^1NH_2$$

When the above reaction is carried out in a hydrogen atmosphere in the presence of a Raney–Nickel catalyst, hydrogenation of the nitrile to azomethine takes place as a first step, followed by addition of dialkyl H-phosphonate [73].

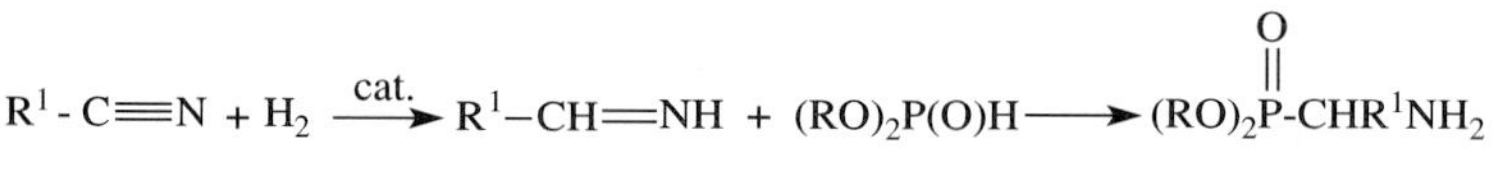

$$R^1\text{-}C\equiv N + H_2 \xrightarrow{cat.} R^1\text{-}CH=NH + (RO)_2P(O)H \longrightarrow (RO)_2\overset{\overset{O}{\|}}{P}\text{-}CHR^1NH_2$$

Cyclic phosphonates easily undergo Markovnikov's addition to enamines of the type **I**, yielding adducts with α-aminophosphonate structure **II** [74, 75].

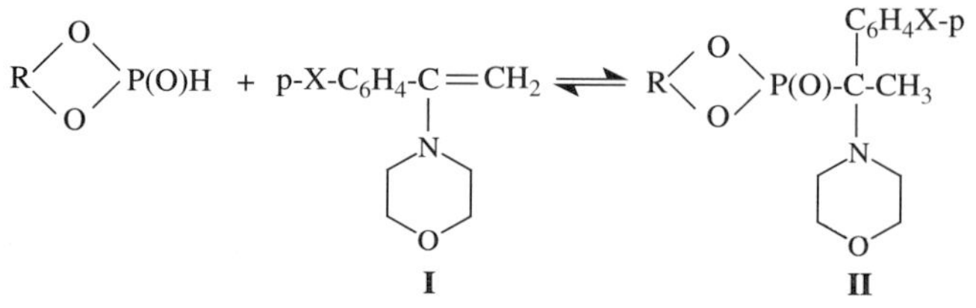

4.1 Aminophosphonates and Aminophosphonic Acids

1,3,5-Trisubstituted-hexahydro-s-triazines reacted with 3 equivalents of diethyl H-phosphonates at 100 °C for 6 h to afford aminomethyldiethylphosphonate [76].

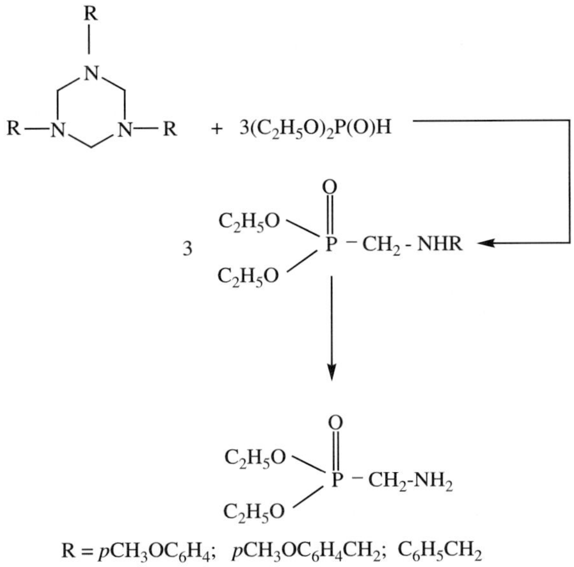

$R = p\text{CH}_3\text{OC}_6\text{H}_4;\ p\text{CH}_3\text{OC}_6\text{H}_4\text{CH}_2;\ \text{C}_6\text{H}_5\text{CH}_2$

Hydrogenolytic dibenzylation gave aminomethyldiethylphosphonate in near quantitative yield. The latter compound is a key intermediate for the synthesis of cephalosporins.

Despite the high level of interest in the asymmetric synthesis of α-aminophosphonic acids, not much is known about the cyclic α-aminophosphonates. The pharmaceutically interesting cyclic α-aminophosphonates are synthesized by the hydrophosphonylation of cyclic imines (thiazolines) in the presence of chiral lanthanoid–potassium–binaphtoxide complexes (LnPB) (Ln = lantanoid metal, P = potassium, B = binaphtol) [77].

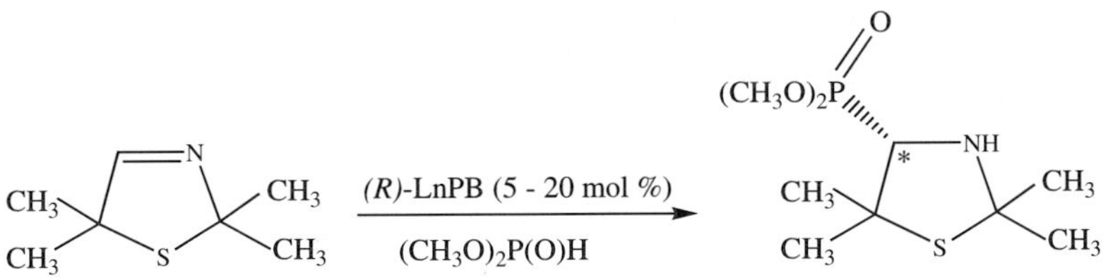

The chiral thiazolidinylphosphonate was obtained in excellent optical purity of up to 98% and high chemical yield up to 97%.

Dialkyl 1-hydroxyalkylphosphonates, which are conveniently available by addition of aldehydes to dialkyl H-phosphonates, are used as starting compounds for the preparation of α-aminoalkanephosphonic acids [78].

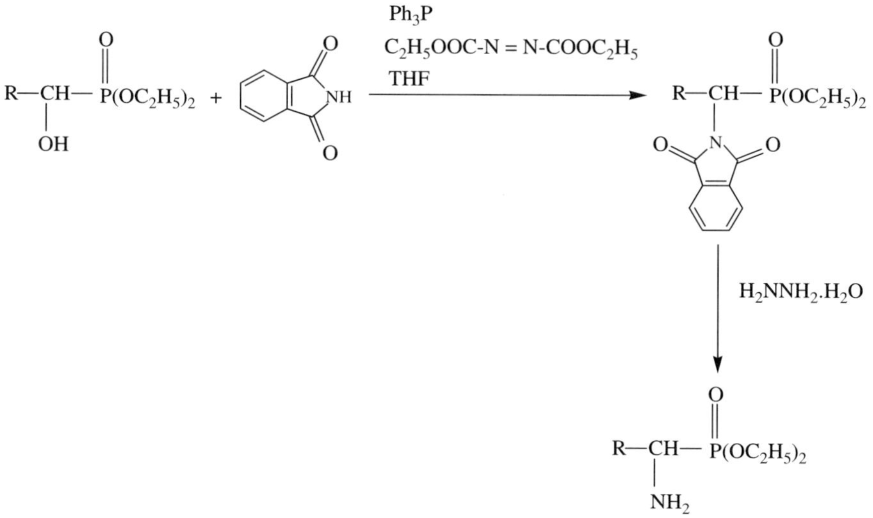

$$R\text{-}CH(OH)\text{-}P(O)(OC_2H_5)_2 + \text{phthalimide} \xrightarrow[\text{THF}]{\substack{Ph_3P \\ C_2H_5OOC\text{-}N=N\text{-}COOC_2H_5}} R\text{-}CH(NPhth)\text{-}P(O)(OC_2H_5)_2 \xrightarrow{H_2NNH_2 \cdot H_2O} R\text{-}CH(NH_2)\text{-}P(O)(OC_2H_5)_2$$

Heating aminoalkylphosphonates leads to the formation of zwitterionic salts resulting from internal alkylation of the amino group [79].

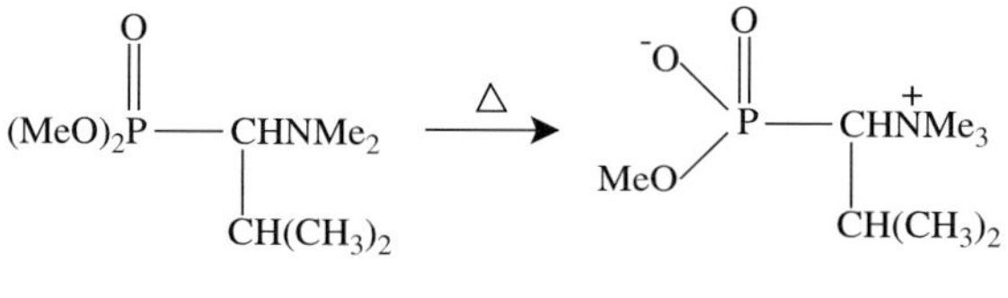

$$(MeO)_2P(O)\text{-}CH(CH(CH_3)_2)\text{-}NMe_2 \xrightarrow{\Delta} MeO\text{-}P(O)(O^-)\text{-}CH(CH(CH_3)_2)\text{-}\overset{+}{N}Me_3$$

N-(Phosphonomethyl) glycine

The most important α-aminophosphonic acid is *N*-(phosphonomethyl) glycine and its derivatives. It has been found that *N*-(phosphonomethyl) glycine is effective as a herbicide and plant growth regulator when used at low rates. This α-aminophosphonic acid is biodegradable within a relatively short time after its application. Due to the importance of *N*-(phosphonomethyl) glycine and certain salts as a herbicide, different methods of making the compounds are constantly being sought in order to provide improved and alternate methods of manufacture. *N*-(phosphonomethyl) glycine can be made in a number of different ways. A Belgian patent describes the preparation of *N*-(phosphonomethyl) glycine from glycine and chloromethylphosphonic acid [80]. The chlorine atom in this acid is of low reactivity, and under the forcing conditions required for the reaction to take place, by-products are

4.1 Aminophosphonates and Aminophosphonic Acids

formed. A U.S. patent describes the oxidation of glycinemethylene phosphinic acid with mercuric chloride to form N-(phosphonomethyl) glycine [81]. In view of the problems of environmental pollution associated with the use of mercury compounds, this method is obviously unsuitable for large-scale manufacturing use. According to U.S. patent 4,008,296, N-(phosphonomethyl) glycine can be prepared from 1,3,5-tricyanomethylhexahydro-1,3,5-triazine and phosphonic acid diesters [82]. The process may be illustrated by the following reaction scheme:

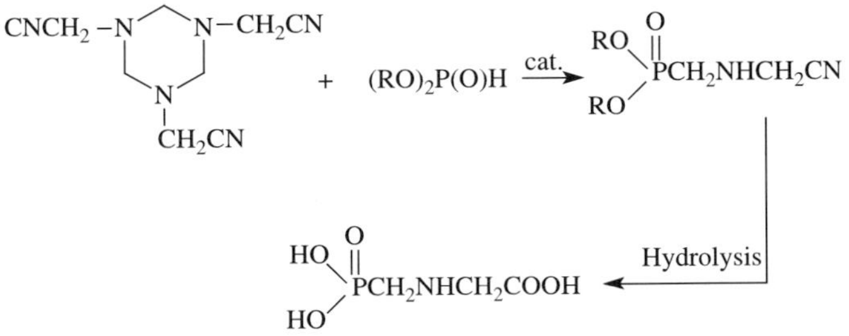

$$CNCH_2-N \underset{CH_2CN}{\overset{N-CH_2CN}{\bigcirc}} + (RO)_2P(O)H \xrightarrow{cat.} (RO)_2P(O)CH_2NHCH_2CN$$

$$(HO)_2P(O)CH_2NHCH_2COOH \xleftarrow{Hydrolysis}$$

The catalyst used in the process is preferably hydrogen chloride or hydrogen bromide. The hydrolysis of N-(phosphonomethyl) glycinonitrile is carried out with a concentrated aqueous solution of a mineral acid. The hydrolysis may also be carried out with an aqueous, aqueous alcoholic, or alcoholic solution of an alkali, for example, sodium or potassium hydroxide.

N-(phosphonomethyl) glycine is obtained in a good yield by reacting the sodium salt of tris-(carboxymethyl)-hexahydro-s-triazine with phosphonic acid diesters [83].

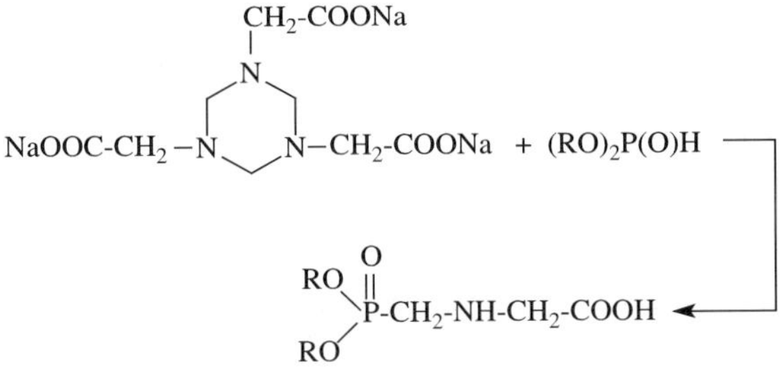

$$NaOOC-CH_2-N \underset{\underset{CH_2-COONa}{N}}{\overset{N-CH_2-COONa}{\bigcirc}} + (RO)_2P(O)H \longrightarrow (RO)_2P(O)-CH_2-NH-CH_2-COOH$$

Here, we will discuss a few widely utilized methods. The most important method for the preparation of N-(phosphonomethyl) glycine is from glycine, formaldehyde, tertiary base, and diethyl H-phosphonate in alcoholic solution (see Appendix) [84].
In the presence of small amounts triethylamine, paraformaldehyde dissolves in methyl alcohol already at 45–50 °C. Under the combined action of triethylamine and CH_2O,

glycine dissolves in methyl alcohol at room temperature. The dissolution reaction of glycine is exothermic. The following equilibrium reactions occur:

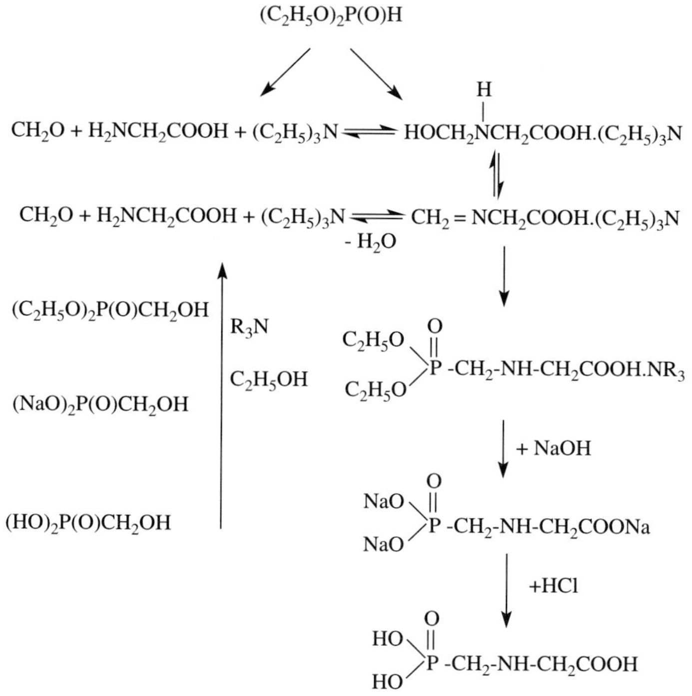

$$(C_2H_5O)_2P(O)H$$

$$CH_2O + H_2NCH_2COOH + (C_2H_5)_3N \rightleftharpoons HOCH_2\underset{H}{N}CH_2COOH \cdot (C_2H_5)_3N$$

$$CH_2O + H_2NCH_2COOH + (C_2H_5)_3N \underset{-H_2O}{\rightleftharpoons} CH_2=NCH_2COOH \cdot (C_2H_5)_3N$$

$(C_2H_5O)_2P(O)CH_2OH \quad\quad R_3N$

$(NaO)_2P(O)CH_2OH \quad\quad C_2H_5OH$

$(HO)_2P(O)CH_2OH$

$$\underset{C_2H_5O}{\overset{C_2H_5O}{>}}\!\!P(O)\text{-}CH_2\text{-}NH\text{-}CH_2COOH \cdot NR_3$$

$\downarrow + NaOH$

$$\underset{NaO}{\overset{NaO}{>}}\!\!P(O)\text{-}CH_2\text{-}NH\text{-}CH_2COONa$$

$\downarrow + HCl$

$$\underset{HO}{\overset{HO}{>}}\!\!P(O)\text{-}CH_2\text{-}NH\text{-}CH_2COOH$$

Regarding the by-products, it should be noted that there is no formation of any secondary derivative of glycine like *N*-diphosphonomethyl glycine, but only of hydroxymethyl phosphonic acid, which can be reclaimed in the form of its calcium salt. The yield calculated on the base of glycine is about 58–63%.

N-(phosphonomethyl) glycine has been obtained from diketopiperazine, formaldehyde, and H-phosphonic acid diesters followed by hydrolysis [85].

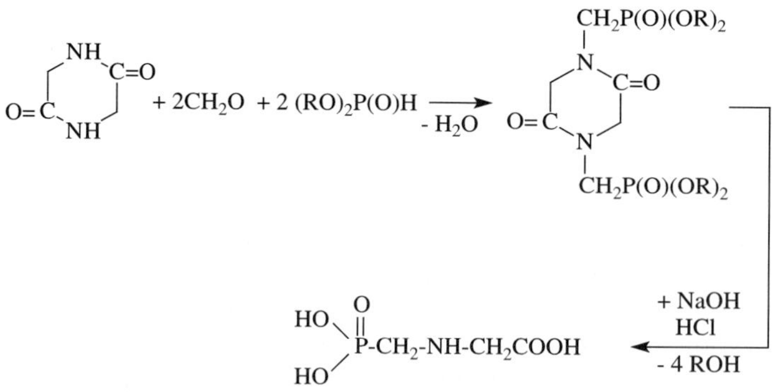

[diketopiperazine] $+ 2CH_2O + 2(RO)_2P(O)H \xrightarrow{-H_2O}$ [N,N'-bis(phosphonomethyl) diketopiperazine]

$\xrightarrow[-4ROH]{+NaOH, HCl}$ $\underset{HO}{\overset{HO}{>}}\!\!P(O)\text{-}CH_2\text{-}NH\text{-}CH_2COOH$

4.1 Aminophosphonates and Aminophosphonic Acids

The reaction of diethyl H-phosphonate with sodium salt of *N*-carboxyglycine resulted in the formation of *N*-(phosphonomethyl) glycine [86].

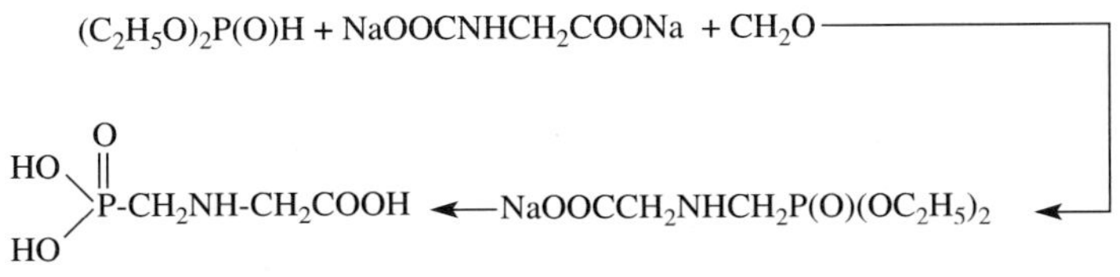

$$(C_2H_5O)_2P(O)H + NaOOCNHCH_2COONa + CH_2O \longrightarrow$$

$$\underset{HO}{\overset{HO}{>}}P(O)\text{-}CH_2NH\text{-}CH_2COOH \longleftarrow NaOOCCH_2NHCH_2P(O)(OC_2H_5)_2 \longleftarrow$$

N-(Phosphonomethyl) glycine derivatives

Searching for new synthetic routes for the preparation of *N*-phosphonomethyl glycine derivatives has been the focus of efforts of scientists because it has been found that the *N*-phosphonomethyl glycine derivatives are especially effective in suppressing the growth of cancer, tumors, viruses, or bacteria.

Heating various H-phosphonic acid diesters with a hydroxyalkylcarbamate mixture [87,88] yielded 3-ethyl-2-hydroxy-2-oxo-1,4,2-oxazaphosphorinane **1**, which is water-soluble and hydrolytically stable.

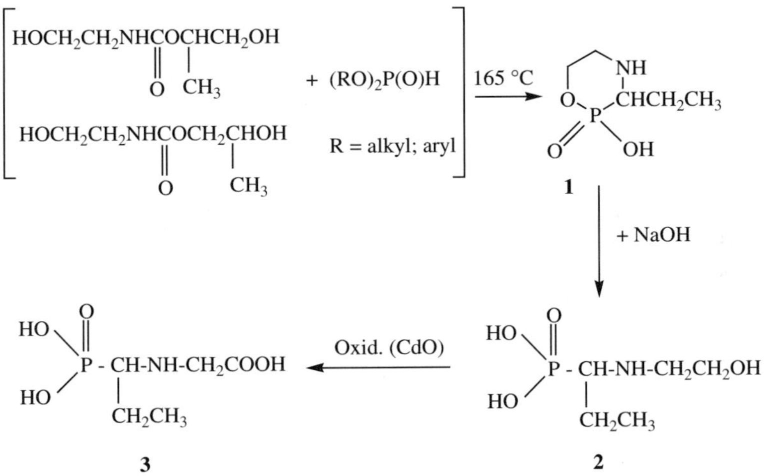

The treatment of **1** with an aqueous solution of NaOH yields α-ethyl-α-*N*-(hydroxyethylamino)methyl phosphonic acid **2** after treatment with Dowex 50WX8-200 [89]. α-Ethyl-*N*-(phosphonomethyl) glycine **3** is obtained by oxidation of **1** or **2** [90]. The cytotoxicity of the new aminophosphonic acids has been studied in continuous cell line and was expressed as a concentration-depending reduction of the uptake of the vital dye Neutral Red (IC_{50}) (Table 4.1) [91]. The results obtained showed that 4-ethyl-2-hydroxyl-2-oxo-1,4,2-dioxaphosphorinane and α-ethyl-α-(hydroxyethylamino)methyl phosphonic acid are toxic compounds. The sodium salt of the

Table 4.1

IC$_{50}$ of some aminophosphonic acids and their salts

Aminophosphonic acids and their salts	IC$_{50}$ (mg/mL)
3-Ethyl-2-hydroxy-2-oxo-1,4,2-oxazaphosphorinane	0.00022
Sodium salt of 3-ethyl-hydroxy-2-oxo-1,4,2-oxazaphosphorinane	0.00173
α-Ethyl-α-N-(hydroxyethylaminomethyl)phosphonic acid	0.00024
α-Ethyl-N-(phosphonomethyl)glycine	0.82272
Monoisopropyl ammonium salt of α-ethyl-N-(phosphonomethyl) glycine	1.90188

4-ethyl-2-hydroxyl-2-oxo-1,4, 2-dioxaphosphorinane has lower toxicity compared to the acid form. The toxicity of α-ethyl-N-(phosphonomethyl) glycine **3** is three orders lower with respect to **1** and **2**.

Phosphonylmethylaminocyclopentane-1-carboxylic acid **1** was obtained in a good yield of 86% by reacting aminocyclopentane-1-carboxylic acid, cycloleuicin, paraformaldehyde, and diethyl H-phosphonate [92].

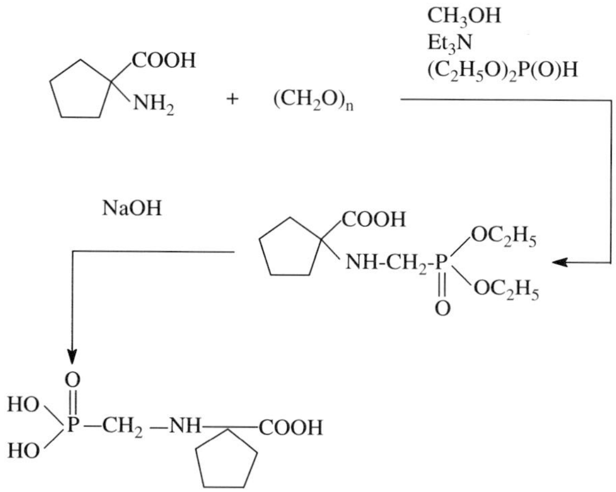

4.1.2 Hydrolytic cleavage of phosphorus–carbon bond

Although the carbon–phosphorus bond is believed to be extremely stable, several examples of the hydrolytic cleavage of the P–C bond in 1-aminoalkanephosphonic acid and their esters, in acidic and basic media, were described recently [24,30,93]. Hydrolysis of

4.1 Aminophosphonates and Aminophosphonic Acids

2- and 4-pyridylmethyl(amino)phosphonates (**1** and **2**) by means of 20% aq. HCl led to decomposition of the phosphonates with a cleavage of the P–C bond and formation of the corresponding amines [24] in 50% yield.

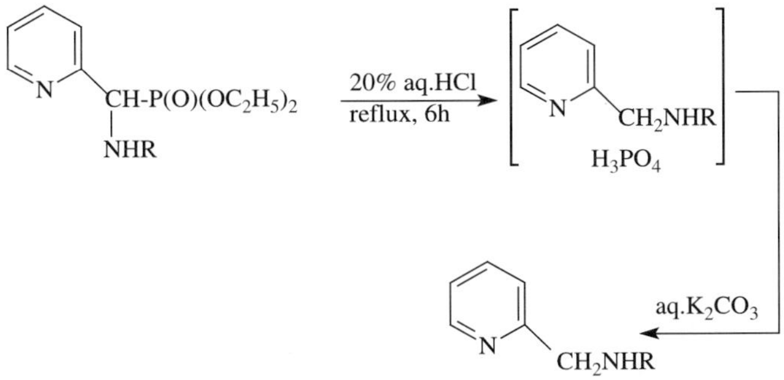

The P–C cleavage is not only characteristic for aminophosphonate esters. The corresponding aminophosphonic acids also undergo a similar cleavage.

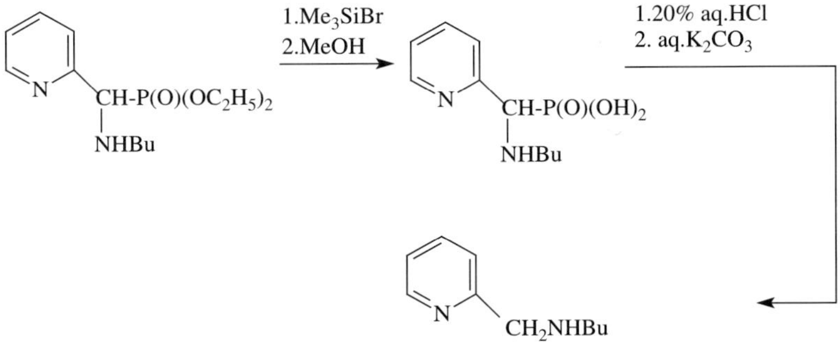

Obviously, the chemical character and electronic structure of 2- and 4-pyridyl moieties is responsible for acidic cleavage of 2- and 4-pyridylmethy(amino)phosphonates. The protonation of pyridine nitrogen in the phosphonate molecule (structure **A**) postulated a delocalization of a positive charge on the 2- and 4- positions of the molecule. It is accompanied by the formation of conjugate double bonds between the C–H group and the pyridine ring (pyridone or enamine-like structure **B**). The formation of species **B** is facilitated and possible when the phosphorus moiety is removed as a positively charged group. It should be pointed out that the protonation of the amino group is an additional force for the breaking of the bond between carbon and phosphorus.

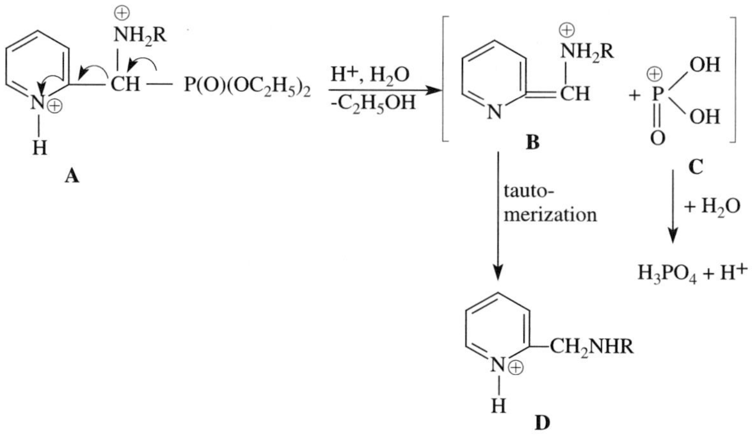

Accumulation of positive charges in the molecule facilitates cleavage of the P–C bond. This observation has been confirmed by the fact that there was no cleavage of the P–C bond in the case of pyridylmethyl(hydroxy)phosphonates.

Treatment of 1-amino-2′-nitrobenzylphosphonic acids with aqueous sodium hydroxide caused a P–C bond cleavage, with the formation of 3-amino-2,1-benzisoazole derivatives [30].

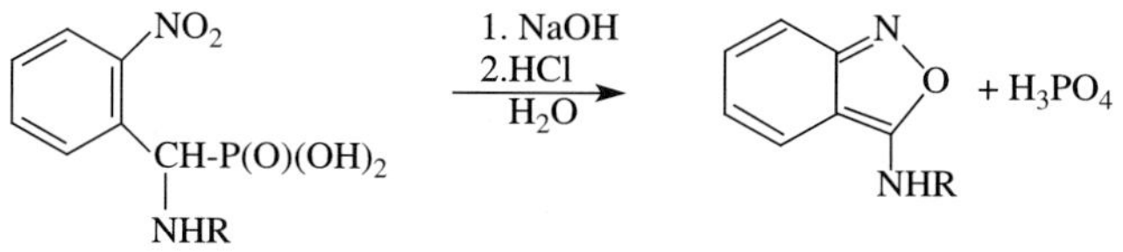

The basic cleavage of 1-amino-2′-nitrobenzylphosphonic acids proceeded via the formation of *aci*-nitro species, in an ionizable form [30].

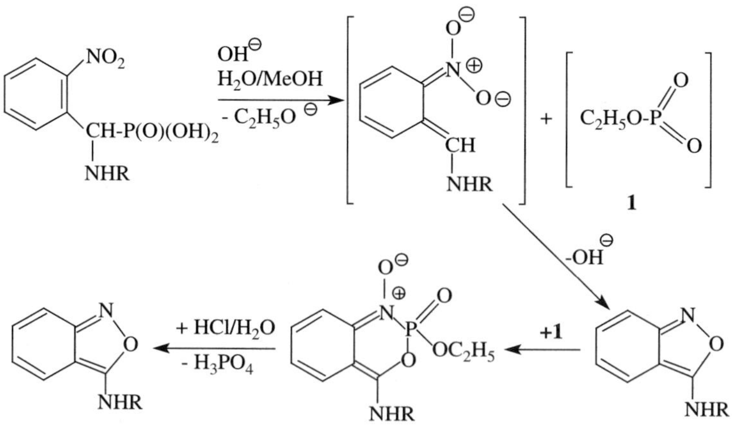

The cleavage of the P–C bond in some aminophosphonates is an unusual reaction, because the strength of a single P–C bond is comparable with the strength of a single C–C bond (the energy value for the P–C bond is about 63 kcal/mol).

4.1 Aminophosphonates and Aminophosphonic Acids

4.1.3 Synthetic application

The most recent developments in drug research have shown that peptides play a key role as endogenous regulators in neurology, immunology, heart, and blood circulation. The replacement of the proteinogenic amino acid residues by aminophosphonic acid analogues to give peptidomimetics as potential antimetabolites of the parent peptides is one the most important synthetic applications of aminophosphonic acids.

In 1975, Marthell *et al.* introduced the term 'phosphonopeptides' for peptide analogues in which an aminocarboxylic residue is replaced by an aminophosphonate at any position in the peptide chain [94]. There is an excellent review by P. Kafarski and B. Lejczak [95] devoted to the synthesis of phosphono- and phosphinopeptides. Here, only references connected with the synthesis of biologically active phosphonopeptides using aminophosphonates based on H-hosphonate diesters have been included.

Synthesis of phosphonopeptides

Peptides containing a P-terminal aminophosphonic acid have been prepared by coupling protected amino acids with dialkyl or diaryl esters of aminoalkanephosphonic acids or free acids. Protection of the amino phosphonic acid groups is an important step in the phosphonopeptide synthesis. Aminophosphonates based on the H-phosphonate diesters are used directly for the preparation of phosphonopeptides. Diphenyl aminoalkanephosphonates are attractive starting materials for phosphonopeptide synthesis because they are readily available, and efficient coupling is easily achieved by most methods used in peptide chemistry.

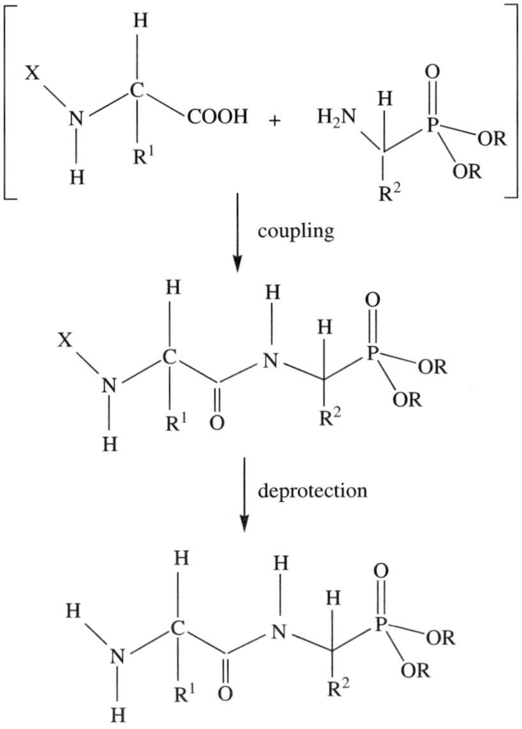

The partial protection of the amino group of the aminophosphonate, as a result of the intramolecular hydrogen bonding with the phosphoryl group, results in a decrease in the

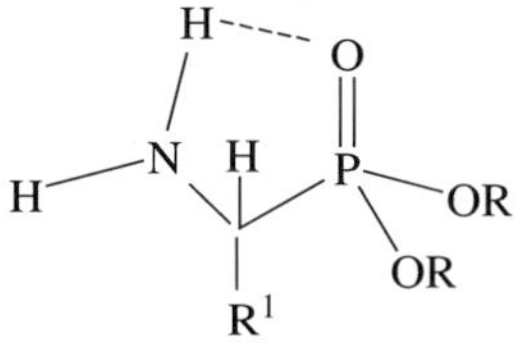

rate of the acylation reaction of the diethyl and diphenyl aminophosphonates when compared to the corresponding esters of amino acids [96].

1-Amino-1-cyclohexylmethylphosphonic acid diethyl ester was utilized for the preparation of hapten **1**, recently employed to induce the first monoclonal catalytic antibodies capable of catalyzing peptide bond formation[97]. The key steps of the synthesis of **1** include (a) preparation of scalemic α-aminophosphonate, (b) generation of the *p*-nitrobenzyl ester, (c) installation of a suitable second phosphonate ester linkage, (d) *N*-acylation with glutaric anhydride.

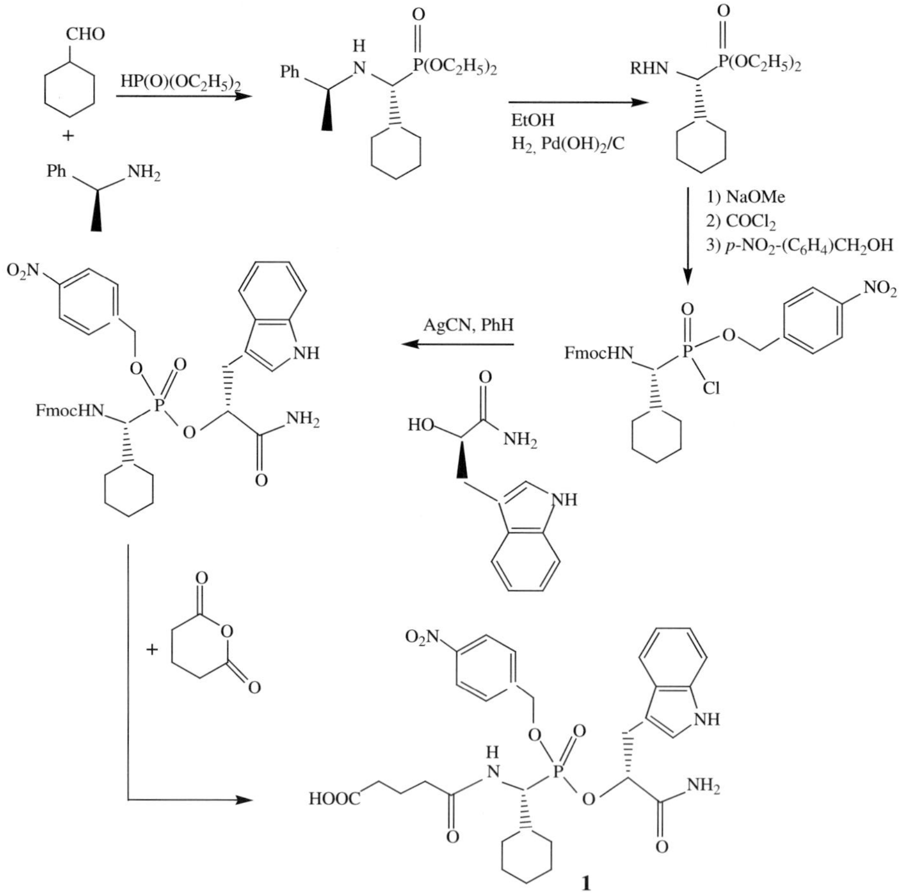

4.1 Aminophosphonates and Aminophosphonic Acids

Synthesis of α-aminophosphonate nucleosides

Honek *et al.* have successfully synthesized the α-aminophosphonate AdoHcy analogue [98].

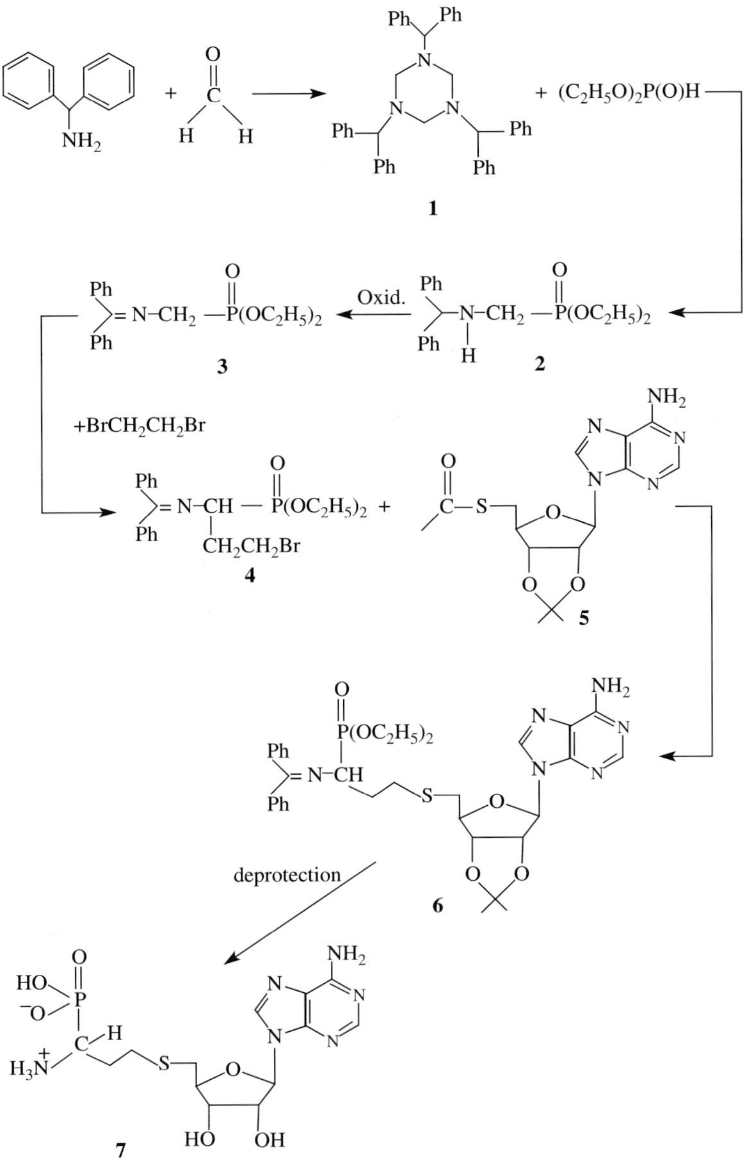

Diphenylmethylamine was reacted with formalin to give 2,4,6-tris(benzydryl)hexahydro-1,3,5-triazine **1**. Addition of diethyl H-phosphonate to **1** gave diethyl N-(diphenylmethyl)methylphosphonate **2**. The oxidation of **2** furnished the imine **3**, which was alkylated with 1,2-dibromoethane to give **4**. The thiolate anion of 5′-thio nucleoside analogue **5** reacts with **4** to give AdoHcy analogue **6**. Full deprotection of **6** was achieved with trimethylsilyl iodide to generate the desired compound **7**. This is the first evaluation of a phosphonate nucleoside as a probe for the enzyme AdoHcy hydrolase.

The phosphonic acid moiety has been shown to be inhibitors of numerous metabolic processes [99,100]. In recent years, S-adenosyl-L-homocystein hydrolase(AdoHcy) has become an attractive target for drug design since its inhibitors have been shown to exhibit antiviral [101], antiparasitic [102], antiarthritic [103], and immunosuppressive effects [104]. Importantly, α-aminophosphonate diesters are more attractive as intermediates for multistep synthesis than the corresponding phosphonic acids. The insolubility of the latter in both organic and neutral aqueous media complicates derivation of both the amine and acid functionalities.

4.2 BISPHOSPHONATES

The first bisphosphonates (originally called diphosphonates) were synthesized in the 19th century. Geminal bisphosphonates **1** are metabolically stable analogues of the naturally occurring inorganic pyrophosphates **2**, in which the P–O–P backbone of pyrophosphate has been replaced with P–C–P. Fleisch and coworkers have shown that bisphosphonates impair the formation and dissolution of calcium phosphate crystals *in vitro* [105].

Substitution of oxygen to carbon makes them resistant to rapid hydrolysis by phosphatase, an enzyme with widespread distribution in the body. More importantly, the replacement of the oxygen atom between the two phosphonic acid moieties of pyrophosphate by a carbon atom opened up the possibility of attaching side chains. The nature of the groups attached to the central carbon atom is the key to the optimization of bisphosphonates as potent inhibitors of osteoclastic bone resorption. The action of the bisphosphonates on calcium salts is similar to the action of pyrophosphates. They exhibit the following actions:

- inhibiting the precipitation of calcium phosphate from solutions;
- blocking the transformation of amorphous calcium phosphate into crystalline form without, however, inhibiting the formation of the initial phase;
- blocking the aggregation of crystals of hydroxyapatite;
- retarding the degree of dissolution of crystals of hydroxyapatite after the latter have absorbed the bisphosphonate from the solutions.

4.2 Bisphosphonates

These compounds are known to possess a number of dental and medical applications: dental—toothpastes and mouthwashes [106]; medical—drugs against rheumatoid arthritis [107,108], inhibitors of bone resorptive processes such as osteoporosis and Paget's diseases [106a,109–111], in breast cancer therapy [112], antiviral [113,114], antiinflammatory [115,116] and antirheumatismal agents [117], and pain-easing activities [115]. They are also used as antibacterial agents [118], plant-growth regulators [119], herbicides [120,121], pesticides [118], in the nuclear industry [122–124], as flame retardants [120], and chelators in water treatment [120]. Geminal bisphosphonates and geminal bisphosphonic acids can readily form complexes with calcium and magnesium, and this behaviour is one of the principles sustaining their use as drugs [106a,114]. It is well established that bisphosphonates bind avidly to mineralized bone where they potently inhibit osteoclast-mediated bone resorption [125]. Biological activities of bisphosphonates appear to be related to their ability to chelate metal ions. Bisphosphonates are remarkably nontoxic to other organs and are very well tolerated clinically. This may be explained by the rapid accumulation of bisphosphonate in bone, resulting in a very short plasma half-life and low exposure of visceral tissue [126]. Bisphosphonates also exert antitumor effects. They exert direct cytostatic effects on a variety of human tumor cells *in vitro*. Bisphosphonates (zoledronic acid, pamidronate) decrease cell proliferation and induce apoptosis in a time- and concentration-dependent manner.

All compounds containing the geminal bisphosphonate moiety may be divided into two subgroups on the basis of whether or not they possess a nitrogen-containing substituent. The nitrogen-containing bisphosphonates are the most potent compounds. They inhibit the mevalonate pathway, thereby reducing the prenylation of small GTP-binding proteins such as Ras and Rho, which are essential for vesicular trafficking and the maintenance of cytoskeletal integrity[127].

4.2.1 Methods for preparation of bisphosphonates

F. Lancas *et al.* have published an excellent review devoted to the major synthetic routes of bisphosphonates [128].

Dialkyl H-phosphonates react with α-amino-α-alkoxyalkanes to form amino-containing bisphosphonic acid tetraesters as final products [129,130].

Substituted aminomethylenebisphosphonates were prepared by reaction of dialkyl H-phosphonates with tri-Et orthoformate and the corresponding heteryl- or arylamine [131].

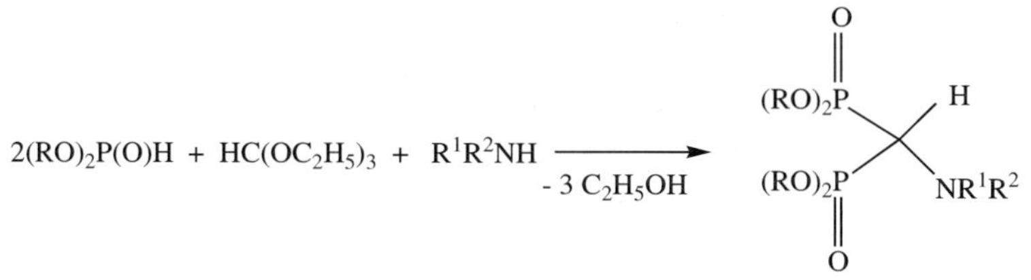

$$2(RO)_2P(O)H + HC(OC_2H_5)_3 + R^1R^2NH \xrightarrow{-3 C_2H_5OH} (RO)_2P(O)\text{-}C(H)(NR^1R^2)\text{-}P(O)(OR)_2$$

Bisphosphonates are obtained by reacting acetic anhydride with H-phosphonic acid [132].

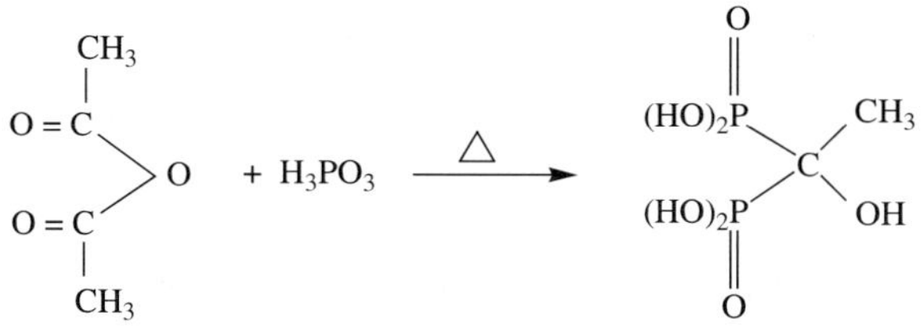

$$(CH_3CO)_2O + H_3PO_3 \xrightarrow{\Delta} (HO)_2P(O)\text{-}C(CH_3)(OH)\text{-}P(O)(OH)_2$$

1-Hydroxyalkyl-1,1-bisphosphonic acids are obtained reacting acylating agents with H_3PO_3 or carbonic acids with PCl_3 and calculated amounts of water [133].

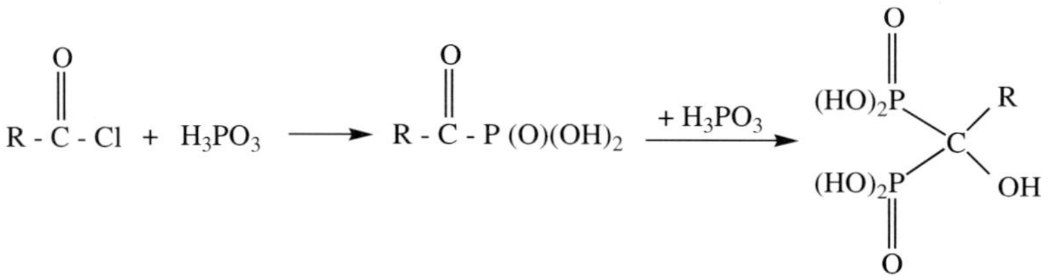

$$R\text{-}C(O)\text{-}Cl + H_3PO_3 \longrightarrow R\text{-}C(O)\text{-}P(O)(OH)_2 \xrightarrow{+ H_3PO_3} (HO)_2P(O)\text{-}C(R)(OH)\text{-}P(O)(OH)_2$$

Bisphosphonates have been prepared by the reaction of a carboxylic acid with H-phosphonic acid and phosphorus trichloride [134].

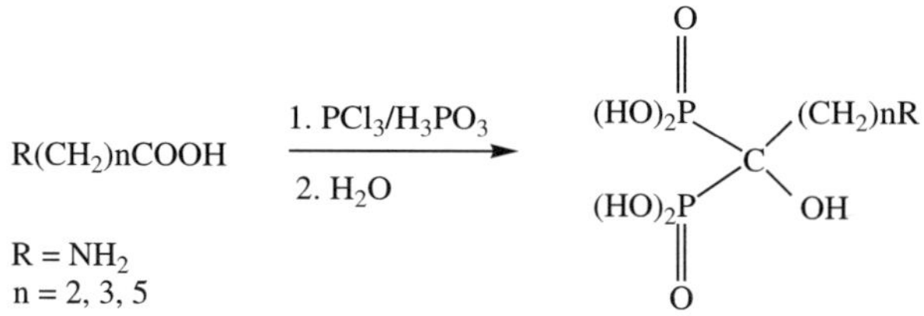

$$R(CH_2)_nCOOH \xrightarrow[\text{2. } H_2O]{\text{1. } PCl_3/H_3PO_3} (HO)_2P(O)\text{-}C((CH_2)_nR)(OH)\text{-}P(O)(OH)_2$$

$R = NH_2$
$n = 2, 3, 5$

The reaction can be carried out in the presence of a diluent, especially chlorinated hydrocarbons, particularly chlorobenzene, which does not solubilize the reaction components and

4.2 Bisphosphonates

serves only as a heat carrier. The reaction mixture starts as a two-phase system, in which the melted phase gradually thickens into a nonstirrable mass. The semisolid sticky mass finally turns into a hard, rigid material, coating the walls of the reaction vessel.

By running the reaction in methanesulfonic acid, which was used to solubilize the reaction components, the reaction remains fluid, thus allowing complete conversion of the carboxylic acid, providing excellent yields and purity of 1-hydroxy-1,1- bisphosphonates [135,136]. It was shown that at 55 °C, the reaction takes 3 days while at 65 °C, the reaction is complete after an overnight age (see Appendix). Methanesulfonic acid reacts with phosphorus trichloride and the reaction becomes self-heating at 85 °C and an uncontrolled exotherm occurs at reaction temperatures >140 °C. It was found that the reaction can be realized in the presence of poly(ethylene glycol) [137]. However, it was reported that large quantities of poly(ethylene glycol) are required. Surprisingly, it was found that the solidification and reactor-fouling problem can be eliminated if the reaction between carboxylic acid, phosphorus acid, and, for example, $POCl_3$ is carried out in a diluent that is an aromatic hydrocarbon (e.g.; toluene) or silicone fluid, especially poly(dimethylsiloxane) in the presence of a heterogeneous solid support such as fumed silica, silica xerogels, and diatomaceous silica [138].

Substituted carboxylic acids have been used for the preparation of 1-hydroxy-1,1-bisphosphonic acids by reaction with H_3PO_3 [139].

A new type of bisphosphonates **1** with various aromatic rings on the side chains linked by amide bonds is synthesized [140]. The driving force for these studies is the assumption that the introduction of a cyclic group on the side chain may increase the antiresorptive potency of the bisphosphonates.

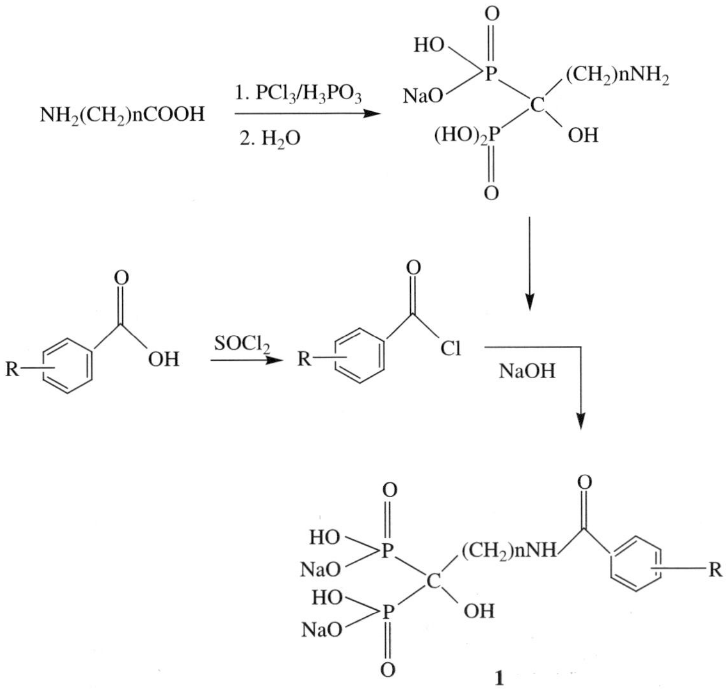

Bisphosphonates were prepared and from naturally occurring 1-amino acids [141]. In this case, initial protection of the amino acid–amine moiety was required. Insertion of a phosphorus atom was divided into two steps: initial reaction with a suitable phosphorus reagent to produce the acylphosphonate, and a second phosphorus attack to form the bisphosphonate. Since conversion to acid chloride served as the acid activation process, the amine-protecting group had to be either Fmoc or phtalimide, both of which are stable in acidic conditions. Prolonged reaction times always resulted in partial rearrangement and formation of by-products.

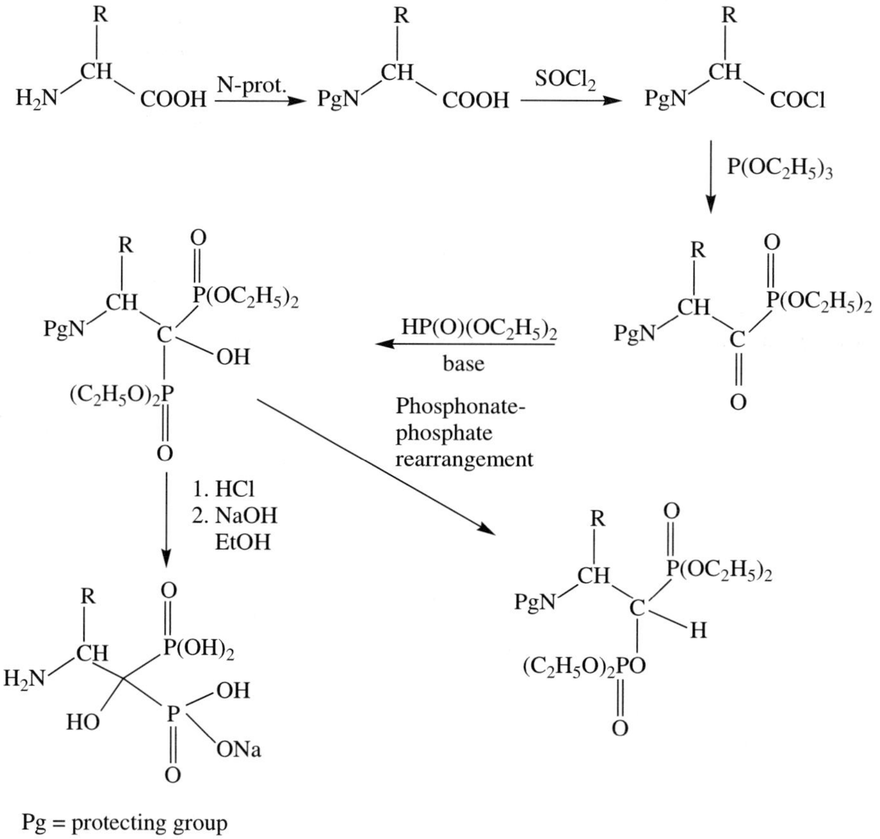

Pg = protecting group

2-Amino-1-hydroxyethylene-1,1-bisphosphonic acid **2** has been synthesized from N-phthaloylglycine [142]. Treatment of N-phthaloylglycine with thionyl chloride gave the corresponding acid chloride, which on reaction with trimethyl phosphite, furnished α-ketophosphonate **1**. Addition of dimethyl H-phosphonate to the carbonyl group of **1** gave **2**.

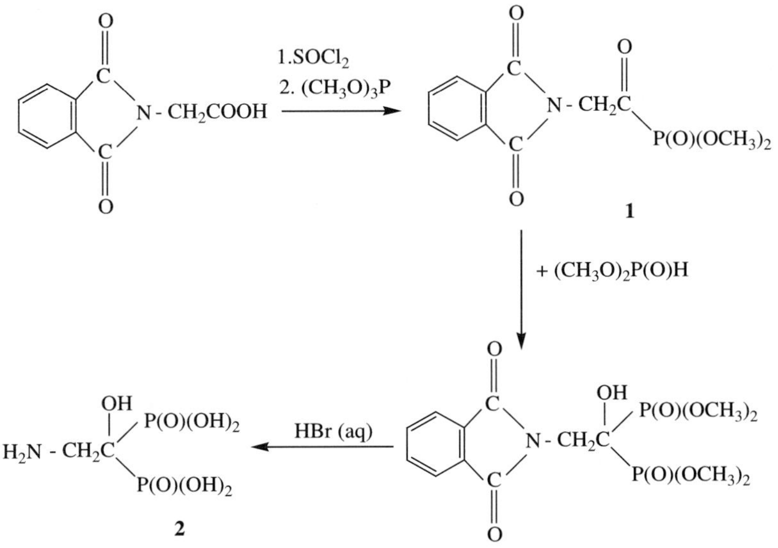

4.2 Bisphosphonates

Pyridinium-1-yl-hydroxy-bisphosphonates have been synthesized using the reaction pathway given below [143]. In the first step, substituted pyridines were obtained by reacting arylmetallic compounds with bromopyridines. The substituted pyridines were then alkylated by using bromoacetic acid, and the resulting pyridinium-1-yl acetic acids were converted into the corresponding bisphosphonates by using $H_3PO_3/POCl_3$ [144].

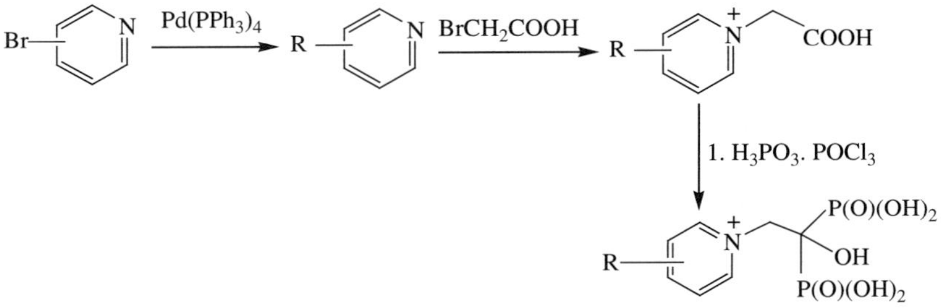

Conversion of the carboxylic group into the diphosphonic acid group takes place by a complex reaction mechanism. The phosphorus–halogen compound acts primarily as a dehydrating agent, although the hydrolysis products of the phosphorus–halogen component that accumulates may also participate in the substitution reaction.

A new class of bisphosphonates containing nitrooxy NO-donor functions has been synthesized [145].

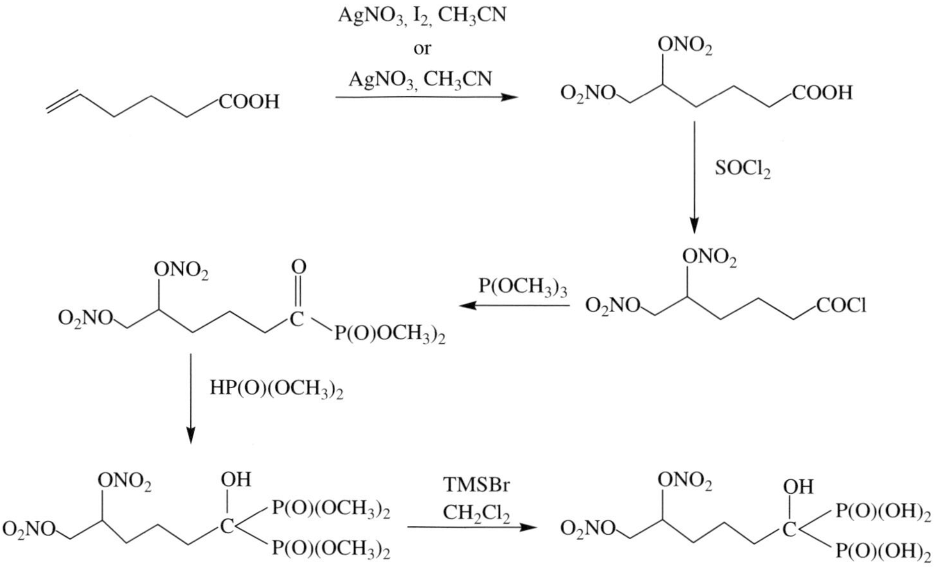

Nitric oxide (NO) is a key signalling molecule involved in the regulation of many physiological processes such as vascular relaxation, neurotransmission, platelet aggregation, and events of the immune system [146,147].

Beckmann rearrangement of oximes in the presence of phosphorus nucleophiles such as $(C_2H_5O)_3P$ or $HP(O)(OC_2H_5)_2$ resulted in the formation of aminomethylene *gem*-diphosphonates [148].

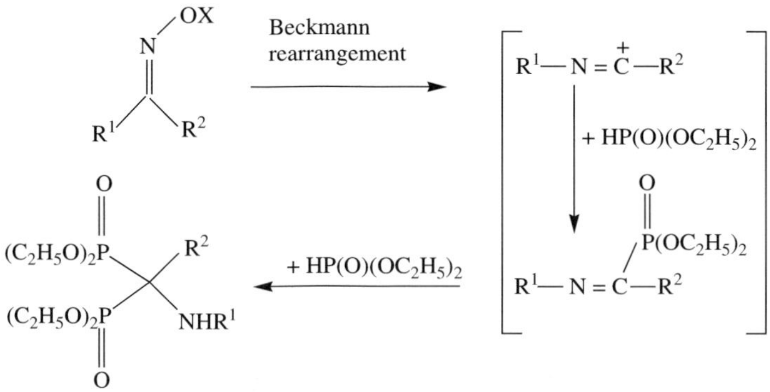

It was found that in the case of diphosphonylation reactions with cyclic aliphatic oximes, diethyl H-phosphonate is the better phosphorus reagent. All bisphosphonates thus obtained show diagnostic triplets ($J \approx 140$ Hz) at 57.8–63.6 ppm due to carbons bearing *gem*-diphosphonate groups in their $^{13}C\{H\}$ NMR spectra.

4.2 Bisphosphonates

The reaction of dialkyl H-phosphonates with activated alkenes or acetylenes, in the presence of a suitable base/catalyst or upon heating, leads to a variety of very useful phosphonates, often with one or more chiral centers [149]. For example, in the reaction of $(RO)_2P(O)H$ with $R^1C \equiv CR^2$, three types of products have been obtained, among which **2** has two chiral centers [150].

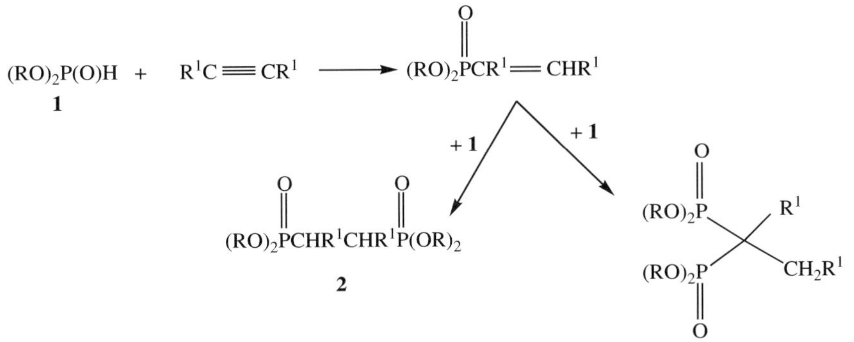

$$(RO)_2P(O)H + R^1C \equiv CR^1 \longrightarrow (RO)_2PCR^1 = CHR^1$$
1

$$(RO)_2PCHR^1CHR^1P(OR)_2$$
2

A series of functional vicinal bisphosphonates have been synthesized in good yield via new palladium-catalyzed bishydrophosphorylation reactions of terminal alkynes and dialkyl H-phosphonates [151]. The terminal alkynes were synthesized in excellent yields by the Heck reaction [152]. The resulting bisphosphonates have been characterized by ^1H, $^1H\{^{31}P\}$, $^{31}P\{H\}$, and $^{13}C\{H\}$ NMR spectroscopy. The characteristic signals for the protons on the alkane backbone appear as three multiplets at $\delta = 3.7, 2.9$ and 2.4 ppm. On decoupling the ^{31}P nuclei, these three proton signals collapse into a doublet,

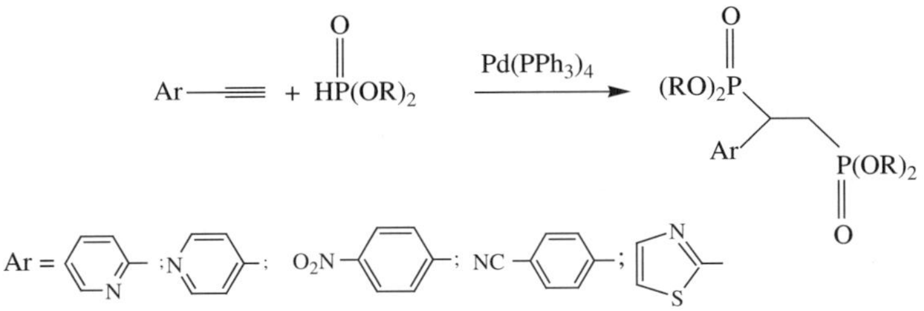

$$Ar-\equiv + HP(OR)_2 \xrightarrow{Pd(PPh_3)_4} (RO)_2P\text{-CH(Ar)-CH}_2\text{-}P(OR)_2$$

Ar = pyridyl, pyridyl, O_2N-C$_6H_4$, NC-C$_6H_4$, thiazolyl

doublet of doublet, and a doublet with a $^2J(H,H) = 15.5$ Hz and $^3J(H,H) = 12.2$ Hz. In the $^{31}P\{^1H\}$ NMR spectra of these vicinal bisphosphonates, two doublets appear at $\delta = 58$ ppm and $\delta = 55$ ppm with $^3J(P,P) = 77$ Hz. The above palladium-catalyzed bishydrophosphorylation reactions proceed only with electron-deficient terminal alkynes. According to the proposed mechanism, the active catalyst is the palladium hydride **1** obtained via oxidative addition of dialkyl H-phosphonate to a palladium [153].

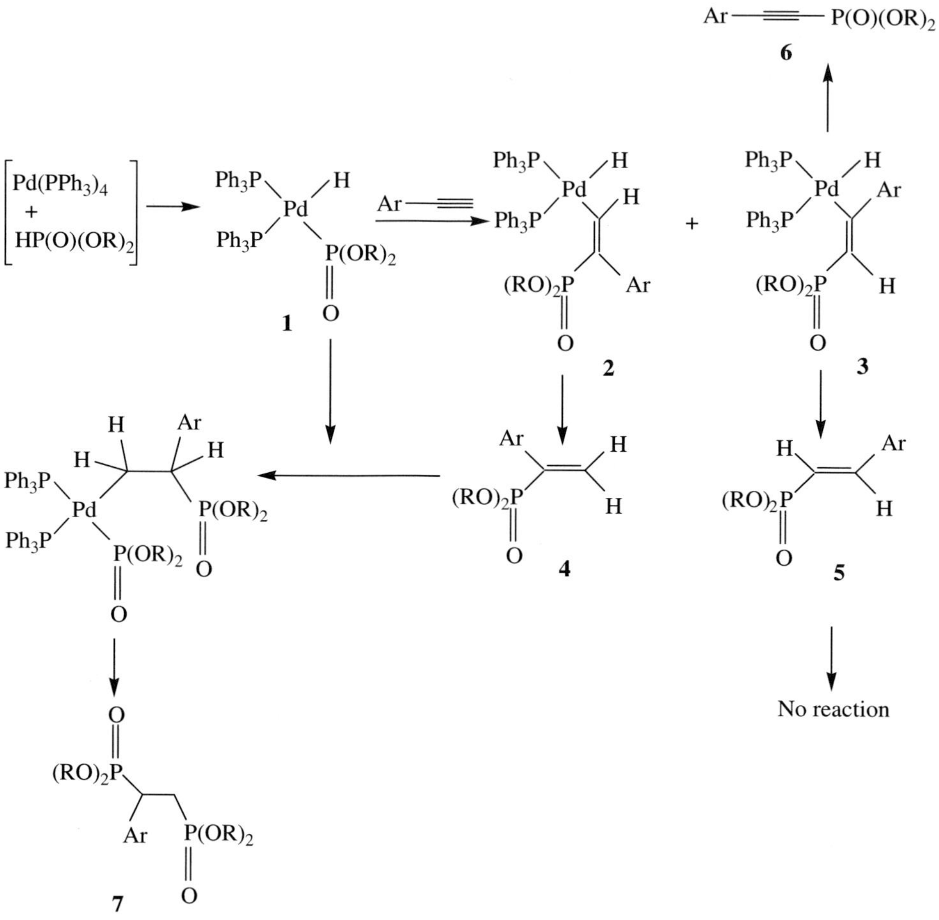

Insertion of alkynes into Pd–P bonds give alkenylpalladium intermediates **2** and **3**. Photonolysis of these intermediates by dialkyl H-phosphonate results in alkenylphosphonates **4** and **5** and regenerates the active catalyst **1**. Indirect evidence for the insertion of alkynes into Pd–P bonds comes from the isolation of trace amounts of alkynyl phosphonate **6**, which can only result from reductive elimination of the alkenylpalladium intermediate **3**. The second hydrophosphorylation reaction proceeds via the insertion of geminal alkenylphosphonate **4** into the Pd-H bonds, followed by reductive elimination of the product **7**. This reaction pathway can explain the failure of **5** to react with dialkyl H-phosphonate because of the steric hindrance of the potential insertion intermediate. These bisphosphonates are evaluating as ligands for radionuclides such as ^{99m}Tc and $^{186/188}$Re for imaging and therapeutic applications.

Vicinal bisphosphonates derived from cyclic H-phosphonates and dialkyl acetyledendicarboxylates are synthesized at room temperature [154].

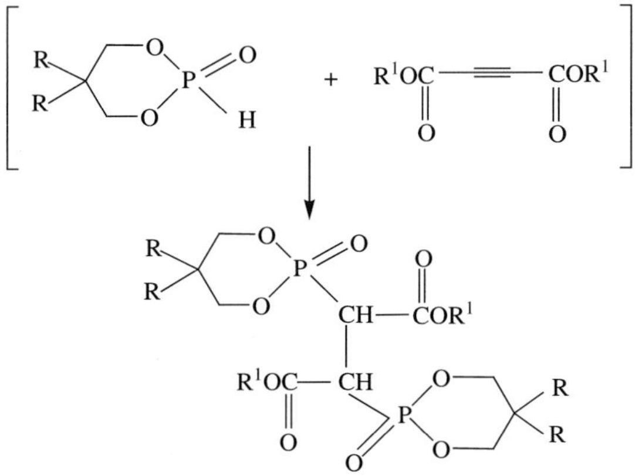

4.2 Bisphosphonates

Aminoalkylbisphosphonates are also formed by addition of dialkyl H-phosphonates to nitriles in the presence of hydrogen chloride [155].

$$2(RO)_2P(O)H + R^1-C\equiv N \longrightarrow \left[(RO)_2\overset{O}{\underset{\|}{P}}\right]_2 CR^1NH_2$$

Aminoalkylbisphosphonates **1** are synthesized reacting nitriles, H_3PO_3, and PBr_3 [156].

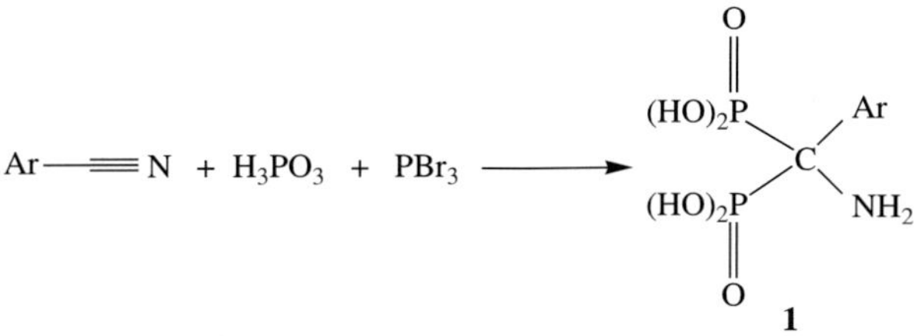

Bisphosphonates have been prepared starting either from the corresponding nitriles or from the respective amides [157].

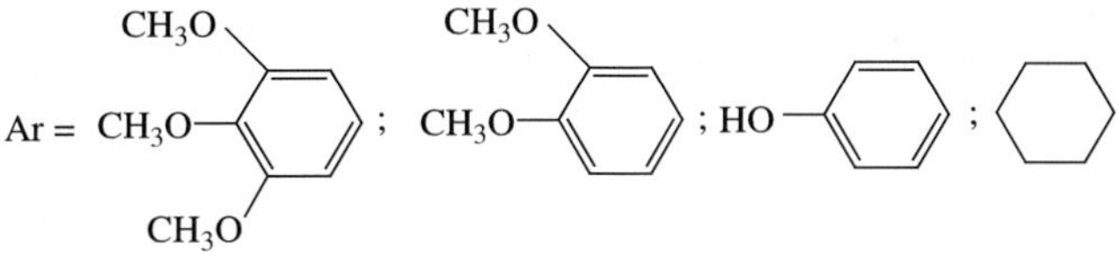

1-Aminoalkylbisphosphonates of the structure **1** can be converted into the corresponding hydroxyalkylbisphosphonates **2** in high yield by the reaction with nitrous acid [158].

When R = Ar, the reaction of **1** often gives the chloroalkylbisphosphonates **3** in concentrated hydrochloric acid.

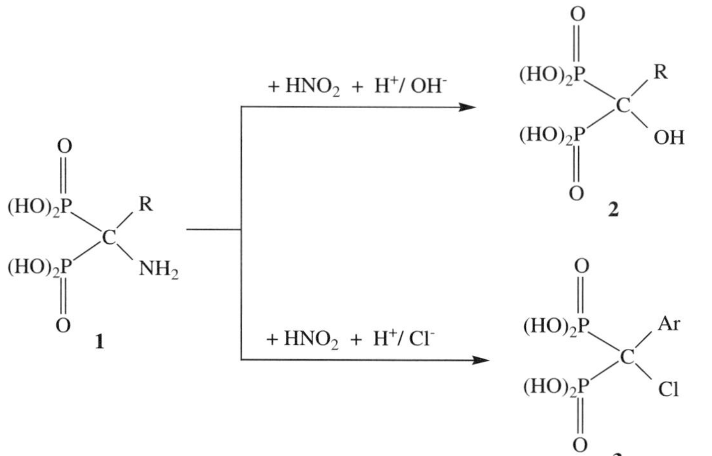

Treatment of several carbonyl compounds with excess strong base and diethyl phosphorochloridite and the oxidation of the reaction mixture with H_2O_2 results in the formation of bisphosphonates in a very good yield (see Appendix) [159]. The appearance of triplets for

$$(C_2H_5O)_2P(O)H + PCl_5 \text{ or } SOCl_2 \longrightarrow (C_2H_5O)_2PCl$$

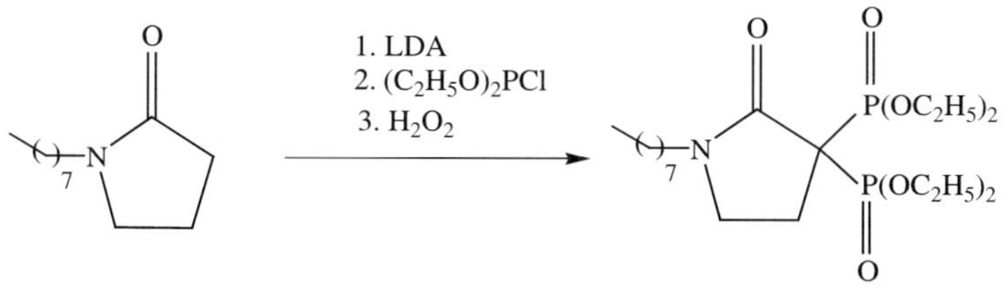

the carbonyl carbon (^2J(P,C) = 4.6 Hz) and for the α-carbon atom (^1J(P,C) = 135.8 Hz) in the $^{13}C\{^1H\}$ NMR spectrum supported the proposed structures with two phosphonate groups attached to the same carbon. The process gives moderate to good yields with *N*-alkyl lactams, but more varied results have been obtained with other carbonyl compounds.

(50 %) (18%)

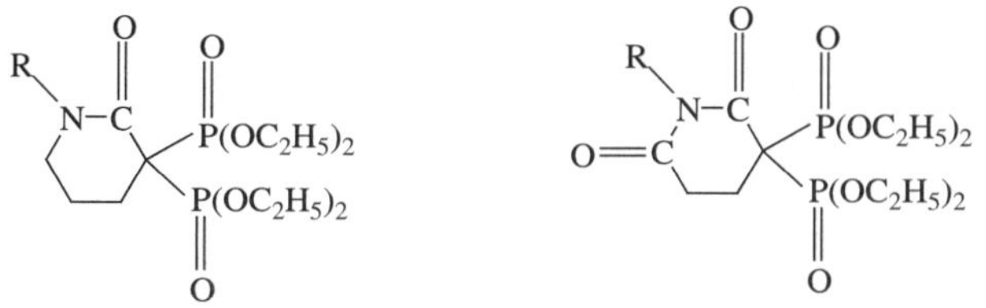

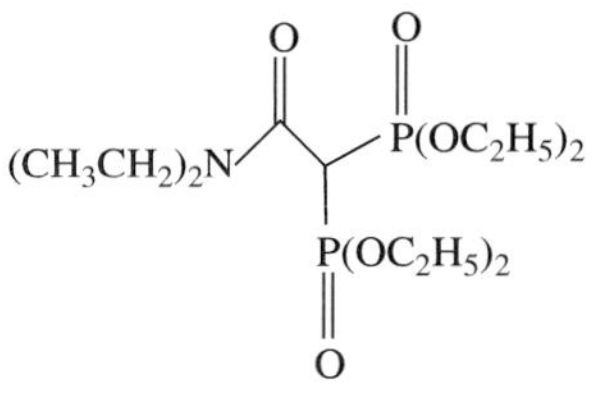

4.2 Bisphosphonates

(72%)

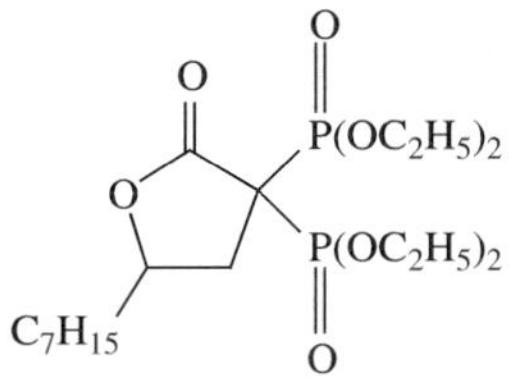

(67%)

Bisphosphonate conjugates, with potential as chemotherapy agents and potential carriers of cytotoxic radionuclides[160], are synthesized according to the following reaction scheme:

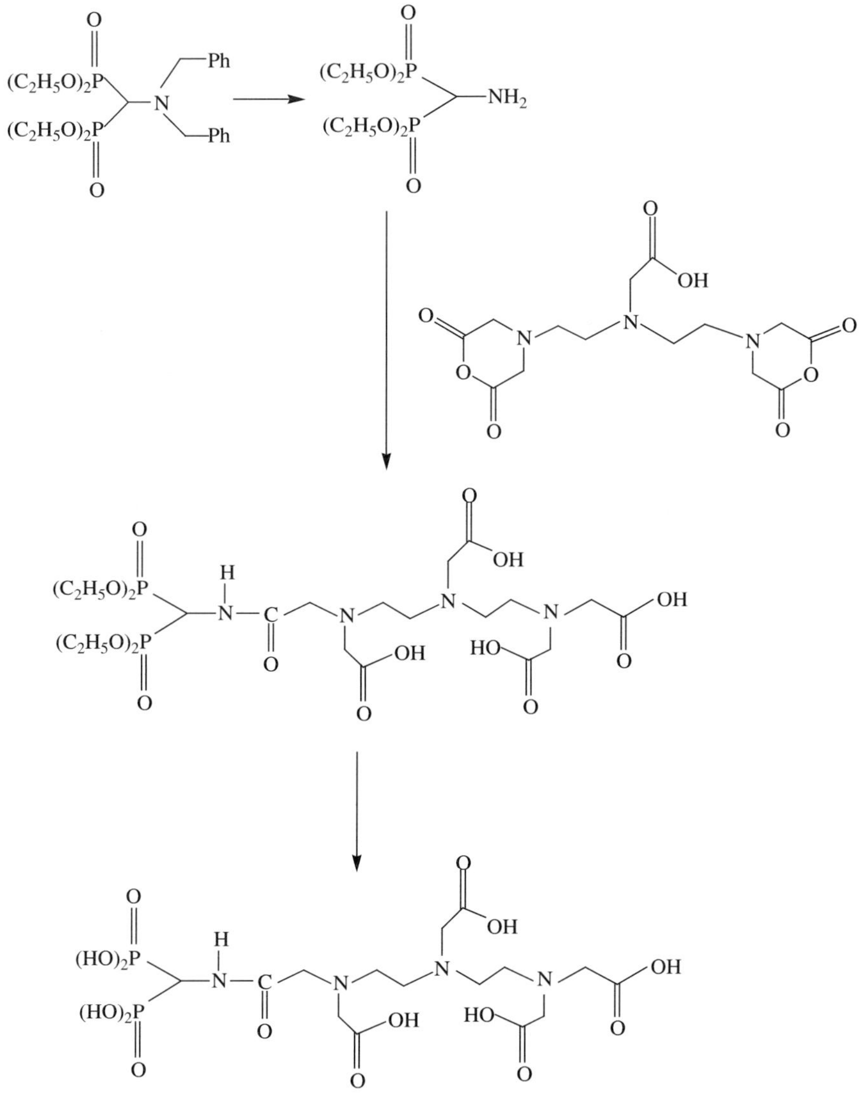

1

[(5-Fluoro-2,4-dioxo-3,4-dihydro-2H-pyrimidine-1-carbonyl)-amino]-methylenebisphosphonic acid **2** was prepared from aminobisphosphonate tetraethyl ester, triphosgene, and 5-fluorouracil [160].

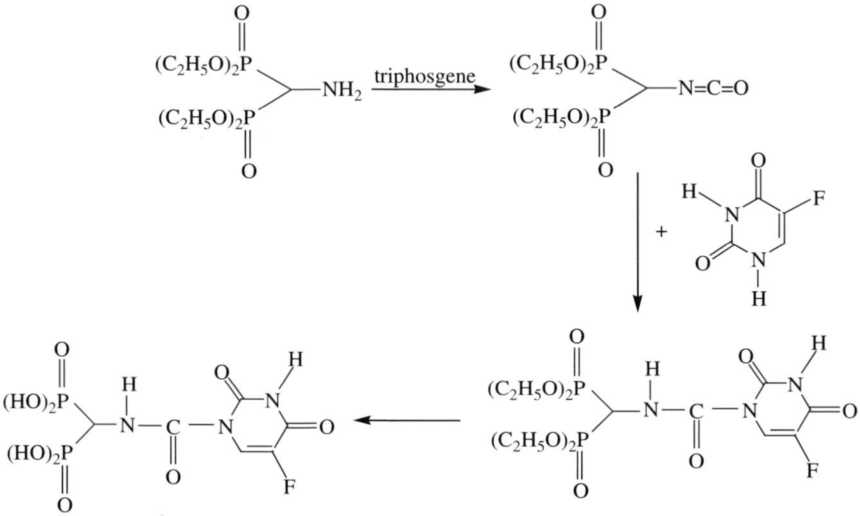

Compounds **1** and **2** are hypothesized to be able to deliver either high doses of radiation or a high concentration of chemotherapy agents at sites of increased osteoclastic activity in patients with bony metastases, while exhibiting minimal toxicity to normal tissues.

Bisphosphonates used currently in the treatment of osteopthia exhibit some quite serious drawbacks with respect to the degree of toxicity in animals and the tolerability, or the inducement of negative collateral side-effects, in humans. It has been found that bisphosphonic acid, obtained by reacting dibromo-difluoromethane with trialkylphosphite and sodium salt of dialkyl phosphonate [161] exhibits high activity that is not accompanied by side effects.

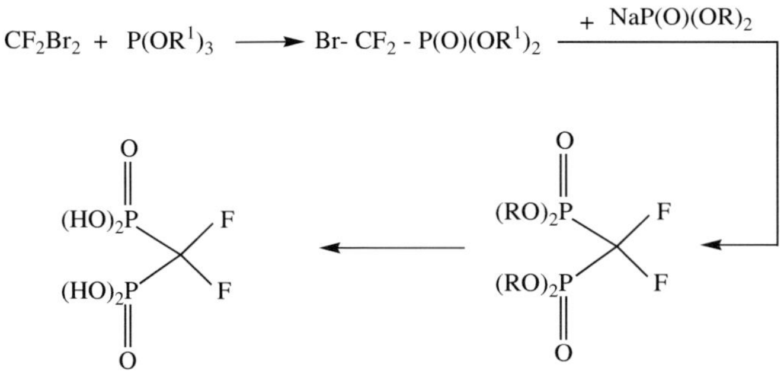

Methylenebis(phosphonic acid) derivatives are of great interest because of their potential biological activity, including antivirus activity, inhibiting osteoclastic bone resorption, and their application as ligands for ^{99}Tc adiopharmaceutical.

4.2 Bisphosphonates

1,1-Cyclopropanediylbis(phosphonic acid) has been synthesized by an intramolecular cyclization of the thallium salt of tetraisopropyl (3-iodo-1,1-propanediyl)bis(phosphonate) [162].

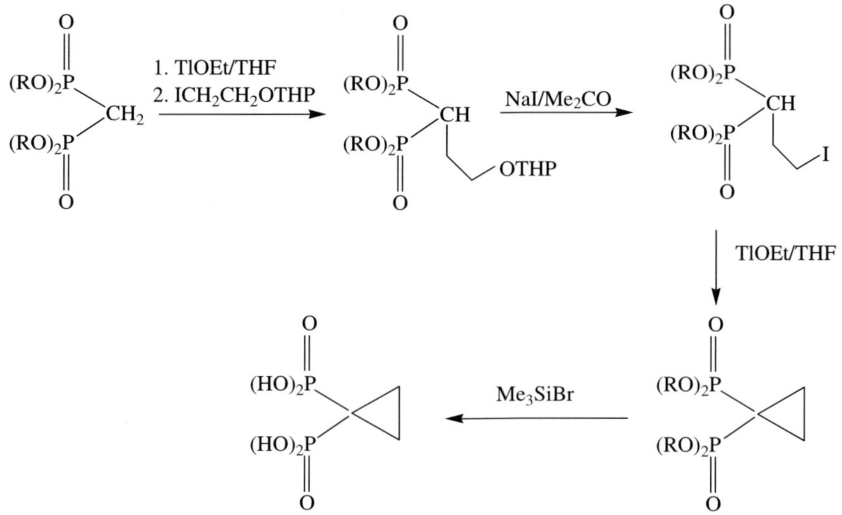

Reaction of bromomethylenebis(phosphonates) with electron-deficient alkenes as Michael acceptors in the presence of thallium ethoxide furnished 2-substituted 1,1-cyclopropaneiylbis(phosphonates) in good yield [163].

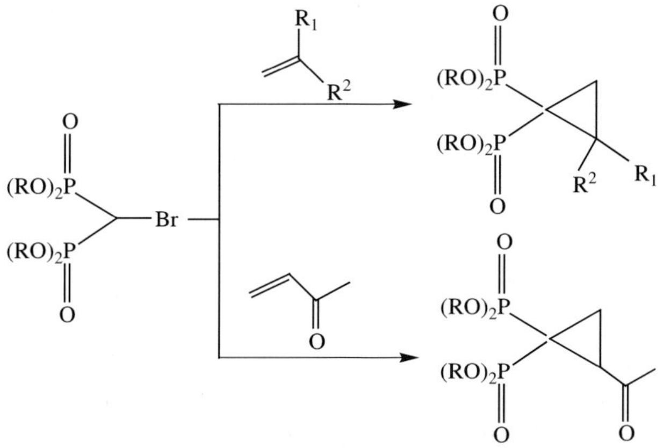

Tetraethyl 2-aminocyclopropylidene-1,1-bisphosphonate **1** was obtained via Michael-type addition of lithiocarbanion of ethyl 2-bromoacetate onto the tetraethyl vinylidene-1,1-bisphosphonate [164].

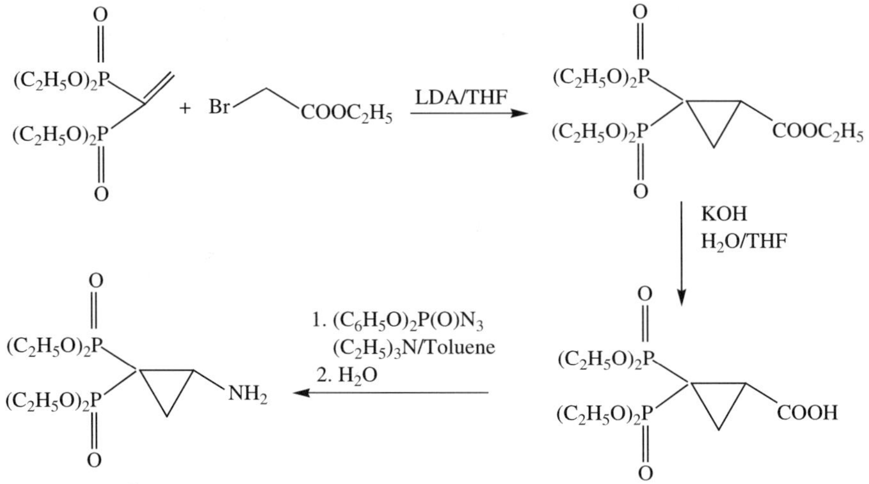

The tetraethyl (2-aminoethyl)cyclopropylidene-1,1-bisphosphonate **2** was prepared from 4-ene-1,1-bisphosphonate.

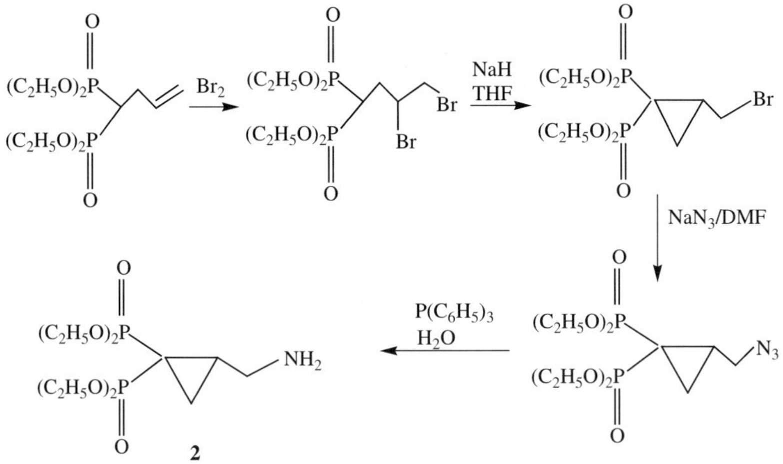

These compounds with three-membered ring are likely to present interesting biological properties by themselves or as new conjugates for bone delivery-targeting of drugs.

Ethenylidenebis(phosphonic acid) **1** and its esters have found utility as sequestering agents, in the development of polymeric flame retardants, and in certain pharmaceutical applications. Ethenylidenebis(phosphonic acid) is prepared via the thermal dehydration of tetrasodium (1-hydroxyethylidene)bis(phosphonate) [165]. Disadvantages of this time-consuming process include the need for precise control of temperature during the dehydration.

4.3 Nucleoside H-Phosphonates 159

$$(NaO)_2P(O)-C(OH)(CH_3)-P(O)(NaO)_2 \xrightarrow[-H_2O]{400\,°C,\ H^+} (HO)_2P(O)-C(=CH_2)-P(O)(HO)_2$$
1

Degenhardt *et al.* [166] have developed a new method for the preparation of tetraalkyl ethenylidenebis(phosphonates) **1**, which involves the base-catalyzed reaction of a methylenebis(phosphonate) ester with paraformadehyde, followed by acid-catalyzed elimination of methanol.

$$(RO)_2P(O)-CH_2-P(O)(RO)_2 \xrightarrow[-CH_3OH]{(CH_2O)_n} (RO)_2P(O)-C(=CH_2)-P(O)(RO)_2$$
1

R = C$_2$H$_5$; CH$_3$; *i*-C$_3$H$_7$

Tetraalkyl ethenylidenebis(phosphonates) can be polymerized by carrying out the polymerization by methods such as heating, and/or using as a catalyst ultra-violet light or a free-radical initiator. Copolymers have been prepared by polymerizing tetraalkyl ethenylidenebis(phosphonates) with an ethylenically unsaturated polymerizable monomer. [167].

4.3 NUCLEOSIDE H-PHOSPHONATES

Rapid development of molecular medicinal diagnostics in recent years has created a high demand for molecular probes enabling detection of specific gene sequences [168]. Since Zamecnik and Stephenson [169] first demonstrated virus replication inhibition by synthetic oligonucleotides, great interest has been generated in oligonucleotides as therapeutic and diagnostic agents. Due to some inherent problems connected with the handling of radioactive tracers, synthetic oligonucleotides equipped with various reporter groups detectable by fluorescent or enzymatic method are attracting interest [170]. Nonradioactive systems of labeling oligonucleotides are preferred due to their safety, ease of handling, and variety of modes of detection.

The backbone of a nucleic acid is a polymer of ribofuranoside rings linked by phosphate ester groups. Each ribose unit carries a heterocyclic base that provides part of the information needed to specify a particular amino acid in protein synthesis. Nucleic acids carry an organism's genetic information.

Nucleoside 3'-H-phosphonates **1** (ribonucleosides) and **2** (deoxyribonucleosides) seem to be the most attractive as starting materials in the chemical synthesis of DNA and RNA fragments. They entirely dominate the field of oligonucleotide synthesis. Nucleoside

H-phosphonates are useful as starting materials for the novel synthesis of the phosphoramidate and thiophosphate.

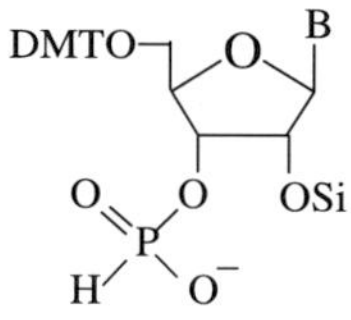

1 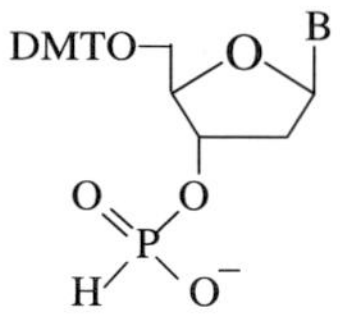 **2**

Nucleosides and nucleotides have demonstrated widespread utility as antiviral agents in the treatment of AIDS and AIDS-related complex and as anticancer therapeutics [171].

Nucleosides 5′-H-phosphonates **3** have been designed with the objectives of improving anti-HIV activity, enhancing blood–brain barrier penetration, modifying pharmacokinetic properties to increase plasma half-life, and improving drug delivery with respect to site-specific targeting or drug localization.

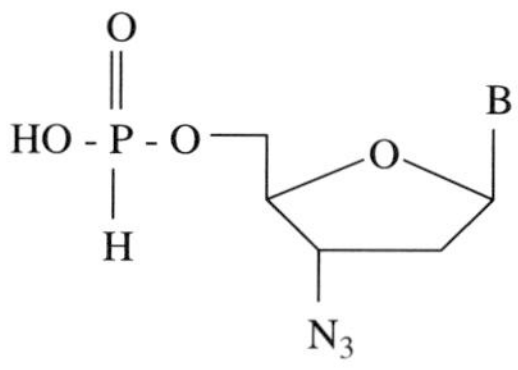

3

4.3.1 Methods of preparation

Nucleoside 3′-H-phosphonates

An important step in the design of oligonucleotide synthesis is the proper choice of the 2′-hydroxyl protecting group for ribonucleosides.

Ribonucleoside 3′-H-phosphonates **1** were synthesized from protecting ribonucleosides 2 in 75–90% yield using the PCl$_3$–imidazole procedure [172].

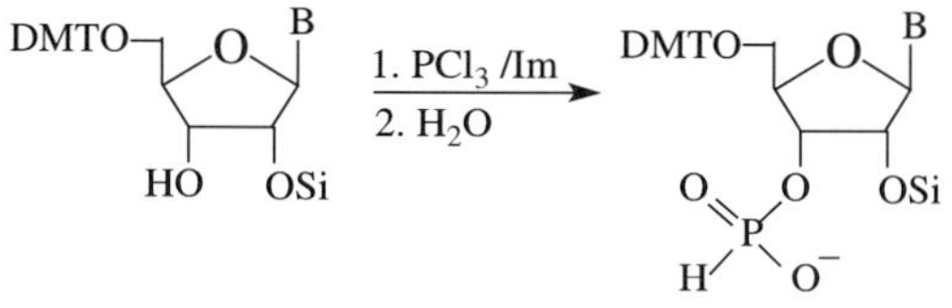

DMT – 4,4′-dimethoxytriphenylmethyl group; Si – t-butyldimethylsilyl group; Im – imidazole; B – uracil

Ribonucleoside 3′-H-phosphonates **1** are easy to prepare, purify, and handle.

When 5′-O-(p-monomethoxytrytil)thymidine was allowed to react with phosphorous acid in the presence of a coupling reagent in dry pyridine, 5′-O-(p-monomethoxytrytil)thymidine 3′-phosphonate (deoxyribonucleoside) was obtained in a high yield [173].

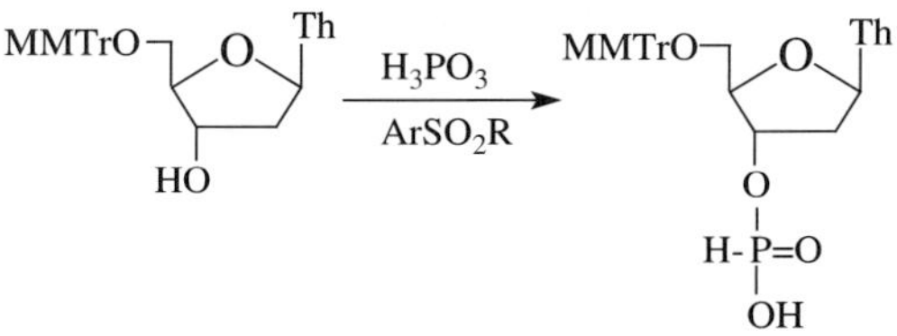

4.3 Nucleoside H-Phosphonates

[Scheme: 5′-O-MMTr-thymidine + H$_3$PO$_3$ / ArSO$_2$R → 3′-H-phosphonate]

Arylsulfonylamides were found to be the most suitable coupling reagents for this reaction. After the reaction, 5′-O-(*p*-monomethoxytrytil)thymidine 3′-phosphonate was extracted with chloroform and the chloroform layer was evaporated to dryness.

A new method is provided for the synthesis of nucleoside H-phosphonate monomers. The method comprises contacting a mononucleoside having a 3′-hydroxyl moiety with triphosgene and an excess of phosphorous acid [174].

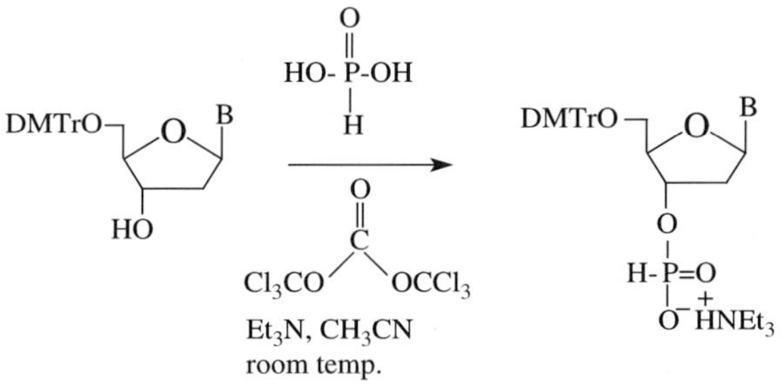

[Scheme: 5′-O-DMTr nucleoside + H$_3$PO$_3$ + triphosgene (Cl$_3$CO-CO-OCCl$_3$), Et$_3$N, CH$_3$CN, room temp. → 3′-H-phosphonate triethylammonium salt]

^{31}P NMR revealed that the phosphorous acid and triphosgene first produce a pyrophosphonate intermediate, which then reacts with the mononucleoside 3′-hydroxyl to furnish in a high yield the desired mononucleoside H-phosphonate product. The reaction is exothermic and proceeds at room temperature in solvent. 5′-Dimethoxytrityl -3′-thymidine H-phosphonate was characterized by ^{31}P NMR (δ = –0.27 ppm, ^1J(P,H) = 605 Hz). The ^{31}P NMR spectrum was obtained in an anhydrous Pyr/CH$_3$CN (1:1) solution.

5′-O-Tritylthymidine 3′-H-phosphonate was prepared in nearly quantitative yield without any phosphate derivatives when 5′-O-tritylthymidine was treated with tris(dimethylamino)phosphine in pyridine at room temperature for 24 h and then hydrolyzed by addition of water [175]. This method was quite useful but limited to the preparation of

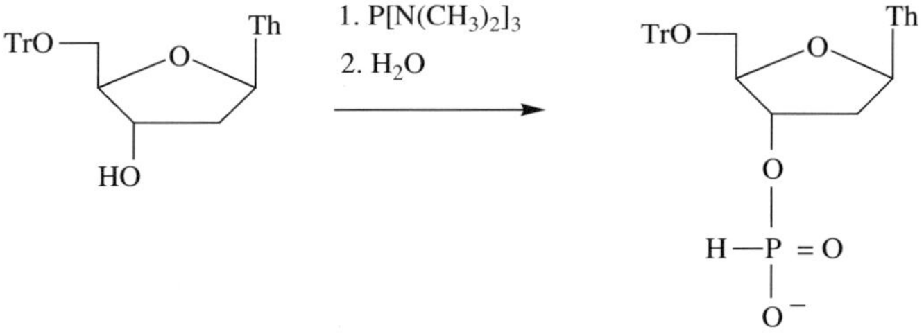

[Scheme: 5′-O-Tr-thymidine, 1. P[N(CH$_3$)$_2$]$_3$, 2. H$_2$O → 3′-H-phosphonate]

alkali-stable nucleoside phosphonates since one equimolar dialkylamine, which was accumulated as the reaction proceeded, caused acyl protecting groups to be taken off.

Tris(1,1,1,3,3,3-hexafluoro-2-propyl)phosphite **1**, easily prepared in 73% yield by treatment with PCl$_3$ and 1,1,1,3,3,3-hexafluoro-2-propanol [176], was used as a phosphonylating reagent for the preparation of nucleoside 3′-H-phosphonate units [177].

$$PCl_3 + (F_3C)_2CHOH \xrightarrow[THF]{Et_3N} [(F_3C)_2CHO]_3P$$
$$\mathbf{1}$$

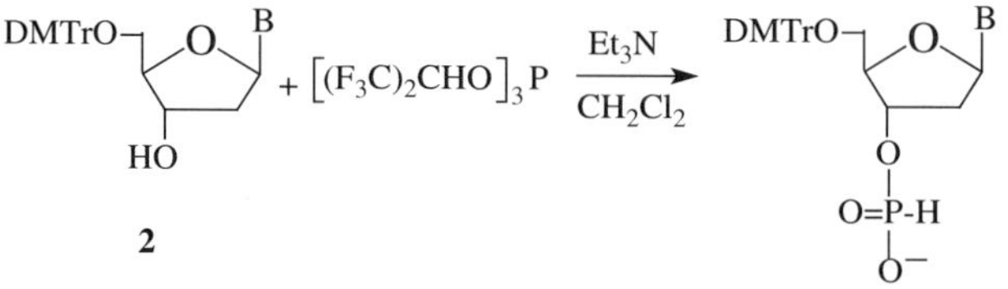

DMTrO—[sugar]—B, HO + $[(F_3C)_2CHO]_3P$ $\xrightarrow[CH_2Cl_2]{Et_3N}$ DMTrO—[sugar]—B, O—P(=O)(H)(O⁻)

2 **3**

The required deoxyribonucleoside 3′-H-phosphonate building blocks **3** were readily prepared by allowing the corresponding *N*-acyl-5′-O-(dimethoxytrityl)-deoxyriboucleosides **2** to react with 1,1 molar equivalent of **1** in the presence of a catalytic amount of tryethylamine in CH_2Cl_2 at room temperature for 5 min. Yields of H-phosphonate units **3** ranged from 87–95%. The ^{31}P NMR spectra data ($CDCl_3$/85% H_3PO_4) were as follows: B = T, 2.81 ppm; B = bzA, 2.71 ppm; B = ibuG, 2.89 ppm; B = ibuC, 2.71 ppm. The results obtained indicate that the use of **1** as a phosphonylating reagent considerably shortens the reaction time for the preparation of protected deoxyribonucleoside 3′-H-phosphonates. It was shown that tris(1,1,1,3,3,3-hexafluoro-2-propyl)phosphite **1** is more effective for the preparation of H-phosphonate units than the bis(1,1,1,3,3,3-hexafluoro-2-propyl)phosphonate [178].

Nucleoside 3′-H-phosphonates were prepared by reacting 5′-dimethoxytrityl thymidine with PCl_3/1,4,2-triazole [179].

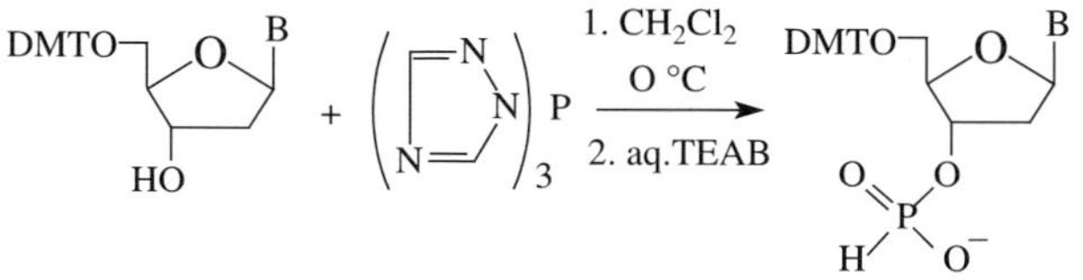

DMTO—[sugar]—B, HO + (triazole)$_3$P $\xrightarrow[\text{2. aq.TEAB}]{\text{1. }CH_2Cl_2,\ 0\ ^\circ C}$ DMTO—[sugar]—B, O—P(=O)(H)(O⁻)

Nucleoside 3′-H-phosphonates can be prepared by reacting deoxycytidine with PCl_3/imidazole mixture [180].

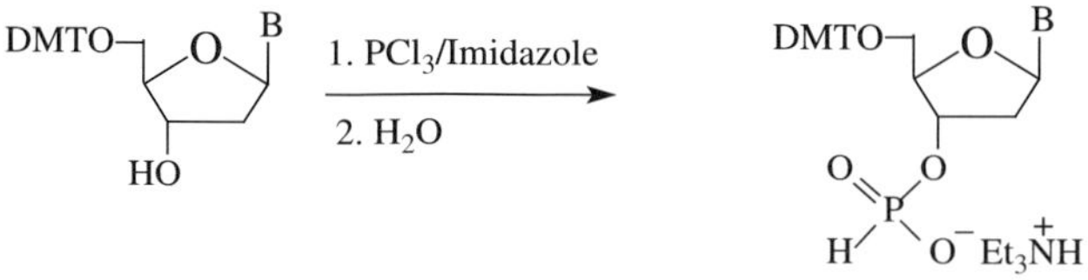

DMTO—[sugar]—B, HO $\xrightarrow[\text{2. }H_2O]{\text{1. }PCl_3/\text{Imidazole}}$ DMTO—[sugar]—B, O—P(=O)(H)(O⁻ $Et_3\overset{+}{N}H$)

B = cytosine

Aryl H-phosphonates diesters are reactive enough to undergo fast and quantitative reaction with various nucleophiles at room temperature. It was found that diphenyl H-phosphonate

4.3 Nucleoside H-Phosphonates

underwent rapid transesterification reaction with suitably protected nucleosides affording nucleoside aryl and dinucleoside H-phosphonates [181].

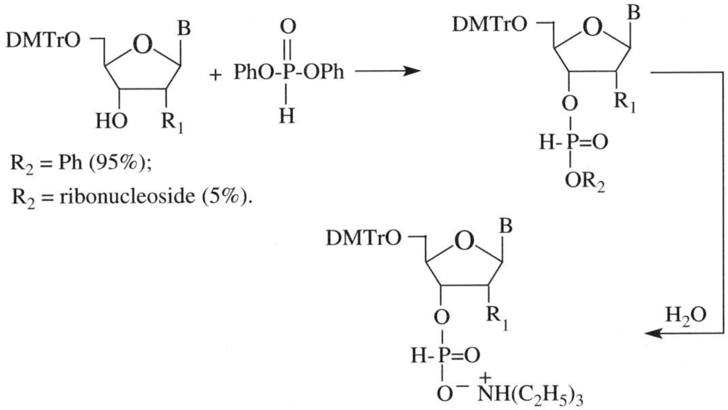

R_2 = Ph (95%);
R_2 = ribonucleoside (5%).

Formation of the undesired symmetrical product (R_2 = ribonucleoside) can be completely eliminated by using three molar excess of diphenyl phosphonate and the reaction went to completion in 15 min under these conditions. Lower reactivity of the ribonucleoside phenyl H-phosphonates (R_2 = phenyl) has been manifested in their slower hydrolysis to the nucleoside H-phosphonates upon addition of water. The hydrolysis was found to be substantially faster in the presence of a base.

Diphenyl H-phosphonate represents an inexpensive, commercially available reagent suitable for the convenient and efficient conversion of partially protected deoxyribo- and ribonucleotides into the corresponding 3′-phosphonate monoesters. The reagent is stable, easy to handle, and affords H-phosphonate monoesters of purity usually better than 95% even without column chromatography. Considering that phosphonylation with diphenyl phosphonate occurred effectively in rather mild conditions, it could be expected that this procedure will find applications outside the nucleotide field, for example, peptides, carbohydrates, etc.

The nucleoside 3′-H-phosphonate monoester, when reacted in pyridine with phenol in the presence of pivaloyl chloride, produced the desired nucleoside phenyl H-phosphonate [182].

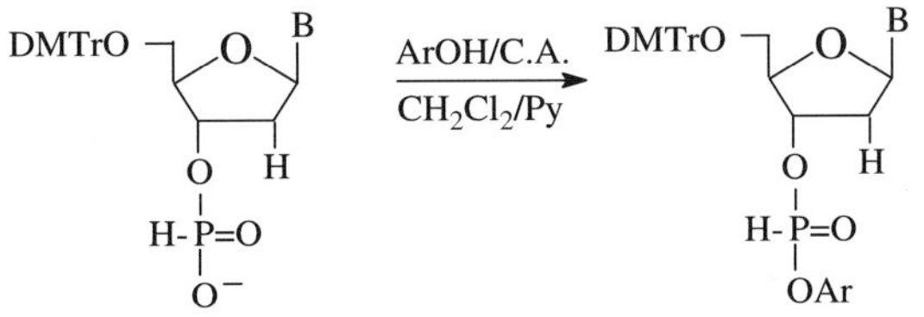

Ar = phenyl; Ar = 4-chlorophenyl ;Ar = 2,4-dichlorophenyl ;Ar = 4-nitrophenyl
Ar = 2,4,6-trichlorophenyl; C.A. – coupling agent: diphenyl phosphorochloridate

The main advantages of aryl H-phosphonates as synthetic intermediates stem from the fact that these compounds possess only one electrophilic center located on the phosphorus atom, and their reactivity can be modulated by changing electronic and/or steric properties of substituents on the aromatic ring of an aryl moiety. It should also be kept in mind that electronic properties of the aryloxy groups may affect the phosphonate–phosphite equilibria in these compounds, and consequently, the phosphorus center in aryl H-phosphonates can change its character from electrophilic (in the H-phosphonate form) to nuclephilic (in the phosphite form), with all the chemical implications of this fact.

Kraszewski et al. [183] have found that the most important factors in the synthesis of nucleoside aryl H-phosphonates from nucleoside H-phosphonates and the corresponding phenols are (i) acidity (pK_a) of the phenols, (ii) the nature of the coupling agent used for the condensation, and (iii) basicity of the reaction medium. These factors affect the yields of the condensation to produce nucleoside aryl H-phosphonates. In pyridine, in the presence of different condensing agents, these affect the extent of various side-reactions (subsequent reactions of the desired aryl H-phosphonate with condensing agents, or disproportionation of the produced aryl H-phosphonate). These problems can be alleviated when synthesis of aryl H-phosphonates is carried out in methylene chloride contaning a limited amount of pyridine in the presence of diphenyl phosphorochloridate as a condensing agent. Under these conditions, coupling of nucleoside H-phosphonate with phenols is clean, relatively fast, and the produced aryl H-phosphonates do not undergo any detectable changes.

Nucleoside 5'-H-phosphonates

3'-Azido-2',3'-dideoxythymidine-5'-H-phosphonate was obtained by reacting 3'-azido-2',3'-deoxythymidine with PCl_3 and subsequent hydrolysis (see Appendix) [184–186].

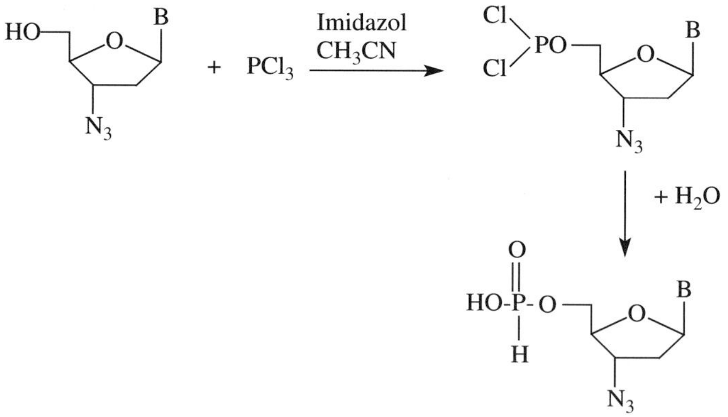

5',5'-O-Dinucleoside H-phosphonate **1** has been synthesized by Meier by reacting 2',3'-dideoxythymidine with diisopropyldichlorophosphine in the presence of diisopropylethylamine, and subsequent hydrolysis [187].

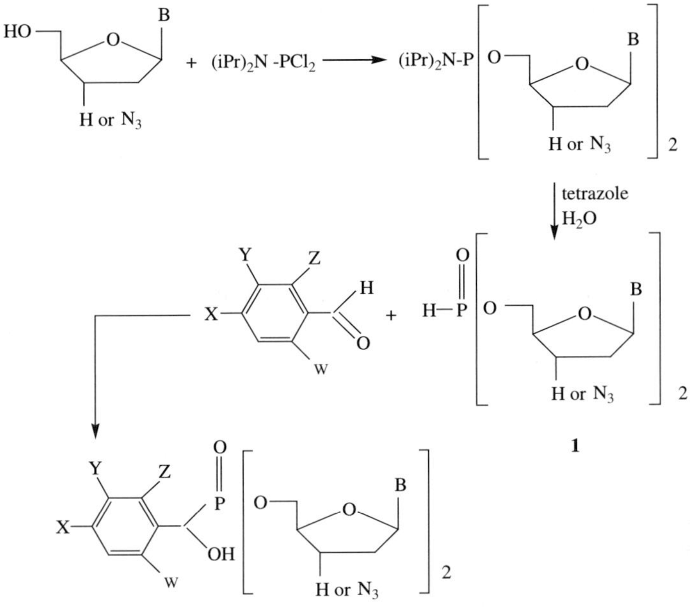

2

5′,5′-O-Dinucleoside-α-hydroxybenzylphosphonates **2** in up to 90% yield were synthesized by reacting the symmetric dinucleoside H-phosphonates **1** with benzaldehyde derivative in the presence of catalytic amounts of a tertiary base like diisopropylethylamine (DIPEA), or triethylamine (TEA). Dinucleoside-α-hydroxybenzylphosphonates **2** are considered uncharged prodrugs of 5′-nucleoside H-phosphonates **3** and 5′-nucleoside monophosphates **4**. These two compounds are released by controlled hydrolysis via two different pathways: the phosphonate–phosphate rearrangement and the direct cleavage mechanism [188].

a) the base-catalysed phosphonate-phosphate rearrangement

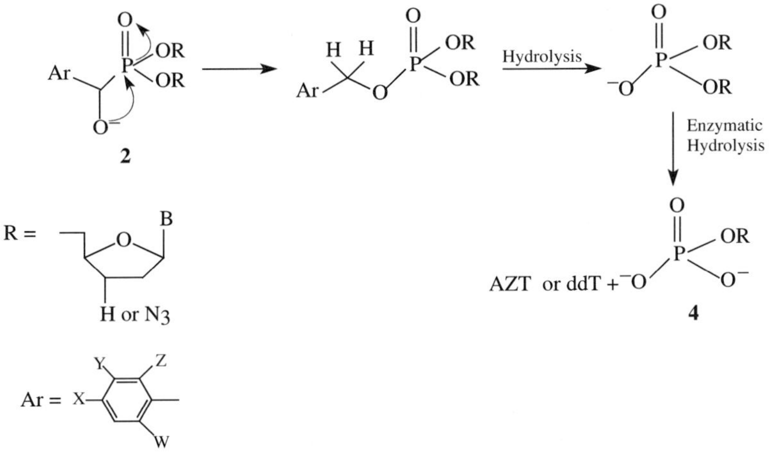

b) the base-induced direct-cleavage mechanism

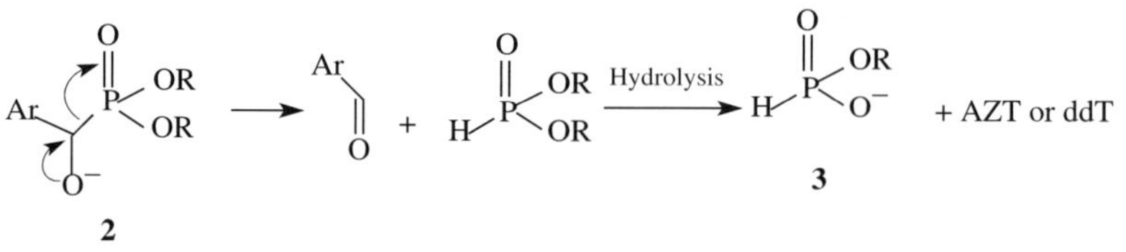

4.3.2 Reactivity of nucleoside H-phosphonates

Reaction with O-nucleophiles (transesterification reaction)

Todd *et al.* used nucleoside 3'-H phosphonates to prepare 3'-5'-internucleotidic bonds for the first time [189]. Diphenyl chlorophosphate was used to activate nucleoside H-phosphonate.

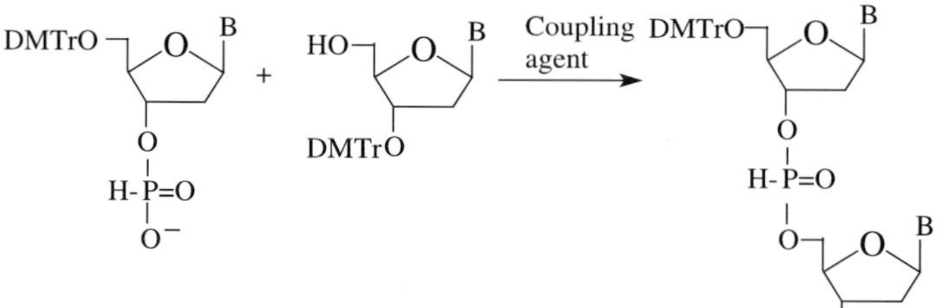

Coupling agent = $(C_6H_5O)_2P(O)Cl$

It was shown that pivaloyl chloride led to dinucleoside H-phosphonates in good yield [190]. Two diastereomers are formed (δ = 8.44 and 7.23 ppm, $J(P,H)$ = 716 Hz).

Dinucleoside H-phosphonates were obtained when deoxyribonucleoside H-phosphonate reacted with 3'-benzoylthymidine in the presence of 1,3-dimethyl-2-chloroimidazolinium chloride-coupling reagent [177]. The $^{31}P\{H\}$ NMR spectrum of the reaction mixture showed that the signal of the nucleoside H-phosphonate at δ = 2.81 ppm completely disappeared and new signals were observed at δ = 8.21 and 7.20 ppm. The chemical shift suggested that nucleoside H-phosphonate was converted into the corresponding dinucleoside H-phosphonate.

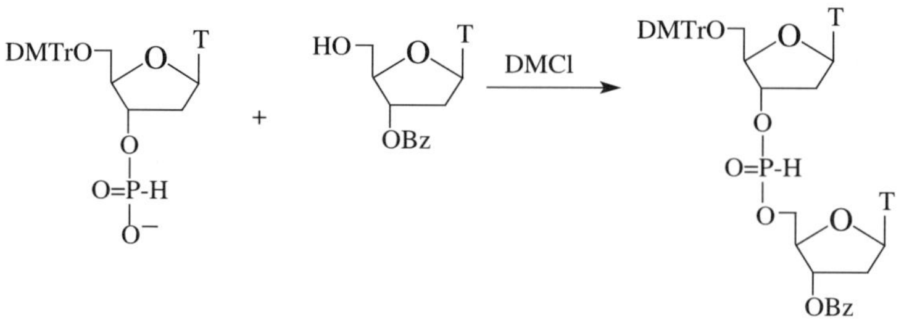

4.3 Nucleoside H-Phosphonates

The transesterification of aryl nucleoside 3'-H-phosphonate with alcohols or with suitably protected nucleosides under mild conditions furnished the corresponding ester or dinucleoside (3'-5') H-phosphonate diesters [182].

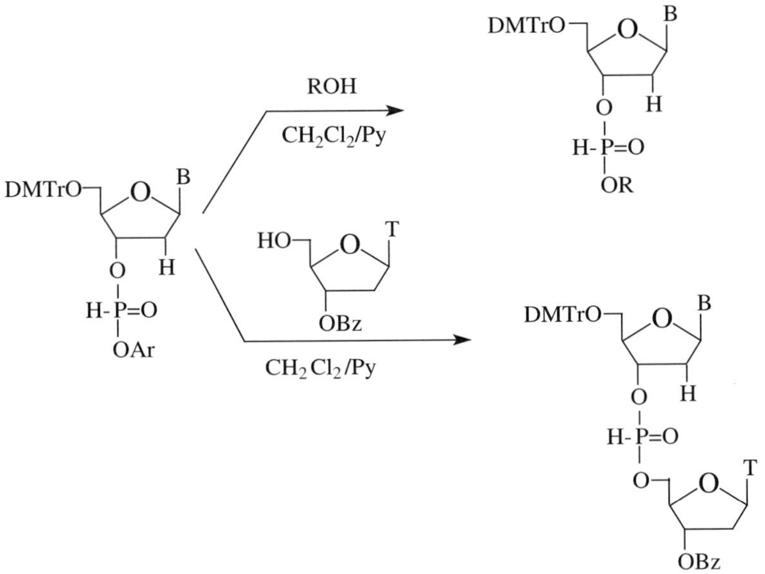

This route provides a new means for the formation of 3'-5'-H-phosphonate internucleotidic linkages. The advantages of this method are (i) the reaction is rapid and efficient, (ii) it can be carried out under mild reaction conditions, and (iii) the starting material, aryl nucleoside H-phosphonates, can be conveniently generated *in situ* from easily accessible corresponding nucleoside H-phosphonate monoesters. The method seems to be a rather general one and thus applicable to the preparation of other phosphorus-containing natural products and their analogues. The most reactive among investigated aryl nucleoside H-phosphonates were found to be those bearing *p*-nitrophenyl and 2,4,6-trichlorophenyl groups. The relative order of reactivity in this reaction, a:b:c:d:e:f:g was found to be [191] 1:4:10:40:350:1100:1100, which parallels that observed for transesterification of aryl nucleoside H-phosphonate with simple alcohols [192]. 4-Nitrophenyl H-phosphonate is at least 30 times more reactive than 4-chlorophenyl H-phosphonate [191].

Reactions of nucleoside H-phosphonates with various diols using different types of condensing agents have been studied [193] because functionalization of oilgonucleotides via attachment of various functionalities enables linking oligonucleotides to other classes of biopolymers or low molecular compounds (e.g., reporter, groups, haptens). Depending on the coupling procedure and the length of a polymethylene chain of the diol, acyclic H-phosphonate diesters or cyclic phosphite triesters were formed.

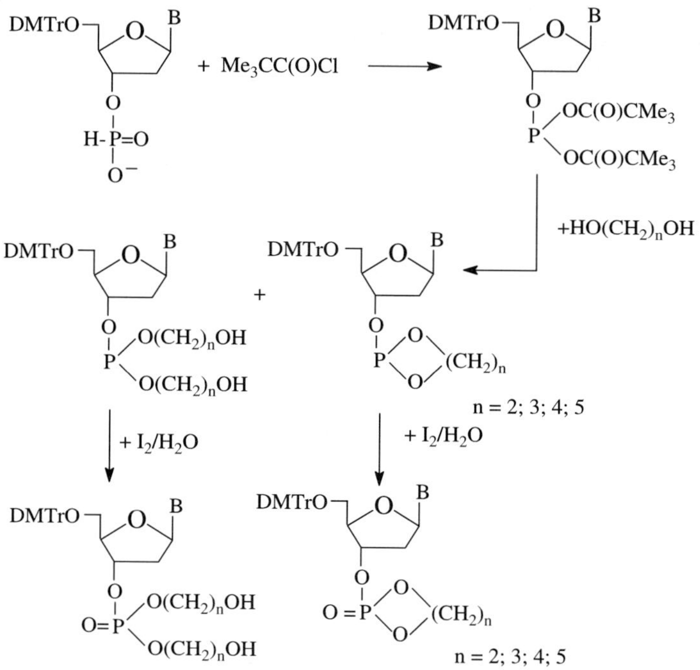

[Scheme showing reactions of DMTrO-nucleoside H-phosphonate with Me₃CC(O)Cl, diols HO(CH₂)ₙOH, and I₂/H₂O, giving acyclic and cyclic phosphite/phosphate products with n = 2; 3; 4; 5]

The exclusive formation of the cyclic species in these reactions indicated a strong tendency to cyclization of the initially formed hydroxyalkyl phosphites. When $n = 4$ and 5, the cyclic phosphites ($n = 4$, $\delta = 132.1$ ppm) and ($n = 5$, $\alpha = 132.0$ ppm) were the major products (>95%) at molar ratio 1:1. Cyclic phosphites were oxidized to the corresponding cyclic phosphates. The progress of the reaction was followed by ^{31}P NMR spectroscopy.

Transesterification of nucleoside H-phosphonates with various diols in pyridine in the presence of 2-chloro-5,5-dimethyl-1,3,2-dioxaphosphorinane (a mild condensing agent) furnished rapid and clean formation of the corresponding hydroxyalkyl nucleoside H-phosphonate [193]. With ethylene glycol (n = 2), the desired hydroxyethyl H-phosphonate was formed as a minor product, only 10% ($\delta p = 8.6$ ppm), while the main phosphorus-containing species was a compound resonating at $\delta p = 23.1$ ppm. The presence

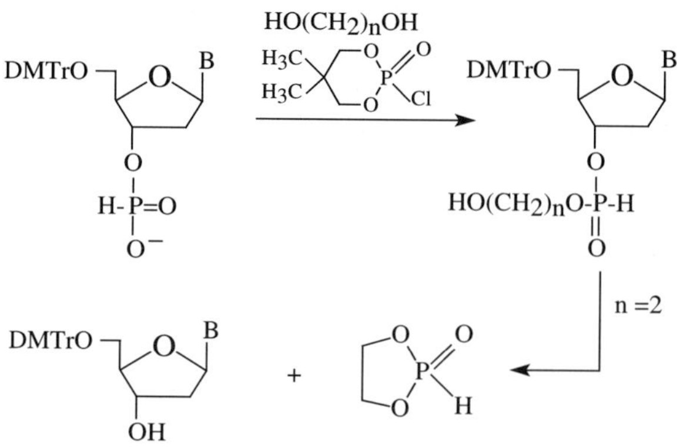

[Scheme showing reaction of DMTrO-nucleoside H-phosphonate with HO(CH₂)ₙOH and 2-chloro-5,5-dimethyl-1,3,2-dioxaphosphorinane, giving HO(CH₂)ₙO-P(H)(=O)-O-nucleoside, and for n = 2, the DMTrO-nucleoside with free OH plus cyclic ethylene H-phosphonate]

4.3 Nucleoside H-Phosphonates

of 5'-O-4,4'-dimethoxytritylthymidine (90%) suggests that the compound resonating at 23.1 ppm is 1-hydro-2-oxo-1,2,3-dioxaphospholane. The strong tendency of 2-hydroxyethyl H-phosphonate diesters to cyclization is explicable in terms of hydrogen bonding of the β-hydroxyl group to the 3'-O atom.

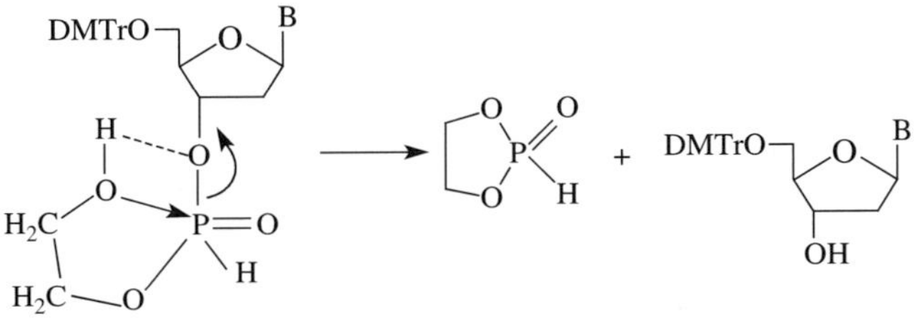

Such hydrogen bonding makes the phosphorus atom a stronger electrophilic center and favors the formation of cyclic H-phosphonate.

Dinucleoside H-phosphonates were obtained by reacting, in the presence of triphosgene, H-phosphonate with a second mononucleoside having a free 5'-hydroxyl group [174].

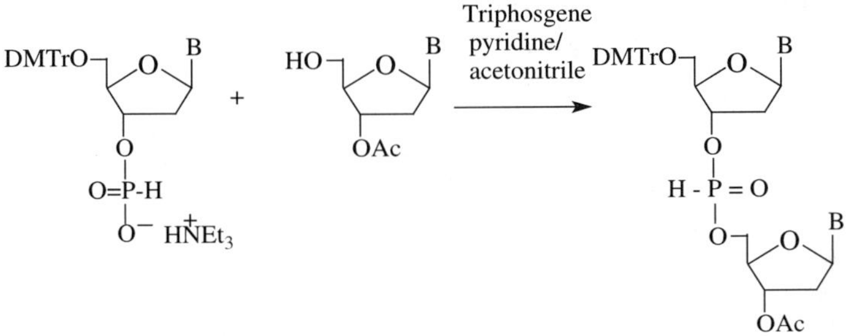

Reaction with N-Nucleophiles

Introduction of modifications at the phosphorus center of natural products may be valuable in elucidating the mechanism of enzymatic reactions [194], for developing new enzyme inhibitors of potential medical value [195], or for the design of natural product analogues [196] that are resistant to enzymatic degradation while keeping some other biological properties. Nucleoside *N*-alkyl-H-phosphonamidates are interesting from this point of view.

Various approaches for the preparation of nucleoside H-phosphonamidates are known, but the most efficient and versatile route to nucleoside *N*-alkyl H-phosphonamidates was found to be aminolysis of the *in situ*–produced aryl nucleoside H-phosphonates with appropriate amines.

Aryl nucleoside H-phosphonates **1** react with primary amines **2** or aminonucleoside **3** to give the desired nucleoside phosphonamidates **4** or **5** in >90% yield [182b,182c, 191,197].

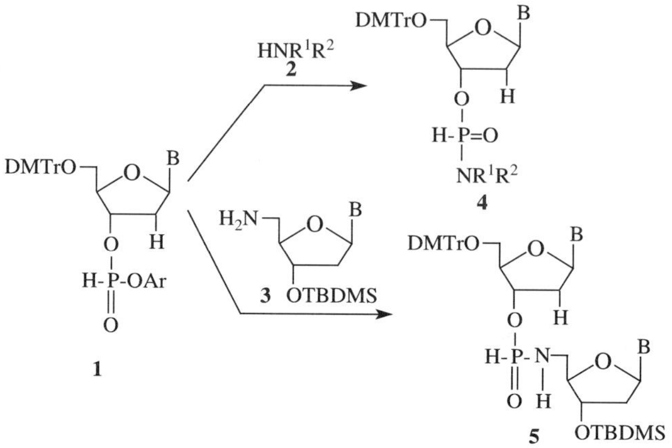

The most important factors affecting the formation of nucleoside H-phosphonamidates of type **4** in reactions promoted by condensing agents are (i) reactivity of amines towards coupling agents, (ii) chemoselectevity of amines towards the reactive species generated during the activation process, and (iii) steric hindrance in the amines. The advantages of using aryl nucleoside H-phosphonates as synthetic intermediates are (i) there is only one electrophilic center in these compounds, which eliminates problems of chemoselectivity in reactions with nucleophiles, and (ii) if necessary, the electrophilicity of the phosphorus atom can be controlled by changing substituents on the aryl ring. To suppress the hydrolysis of aryl H-phosphonate **1** by moisture, the aminolysis with **3** was carried out in the presence of trimethylsilyl chloride. This synthetic pathway can be considered as a general method for the preparation of natural product analogues with the P–N bond in a bridging position of the phosphoramidate linkage.

Reactions with sulfide and hydrogen sulfide

The pharmacological value of oligonucleotides with natural phosphodiester-internucleotide linkages is seriously hampered by their rapid degradation in the presence of cell and serum nucleases [195]. As a result of this, most investigations have been focused on finding chemical modifications that would confer stability to nucleases while preserving the specificity and efficiency of complexation of the modified oligonucleotide with target DNA or RNA sequences. Among nucleic acid analogues with modifications at the phosphorus center, oligonucleoside phosphorodithioates possess most favorable biological and pharmacological properties as potential antisense or antigene agents.

Nucleoside H-phosphonothioates **1** and H-phosphonodithioates **2** represent an interesting class of compounds in the growing field of nucleotide analogues.

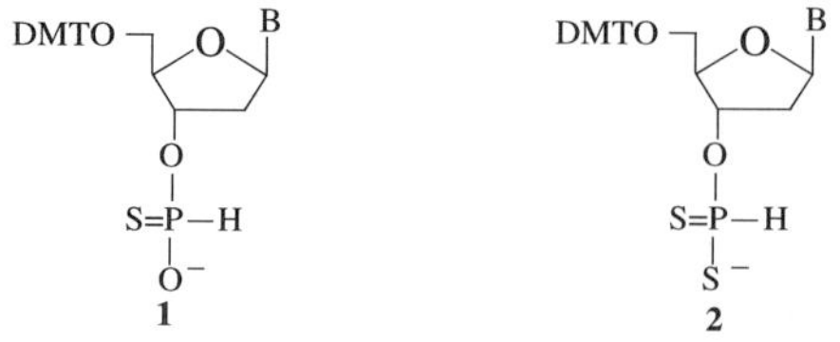

4.3 Nucleoside H-Phosphonates

The potential usefulness of nucleoside H-phosphonothioates stems from the fact that they can be considered as convenient synthons for the introduction of an H-phosphonothioate group into oligonucleotides. This group can in turn be converted into a phosphoromonothioate or a phosphorodithioate function. Phosphorodothioates are known to confer favorable biological properties onto oligonucleotide analogues (e.g.; resistance to nucleases) without significantly impairing the ability of such analogues to form stable complexes with the target DNA or RNA sequences [198, 199].

In contradistinction to nucleoside H-phosphonate monoesters, the corresponding H-phosphonothioates are chiral at the phosphorus center. Since the stereochemistry at the phosphorus center can be controlled [200,201] during oxidation of nucleoside H-phosphonothioates, one may obtain phosphate analogues with the same or with inverted configuration.

Nucleoside H-phosphonothioate monoesters can be converted into the corresponding diesters with an efficiency similar to that of H-phosphonate derivatives. H-Phosphonothioate diesters seem to be promising for the preparation of other phosphate analogues—phosphorothioamidates, alkyl phosphorothioates, and phosphorothioates [202].

The ^{31}P NMR spectrum showed that when pivaloyl chloride (PV-Cl) was added to a reaction mixture containing 5′-O-(dimethoxytrityl)thymidine and triethylammonium phosphinate, the reaction went to completion in ~2 min, producing the nucleoside phosphinate **1** as a single phosphorus-containing species [203,204]. After the addition of 2

equivalents of sulfur, two diastereoisomers **2** are formed. The nucleoside H-phosphonothioate **2** is resistant to further oxidation even after prolonged treatment with sulfur. The main advantages of the method are (i) one-pot synthesis involving two experimentally simple steps, namely, condensation and oxidation with sulfur, (ii) mild conditions, (iii) reasonable high yield.

A convenient method for the preparation of nucleoside 3′-H-phosphonothioate monoesters from suitably protected nucleoside 3′-H-phosphonates has been developed by Stawinski *et al*. [205]. It consists of the activation of nucleoside 3′-H-phosphonate monoesters with pivaloyl chloride followed by treatment with 1,1,1,3,3,3-hexamethyldisilanthiane (HMDTS).

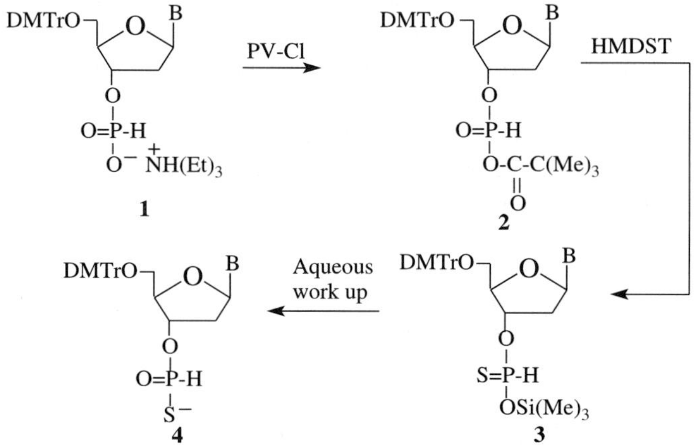

The formation of mixed anhydrides **2** is critical for overall yield of the conversion of H-phosphonate monoesters **1** into H-phosphonothioate **4**. The transformation of an H-phosphonate into an H-phosphonothioate function in the ribo series proved to be a stereoselective process. In contradistinction to this transformation, synthesis of the ribonucleoside 3′-H-phosphonothioate via phosphinate intermediates [202] hardly showed any stereoselectivity.

5′-O-(9-Phenylxanthen-9-yl)thymidine is converted into the triethylammonium salt of its 3′-phosphonodithioate in good yield [206].

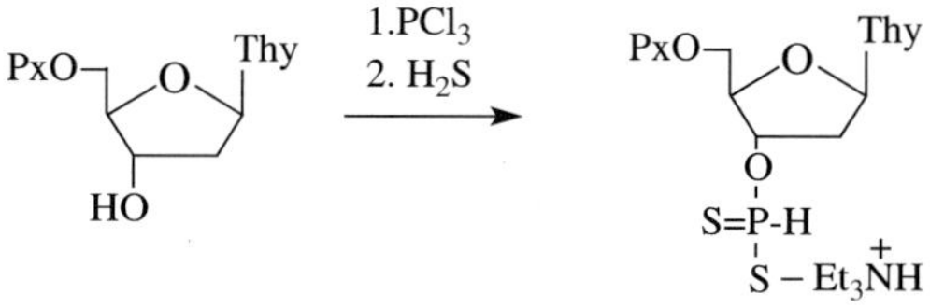

On reacting nucleoside aryl H-phosphonates **1** with hexamethyldisilathiane, various nucleoside H-phosphonothioate monoesters were obtained in high yields (over 90%) [182d, 207].

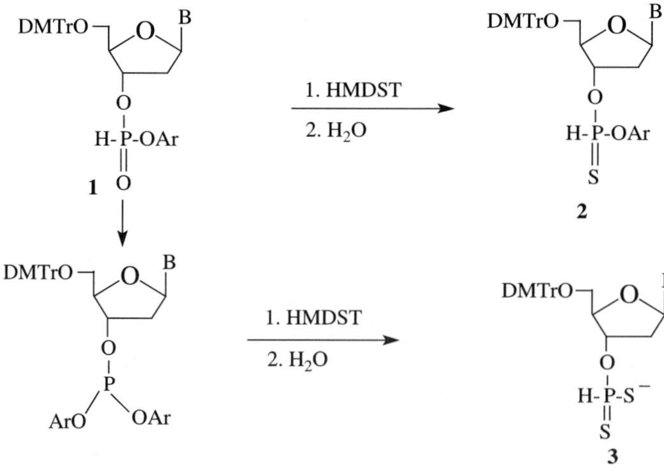

4.3 Nucleoside H-Phosphonates

Kraszewski et al. [208] have developed a new, efficient method for the synthesis of phosphorothioate diester, based on the H-phosphonate chemistry. The reaction of aryl nucleoside H-phosphonate with elemental sulfur furnished the corresponding aryl nucleoside phosphorothioate. The synthesis of phosphorothioates can be carried out as a four-components-one-pot reaction, by allowing nucleoside H-phosphonate to react with phenols in the presence of diphenyl phosphorochloridate to furnish aryl nucleoside H-phosphonates, which react with elemental sulfur.

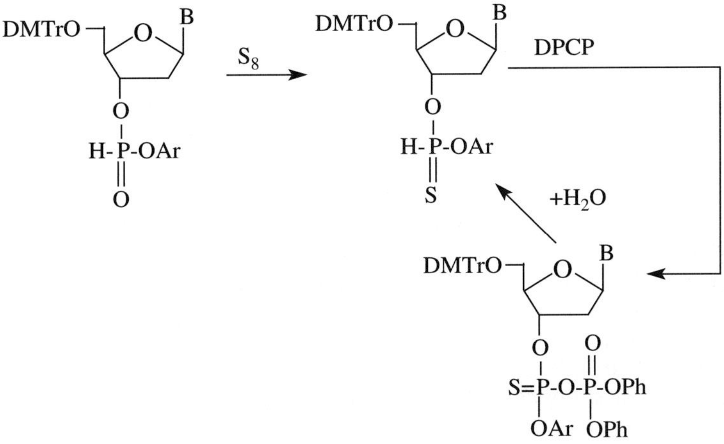

Nucleoside H-phosphorothioates of type **2** were synthesized by reacting unprotected nucleoside H-phosphonates **1** with elemental sulfur, suspended in pyridine containing triethylamine and trimethylsilyl chloride [209]. After 5 min, the starting compounds were completely converted into nucleoside phosphorothioate.

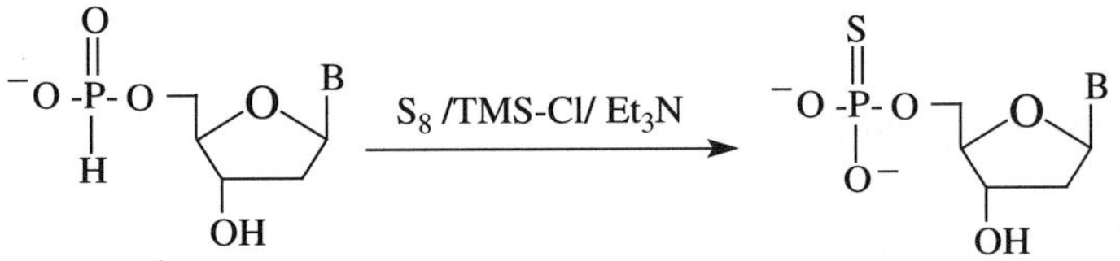

Dinucleoside phosphorothioates are chiral analogues of phosphodiesters in which one of the nonbridging oxygens has been replaced by sulfur. These compounds are important research tools in stereochemical investigations, in mechanistic studies of various enzymatic reactions, and other biochemical studies [210–212]. There have been only a few methods for the preparation of ribonucleoside phosphorothioates [213–216]. The H-phosphonate approach is the best method for the preparation of phosphorothioate diesters [217–219].

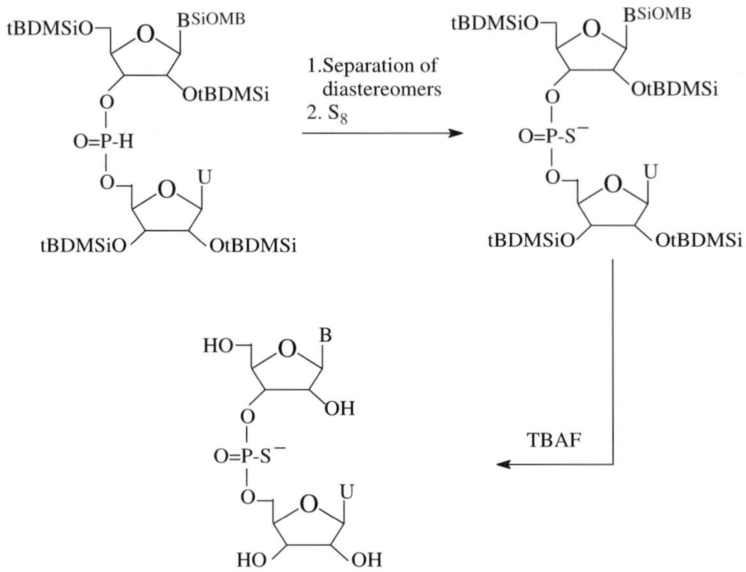

tBDMSi – t-butyldimethylsilyl;
SiOMB – 2-[(tert-butyldiphenylsiloxy)methyl]benzoyl;
TBAF – tertabutylammonium fluoride

A crucial point in the synthesis of optically pure dinucleoside phosphorothioates is the separation of the precursors of the final products that differ only in their configuration around phosphorus. Column chromatography on silica gel is an efficient and convenient way to separate the diastereomers of H-phosphonate diesters. The results obtained reveal that the sulfurization of suitable protected diribonucleoside H-phosphonates with elemental sulfur is a stereospecific reaction, which is most likely proceeding with the retention of the configuration at the phosphorus center.

4.3.3 Synthetic application

Synthesis of oligonucleotides

Simple methods for synthesizing and purifying oligonucleotides are now in great demand due to the utility of synthetic oligonucleotides in a wide variety of molecular biological techniques. The importance of chemically synthesized oligonucleotides is principally due to the wide variety of applications to which oligonucleotides can be directed. For example, oligonucleotides can be utilized in biological studies involving genetic engineering, recombinant DNA techniques, antisense DNA, detection of genomic DNA, probing DNA and RNA from various systems, detection of protein–DNA complexes, detection of site-directed mutagenesis, primers for DNA and RNA synthesis, primers for amplification techniques such as the polymerase chain reaction, ligase chain reaction, etc.

4.3 Nucleoside H-Phosphonates

Oligonucleotide synthesis using nucleoside H-phosphonates has been reported first by Todd et al. [184] and revisited by Hata et al. [220,221] and Garegg et al. [222–226], but became practical only with the recent introduction of pivaloyl chloride (trimethyl acetyl chloride) as the condensing agent. The key step in nucleic acid synthesis via the H-phosphonate method is the specific and sequential formation of internucleotide phosphate linkages between a 5'-OH group of one nucleotide and a 3'-OH group of another nucleotide. Once the entire chain is constructed, the phosphonate diester linkages are oxidazed with aqueous iodine [174, 177,179,190,191] or with 2,4,6-triisopropylbenzenesulfonyl chloride [175]. It has been clearly shown [227] that an improvement of the hydrogen-phosphonate method for oligodeoxynucleotide synthesis has been achieved via an efficient capping reagent, triethylammonium isopropylphosphonate **1**. Activation for capping and coupling is best effected with 1-adamantanecarbonyl chloride. The model reaction revealed that addition of 1-adamantanecarbonyl chloride to the mixture of triethylammonium isopropylphosphonate **1** ($\delta = -0.13$ ppm, $^1J(P,H) = 610$ Hz) and 3'-acetyl thymidine caused essentially immediate conversion to the 5-isopropylphosphonate derivative of 3'-acetyl thymidine **2** ($\delta = 7.13$ ppm, $^1J(P,H) = 702$ Hz).

There have been reports of successful use of this method in both deoxyribonucleotide and ribonucleotide syntheses [224,226]. The H-phosphonate method offers several advantages over the β-cyanoethyl phosphoramidite method: (i) the H-phosphonate method can be accomplished in a shorter cycle time, (ii) the 3' phosphonate monomers have greater stability than the corresponding 3'-phosphoramidiates, (iii) a simple reaction can be used to prepare backbone-modified DNA or RNA. The H-phosphonate methods of Froehler et al. and Garegg are adequate for small-scale synthesis. The main reason is that the methods require 20–30 equivalents of monomer per coupling reaction. An improved process for synthesizing oligonucleotides by condensing a solid support–bound nucleotide with free activated nucleoside H-phosphonate was described by Agrawal et al.[228]. The first nucleotide (monomer 1) is bound on one end (e.g., the 3' hydroxyl group) with the result that only its other end (e.g., the 5' hydroxyl group) is able to react. The next (incoming) nucleotide (monomer 2) is blocked or protected by a chemical group (e.g., the 4,4' dimethoxytrityl (DMT)), only at one end (e.g., the 5' hydroxyl group). The other end (e.g., the 3' hydroxyl group) is the able to react with the first end (e.g., the 5' hydroxyl group) of bound monomer 1. The method can also be used with a solid support to which the first nucleotide is bound by its 5' end. As solid supports, controlled pore glass [CPG], cellulose, and polystyrene, for example, are used. Carriers that are insoluble in the reagents used for the oligonucleotide synthesis are selected, this being for convenience in recovering the polynucleotide after each condensation cycle. Carrier-bound nucleosides are prepared by the procedure of Chow et al. [229]. The process results in a coupling efficiency of greater than 97%, and the process does not require a separate capping step and capping reagent because the activating reagent serves as a self-capping function, thereby preventing elongation of failed sequences.

Coupling of di- or oligo-nucleosides having H-phosphonate at the 3'-position with a second mono- or oligo- nucleoside having a free 5'-hydroxyl group results in the formation of oligonucleosides (n > 2) [174].

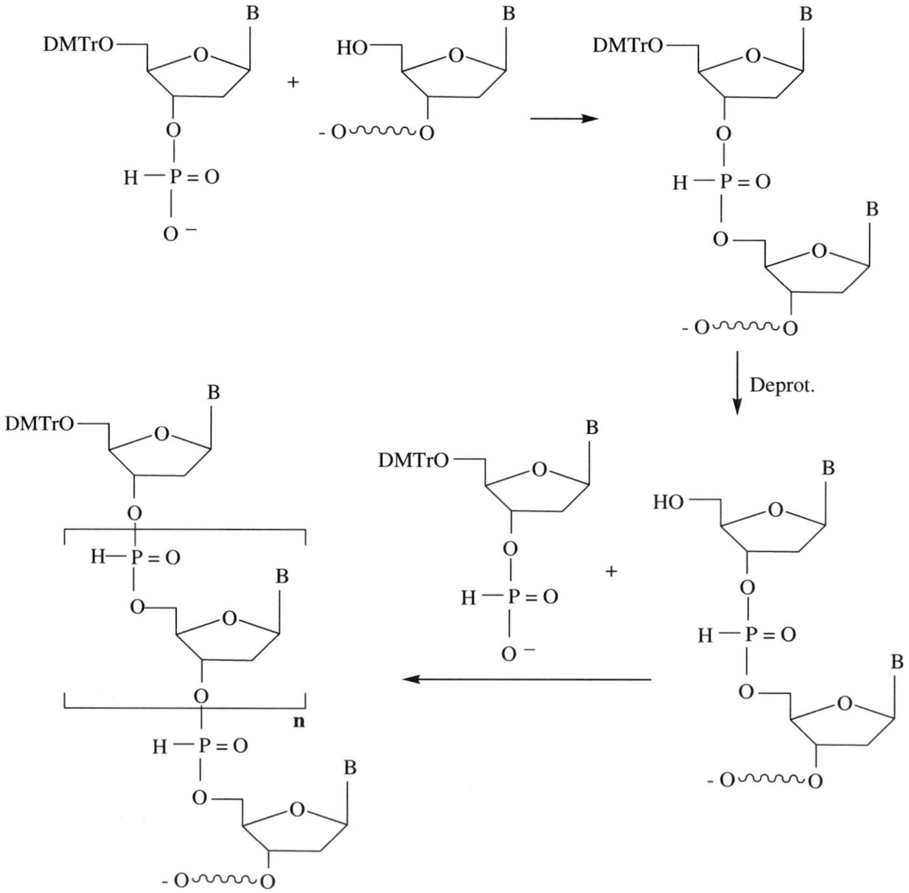

The H-phosphonate approach for the preparation of 3'-5'- internucleotidic bonds shows the following advantages: protection of phosphate is not required; the oxidation reaction is performed by I_2 solution at the end of the synthetic cycles; the coupling reaction is much faster than the phosphoramidate approach; the H-phosphonate units are more stable than

4.3 Nucleoside H-Phosphonates

the phosphoramidate units, and the capping step after each coupling reaction can be omitted [224,230].

Synthesis of phospholipids

The vital role played by phospholipids in many biological processes has stimulated a number of studies concerning their chemistry, biochemistry, and physical properties. Phospholipid analogues were found to be valuable tools in studies concerning elucidation of the mechanism of some enzymatic reactions [231], in probing biomembrane structures [232], and in the preparation of liposomes with the desired properties [233]. Phospholipids have been used as drug carriers [234] or as drugs [233,235].

The most important stage in the chemical synthesis of phospholipids is phosphorylation resulting in the formation of a phosphodiester bond, the major structural element of these compounds. A number of methods have been developed for phospholipid preparation. The most straightforward one, the phosphodiester method for phospholipid synthesis, is inefficient in terms of yield and involves a laborious procedure. Simple experimental procedures and high yield of phosphodiester formation on one hand and easy access to phospholipid analogues make the H-phosphonate approach the method of choice for phospholipid synthesis [236]. The utility of H-phosphonate intermediates in the synthesis of phospholipids was demonstrated by Stawinski *et al.* [236]. In a typical synthesis of phospholipid diesters, 1,2-dipalmitoyl-*sn*-glycero-3-H-phosphonate **2** (^{31}P in ppm, $\delta = 4.60$, $^1J(P,H) = 626$ Hz), prepared by the reaction of 1,2-dipalmitoyl-*sn*-glycerol **1** with the PCl$_3$/imidazole or with salicylchlorophosphite as a phosphylating agent, was rendered anhydrous by repeated evaporation of added pyridine and then condensed with a hydroxylic component in the presence of a coupling agent (PV-Cl **5**).

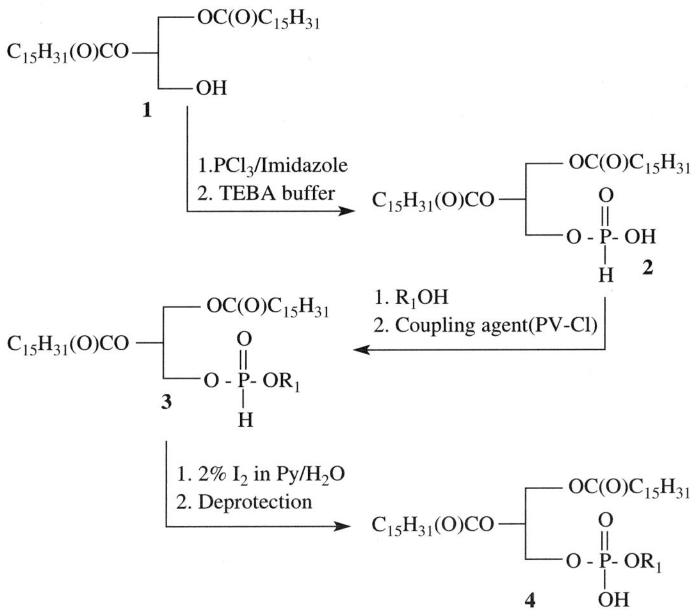

After 5 min, **2** was completely converted into gycero-H-phosphonate diester **3** (^{31}P in ppm, δ = 8.65 and 8.43, ^1J(P,H) = 713 Hz), which was oxidized by addition of iodine in aqueous pyridine to give **4** (^{31}P in ppm, δ = −0.45). The total yield of phospholipids **4** was ca. 80%.

The most convenient condensing reagent seems to be 5,5-dimethyl-2-oxo-2-chloro-1,3,2-dioxaphosphorinane **6**. This chlorophosphate is a stable, crystalline compound, easy to prepare in large quantities [237], and has good solubility in most organic solvents. This coupling agent ensures clean and reasonably fast coupling without dangerous side reactions even if the reaction mixture is left for a long time. Pivaloyl chloride **5**, which is commonly used in oligonucleotide synthesis, seems to be unnecessarily reactive for the purpose of phospholipid synthesis. It may also lead to the formation of side products if the reaction mixture is not quenched when the reaction is over.

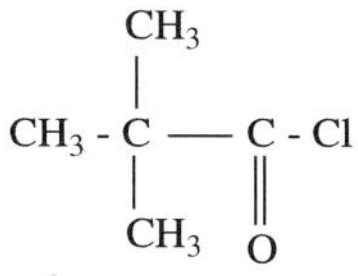

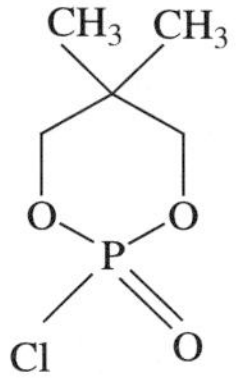

The H-phosphonate approach appears to be an efficient and experimentally simple method for phospholipid synthesis. The distinctive features of the method are (i) easy preparation of the key intermediate **2**, which can be stored for several months, (ii) coupling reactions are fast and clean, (iii) the possibility of isolating intermediates or carrying out the synthesis as a 'one-pot' reaction, (iv) the lack of a protecting group at the phosphorus center simplifies the deprotection procedure, and (v) the possibility of synthesizing various phospholipid analogues, including isotope labeling, via changing the oxidation procedure.

H-Phosphonate diesters of type **3** can be used in the preparation of phospholipid analogues **7** and **8** having a modified phosphorus center.

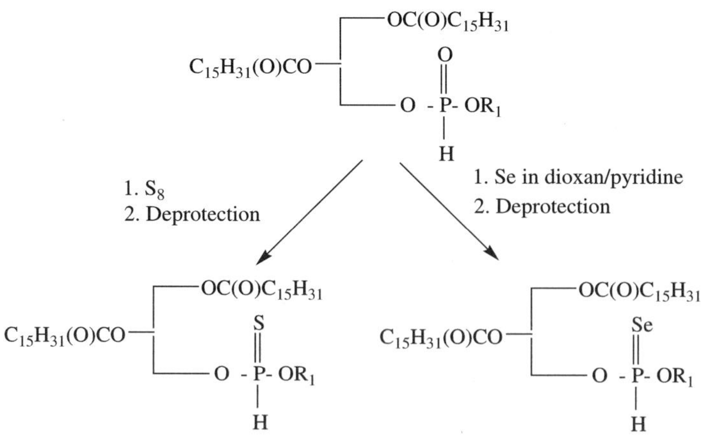

4.3 Nucleoside H-Phosphonates

Synthesis of 5'-O-hydrogen phospholipids

5'-O-hydrogen phospholipids were synthesized in good yield by tandem transesterification of diphenyl H-phosphonate with 3'-azido-2', 3'-dideoxythymidine and a long-chain alcohol [238].

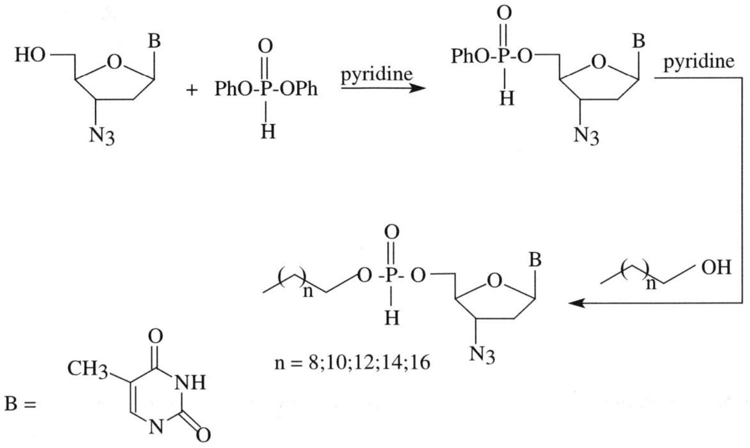

These hydrogen phospholipids present prodrugs of nucleosides such as 2',3'-didehydro-3'-deoxythymidine (d4T) and 2',3'-dideoxyinosine (ddI), RT inhibitors approved for clinical use to treat human immunodeficiency virus infection.

Construction of a nonradioactive hybridization probe

In order to serve as a hybridization probe, a synthetic oligonucleotide usually has to be equipped with a radioactive tracer or a reporter group that facilitates its localization. The growing usage of hybridization probes in molecular diagnostics makes nonradioactive systems of labeling of oligonucleotides the preferred ones, due to their safety, ease of handling, and a variety of modes of detection [168,170].

The most common approach to constructing a nonradioactive hybridization probe consists of functionalization of an oligonucleotide, that is, the introduction of a new functional group to which suitable reporter groups can be attached [239, 240]. Most commercially available reporter groups are active aryl carboxyesters or isothiocyanates, which can be conveniently attached to oligonucleotides via appropriate linkers carrying a primary amino group [182c].

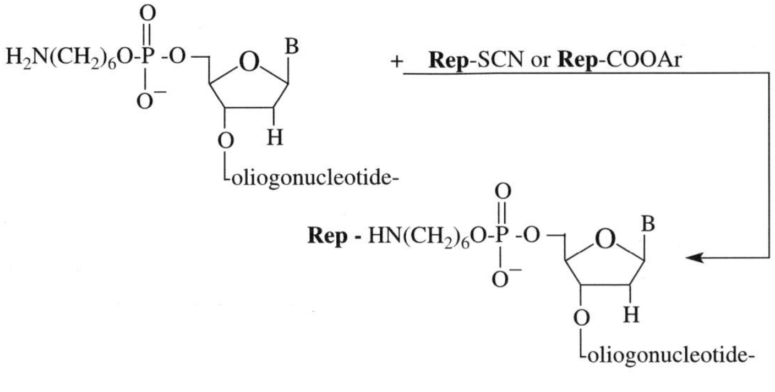

Using this method, several fluorescein- and biotin-tagged oligonucleotidic hybridization probes (a probe for the detection of genes encoding cysteine proteases of *Fasciola hepatica*, a probe for the detection of human papillomavirus (type 16)) have been synthesized and successfully used in biological studies.

4.4 DIALKYL EPOXYALKYLPHOSPHONATES

The discovery of the antibiotic fosfonomycin [241] [(−)(1R,2S)-(Z)-1,2-epoxypropylphosphonic acid] in 1969 generated an increased interest in the chemistry of dialkyl 1,2-epoxyalkylphosphonates. Its mode of action, an analogue of phosphoenol pyruvate in its inhibition of the enzyme pyruval transferase, has given epoxyphosphonates biochemical significance [242]. Dialkyl epoxyalkylphosphonates are of interest because of their use as intermediates in the synthesis of bioactive substances [243–246], and as modifiers of natural and synthetic polymers [247–249].

4.4.1 1,2-Epoxyalkylphosphonates

Methods of preparation

Different methods are known for the synthesis of dialkyl 1,2- epoxyalkylphosphonates. There are excellent reviews by Redmore [250], Yamamoto [251], and Ioga [252] describing the present methods for the synthesis of epoxyphosphonates. In this book, the following methods for the preparation of epoxyphosphonates based on diesters of H-phosphonic acid and their derivatives will be included: (i) reaction of sodium dialkyl H-phosphonates with α-halo ketones, (ii) reaction of dialkyl halohydrinphosphonates with bases, (iii) the Darens reaction of dialkyl chloromethylphosphonates with carbonyl compounds, (iv) oxidation of 1,2-unsaturated phosphonates with a peroxide.

Reaction of sodium dialkyl phosphonates with a-halo ketones. Alkali metal derivatives of dialkyl H-phosphonates react with α-halo ketones to form an alkoxide anion, which displaces the β-halogen, resulting in the formation of the dialkyl 1,2-epoxyalkylphosphonates with yields in the range 60–70% [253].

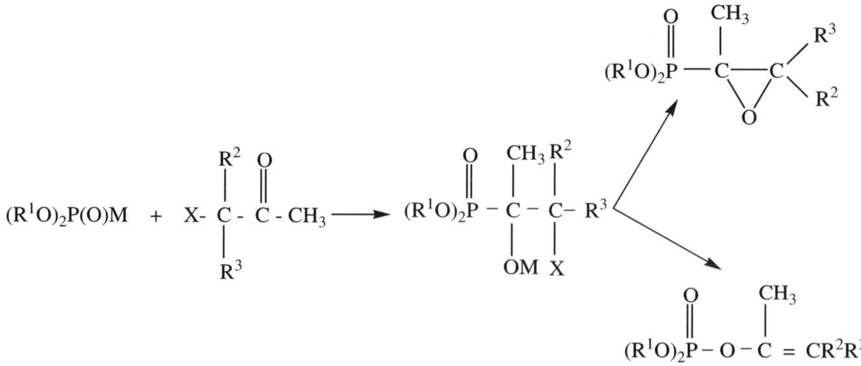

4.4 Dialkyl Epoxyalkylphosphonates

Reaction of dialkyl halohydrinphosphonates with bases. Dialkyl-1,2-epoxyphosphonates are exclusively formed in the reaction between dialkyl H-phosphonates and α-halo ketones in the presence of equimolar quantities of sodium alkoxides in methanol at room temperature [254]. A two-step mechanism of the above reaction has been proposed based on ^{1}H NMR data, which includes the intermediate formation of the corresponding halohydrin.

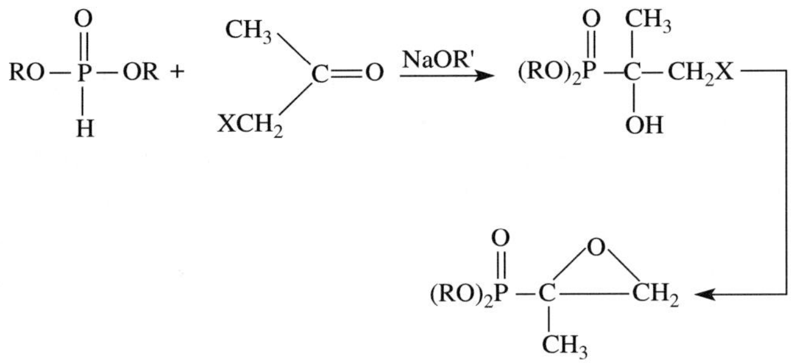

$$RO-\overset{O}{\underset{H}{\overset{\|}{P}}}-OR + \underset{XCH_2}{\overset{CH_3}{>}}C=O \xrightarrow{NaOR'} (RO)_2\overset{O}{\overset{\|}{P}}-\underset{OH}{\overset{CH_3}{\underset{|}{C}}}-CH_2X \longrightarrow$$

$$(RO)_2\overset{O}{\overset{\|}{P}}-\underset{CH_3}{\overset{}{\underset{|}{C}}}\overset{O}{\underset{}{\diagdown}}CH_2$$

The reaction between diethyl H-phosphonate and α-chloroketone in the presence of KF resulted in the formation of diethyl-1,2-epoxy-2-propylphosphonate [255].

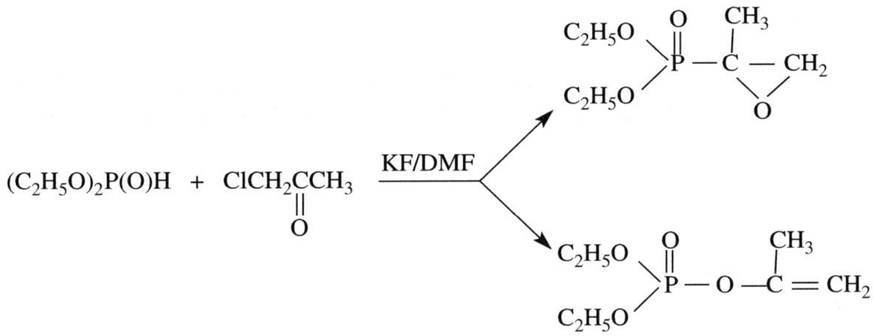

$$(C_2H_5O)_2P(O)H + ClCH_2\overset{O}{\overset{\|}{C}}CH_3 \xrightarrow{KF/DMF}$$

giving:
- $(C_2H_5O)_2P(O)-C(CH_3)(CH_2)O$ (epoxide)
- $(C_2H_5O)_2P(O)-O-C(CH_3)=CH_2$

An alternative procedure for the preparation of dialkyl-1,2-epoxyalkylphosphonates includes reaction of diethyl H-phosphonate in methanol with α-tosyloxy ketones in the presence of DBU (1,8-diazobicyclo[5.4.0]undec-7-ene) [256].

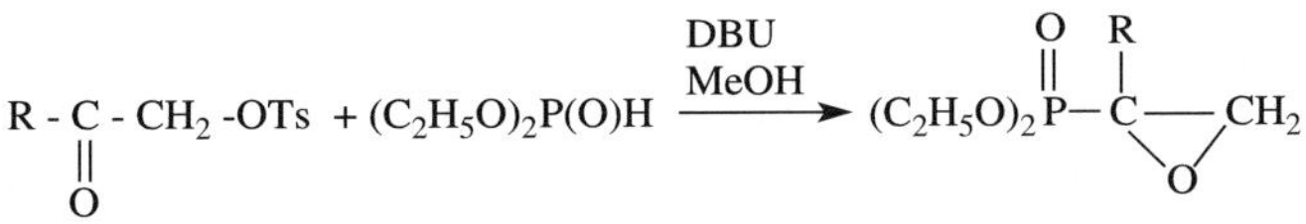

$$R-\overset{O}{\overset{\|}{C}}-CH_2-OTs + (C_2H_5O)_2P(O)H \xrightarrow[MeOH]{DBU} (C_2H_5O)_2\overset{O}{\overset{\|}{P}}-\underset{}{\overset{R}{\underset{|}{C}}}\overset{}{\diagdown}CH_2$$

This method is a useful modification having a great synthetic utility since the living group is introduced in mild conditions [256c]. The same reaction is carried out in the two-phase system CH$_2$Cl$_2$/NaOH 50% TBAC at room temperature [257]. The phase-transfer conditions simplify and accelerate numerous reactions traditionally performed in nonaqueous media. The most recent development in the field has featured the use of phase-transfer catalyst conditions in the preparation of dialkyl-1,2-epoxyalkylphosphonates from dialkyl H-phosphonates and α-haloketones [258].

$$(RO)_2P(O)-H + ClCH(R^1)\overset{O}{\underset{\parallel}{C}}-CH_3 \xrightarrow[-KCl;\ KHCO_3]{K_2CO_3/NaHCO_3 \atop Bu_4\overset{+}{N}\overset{-}{Br}} (RO)_2\overset{O}{\underset{\parallel}{P}}-\underset{\underset{O}{\diagdown\ \diagup}}{C}-\overset{R^1}{\underset{H}{C}}$$

$R = CH_3;\ C_2H_5;\ C_4H_9;\ i\text{-}C_3H_7;$

The structure of the final product has been verified by a combination of ^{1}H, ^{31}P, and ^{13}C NMR spectroscopy. The phosphorus chemical shift appears at 21.08 pp (R = CH$_3$). The signal for the carbon atom bounded to the phosphorus appears at 50.70 ppm as a doublet with ^1J(P,C) = 203.0 Hz (R = CH$_3$). A two-step mechanism involving deprotonation of the dialkyl H-phosphonates by [Bu$_4$NHCO$_3^-$]$^+$, followed by nucleophilic attack of the dialkyl phosphonate anion (RO)$_2$P(O)$^-$ at the carbonyl group of the ketone, is presented. The oxirane formation during the second stage of the reaction involves displacement of the chloride ion.

$$(RO)_2P(O)H + Bu_4\overset{+}{N}\,\overset{-}{HCO_3} \longrightarrow (RO)_2P(O)^-\,Bu_4\overset{+}{N}$$

$$+ ClCH_2\overset{O}{\underset{\parallel}{C}}CH_3$$

$$(RO)_2\overset{O}{\underset{\parallel}{P}}-\overset{CH_3}{\underset{\diagdown O \diagup}{C}}-CH_2 \xleftarrow{-Bu_4\overset{+}{N}\,Cl^-} (RO)_2\overset{O}{\underset{\parallel}{P}}-\overset{CH_3}{\underset{O^-\,Bu_4\overset{+}{N}}{C}}-CH_2Cl$$

Side reactions

$$(RO)_2\overset{O}{\underset{\parallel}{P}}-O-\overset{CH_3}{\underset{}{CH}}-CH_2Cl \xleftarrow{} (RO)_2\overset{O}{\underset{\parallel}{P}}-\overset{CH_3}{\underset{OH}{C}}-CH_2Cl$$

$\Big|\ -HCl$

$$(RO)_2\overset{O}{\underset{\parallel}{P}}-O-\overset{CH_3}{\underset{}{C}}=CH_2$$

On the basis of the results obtained, it can be assumed that the reaction proceeds either on the phase boundary or in the organic phase (the starting reagents form the liquid phase). The established inhibition of the formation of the 1,2-epoxyphosphonates in the presence of the CO$_2$ leads to the conclusion that the interaction proceeds in the organic phase. It is likely, therefore, that the deprotonation of the dialkyl H-phosphonate occurs in the organic phase by the reaction with small amounts of potassium hydrogen carbonate. This method does not improve the yields of dialkyl-1,2-epoxyalkylphosphonates, but the absence of undesired side products, which usually accompany this reaction, renders this methodology a viable alternative to methodsthat require the use of a solvent.

The Darzens reaction. The synthesis of glycidic esters by the condensation of carbonyl compounds with α-halo esters, known as Darzens reaction [259], has been used in phosphorus

4.4 Dialkyl Epoxyalkylphosphonates

chemistry for the preparation of dialkyl 1,2-epoxyalkylphosphonates. This method involves the reaction of dialkyl chloromethylphosphonates with carbonyl compounds. The first report for the preparation of epoxyphosphonates in 31% yield by the Darzens route is described by Martinov et al. [260].

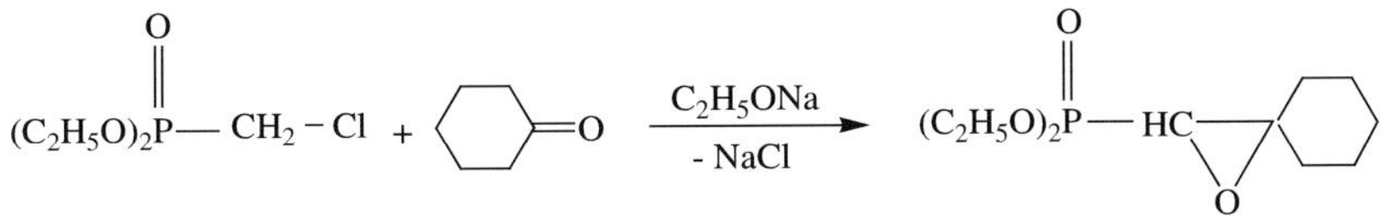

$$(C_2H_5O)_2P(O)-CH_2-Cl + \text{cyclohexanone} \xrightarrow[-NaCl]{C_2H_5ONa} (C_2H_5O)_2P(O)-HC\underset{O}{\overset{}{\diagdown}}\text{(cyclohexyl epoxide)}$$

Best yields were obtained when dialkyl chloromethanephosphonates were metallated by treatment with BuLi in THF [261].

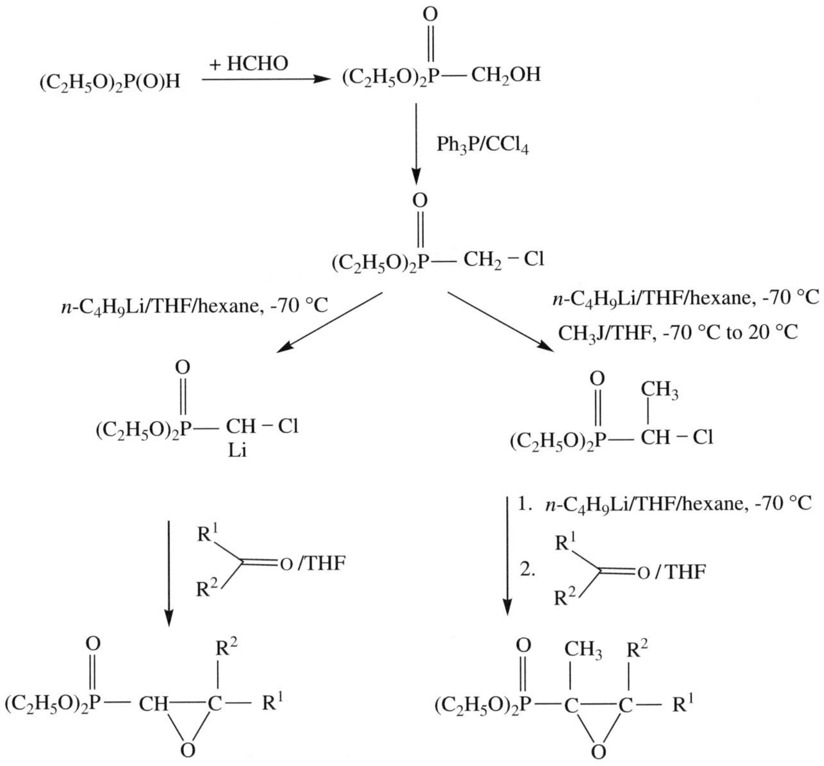

$$(C_2H_5O)_2P(O)H \xrightarrow{+HCHO} (C_2H_5O)_2P(O)-CH_2OH \xrightarrow{Ph_3P/CCl_4} (C_2H_5O)_2P(O)-CH_2-Cl$$

Left branch: $n\text{-}C_4H_9Li/THF/hexane, -70\,°C$ gives $(C_2H_5O)_2P(O)-CHCl(Li)$, then $R^1R^2C=O/THF$ gives $(C_2H_5O)_2P(O)-CH\underset{O}{\overset{}{\diagdown}}CR^1R^2$.

Right branch: $n\text{-}C_4H_9Li/THF/hexane, -70\,°C$; $CH_3I/THF, -70\,°C$ to $20\,°C$ gives $(C_2H_5O)_2P(O)-CH(CH_3)-Cl$; then 1. $n\text{-}C_4H_9Li/THF/hexane, -70\,°C$; 2. $R^1R^2C=O/THF$ gives $(C_2H_5O)_2P(O)-C(CH_3)\underset{O}{\overset{}{\diagdown}}CR^1R^2$.

Addition of a carbonyl compound results in the almost immediate formation of a phosphonates chlorohydrin. After a few hours, only the signals for the corresponding epoxyalkanephosphonates can be detected in the NMR spectra. This procedure has led to almost quantitative yields of the epoxyalkanephosphonate (see Appendix). This method is applicable to (i) a wide range of aliphatic and aromatic aldehydes and ketones, (ii) to the preparation of 1-substituted 1,2-epoxyalkanephosphonates.

1-Substituted -1,2-epoxyalkylphosphonates are obtained via metallation of dialkyl phosphonates with BuLi in THF followed by electrophilic chlorination with benzenesulfonyl chloride [262].

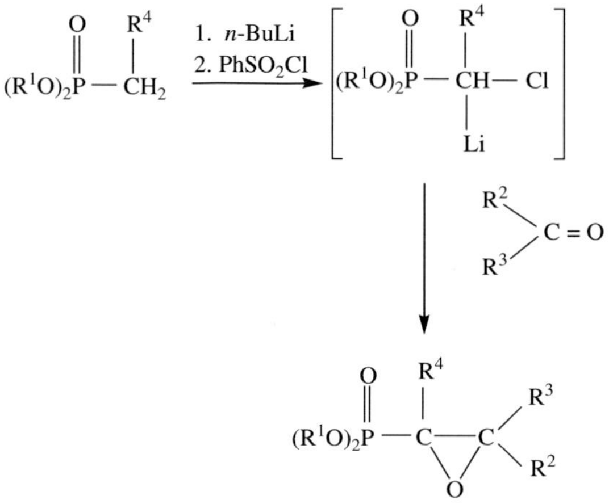

$$(R^1O)_2\overset{\overset{O}{\|}}{P}-\overset{\overset{R^4}{|}}{CH_2} \xrightarrow{\substack{1.\ n\text{-BuLi}\\ 2.\ PhSO_2Cl}} \left[(R^1O)_2\overset{\overset{O}{\|}}{P}-\overset{\overset{R^4}{|}}{\underset{\underset{Li}{|}}{CH}}-Cl\right]$$

$$\xrightarrow{\underset{R^3}{\overset{R^2}{>}}C=O} (R^1O)_2\overset{\overset{O}{\|}}{P}-\overset{\overset{R^4}{|}}{\underset{}{C}}\underset{\underset{O}{\diagdown\diagup}}{-}C\overset{R^3}{\underset{R^2}{<}}$$

Fosfomycin has been prepared from chloromethylphosphonate by nucleophilic substitution of the chlorine with dimethyl sulfide [263].

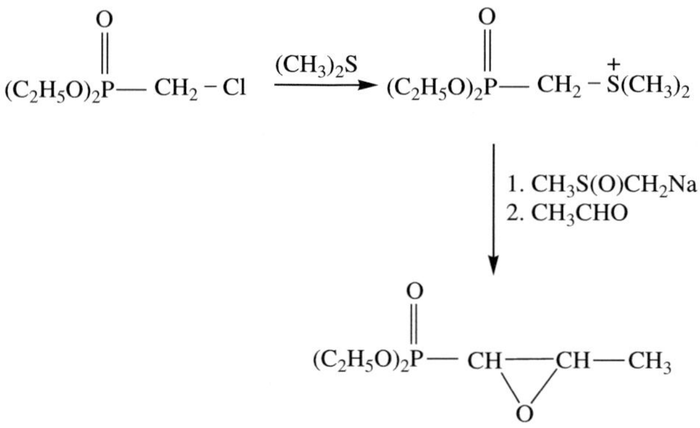

$$(C_2H_5O)_2\overset{\overset{O}{\|}}{P}-CH_2-Cl \xrightarrow{(CH_3)_2S} (C_2H_5O)_2\overset{\overset{O}{\|}}{P}-CH_2-\overset{+}{S}(CH_3)_2$$

$$\xrightarrow{\substack{1.\ CH_3S(O)CH_2Na\\ 2.\ CH_3CHO}} (C_2H_5O)_2\overset{\overset{O}{\|}}{P}-CH\underset{\underset{O}{\diagdown\diagup}}{-}CH-CH_3$$

Oxidation of 1,2-unsaturated phosphonates with a peroxide. The direct oxidation of dialkyl vinylphosphonates appears to be the most attractive synthetic route for the preparation of dialkyl 1,2-epoxyalkylphosphonates [264].

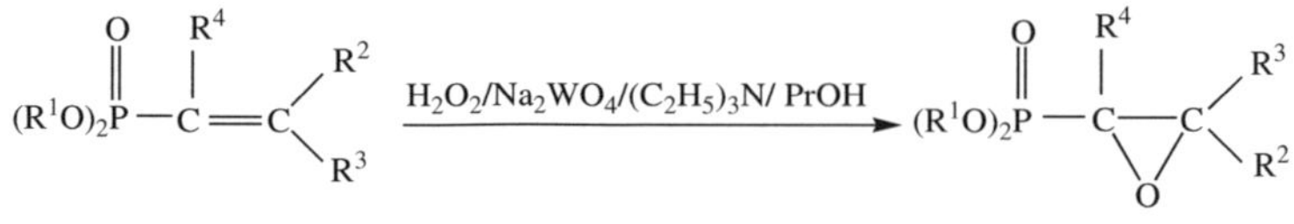

$$(R^1O)_2\overset{\overset{O}{\|}}{P}-\overset{\overset{R^4}{|}}{C}=C\overset{R^2}{\underset{R^3}{<}} \xrightarrow{H_2O_2/Na_2WO_4/(C_2H_5)_3N/\ PrOH} (R^1O)_2\overset{\overset{O}{\|}}{P}-\overset{\overset{R^4}{|}}{C}\underset{\underset{O}{\diagdown\diagup}}{-}C\overset{R^3}{\underset{R^2}{<}}$$

4.4 Dialkyl Epoxyalkylphosphonates

4.4.2 2,3-Epoxyalkylphosphonates

Dialkyl H-phosphonates react with epoxyketones in the presence of a catalytic amount of sodium dialkyl phosphonate (pathway **A**) at room temperature, giving dialkyl esters of 1-hydroxyl-1,3-dimethyl-2,3-epoxybutylphosphonic acid in 40–99% yield [265].

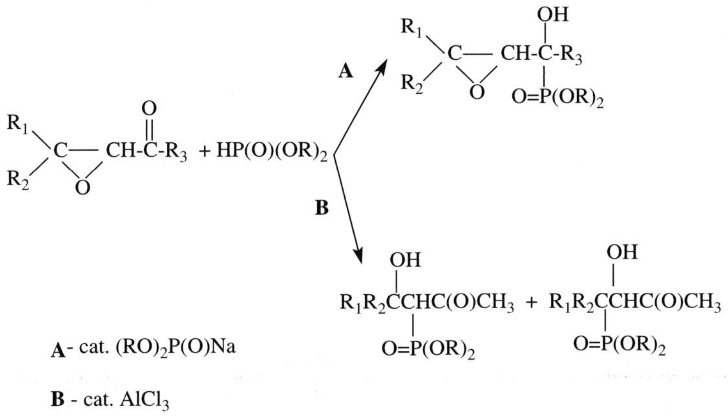

A - cat. $(RO)_2P(O)Na$

B - cat. $AlCl_3$

In the presence of $AlCl_3$, ring opening (pathway **B**) occurs, resulting in the formation of a mixture of dimethyl ester of 2-hydroxyl-2-methyl-1-acetopropylphosphonic acid and 1-hydroxyl-2-methyl-2-acetoisopropylphosphonic acid. ^{31}P NMR spectra showed signals at 23–26 ppm, which is characteristic for the phosphonate structure.

4.4.3 Reactivity of 1,2-epoxyalkylphosphonates

It was shown that diethyl 2,2-disubstituted -1,2-epoxyethylphosphonates undergo thermal rearrangement to yield diethyl 1,1-disubstituted -1-formylmethylphosphonate [266].

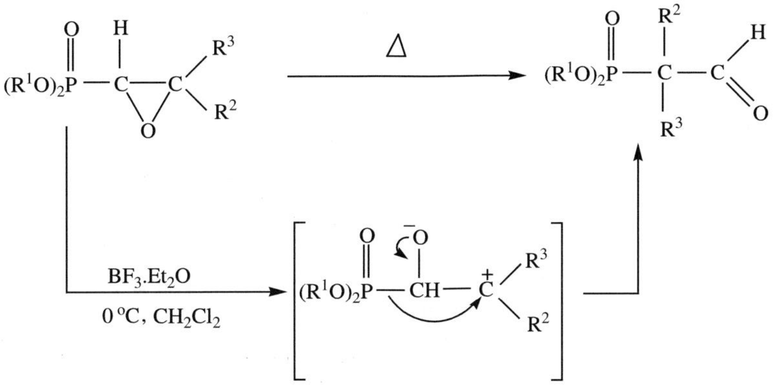

This is the first example of thermal and acid-catalyzed rearrangement. The preferred catalyst for this rearrangement was found to be $BF_3 \cdot Et_2O$.

The opening of the epoxide with methanol- or ethanol-containing H_2SO_4 resulted in the formation of 1-methoxy-2-hydroxy- or 1-ethoxy-2-hydroxyethylphosphonate [267].

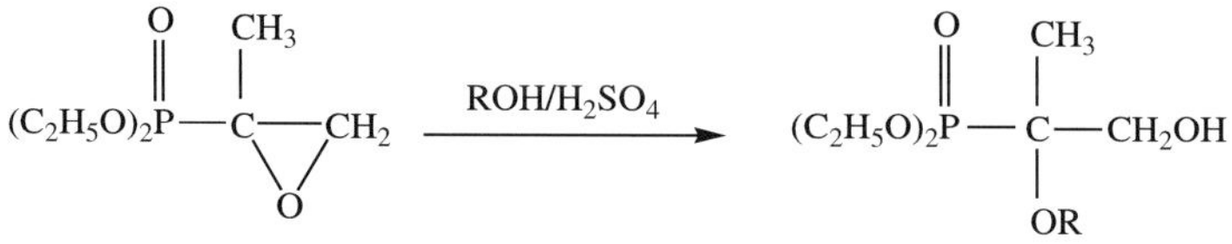

$$(C_2H_5O)_2P(O)-C(CH_3)(-CH_2-)O \xrightarrow{ROH/H_2SO_4} (C_2H_5O)_2P(O)-C(CH_3)(OR)-CH_2OH$$

Acid-catalyzed hydration of diethyl 1,2-epoxyethylphosphonate with aqueous H_2SO_4 leads to the formation of diethyl 1,2-dihydroxyethylphosphonate [268].

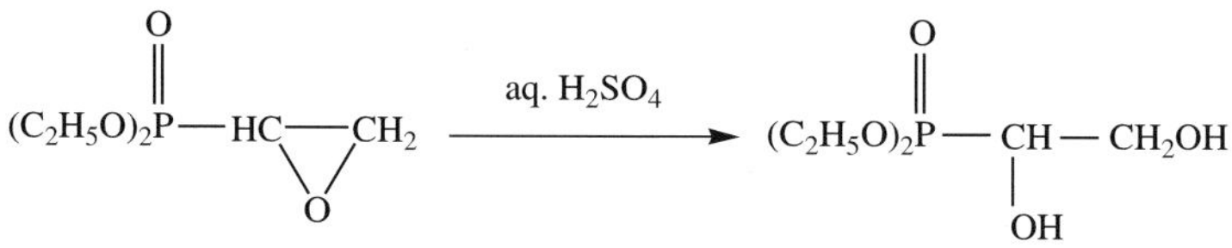

$$(C_2H_5O)_2P(O)-HC(-CH_2-)O \xrightarrow{aq.\ H_2SO_4} (C_2H_5O)_2P(O)-CH(OH)-CH_2OH$$

Treatment of diethyl 1,2-epoxyethylphosphonate with metanolic or ethanolic sulfuric acid gave diethyl 1-hydroxy-2-alkoxyethylphosphonates.

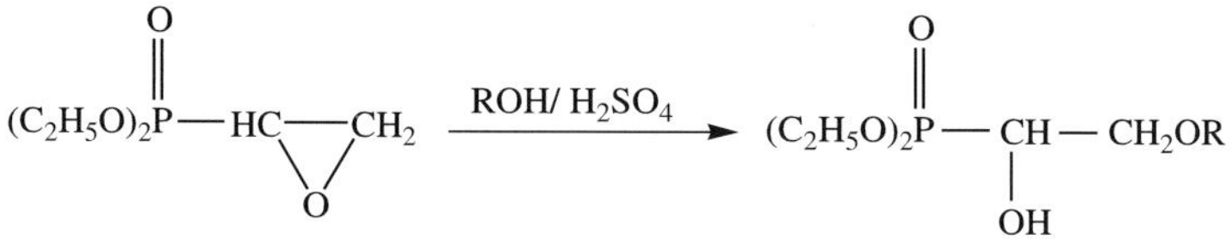

$$(C_2H_5O)_2P(O)-HC(-CH_2-)O \xrightarrow{ROH/H_2SO_4} (C_2H_5O)_2P(O)-CH(OH)-CH_2OR$$

Phosphorylated amino alcohols were obtained by reacting 1,2-epoxyphosphonates with amines.

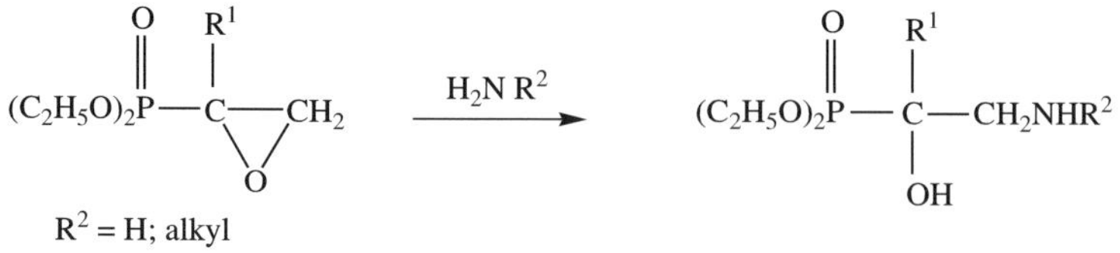

$$(C_2H_5O)_2P(O)-C(R^1)(-CH_2-)O \xrightarrow{H_2NR^2} (C_2H_5O)_2P(O)-C(R^1)(OH)-CH_2NHR^2$$

R^2 = H; alkyl

It was shown that sodium dialkyl phosphonates react in an alcoholic medium with diethyl 1,2-epoxyalkanephosphonates to give dialkyl 2-oxoalkanephosphonates [269]. The sodium dialkyl phosphonate attacks the β-carbon atom

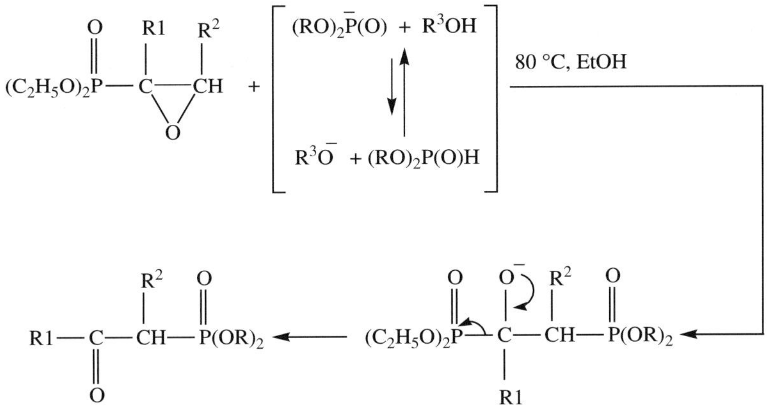

4.4 Dialkyl Epoxyalkylphosphonates

[Reaction scheme showing diethyl epoxyalkylphosphonate reacting with R³O⁻/(RO)₂P(O)H equilibrium at 80 °C in EtOH, proceeding through an intermediate to give the final dialkyl 2-oxoalkylphosphonate product]

of the epoxide ring to give unstable sodium α-hydroxyphosphonates, which eliminate sodium dialkyl phosphonate to produce dialkyl 2-oxoalkylphosphonates in 50–85% yield.

α-Metallated 1,2-epoxyethylphosphonate can serve as a precursor for the preparation of 1,2-epoxyethylphosphonate derivatives [270]. Reaction of diethyl 1,2-epoxyethylphosphonate with LDA (lithium diisopropylamine) in THF at low temperatures (−115 °C) resulted in the deprotonation of the α-carbon atom, giving a high yield of diethyl α-lithioepoxyethylphosphonate, which reacts with alkyl halides (electrophiles) to give diethyl 1-alkyl-1,2-epoxyethylphosphonates.

[Reaction scheme: diethyl 1,2-epoxyethylphosphonate + LDA at −115 °C in THF gives the α-lithio derivative, which on treatment with RX gives diethyl 1-alkyl-1,2-epoxyethylphosphonate]

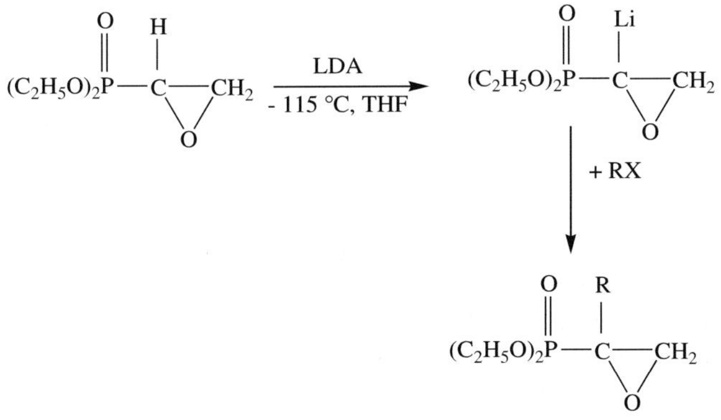

When dialkyl 1-alkyl-1,2-epoxyethylphosphonates were treated with an excess of thiourea in MeOH at reflux, dialkyl 1-alkylvinylphosphonates were produced in good yields [271].

$$(RO)_2P(O)-C(R^1)-CH_2 \text{ (epoxide)} \xrightarrow[\text{r.t., MeOH}]{\text{thiourea}} (RO)_2P(O)-C(R^1)-CH_2 \text{ (episulfide)}$$

$$\xrightarrow[\text{MeOH}]{\text{thiourea reflux}} (RO)_2P(O)-C(R^1)=CH_2$$

4.5 POLY(ALKYLENE H-PHOSPHONATE)S

Polymer chemistry has contributed in various ways to the present progress in biology, biochemistry, and medicine, providing new, highly specified materials. One of these ways is the synthesis of new polymers, structurally related to the natural biopolymers with poly(alkylene phosphate) main chains. Biopolymers with polyphosphate backbones belong to the particularly studied areas in chemistry, biochemistry, and biology. Poly(alkylene phosphate)s can readily be prepared from poly(alkylene H-phosphonate)s. Poly(alkylene H-phosphonate)s are an interesting class of phosphorus-containing polymers because both the polymer backbone and phosphorus substituents can be modified.

$$RO-\overset{O}{\underset{H}{P}}-O-R^1 {\left[O-\overset{O}{\underset{H}{P}}-O-R^1 \right]}_n O-\overset{O}{\underset{H}{P}}-OR$$

Poly(alkylene H-phosphonate)s are particularly interesting due to the fact that P–H groups can be converted into a number of interesting functional groups. Oxidative chlorination of these polymers, using Atherton–Todd reaction conditions followed by reaction with alcohols and amines yields, the corresponding polymeric phosphate esters or amides. Oxidation with N_2O_4 furnishes polymeric phosphoric acids. Poly(alkylene H-phosphonate)s are a relatively new family of biodegradable polymers that are being actively investigated for pharmaceutical and biomedical applications such as drug delivery and tissue engineering [272–274]. These polymers are biodegradable through hydrolysis, and possibly enzymatic cleavage of the P–O–C bonds under physiological conditions. The application of these polymers for pharmaceutical and biomedical applications largely depends on their high purity and polydispersity. A number of synthetic routes have been explored for the synthesis of these polymers, including ring-opening [272], bulk polymerization [273], and enzymatic polymerization [275]. Bulk polycondensation is often used as a preferred route for large-scale production of poly(alkylene

4.5 Poly(alkylene H-phosphonate)s

H-phosphonate)s. The advantages of this method are the short reaction time, minimal purification steps, and feasibility for scale-up [276].

4.5.1 Methods of preparation

Poly(alkylene H-phosphonate)s can be synthesized by polycondensation and polymerization methods.

By polycondensation

Commercially available low-cost H-phosphonate diesters are used as starting compounds for the preparation of poly(alkylene H-phosphonate)s. H. Coates [277] has used for the first time transesterification of dialkyl H-phosphonates with hydroxyl-containing compounds in the molar ratio 1:1 for the synthesis of poly(alkylene H-phosphonate)s. Petrov et al. [278] claimed formation of a highmolecular–weight poly(1,6-hexane H-phosphonate) by transesterification of dimethyl H-phosphonate with 1,6-hexane diol in the molar ratio 1:1. Vogt was the first who has studied in detail the transesterification of diethyl H-phosphonate with different diols, showing the reasons that restricted the preparation of poly(alkylene H-phosphonate)s with high molecular weight and describing the conditions—temperature, molar ratio between starting compounds, catalyst, pressure—leading to high molecular–weight products [279]. Experimental results [279–284] showed that poly(alkylene H-phosphonate)s with high molecular weight can be obtained when the polycondensation is carried out in excess of dialkyl H-phosphonate in two stages. In the first stage, called the 'low temperature stage', dialkyl H-phosphonate in an excess reacted with hydroxyl-containing compounds at 120–125 °C. Oligomers with end alkyl phosphonate groups were obtained.

$$m\ RO-\underset{H}{\underset{|}{P}}(=O)-OR + n\ HO-R^1-OH \xrightarrow{120-125\ °C} RO-\underset{H}{\underset{|}{P}}(=O)-O-R^1\left[O-\underset{H}{\underset{|}{P}}(=O)-O-R^1\right]_x O-\underset{H}{\underset{|}{P}}(=O)-OR$$

$$\downarrow 160\text{-}180\ °C$$
$$-(m-n)\ (RO)_2P(O)H$$

$$RO-\underset{H}{\underset{|}{P}}(=O)-O-R^1\left[O-\underset{H}{\underset{|}{P}}(=O)-O-R^1\right]_n O-\underset{H}{\underset{|}{P}}(=O)-OR$$

In the second stage, called the 'high temperature stage', the oligomer obtained at the first stage is heated at temperatures 160–180 °C. The formation of a high molecular–weight poly(alkylene H-phosphonate) is accompanied by dialkyl hydrogen phosphonate elimination (see Appendix). The two-stage process allows formation of high molecular–weight poly(alkylene H-phosphonate)s because at the second stage hydroxyl groups did not exist

190   4. Important Classes of Compounds

and their dehydration or other side reactions are mostly eliminated. Poly(alkylene H-phosphonate)s are hydrolytically unstable. They are soluble in $CHCl_3$, CH_2Cl_2, and CH_3OH. The structure of the poly(alkylene H-phosphonate)s was elucidated by means of ^1H, ^{13}C, and ^{31}P NMR spectroscopy.

The ^1H NMR spectrum of poly(oxyethylene H-phosphonate)s (Figure 4.1) reveals two types of P–H protons, which appears as doublets at δ = 6.87 ppm ($^1J(P,H)$ = 709.1 Hz) and 6.95 ppm ($^1J(P,H)$ = 715.88 Hz). These signals have to be assigned to the P–H proton bonded to the phosphorus atom in the end group and in the repeating units, respectively. The multiple at δ = 4.15–4.25 ppm is characteristic for P–OCH$_2$ protons. This group is formed as a result of the transesterification of the methoxy group of dimethyl H-phosphonate with the hydroxyl group of PEG. In the $^{31}P\{H\}$NMR spectrum (Figure 4.2) of the reaction product obtained after heating dimethyl H-phosphonate with PEG, there are two signals at 11.17 and 10.46 ppm. The signal at 11.17 ppm in the ^{31}P NMR spectrum appears as a doublet of sextets with $^1J(P,H)$ = 709.0 Hz and $^3J(P,H)$ = 11.0 Hz and can be assigned to the phosphorus atom in the end group of poly(oxyethylene H-phosphonate)s bonded with OCH_3 and

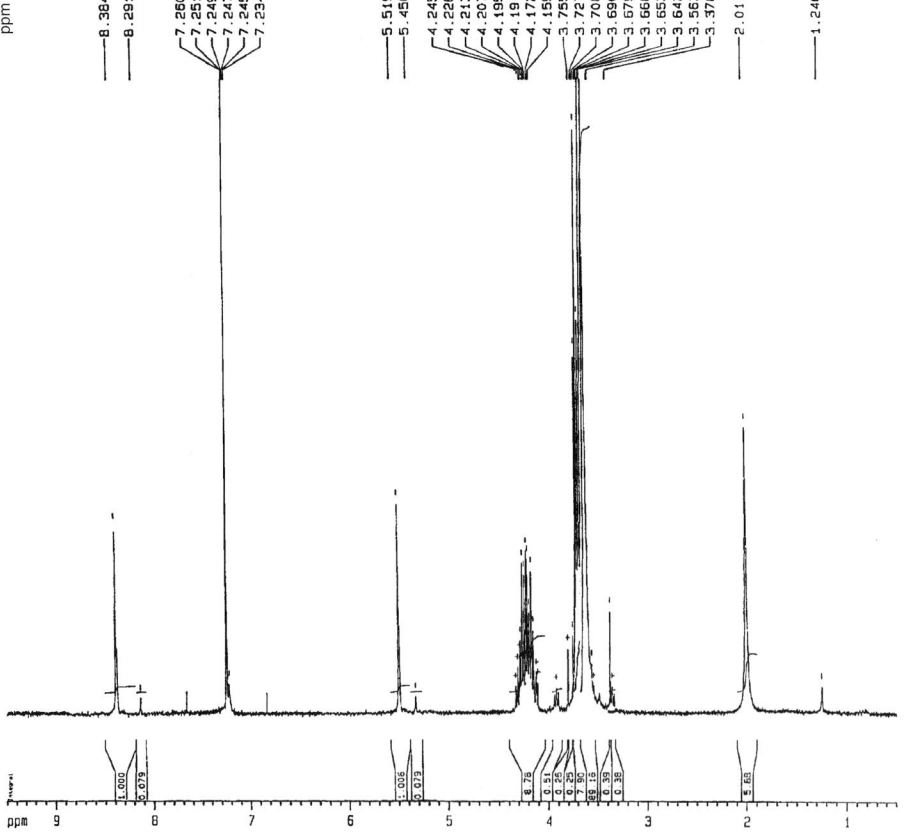

Figure 4.1 ^1H NMR spectrum of poly(oxyethylene H-phosphonate) based on dimethyl H-phosphonate and PEG.

4.5 Poly(alkylene H-phosphonate)s

OCH$_2$ groups. The signal at 10.46 ppm in the ^{31}P NMR spectrum appears as a doublet of quintets with ^1J(P,H) = 716.8 Hz and ^3J(P,H) = 9.8 Hz, and can be assigned to the phosphorus atom in the repeating unit of poly(oxyethylene H-phosphonate)s bonded with two OCH$_2$ groups.

The ^{13}C{H} NMR spectrum (Figure 4.3) also confirms the proceeding of the transesterification reaction. The doublets at 64.57 and 70.04 ppm with ^2J(P,C) = 6.2 Hz and ^3J(P,C) = 5.4 Hz, respectively, can be assigned to the POCH$_2$ and POCH$_2$CH$_2$O carbon atoms. The signal at 70.43 ppm can be assigned to the carbon OCH$_2$CH$_2$O atoms.

The molecular weight can be calculated from the ratio between the integral intensity of the P–H protons in the repeating unit at 6.95 ppm and the integral intensity of the P–H protons in the end groups at 6.87 ppm (Figure 4.1) divided by 2 (there are two end P–H protons). This gave the number of the repeating units, respectively, M_n [284]. This calculation was made on the fact that the intensity of a proton magnetic resonance line is proportional to the concentration of the relevant proton in the sample [285]. The number average degree of polymerization (DP) could also be estimated by ^{31}P{H} NMR spectroscopy. The ratio between the integral intensity of phosphorus atoms in the repeating unit at 10.46 ppm and the integral intensity of the phosphorus atom in the end groups at 11.17 ppm divided by 2 (there are two end phosphorus atoms) gave the number of the repeating units, respectively, M_n. The polymeric character of the poly(oxyethylene H-phosphonate)s was confirmed by size-exclusion chromatography (SEC).

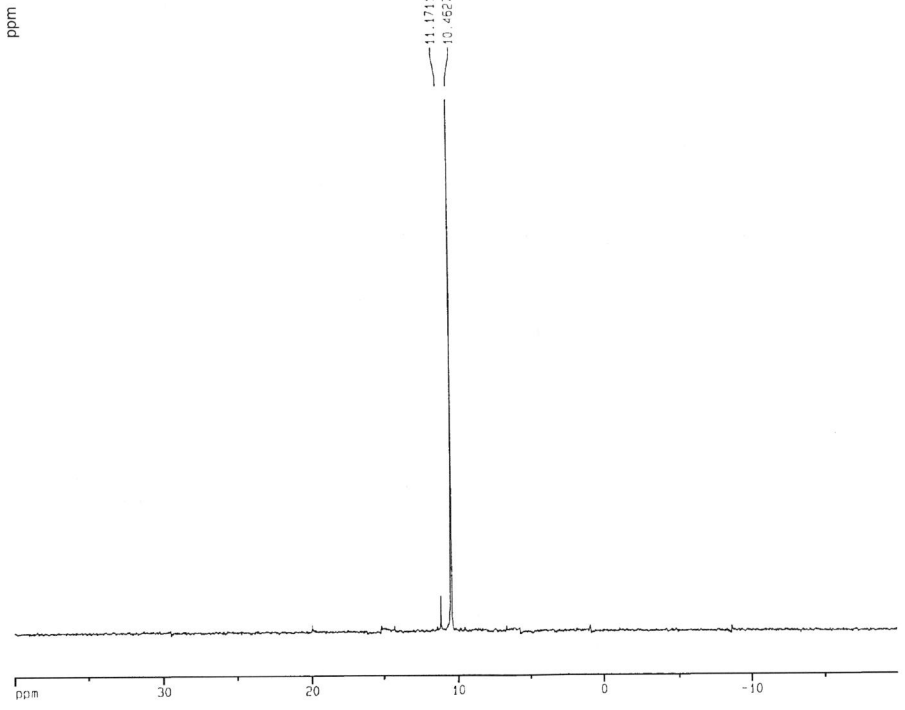

Figure 4.2 ^{31}P{H} NMR spectrum poly(oxyethylene H-phosphonate) based on dimethyl H-phosphonate and PEG.

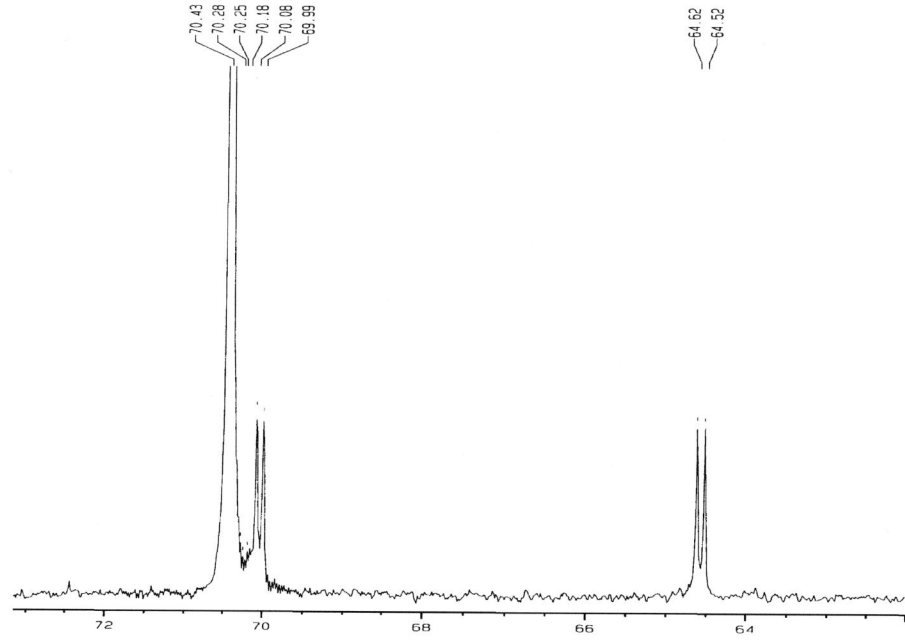

Figure 4.3 $^{13}C\{H\}$ NMR spectrum of poly(oxyethylene H-phosphonate).

Transesterification of dimethyl H-phosphonate is accompanied by side reactions, which represent a transfer of the methyl group of the methyl phosphonates end group to the hydroxyl-containing compound to form phosphonic acid end groups and ether compounds. The formation of these side products is due to the nucleophilic attack of the

4.5 Poly(alkylene H-phosphonate)s

the α-carbon atom, the second electrophlic center in the molecule of dimethyl H-phosphonate. The phosphonic end group and ether group do not participate in the transesterification reaction and prevent the formation of high molecular weight–polymers. The side reaction can be eliminated when diphenyl H-phosphonate is used as a starting compound [286] because in its molecule, the α-carbon atom does not play the role of an electrophilic center. Transesterification of diphenyl H-phosphonate

$$n\, C_6H_5\text{-O-P(O)(H)-O-}C_6H_5 + n\, HO\text{-R-}OH \xrightarrow{-(2n-1)\, C_6H_5\text{-OH}} C_6H_5\text{-O-}[P(O)(H)\text{-O-R-O}]_n\text{-P(O)(H)-O-R-OH}$$

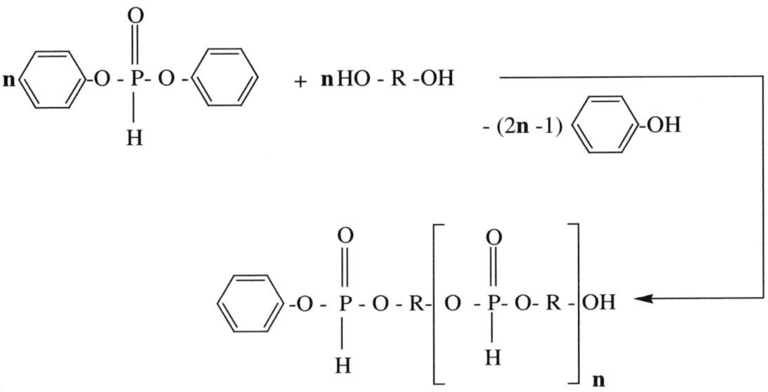

with hydroxyl-containing compounds is practically irreversible and can be performed without removing phenol. However, removing phenol enhances formation of poly(alkylene H-phosphonate)s with high molecular weight. Diphenyl H-phosphonate is more hydrolytically unstable than dialkyl esters of phosphonic acid.

The phosphonic end groups can be converted into reactive methyl phosphonate groups by treatment with diazomethane [287]. This allows the preparation of poly(alkylene H-phosphonate)s with number-average molecular weights greater than 10^4 Da. An alternative route for the preparation of high molecular weight poly(alkylene H-phosphonate)s avoiding the use of explosive and toxic diazomethane is to carry out the transesterification of dialkyl H-phosphonates with diols in the presence of Na_2CO_3 [288].

Poly(alklylene H-phosphonate) containing a nitrogen base in the side chain was synthesized by polycondensation of dialkyl H-phosphonate with 1-(2′,3′-dihydroxypropyl)imidazole [289].

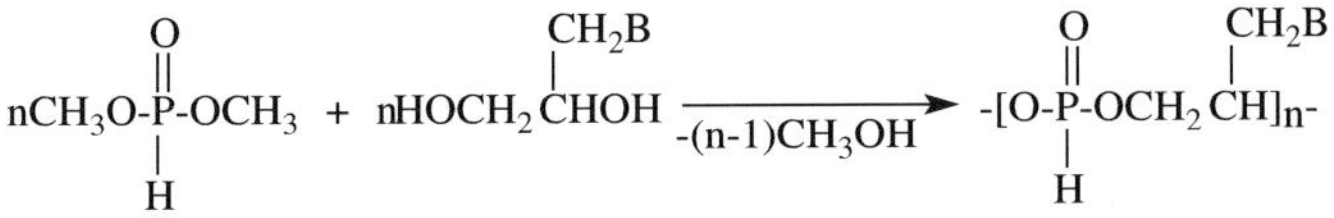

$$n\, CH_3O\text{-P(O)(H)-}OCH_3 + n\, HOCH_2CH(CH_2B)OH \xrightarrow{-(n-1)CH_3OH} -[O\text{-P(O)(H)-}OCH_2CH(CH_2B)]_n-$$

B – imidazole ring

Poly(alkylene H-phosphonate)s with a number-average molecular weight of about 3000 Da were obtained by the transesterification of dimethyl H-phosphonate with poly(ethylene glycol) (PEG 400) under microwave irradiation with a very short reaction time (55 min) relative to that of classical thermal heating (9 h) [290].

By polymerization

Polymerization of the 2-hydro-2-oxo-1,3,2-dioxaphosphorinane, initiated anionically or with aluminum alkyls, resulted in poly(alkylene H-phosphonate)s [291].

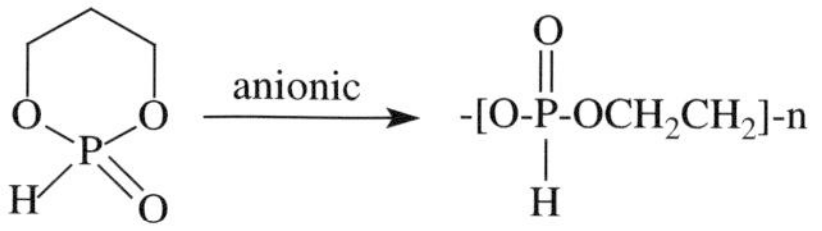

$$\text{[cyclic monomer]} \xrightarrow{\text{anionic}} -[O-\overset{O}{\underset{H}{\overset{\|}{P}}}-OCH_2CH_2]-n$$

Ring-opening polymerization of 4-methyl-2-oxo-1,3,2-dioxapholane yielded the following structures [292]:

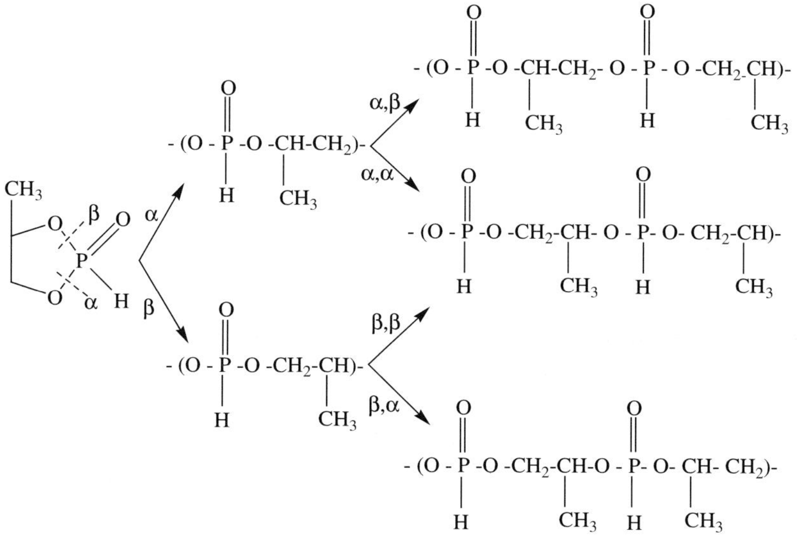

[Scheme showing ring-opening polymerization products with α,β; α,α; β,β; β,α regiochemistry:
- -(O-P(=O)(H)-O-CH-CH₂)- with CH₃
- α,β: -(O-P(=O)(H)-O-CH-CH₂-O-P(=O)(H)-O-CH₂-CH)- with CH₃ groups
- α,α: -(O-P(=O)(H)-O-CH₂-CH-O-P(=O)(H)-O-CH₂-CH)- with CH₃ groups
- -(O-P(=O)(H)-O-CH₂-CH)- with CH₃
- β,β: -(O-P(=O)(H)-O-CH₂-CH-O-P(=O)(H)-O-CH-CH₂)- with CH₃ groups
- β,α: -(O-P(=O)(H)-O-CH₂-CH-O-P(=O)(H)-O-CH-CH₂)- with CH₃ groups]

Teichoic acids are polyesters of phosphoric acid. There is a large variety of these polymers, but the common feature is a phosphoryl group in the backbone.

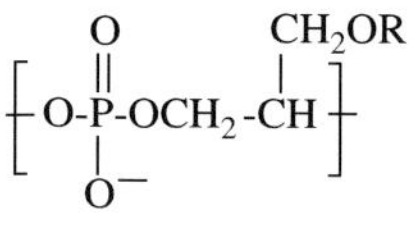

$$\left[O-\overset{O}{\underset{O^-}{\overset{\|}{P}}}-OCH_2-\overset{CH_2OR}{\underset{}{\overset{|}{CH}}} \right]$$

Teichoic acids are major components of cell walls and are responsible for a number of functions, including the flux of ions through the membranes [293]. The 'phosphonate route' has been applied to the synthesis of poly(1,2-glycerol phosphate), the simplest analogue of the natural teichoic acid. Polymerization of 4-acethoxymethyl-2-hydro-2-oxo-1,3,2-dioxaphospholane resulted in the formation of the following poly(alkylene H-phosphonate) [294]. Polymerization of 4-acethoxymethyl-2-hydro-2-oxo-1,3,2-dioxaphospholane is the equilibrium process (70% polymer and 30% starting monomer). The resulting polymer was oxidized by N_2O_4 to the corresponding polyphosphate.

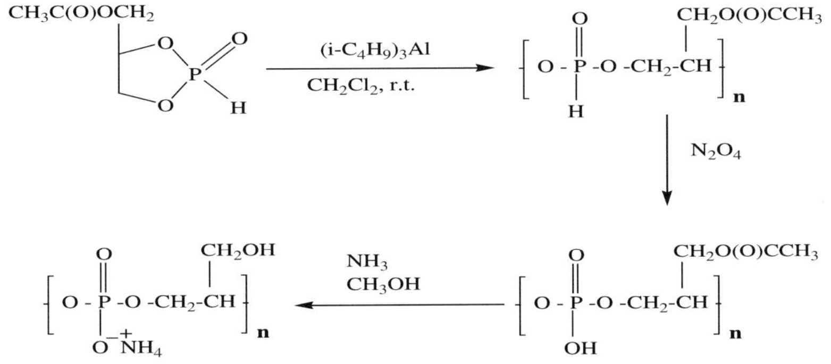

4.5 Poly(alkylene H-phosphonate)s

[reaction scheme showing cyclic acetoxymethyl H-phosphonate polymerized with (i-C$_4$H$_9$)$_3$Al in CH$_2$Cl$_2$, r.t. to give poly(acetoxyethyl H-phosphonate), then N$_2$O$_4$ oxidation and NH$_3$/CH$_3$OH treatment to give the ammonium salt]

The hydrolysis of the acetoxy group proceeds in one step to furnished poly(1,2-glycerol phosphate ammonium salt), the simplest analogue of the natural teichoic acid.

4.5.2 Reactivity of poly(alkylene H-phosphonate)s

The P–H group in poly(alkylene H-phosphonate)s is highly reactive in various reactions. The most important reaction are (i) hydrolysis, (ii) oxidation, and (iii) addition reactions to the double bonds and to carbonyl group.

Hydrolysis

Poly(alkylene H-phosphonate)s are hydrolytically unstable. A small amount of water easily could irreversibly break the main chain. The rate of hydrolysis was determined by direct titrimetric and ^{31}P NMR spectroscopic analyses. In the ^{31}P{H}NMR spectrum (Figure 4.4) of the partially hydrolyzed poly(oxyethylene H-phosphonate) appears a new signal at 8.37 ppm, which can be assigned to the phosphorus atom bonded to the OH group. It was established [279] that on increasing the pH of the medium, the rate of the hydrolysis increased but in all cases the degree of the hydrolysis was about 29%. This fact was explained by the formation of micelles, inside which the hydrolytically unstable ester bonds were protected from the water.

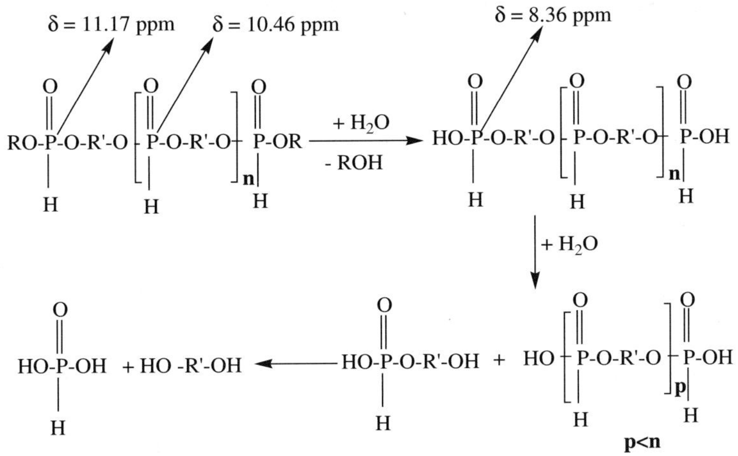

[scheme showing hydrolysis of poly(alkylene H-phosphonate) with chemical shifts δ = 11.17 ppm, δ = 10.46 ppm, and δ = 8.36 ppm assigned; + H$_2$O / – ROH leading to HO-terminated chains, further + H$_2$O giving HO-P(O)(H)-O-R'-OH + HO-[P(O)(H)-O-R'-O]$_p$-P(O)(H)-OH with p<n, and finally HO-P(O)(H)-OH + HO-R'-OH]

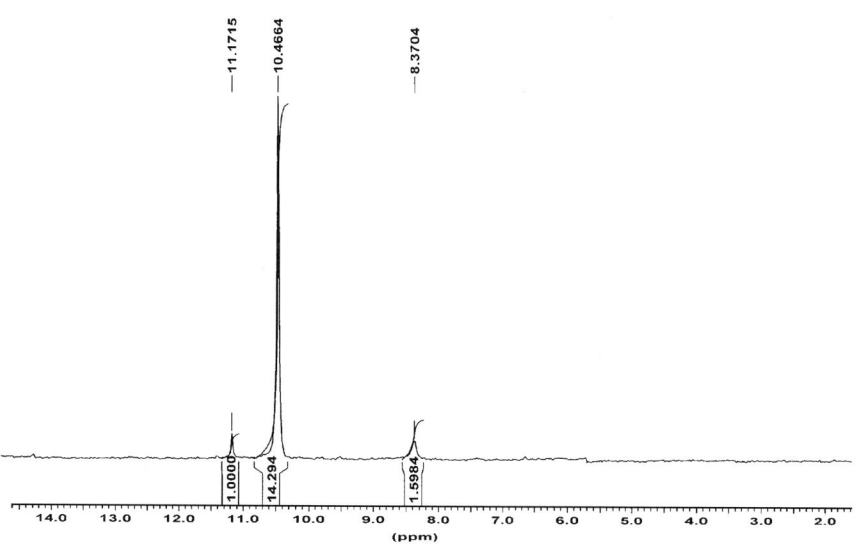

Figure 4.4. 31P{H}NMR spectrum of the partially hydrolyzed poly(oxyethylene H-phosphonate).

The initial stage of the hydrolysis involved cleavage of the alkoxy end groups. The rate of the hydrolysis of the end alkoxy groups was higher compared to that of P–O–C bonds in the repeating unit. The ^{31}P{H} NMR study of the hydrolytic stability of poly(oxyethylene H-phosphonate) showed that the degree of hydrolysis after 6 h at acid pH is about 20% and at basic pH, 36%. This result could be explained by the aggregation of poly(oxyethylene-H-phosphonate), which prevents further hydrolysis of the polymer. Aqueous size-exclusion chromatography study of poly(oxyethylene-H-phosphonate)s showed predominant self-assembly of these polymers in water [295]. Obviously, the acidic P–OH groups that are formed as a result of the hydrolysis of P–O–C bond participate in hydrogen bonding with P=O groups resulting in the formation of aggregates. At elevated temperatures, the final products of the hydrolysis were H-phosphonic acid and the starting hydroxyl-containing compound.

Oxidation

The Atherton–Todd reaction was used to transform poly(alkylene H-phosphonate)s into poly(alkylene phosphate)s [279].

4.5 Poly(alkylene H-phosphonate)s

The structure of poly(alkylene phosphate)s was confirmed by ^{1}H and ^{31}P NMR spectroscopy.

In the ^{31}P{H}NMR spectrum of the reaction product obtained after treatment of poly(oxyethylene H-phosphonate) with methanol in the presence of CCl$_4$ and triethyl amine (Figure 4.5), there are two signals at δ = 0.59 and 1.28 ppm, which confirm the phosphate structure of the product. The signal at 1.28 ppm represents an octet with ^3J(P,H) = 11.28 Hz and ^3J(P,H) = 8.61 Hz, which can be assigned to the phosphorus atom in the repeating unit. The absence of any resonance in the ^{1}H NMR spectrum for P–H protons is a direct proof that the reaction proceeds quantitatively. The signal at 0.59 ppm appears as a multiplet.

Poly(alkylene H-phosphonate)s were converted into the corresponding poly(alkylene phosphate)s using N$_2$O$_4$ in CH$_2$Cl$_2$ as an oxidizing agent [296].

$$RO\left[\!\!\begin{array}{c}O\\\|\\-P\text{-}O\text{-}R_1\text{-}O\text{-}\\|\\H\end{array}\!\!\right]_n H \xrightarrow{N_2O_4} RO\left[\!\!\begin{array}{c}O\\\|\\-P\text{-}O\text{-}R_1\text{-}O\text{-}\\|\\OH\end{array}\!\!\right]_n H$$

The hydrolysis of poly(oxyalkylene phosphate)s have been studied by Penczek et al. [297]. The rate of hydrolysis was determined by direct titrimetric and NMR methods. ^{1}H NMR gave quantitative information on the rate constants of hydrolysis of the main chain (k_m) and the side group (k_s). The rate constants for poly(methyl ethylene phosphate) **1** have been found to be close to those measured for the corresponding low molecular–weight models.

$$\left[\!\!\begin{array}{c}O\\\|\\\text{-}CH_2CH_2O\text{-}P\text{-}O\text{-}\\|\\CH_3O\end{array}\!\!\right]_q \begin{array}{c}O\\\|\\\text{-}CH_2CH_2O\text{-}P\text{-}OH\\|\\OCH_3\end{array} + H\!\!\left[\!\!\begin{array}{c}O\\\|\\\text{-}OCH_2CH_2O\text{-}P\text{-}\\|\\CH_3O\end{array}\!\!\right]_p$$

$$\left[\!\!\begin{array}{c}O\\\|\\\text{-}CH_2CH_2O\text{-}P\text{-}O\text{-}\\|\\OCH_3\end{array}\!\!\right]_n + H_2O \quad\begin{array}{c}\nearrow k'_m\\\\\searrow k'_s\end{array}\quad \left[\!\!\begin{array}{c}O\\\|\\\text{-}CH_2CH_2O\text{-}P\text{-}O\text{-}\\|\\OH\end{array}\!\!\right]_n$$

1

The following values for the pseudo first-order constant in s^{-1} have been measured for **1** at 45°C: k'_s: 3.18 × 10^{-5} at pH = 12.30; 3.76 × 10^{-8} at pH = 7.30; 5.45 × 10^{-7} at pH = 1.50. k'_m: 3.96 × 10^{-5} at pH = 12.30; 1.27 × 10^{-8} at pH = 7.30; 2.01 × 10^{-8} at pH = 1.50.

In acidic conditions, the methyl group hydrolyzes faster, whereas in basic conditions, both the methyl group and the main chain depart with approximately similar rates. This provides an interesting system in which side groups can be removed first and then the macromolecules will eventually hydrolyze at a lower rate.

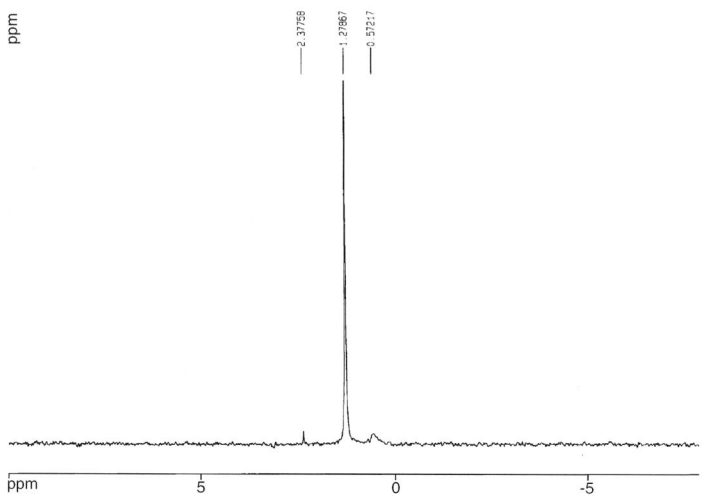

Figure 4.5. ^{31}P{H}NMR spectrum of the poly(oxyethylene phosphate).

For poly(trimethylene phosphate) **2**, following values have been measured at 70 °C: k_2 (in mol/L/s): 5.94×10^{-6} at pH = 11.70; k'_2 (pseudo second-order constant in s^{-1}): 4.60×10^{-9} at pH = 7.32; 1.14×10^{-8} at pH = 1.82.

$$\left[-(CH_2)_3O-\overset{\overset{O}{\|}}{\underset{OH}{P}}-O-\right]_n$$

2

$$\downarrow + H_2O$$

$$\left[-(CH_2)_3O-\overset{\overset{O}{\|}}{\underset{OH}{P}}-O-\right]_m(CH_2)_3O-\overset{\overset{O}{\|}}{\underset{OH}{P}}-OH + H-\left[O(CH_2)_3O-\overset{\overset{O}{\|}}{\underset{OH}{P}}-\right]_p$$

The results obtained show that the hydrolysis of the first group proceeds faster than the second one and the rate constant strongly depends on pH.

Poly(alkylene H-phosphonate)s were converted into the poly(alkylene phosphoramidate) through the Atherton–Todd reaction with amines [298].

$$RO\left[-\overset{\overset{O}{\|}}{\underset{H}{P}}-O-R_1-O-\right]_n H \xrightarrow{\underset{CCl_4/CH_2Cl_2}{R_2NH_2/Base}} RO\left[-\overset{\overset{O}{\|}}{\underset{NHR_2}{P}}-O-R_1-O-\right]_n H$$

4.5 Poly(alkylene H-phosphonate)s

Pharmacologically active ethyl 4-aminobenzoate or phenethylamine were used as amines.

Addition reactions

Addition to C=C double bond. Biodegradable polymers having pendant functional groups are of particular interest since they are capable of covalent prodrug formation and further functionalizations. The oxirane group is used for conjugation of bioactive substances due to the high reactivity of this group towards the amino function under mild conditions [299]. The presence of the highly reactive P–H group in the repeating unit of poly(alkylene H-phosphonate)s toward the C=C double bond makes possible the synthesis of poly(oxyethylene phosphonate)s bearing oxirane groups in the side chain by the addition of epoxyalkenes to poly(oxyethylene H-phosphonate)s [300].

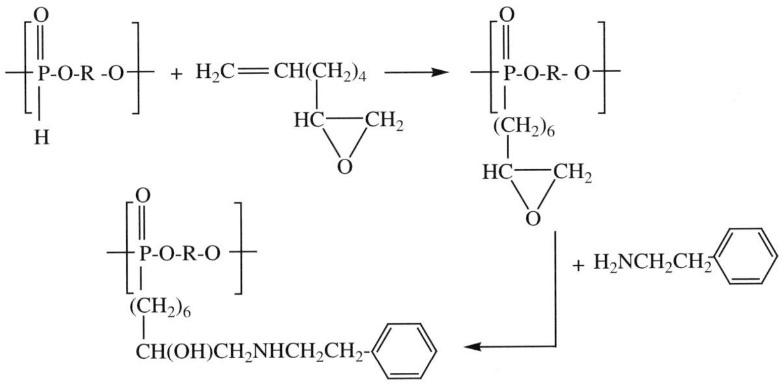

R = PEG

2-Phenethylamine was immobilized onto poly(oxyethylene 7,8-epoxy-1-octylphosphonate). The structure of the polymer was confirmed by means ^{1}H and ^{31}P NMR spectroscopy.

Addition to carbonyl group. Polymers bearing a hydroxyl group in their side chains have been widely used for biomedical application since the hydroxyl group makes such polymers reactive, hydrophilic, and biocompatible [301].

The reaction of poly(oxyethylene H-phosphonate) **1** with chloroacetone under phase-transfer catalysis conditions resulted in poly(alkylene phosphonate) bearing hydroxyl or oxirane groups. The reaction proceeds in two stages [302] (Scheme 4.1). In the first stage performed at 40°C the reaction results mainly in the formation of poly(oxyethylene α-hydroxyl phosphonates) **2**; at 60 °C the main product is poly(oxyethylene phosphonates) bearing oxirane groups **3**. The structure of the products **2** and **3** was proved by ^{31}P, ^{1}H, and ^{13}C{H} NMR spectroscopy. The ^{31}P{H} NMR spectrum of the reaction product shows resonances for three types of phosphorus atoms at

(i) δ = 24.82 ppm, a multiplet (from ^{31}P NMR) that can be assigned to the phosphorus atom in the α-hydroxyphosphonate group with a calculated content of 86.22%,

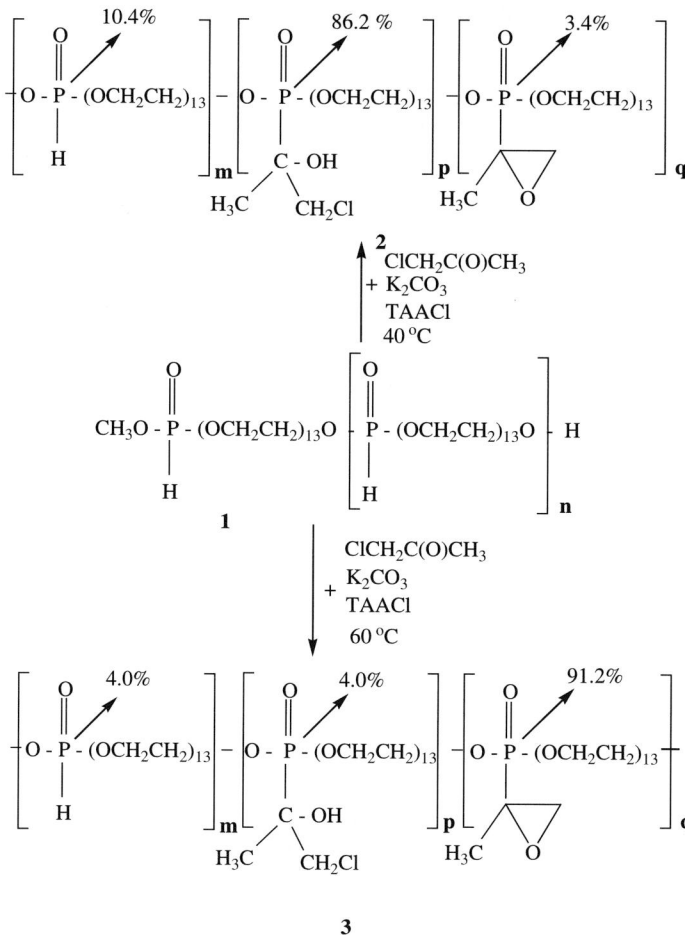

Scheme 4.1 Reaction of poly(oxyethylene H-phosphonate) with chloroacetone.

(ii) δ = 10.42 ppm, which is a doublet of quintets with coupling constants $^1J(P,H) = 715.8$ Hz and $^3J(P,H) = 9.9$ Hz, characteristic for the phosphorus atom in the repeating unit with a content of 10.4%, and

(iii) δ = 22.89 ppm, a multiplet assigned to the phosphorus atom connected with the oxirane group.

The calculated IC_{50} values are for poly(oxyethylene phosphonate) **1**, 2.96 mg/mL; for poly(oxyethylene phosphonate) bearing hydroxyl groups **2**, 0.012 mg/mL; and for poly(oxyethylene phosphonate) bearing oxirane groups **3**, 1.24 mg/mL.

Poly(oxyethylene phosphonate)s bearing α-hydroxyl and oxirane groups are of interest as carriers of low-molecular mass bioactive substances and as polymers with their own bioactivity.

4.5 Poly(alkylene H-phosphonate)s

4.5.3 Application of poly(alkylene H-phosphonate)s

Polymer–drug conjugates

Polymer–drug conjugates are designed to solve major problems in human medicine, namely, (i) the toxic side effects of the drugs, and (ii) the duration of drug action. The toxic side effects of the drugs to normal cells limit the dose of drugs that can be given to patients. Much research has focused on the development of more specific therapeutic strategies to reduce toxicity to normal cells. One of the promising strategies to suppress the toxicity of the drugs involves their conjugation with polymers [303–309]. The macromolecular approach to improve some characteristics of widely used low molecular–mass bioactive compounds enables the formation of unique types of therapeutics. The polymeric drugs are macromolecules that contain a drug unit attached to the backbone chain, or which exhibit drug action in the absence of such unit. One of the major challenges in polymeric drugs is the possibility of targeting the delivery of a therapeutic agent to a specific disease site or organ in the body. Polymer–drug conjugates are designed to improve not only drug localization in the target tissue and diminish drug exposure in potential sites of toxicity, but also to optimize drug-release rate. An interesting model showing the potentials of polymeric drug combinations was presented by Ringsdorf [303]. The properties of the polymer are directly responsible for defining the circulation

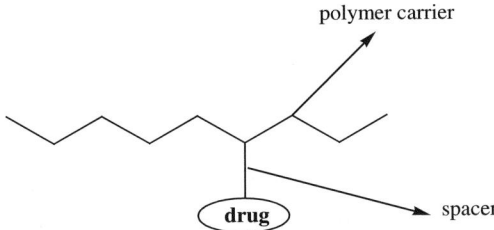

half life, rate of cellular uptake, minimizing toxicity of potent cytotoxic drugs, and imparting favorable physicochemical properties (e.g.; increasing solubility of lipophilic drugs). Unlike most low molecular–weight compounds, which tend to be systematically distributed and rapidly cleared from circulation, high molecular–weight compounds, due to of their size, exhibit prolonged circulation. The smaller ability of polymers to penetrate the cell membranes and to overcome various biological barriers limits their spreading over the human body. As a result, the pharmacokinetics of the conjugated drug changes entirely. Meanwhile, its ability to react with substrates and receptors is preserved. That is why conjugated drugs possess lower toxicity and a quite retarded activity compared to the low molecular–weight free drugs. The retarded effect results from the fact that the macromolecules are trapped by the cells in the reticule-endothelial system. An increasing number of soluble polymers has been used as macromolecular partners for pendent chain drug conjugation. These polymers can be organized into three broad classes: (i) nondegradable synthetic polymers, (ii) potentially biodegradable synthetic polymers, and (iii) natural polymers. Biodegradable polymers have garnered much of the recent attention and development in drug delivery systems because nonbiodegradable polymers need retrieval or further manipulation after introduction in the body.

Gebelein suggests that there are five methods of immobilizing biologically active substances; three are physical methods and two are chemical binding [310]. The three physical methods are (i) adsorption, (ii) matrix entrapment, and (iii) encapsulation. The methods by which these systems release the biologically active substance are (i) desorption, (ii) diffusion, and (iii) polymer degradation or destruction. The two chemical methods are (i) bonding in chain and (ii) bonding through crosslinking. The most common linkages are esters, anhydrides, and acetals [310]. The chemical degradation will most likely take place through a hydrolysis reaction, the anhydride being the easiest to hydrolyze. A cross-linking polymer will afford more resistance to degradation than linear polymers.

Historically, polymer–drug combinations have been achieved by

1. Linking the drug to the polymer via ionic bond–ionic polymer drug combinations. Ionic polymers used as drug carriers include soluble as well as insoluble (crosslinked) polymer systems. Different ionic macromolecules, ion exchange resins, have been investigated. The use of ionic polymer to prolong the effect of drugs is based on the principle that positively or negatively charged pharmaceuticals combined with appropriate polymer yield poly-salt:

$$\text{(R)-SO}_3^-\text{H}^+ + \text{H}_2\text{N-A} \rightleftharpoons \text{(R)-SO}_3^- \text{H}_3\overset{+}{\text{N}}\text{-A}$$

$$\text{(R)-}\overset{+}{\text{NH}}_3\text{OH}^- + \text{HOOC-B} \rightleftharpoons \text{(R)-}\overset{+}{\text{NH}}_3 \text{OOC-B} + \text{H}_2\text{O}$$

where H$_2$N-A and HOOC-B represent a basic and an acidic drug, respectively, and R-SO$_3^-$H+ and R-NH3+OH$^-$, a cationic and an anionic polymer.

2. Linking the drug to the polymer via covalent bond–covalent polymer drug combinations. In this type, the drug can be incorporated in the macromolecule (i) without linker **1**; (ii) with linker **2**; (iii) as a comonomer unit in the backbone **3**.

Poly(ethylene glycol) (PEG) is a nondegradable synthetic polymer that has been extensively studied as a polymer-drug carrier. PEG is hydrophilic and is well tolerated in human. The main disadvantage of PEG is that the polymer backbone is not biodegradable *in vivo*. PEG has been used to conjugate anticancer drugs such as doxorubicin [311], camptothecin

4.5 Poly(alkylene H-phosphonate)s

[312–314], and paclitaxel [315,316]. PEG–drug conjugates can be prepared by linking via one terminus or both termini of an end-functionalized PEG. Usually, the PEGs selected for protein conjugation have a molecular weight of 3400–5000 Da. The major disadvantage of PEG is the low drug-loading capacity, which is limited by the availability of only two attachment sites at the termini of the linear PEG macromolecule. However, PEG chains can be used as building blocks to construct functionalized polymers with low toxicity and reduced immunoreactivity. Branched PEG polymers [311], alternating PEG polymers (e.g.; PEG-lysine [317,318]), and copolymers (e.g.; PEG-oligopeptide-PEG [319,320]) with pendent chains are capable of conjugating a drug along the polymer backbone. PEG block copolymers that can form polymeric micelles have recently been prepared [321–324]. PEG was used for the preparation of poly(oxyethylene H-phosphonate)s. The incorporation of phosphonate units in the backbone of a polymer built of PEG guarantees solubility and degradation under physiological conditions, as well as feasible modification at the P-center.

Poly(oxyalkylene H-phosphonate)s have the following advantages: (i) they are water-soluble, (ii) the extent of polymer drug loading is not limited by the availability of only two reaction sites at the termini of the linear PEG molecule. The P–H, OH, COOH, NH_2, oxirane group, and P=O group in the repeating unit of the poly(oxyalkylene phosphonate)s determine various chemical functionalities, (iii) these polymers can be used to prepare amphiphilic polymers either with the hydrophobic main chain and hydrophilic side chain or vice versa with the exactly controlled length of hydrophobic and hydrophilic units, (iv) these polymers can be used to prepare gels, (v) they can be administered over a wider molecular weight range because after hydrolysis, the individual segments (low molecular PEG) will be safely excreted—the most important potential advantage of poly(oxyalkylene phosphonate)s, (vi) they can be regarded as degradable synthetic polymers. An important feature of these polymers is the exactly controlled polymer-drug ratio.

The Atherton–Todd reaction was used for the preparation of a new class of hydrophilic drug-carriers with an organic polyphosphate main chain. Pharmacologically active ethyl-4-aminobenzoate and phenethylamine were attached to poly(propylene H-phosphonate) and poly(3,6,9-trioxaundecamethylene H-phosphonate) [325].

$$\left[\begin{array}{c} O \\ \| \\ -P\text{-O-R-O}- \\ | \\ H \end{array} \right] \xrightarrow[CCl_4/CH_2Cl_2]{RNH_2/base} \left[\begin{array}{c} O \\ \| \\ -P\text{-O-R-O}- \\ | \\ NHR^1 \end{array} \right]$$

R = $CH_2CH(CH_3)CH_2$; $(CH_2CH_2O)_4$

R^1 = ⟨phenyl⟩-CH_2CH_2-; ⟨phenyl⟩-$\underset{\underset{O}{\|}}{C}OC_2H_5$

$^{31}P\{H\}$ NMR spectra of the final polymers indicate the formation of the P(O)NHR structure due to the presence of the signals around $\delta = 2$ ppm (aromatic phosphoramidates) and $\delta = 9$ ppm (aliphatic phosphoramidates).

2-Phenethylamine was immobilized on poly(oxyethylene phosphate)s, obtained by oxidation of poly(alkylene H-phosphonate)s, via ionic bonds [284] by alkylation reaction or by dealkylation reaction.

$$\left[\begin{array}{c} O \\ \| \\ P\text{-}O\text{-}R\text{-}O \\ | \\ H \end{array}\right] \xrightarrow[\text{CH}_3\text{OH}]{\text{oxidation}} \left[\begin{array}{c} O \\ \| \\ P\text{-}O\text{-}R\text{-}O \\ | \\ OCH_3 \end{array}\right]$$

$$\swarrow + RCH_2NH_2 \cdot HBr \qquad + RNH_2 \searrow$$

$$\left[\begin{array}{c} O \\ \| \\ P\text{-}O\text{-}R\text{-}O \\ | \\ O^- \ H_3\overset{+}{N}R \end{array}\right] \qquad \left[\begin{array}{c} O \\ \| \\ P\text{-}O\text{-}R\text{-}O \\ | \\ O^- \ H_2\overset{+}{N}R \\ | \\ CH_3 \end{array}\right]$$

R = ⟨phenyl⟩-CH$_2$CH$_2$-

The presence of methoxy groups in the poly(oxyethylene phosphate)s predetermines two possible ways for immobilization through an ionic bond of amine-containing biologically active substances: (i) alkylation reaction, (ii) dealkylation reaction. Both of them are due to the reactivity of the α-carbon atom of the methoxy group, which acts as an electrophilic center.

Immobilization of cysteamine hydrochloride onto poly(oxyethylene phosphate)s

Human and technological factors continue to contribute the necessity for efficient protection against nuclear radiation. Cysteamine (Cy) hydrochloride is a well-known conventional chemical radio protector. The first clinical applications of this substance were reported by Bacq several decades ago [326]. Cy was administered in single or multiple doses before or immediately after irradiation and it protected against the symptoms of radiation sickness. Its protective effect was manifested mainly by the more rapid recovery after the termination of the therapy of treated patients [327,328]. The need to use near-toxic amounts of radio-protective drugs to achieve adequate protection, however, still represents a major problem in human medicine [329]. If Cy could be immobilized on biodegradable polymers, it would be possible to reduce its toxicity while preserving its good radio-protective capability. One of the possible candidates for an immobilization template is the poly(oxyalkylene phosphates) (PAPh), a family of biodegradable, hydrophilic, and nontoxic polymers that contain suitable multifunctional groups for immobilization. The basic strategy for the synthesis of radio-protective substances involves a 'molecule combination' based on covalent or ionic bonds between the basic components [330,331]. Cy was immobilized on poly(oxyethylene phosphate)s **2**, obtained by oxidation of poly(oxyethylene H-phosphonate) **1**, by dealkylation reaction [307].

4.5 Poly(alkylene H-phosphonate)s

$$\left[\begin{array}{c} \text{O} \\ \| \\ \text{P-O-R-O} \\ | \\ \text{H} \end{array}\right]_n \xrightarrow[\text{CH}_3\text{OH}]{\text{oxidation}} \left[\begin{array}{c} \text{O} \\ \| \\ \text{P-O-R-O} \\ | \\ \text{OCH}_3 \end{array}\right]_n$$

1 → **2**

$+ \text{mHSCH}_2\text{CH}_2\text{NH}_2\cdot\text{HCl}$
$- \text{CH}_3\text{Cl}$

$$\left[\begin{array}{c} \text{O} \\ \| \\ \text{P-O-R-O} \\ | \\ \text{O}^-\ \overset{+}{\text{H}_3}\text{N(CH}_2)_2\text{SH} \end{array}\right]_m \left[\begin{array}{c} \text{O} \\ \| \\ \text{P-O-R-O} \\ | \\ \text{OCH}_3 \end{array}\right]_{(n-m)}$$

3

The results from the ^{31}P{H} NMR spectroscopy revealed that Cy was attached to the poly(oxyethylene phosphate) by an ionic bond. The elemental analysis of the chlorine revealed that the content of the free cysteamine hydrochloride in the products **3** is less than 1%.

The toxicity of the newly synthesized compounds **2**, (Mn = 1900, 3000, and 4700) and **3** were investigated in mice *in vivo*. The acute toxicity of poly(oxyethylene phosphate)s **2**, cysteamine, and of the complex [PAPh-Cy] were monitored up to 48 h after intraperitoneal injections where cysteamine dose corresponds to LD_{50} (265 mg/kg). Poly(oxyethylene phosphate)s **2** are practically nontoxic up to 2000 mg/kg. The investigation of the acute Cy toxicity yields the following data: maximum tolerable dose (LD_0)—167 mg/kg; LD_{50}— 265 mg/kg, as calculated by Probit analysis; and absolutely lethal dose (LD_{100})—377 mg/kg. The toxicity of the complex **3** (**m** = **n**, PEG 600, Mn = 4700 Da) is investigated at three distinct concentrations that would match the Cy toxic doses: 1037 mg/kg PAPh-Cy (167 mg/kg Cy); 1645 mg/kg PAPh-Cy (265 mg/kg Cy), and 2340 mg/kg PAPh-Cy (377 mg/kg Cy). It should be emphasized that for this time span, the mice survival rate is 100% even at the highest dose applied. It could be assumed that the immobilization of Cy on PAPh leads to a significant reduction of its toxicity.

The radioprotective effect of Cy before and after its immobilization on **2** was determined by the dose reduction factor (DRF) [332]. It is seen that the effect is most pronounced at high doses (1/2 LD_0), but it is also statistically significant at the minimal dose. The experiments with *E. coli* show the statistically significant radioprotection efficiency of the [PAPh-Cy] complex. The maximum DRF value attained is 1.996. The effect is distinctly expressed in all three doses investigated (1/2, 1/4, and 1/8 of Cy LD_0) while pure Cy did not show any protective action even after the administration at half the maximum tolerable dose. The calculated values for PF (protection factor) and PI (protection index) revealed that the immobilization of cysteamine on the poly(oxyethylene phosphate)s resulted in a well-expressed increasing of radioprotection. The degree of the immobilization (**m**) of the cysteamine onto the poly(oxyethylene phosphate)s had significant influence on the radio-protective effect of the preparations. Substance **3**, with degree of the immobilization 24.3%, protected 100% of the irradiated animals and therapeutic width of the preparation (RF = 2.00; PI = 19.33) better. The Cy dose applied with it was 8 times less than the optimal one used in radio-protection. A slightly expressed depot effect is observed with all complexes, the effect being clearly visible again with **3** (PEG 600, Mn = 4700 Da; m = 24%).

Poly(5'-AZT oxyethylene phosphate)s

Nucleoside analogues have demonstrated widespread utility as antiviral and anticancer therapeutics [333]. Among the available drugs for anti-HIV therapy, 3'-azido-2',3'-dideoxythimidine (AZT) is still one of the most effective. AZT-based treatment delays HIV progression and improves survival in patients with advanced disease. The efficiency of AZT inhibition of HIV-1 replication is dependent on several factors, the phosphorylation of AZT by cellular kinases to AZT monophosphate being one of the most important. The phosphorylation of AZT is the rate-limiting step for the formation of AZT triphosphate (AZT-TP), which is believed to interfere with retroviral RNA-dependant DNA polymerase [334]. The major therapeutic limitations of AZT are caused by its inherent clinical toxicities, which include bone marrow suppression, hepatic abnormalities, and myopathy [335]. These limitations are the driving force for developing strategies for designing AZT prodrugs. Most prodrug derivatives of AZT have been prepared by derivation of AZT at its 5'-O position [333,336] in order to reduce its toxicity and increase the anti-HIV activity [334,335b,337]. This approach is based on the assumption that the 5'-O-ester prodrugs will be hydrolyzed to AZT or AZT-monophosphate [335b].

Poly(5'-O-AZT oxyethylene phosphate)s (Poly-AZT) have been designed (Scheme 5.1) with the main goal being to decrease toxicity, to improve anti-HIV activity, to modify pharmacokinetic properties, to improve drug localization, and patient acceptability [338].

The complete conversion of **1**(**a**, m = 12; **b**, m = 8) into the corresponding phosphate structure **2a** is confirmed by the absence of any resonance with large $^1J(P,H)$ coupling constant of 700 Hz, characteristic for P–H protons in ^1H NMR spectrum of **2a**.

The solution behavior in aqueous media is an important evaluation factor in any drug screening. In water, the poly-AZT undergoes a well-pronounced self-assembly as shown by size-exclusion chromatography (SEC). Most of the substance (90% w) is organized in a micelle with apparent molecular mass of 94,000 Da, which elutes at 19.17 mL, while the rest exists in a monomolecular form (the peak at 24.84 mL). A preliminary hydrolytic screening is performed at 37.5 °C in buffered aqueous solution at pH = 3.5. Aliquots from the incubated solution are taken at 3-h intervals over a 24-h period and subjected to SEC analysis in THF. The data obtained indicate that the poly-AZT is gradually degrading into PEG oligomers eluting between 24.8 and 28.3 mL, and AZT conjugates (retention volume = 29.7 mL, Figure 4.6).

The cytotoxicity evaluation is performed with *CHO* and *BALB/c 3T3* cells using standard procedures [339]. The results from the comparative study show that the IC_{50} value of the original AZT is 1.63 and 1.35 mg/mL, for *CHO* and *BALB/c 3T3* lines, respectively. Remarkably, the IC_{50} value of Poly-AZT is 4.74 mg/mL for *CHO* and 3.40 mg/mL, when tested in *BALB/c 3T3* clone 31 cell cultures. The results generated by this study strongly indicate that the immobilization of AZT onto poly(oxyethylene H-phosphonate) enables the formation of a potent prodrug substance that has high water solubility, can be controllably

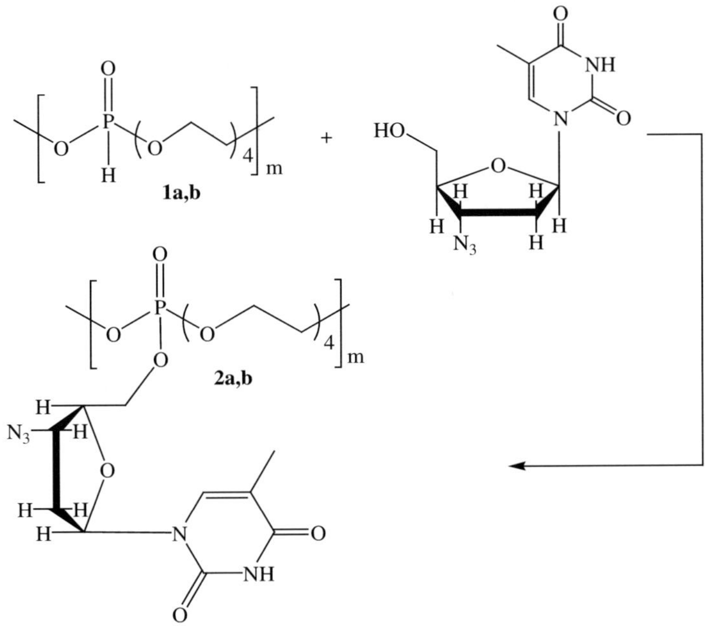

4.5 Poly(alkylene H-phosphonate)s

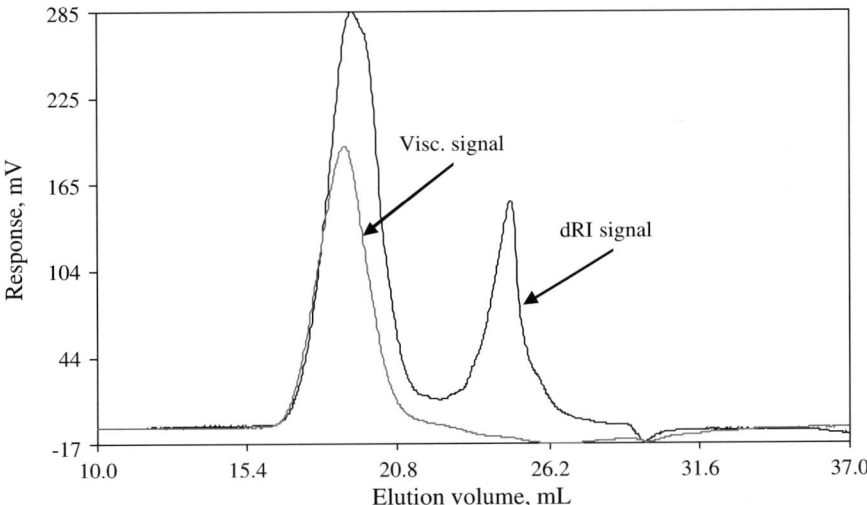

Scheme 5.1 Synthesis of poly(O'-5-AZT oxyethylene phosphate)s **2a** and **2b**.

Figure 4.6 Aqueous SEC of **2a**. Differential refractive index detector (dRI) and differential viscometric detector (Visc.) traces are shown in black and gray, respectively.

hydrolyzed in a medium mimicking the stomach environment, and has significantly decreased cytotoxicity in both cell lines investigated.

Poly(alkylene H-phosphonate)s are used for the preparation of novel biodegradable gene carrier **1** [340].

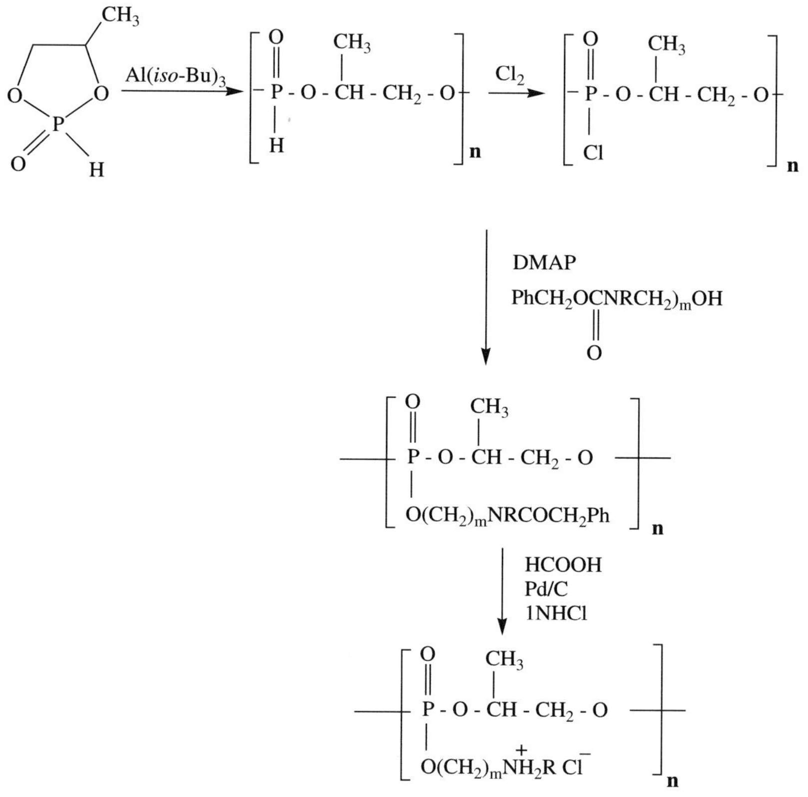

$$\text{cyclic H-phosphonate} \xrightarrow{\text{Al}(iso\text{-Bu})_3} \left[-\overset{\text{O}}{\underset{\text{H}}{\overset{\|}{P}}}-\text{O}-\overset{\text{CH}_3}{\underset{}{\text{CH}}}-\text{CH}_2-\text{O}- \right]_n \xrightarrow{\text{Cl}_2} \left[-\overset{\text{O}}{\underset{\text{Cl}}{\overset{\|}{P}}}-\text{O}-\overset{\text{CH}_3}{\underset{}{\text{CH}}}-\text{CH}_2-\text{O}- \right]_n$$

$$\downarrow \begin{array}{c} \text{DMAP} \\ \text{PhCH}_2\text{OCNRCH}_2)_m\text{OH} \\ \overset{\|}{\text{O}} \end{array}$$

$$\left[-\overset{\text{O}}{\overset{\|}{\underset{\text{O(CH}_2)_m\text{NRCOCH}_2\text{Ph}}{P}}}-\text{O}-\overset{\text{CH}_3}{\underset{}{\text{CH}}}-\text{CH}_2-\text{O}- \right]_n$$

$$\downarrow \begin{array}{c} \text{HCOOH} \\ \text{Pd/C} \\ \text{1N HCl} \end{array}$$

$$\left[-\overset{\text{O}}{\overset{\|}{\underset{\text{O(CH}_2)_m\overset{+}{\text{NH}_2}\text{R } \text{Cl}^-}{P}}}-\text{O}-\overset{\text{CH}_3}{\underset{}{\text{CH}}}-\text{CH}_2-\text{O}- \right]_n$$

1

R = H, m = 2
R = CH$_3$, m = 6

Polymer **1** is designed to have nontoxic building blocks. The ultimate degradation products are expected to be 1,2-propylene glycol, phosphate, and ethanolamine, all with minimal toxicity profiles. The polymer readily forms complexes with plasmid DNA. A unique feature of this system is the capability of controlled release of plasmid from the polymer/DNA complexes, achieved as a result of polymer degradation.

Polyphosphoramidate with a spermidine side chain as a gene carrier was synthesized from poly(1,2-propylene H-phosphonate) by the Atherton–Todd reaction [341].

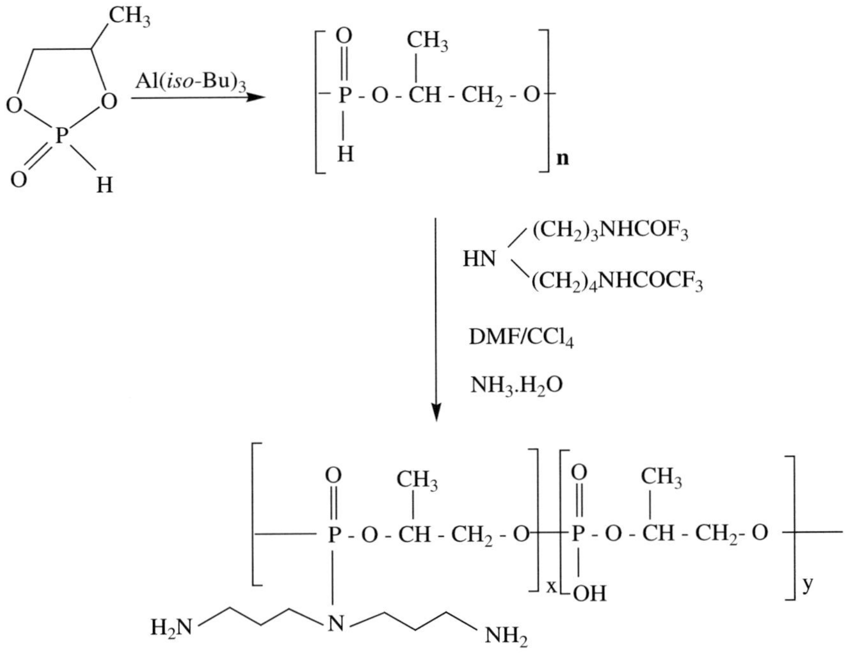

4.5 Poly(alkylene H-phosphonate)s

[Reaction scheme: cyclic H-phosphonate with CH₃ substituent reacts with Al(iso-Bu)₃ to give poly(H-phosphonate) [-P(=O)(H)-O-CH(CH₃)-CH₂-O-]ₙ; then reaction with HN((CH₂)₃NHCOF₃)((CH₂)₄NHCOCF₃) in DMF/CCl₄, followed by NH₃·H₂O, to give a copolymer with two repeating units, one bearing a –N(H)-(CH₂)₃-N(H)-(CH₂)₃-NH₂ (spermidine-like) side chain and the other bearing –OH, with degrees of polymerization x and y.]

The new gene carrier offered significant protection to DNA against nuclease degradation, and showed lower cytotoxicity than poly(-L-lysine) and poly(ethyleneimine).

Hydrogels for biomedical application

Hydrogels are cross-linked polymers that have the ability to swell in water or aqueous solvent systems but will not dissolve regardless of the solvent [342]. Wichterle and Lim developed the first synthetic polymeric hydrogel based on hydroxyethyl methacrylate, for use as biomedical materials [343]. There has been continuing interest in the development of novel types of hydrogel materials for biomedical application. Biodegradable hydrogels have been used quite extensively in the controlled release area. Poly(alkylene phosphate)s have been investigated as biodegradable materials, especially for use in drug delivery and orthopedic applications due to the fact that P–O–C bonds exhibit good hydrolytic degradability [344–346]. On the other hand, due to the recognized biocompatibility of PEG, PEG-containing polymers have been widely explored as biomaterial. PEG-containing hydrogels with biodegradable phosphoester bonds were synthesized through photo-initial cross-linking polymerization of methacryloyl polyphosphate [347].

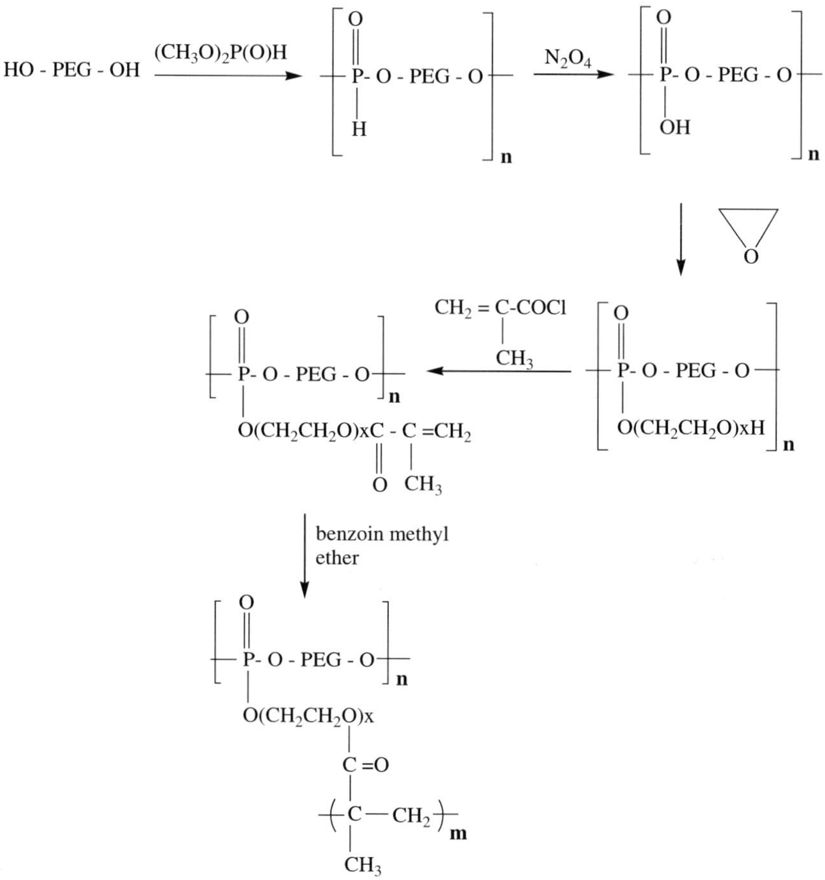

$$HO-PEG-OH \xrightarrow{(CH_3O)_2P(O)H} \left[\begin{matrix} O \\ \parallel \\ -P-O-PEG-O- \\ \mid \\ H \end{matrix} \right]_n \xrightarrow{N_2O_4} \left[\begin{matrix} O \\ \parallel \\ -P-O-PEG-O- \\ \mid \\ OH \end{matrix} \right]_n$$

$$\left[\begin{matrix} O \\ \parallel \\ -P-O-PEG-O- \\ \mid \\ O(CH_2CH_2O)xC-C=CH_2 \\ \parallel \quad \mid \\ O \quad CH_3 \end{matrix} \right]_n \xleftarrow{\begin{matrix} CH_2=C-COCl \\ \mid \\ CH_3 \end{matrix}} \left[\begin{matrix} O \\ \parallel \\ -P-O-PEG-O- \\ \mid \\ O(CH_2CH_2O)xH \end{matrix} \right]_n$$

↓ benzoin methyl ether

$$\left[\begin{matrix} O \\ \parallel \\ -P-O-PEG-O- \\ \mid \\ O(CH_2CH_2O)x \\ \mid \\ C=O \\ \mid \\ -(C-CH_2)_m \\ \mid \\ CH_3 \end{matrix} \right]_n$$

The hydrogel was loaded with 5-fluorouracil (5-FU). It is shown that the load and release of 5-FU can be controlled by the PEG segments molecular weight and the cross-linking density. The biodegradability of phosphate ester bonds contained in hydrogels is potentially useful for the elimination and removal of the drug delivery system after the entire drug is released.

4.6 METAL SALTS OF DIALKYL H-PHOSPHONATES

This chapter describes the various metal derivatives of dialkyl H-phosphonates. Metal salts of H-phosphonate diesters are an important class of compounds that are widely used as reagents, catalysts, stabilizers, and corrosion inhibitors.

Dialkyl H-phosphonates furnish in general two types of metal salts after appropriate treatment. The first type is obtained through deprotonation of the hydrogen atom of the P–H group and has a phosphite structure.

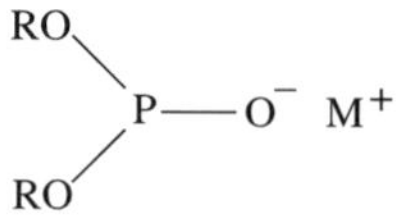

$$\begin{matrix} RO \\ \diagdown \\ P-O^- \; M^+ \\ \diagup \\ RO \end{matrix}$$

4.6 Metal Salts of Dialkyl H-phosphonates

The other type of metal salt is prepared via dealkylation of one or two alkyl substituents of the alkoxy groups of the dialkyl H-phosphonate. These salts retain their phosphonate structure since the immediate surrounding of the tetrahedral phosphorus atom remains unchanged.

$$RO-\overset{\overset{O}{\|}}{\underset{H}{P}}-\overset{-}{O}\;\overset{+}{M} \quad \text{or} \quad \overset{+}{M}\overset{-}{O}-\overset{\overset{O}{\|}}{\underset{H}{P}}-\overset{-}{O}\;\overset{+}{M}$$

4.6.1 Phosphite-type metal salts

In this section, the alkaline metal derivatives of the first type of metal salts of dialkyl H-phosphonates are described, namely, those having phosphite structure. The reactivity of these metal salts has been already discussed in the preceding chapters, since in most of the cases they are generated as intermediates and used *in situ* as phosphorylation reagents.

Alkali metal phosphite derivatives $(RO)_2P–OM$ are obtained from dialkyl H-phosphonates in two ways. The first one includes reaction of dialkyl H-phosphonates with alkali metals, usually in the presence of inert organic solvents such as diethyl ether or THF.

$$(RO)_2P(O)H \xrightarrow[-H_2]{M/THF} (RO)_2P-O^-\;M^+$$

$$M = Li, Na, K$$

Alternatively, these salts may be prepared from dialkyl phosphonates and alkali alkoxides or hydrides. The structure of the alkali metal derivatives of dialkyl H-phosphonates has been studied mainly by spectroscopic methods. IR spectra of the corresponding Li, Na, and K derivatives reveal the disappearance of the intense absorption band in the region 1240–1260 cm^{-1} corresponding to the P=O stretch, as well as the characteristic band in the region 2200–2400 cm^{-1}, resulting from the P–H stretch [348]. It has been assumed that these salts have phosphite structure with a three-coordinated phosphorus atom metal-bonded to the oxygen atom of the former phosphoryl group. ^{31}P NMR studies of the alkali salts of dialkyl phosphonates provide further support for this suggestion, since they reveal chemical shifts in the region 130–150 ppm, corresponding to phosphite-type phosphorus [349].

On the other hand, the moderate solubility of these compounds in organic solvents leads some authors to assume a dimeric structure for these salts where the metal is directly bonded to the phosphorus atom and the latter is four-coordinated [350].

$$\begin{array}{c} RO\diagdown\;\;\;\;O-----M\;\;\;\;\diagup OR \\ P\;\;\;\;\;\;\;\;\;\;\;\;\;\;\;P \\ RO\diagup\;\;\;\;M-----O\;\;\;\;\diagdown OR \end{array}$$

Although this assumption turns out to be correct in case of some other metals like mercury, there is no evidence to support it for the corresponding alkali metal derivatives. Indeed, an *ab initio* conformational analysis of the sodium salt of dimethyl phosphite carried out with the HF/6-31+G* basis set supports the phosphite character of this compound (Figure 4.7) [351].

The phosphorus atom is three-coordinated and tetrahedral with the fourth coordination position being occupied by the phosphorus lone pair. Computational search for a conformer having a four-coordinated phosphorus atom with direct P–Na bond using the same basis set as above does not indicate the presence of an energy minimum structure of that particular type.

Reactivity of metal phosphites

Michaelis–Becker reaction. The Michaelis–Becker reaction is known as a synthetic route for alkylation of dialkyl H-phosphonates. Its standard procedure involves an interaction between alkali salts of dialkyl phosphites and alkyl halides [352].

$$(RO)_2POM + R'X \longrightarrow RO)_2P(O)R' + MX$$

Numerous synthetic applications of this reaction have been already reviewed in the early literature [353]. The Michaelis–Becker reaction takes place as a bimolecular nucleophilic substitution [354]. Direct evidence supporting the S_N2 mechanism of this process is provided by the detected inversion of the configuration at the α-carbon atom of the alkyl halide in the product of the reaction between sodium diethyl phosphonate and methanesulfonates (mesitylates) [355]. On the other hand, it has been established for related phosphoryl compounds that they do not change the configuration at the phosphorus atom in the course of the Michaelis–Becker reaction [356–360]. One of the specifics of this process is the possibility of both O-, and P- alkylation occuring as a result of the ambient character of the dialkyl phosphite anion.

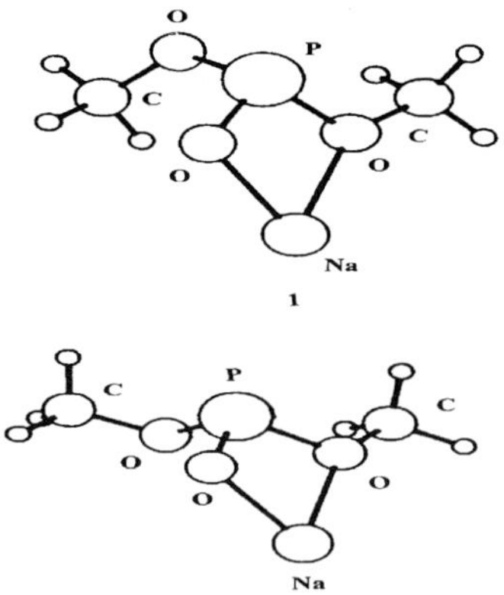

Figure 4.7 HF/6–31+G* optimized geometry of the conformers for the sodium salt of dimethyl phosphite.

4.6 Metal Salts of Dialkyl H-phosphonates

$$(RO)_2P-O^- \longleftrightarrow (RO)_2\overline{P}=O$$

An example illustrating this possibility is the reaction of sodium diethyl phosphite with methanesulfonate of *cis*- and *trans-tert*-butyl-4-cyclohexanols. The products of this interaction are 3-*tert*-butylcyclohexene **I**, (4-*tert*-butyl) cyclohexyl diethylphosphite **II** and (4-*tert*-buty) cyclohexanedialkylphosphonate **III** [355].

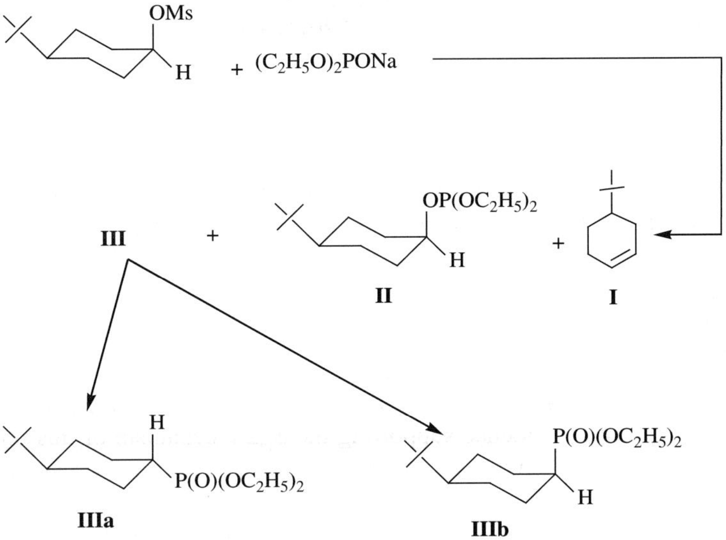

These products result from three types of reactions that take place in this system—elimination, O-alkylation, and P-alkylation. Quantitative O-alkylation takes place in dimethylformamide, while in dioxane and toluene, the yield of the O-alkylated products varies from 56 to 70%. The formation of the two products corresponding to O-alkylation and P-alkylation is explained by the authors in terms of Pearson's theory of 'hard' and 'soft' acids and bases, with phosphorus being the soft and oxygen the hard base. The elimination is suggested to take place as an attack of the oxygen atom of the dialkyl phosphite anion, considered as a 'hard' base toward the H atom in the β position with respect to the mesityl group:

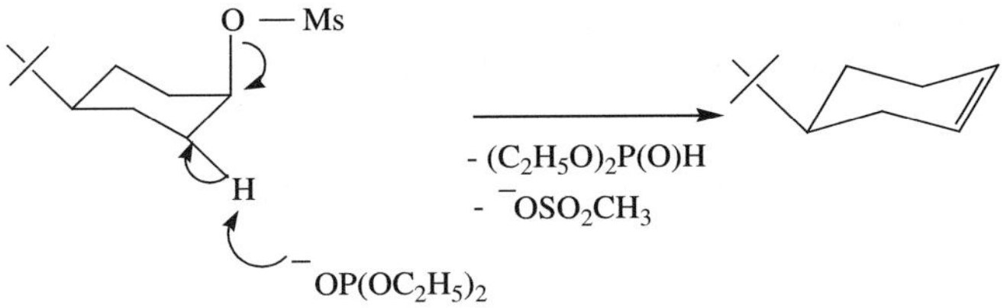

The concept of soft and hard acids and bases, although very useful, is not quantitative and cannot account for some of the processes in solution concerning the type of the ion pairs formed under the above conditions, which may be an important factor for the reaction's outcome.

Halides of some hydrocarbons, containing multiple carbon–carbon bonds as well as other functionalities, have also been phosphorylated via the Michaelis–Becker procedure [361–363]. The modern tendency in the application of the Michaelis–Becker synthesis is to carry out a one-pot synthesis without the necessity of intermediate isolation of the dialkyl phosphite salt. It has been shown that the Michaelis–Becker reaction can be carried out in the presence of potassium carbonate and a phase-transfer catalyst (tetrabutyl ammonium bromide—TBAB) [364].

$$(RO)_2P(O)H + PhCH_2Cl \xrightarrow{K_2CO_3 / TBAB} PhCH_2P(O)(OR)_2$$

This method is based on the fact that potassium carbonate plays the role of an effective base for deprotonation of organic acids such as dialkyl H-phosphonates [365]. The advantage of this method is the possibility of a direct application of dialkyl H-phosphonates instead of their alkaline salts.

It was also established that alkylation of dialkyl H-phosphonates can be carried out also without phase-transfer catalysts in the presence of potassium carbonate [366]. In this case, the difficulties connected with the separation of the phase-transfer catalyst from the reaction product are avoided. This is one of the advantages of this method. This procedure is the only one so far that allows for Michaelis–Becker alkylation of dimethyl H-phosphonate by alkyl halides.

$$(MeO)_2P(O)H + CH_3I \xrightarrow{K_2CO_3} CH_3P(O)(OMe)_2$$

Better results and higher yields have been obtained when the above reaction is carried out in the presence of cesium carbonate instead of potassium carbonate [367]. Thus, O,O-dimethyl-butyl phosphonate has been synthesized from dimethyl H-phosphonate and butyl bromide in 71% yield:

$$(MeO)_2P(O)H + BrCH_2CH_2CH_2CH_3 \xrightarrow{Cs_2CO_3} H_3CCH_2CH_2CH_2P(O)(OMe)_2$$

It was shown that dialkyl arylphosphonates were prepared in good yield by the reaction of aryl iodides or bromides with dialkyl H-phosphonates in the presence of triethylamine and a catalytic amount of tetrakis(triphenylphosphine)palladium [368].

$$ArX + HP(O)(OR)_2 \xrightarrow[Et_3N]{cat.\ Pd(PPh_3)_4} Ar\text{-}P(O)(OR)_2 + Et_3N.HX$$
$$X = Br; I$$

The palladium-catalyzed phosphonylation was successfully applied to stereoselective synthesis of dialkyl vinylphosphonates from vinyl bromides and dialkyl H-phosphonates [368].

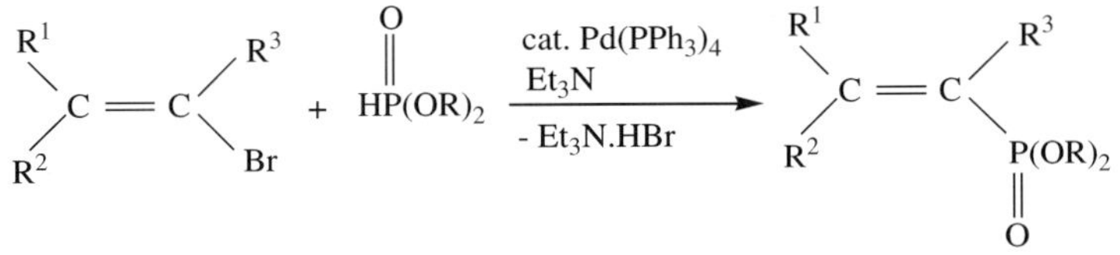

4.6 Metal Salts of Dialkyl H-phosphonates

$$\underset{R^2}{\overset{R^1}{>}}C=C\underset{Br}{\overset{R^3}{<}} + HP(OR)_2 \xrightarrow[- Et_3N.HBr]{\text{cat. Pd(PPh}_3)_4 \atop Et_3N} \underset{R^2}{\overset{R^1}{>}}C=C\underset{\underset{O}{\overset{\|}{P(OR)_2}}}{\overset{R^3}{<}}$$

Dialkyl arylphosphonates are synthesized by the reaction of the corresponding sodium salts of dialkyl phosphites with arylhalides in the presence of copper (I) iodide [369].

$$Ar-X + NaO-P(OR)_2 \xrightarrow{CuI} Ar-P(O)(OR)_2$$

This method is also suitable for the synthesis of steric-hindered dialkyl arylphosphonates. The yields of dialkyl arylphosphonates are, however, lower than in the case where cupric iodide is used as a catalyst.

The Michaelis–Becker reaction between the sodium diethyl phosphite and either of the halofluoro methanes, CF_2Cl_2 or CF_3Br, gives tetraethyl difluoromethylene bisphosphonate [370,371].

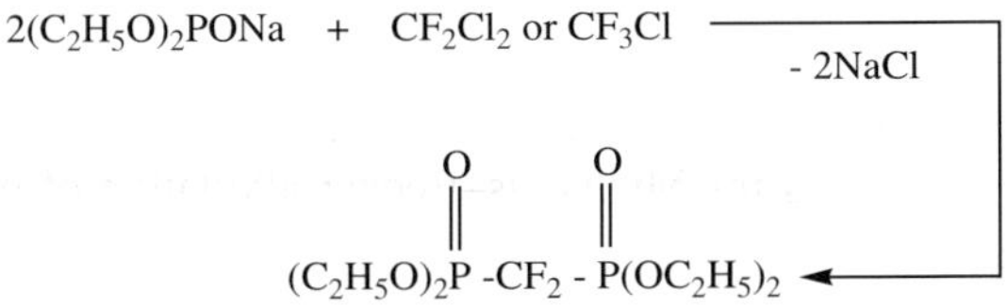

$$2(C_2H_5O)_2PONa + CF_2Cl_2 \text{ or } CF_3Cl \xrightarrow{- 2NaCl} (C_2H_5O)_2\overset{O}{\overset{\|}{P}}-CF_2-\overset{O}{\overset{\|}{P}}(OC_2H_5)_2$$

The sodium salts of dialkyl phosphite, except for the dimethyl homologue, react with dialkyl H-phosphonates to form dialkyl alkylphosphonate and the corresponding sodium salt of monoalkyl H-phosphonic acid [372].

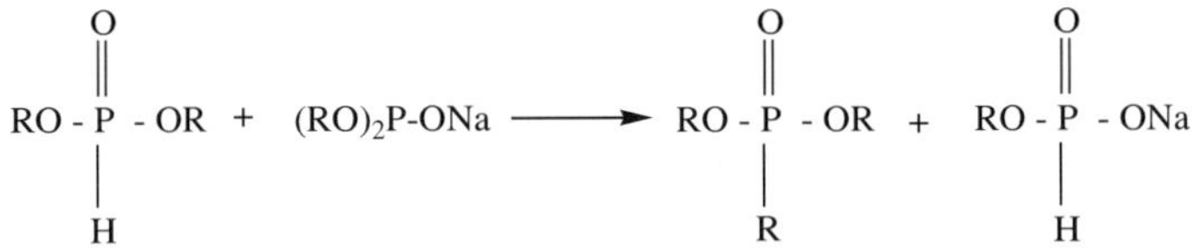

$$\underset{H}{\overset{O}{\overset{\|}{RO-P-OR}}} + (RO)_2P-ONa \longrightarrow \underset{R}{\overset{O}{\overset{\|}{RO-P-OR}}} + \underset{H}{\overset{O}{\overset{\|}{RO-P-ONa}}}$$

The same reaction with dimethyl H-phosphonate occurs in a different way with formation of dimethyl ether, according to the following scheme [372]:

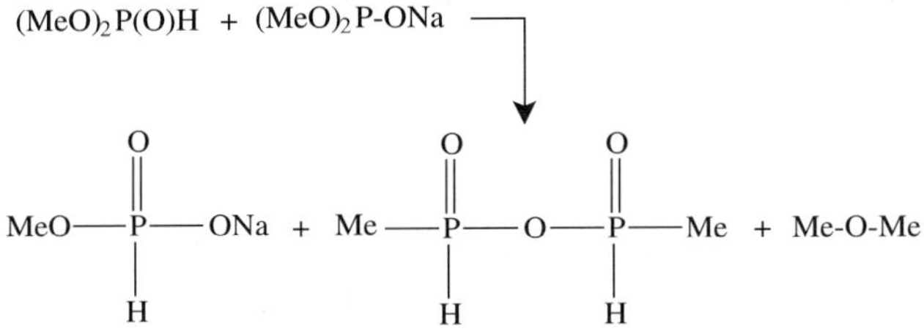

$$(MeO)_2P(O)H + (MeO)_2P-ONa \longrightarrow$$

$$\underset{H}{\overset{O}{\overset{\|}{MeO-P-ONa}}} + Me-\underset{H}{\overset{O}{\overset{\|}{P}}}-O-\underset{H}{\overset{O}{\overset{\|}{P}}}-Me + Me-O-Me$$

Sodium salts of dialkyl phosphite undergo the following thermal changes [373]:

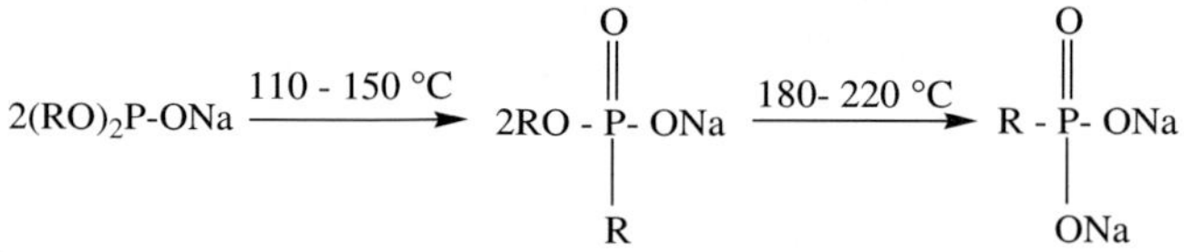

$$2(RO)_2P\text{-}ONa \xrightarrow{110\text{-}150\,°C} 2RO\underset{R}{\overset{O}{\underset{\|}{-P-}}}ONa \xrightarrow{180\text{-}220\,°C} R\underset{ONa}{\overset{O}{\underset{\|}{-P-}}}ONa$$

Reaction of sodium diethyl phosphite and diethyl chlorophosphate leads to a P–P coupling with the formation of the corresponding tetraethyl diphosphonate [374].

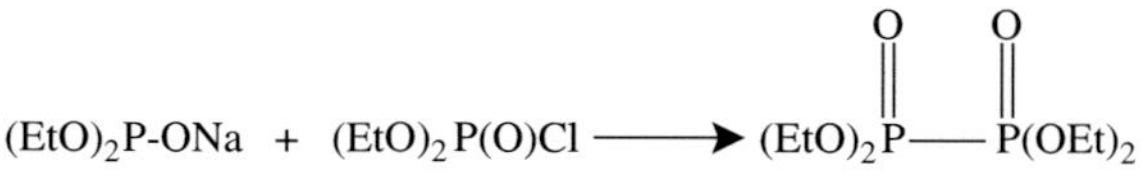

$$(EtO)_2P\text{-}ONa + (EtO)_2P(O)Cl \longrightarrow (EtO)_2\overset{O}{\overset{\|}{P}}\text{—}\overset{O}{\overset{\|}{P}}(OEt)_2$$

Depending on the temperature conditions, carbon disulfide can be phosphorylated by sodium dialkyl phosphites either at the carbon or at one of the sulfur atoms. At relatively low temperatures (2–8 °C), the phosphorus-containing xanthate **I** is obtained [375], whereas at higher temperatures (70–100 °C), the O,O-diethyl ester of S-thioacetyl-thiophosphoric acid **II** is the major product [376].

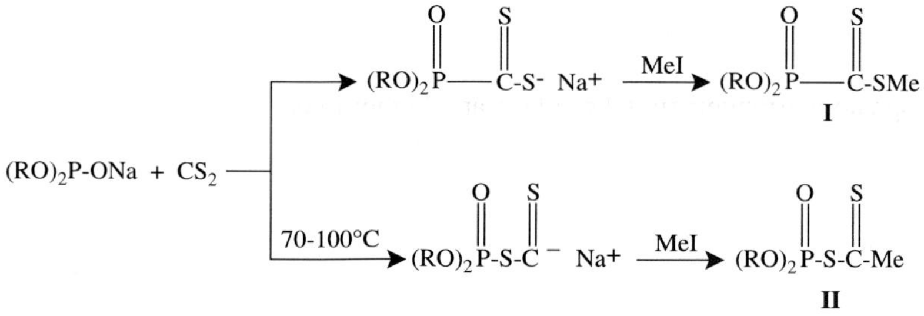

$$(RO)_2P\text{-}ONa + CS_2 \longrightarrow \begin{cases} (RO)_2\overset{O}{\overset{\|}{P}}\text{—}\overset{S}{\overset{\|}{C}}\text{-S}^-\,Na^+ \xrightarrow{MeI} (RO)_2\overset{O}{\overset{\|}{P}}\text{—}\overset{S}{\overset{\|}{C}}\text{-SMe} \quad \mathbf{I} \\ \xrightarrow{70\text{-}100\,°C} (RO)_2\overset{O}{\overset{\|}{P}}\text{-S-}\overset{S}{\overset{\|}{C}}^-\,Na^+ \xrightarrow{MeI} (RO)_2\overset{O}{\overset{\|}{P}}\text{-S-}\overset{S}{\overset{\|}{C}}\text{-Me} \quad \mathbf{II} \end{cases}$$

Treatment of oximes with sodium dialkyl phosphites leads to the formation of N-phosphorylated imines according to the following scheme [377]:

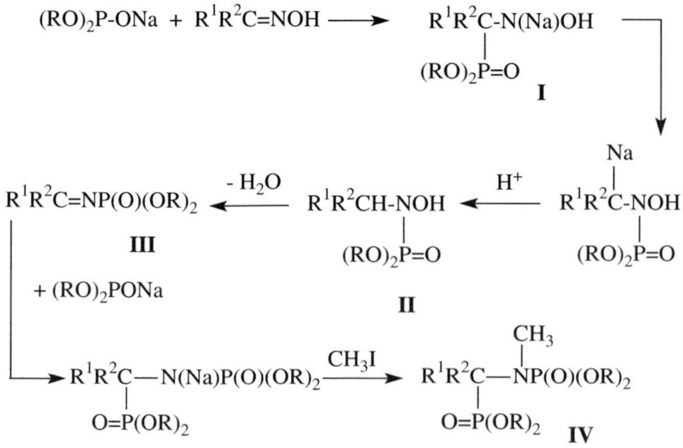

$$(RO)_2P\text{-}ONa + R^1R^2C=NOH \longrightarrow \underset{(RO)_2P=O}{R^1R^2C\text{-}N(Na)OH} \quad \mathbf{I}$$

$$R^1R^2C=NP(O)(OR)_2 \xleftarrow{-H_2O} \underset{(RO)_2P=O}{R^1R^2CH\text{-}NOH} \xleftarrow{H^+} \underset{(RO)_2P=O}{R^1R^2\overset{Na}{C}\text{-}NOH}$$

III + (RO)_2PONa   **II**

$$\longrightarrow \underset{O=P(OR)_2}{R^1R^2C\text{—}N(Na)P(O)(OR)_2} \xrightarrow{CH_3I} \underset{O=P(OR)_2}{R^1R^2C\text{—}\overset{CH_3}{N}P(O)(OR)_2} \quad \mathbf{IV}$$

4.6 Metal Salts of Dialkyl H-phosphonates

The proposed mechanism for this reaction involves initial formation of the corresponding sodium derivative of the *N*-hydroxyamino phosphonate (**I**). The structure of compound (**II**) was proved by ^{31}P NMR spectroscopy. The signal for the phosphorus nuclei appears at $\delta = 9$ ppm, which is characteristic for amidophosphates. This compound undergoes further aminophosphonate–amidophosphate rearrangement yielding the hydroxyamido phosphate (**II**), which transforms to the final phosphorylated imine (**III**) upon water cleavage. Methylamino phosphonates are the only type of compounds known to undergo the aminophosphonate–amidophosphate rearrangement [378]. It is suggested that a P–C bond cleavage and a P–N bond formation in the above scheme take place as a result of an intramolecular attack of the negatively charged nitrogen atom in (**I**) at the electrophilic phosphorus center [375]. It is known that only aminophosphonates or their sodium-N derivatives, which have electroacceptor substituents at the aminocarbon atom [308], undergo aminophosphonate–amidophosphate rearrangement. The aminophosphonate–amidophosphate rearrangement takes place only with sodium or potassium salts of dialkyl phosphites. The corresponding magnesium salts yield a mixture of both products. An isomerization of the N-lithium salt of the aminophosphonate does not take place at all. This difference in the reactivity of the different metal salts is explained in terms of different nucleophilicity of the nitrogen atom in the addition product (**I**) [379]. When M equals H or Li in the above reaction, a rearrangement does not take place because of the low nucleophilicity of the nitrogen atom, while for the Na, K, and Mg salts, the bond ionization is considered to be significantly stronger. This makes the nitrogen atom more nucleophilic and the aminophosphonate–aminophosphate rearrangement takes place at a higher rate.

The sodium diethyl phosphite reacts with aromatic aldehydes yielding *trans*-stilbene-type compounds [380].

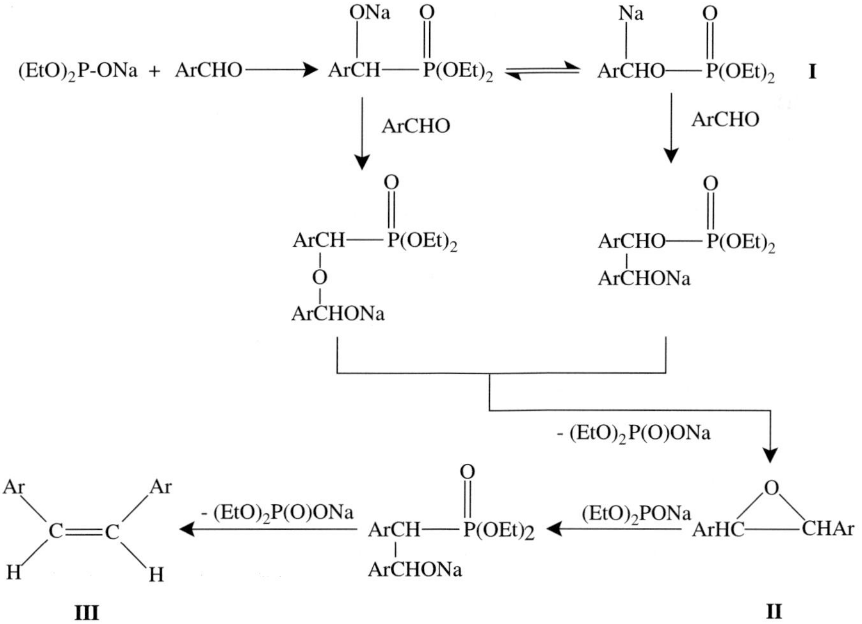

The phosphate carbanion **I** obtained as a result of phosphonate–phosphate rearrangement reacts further with a second molecule of aldehyde, forming 1,2-diphenyloxirane (stilbenic oxide) **II**. The latter undergoes the Horner–Emons reaction to furnish *trans*-stilbene **III**.

Sodium diethyl phosphite reacts with fluorinated alkylketones at temperatures in the −10 °C to 0 °C range, yielding substituted fluoroalkenyl phosphates via phosphonate–phosphate rearrangement of the initially formed β-carbonyl phosphonates [381].

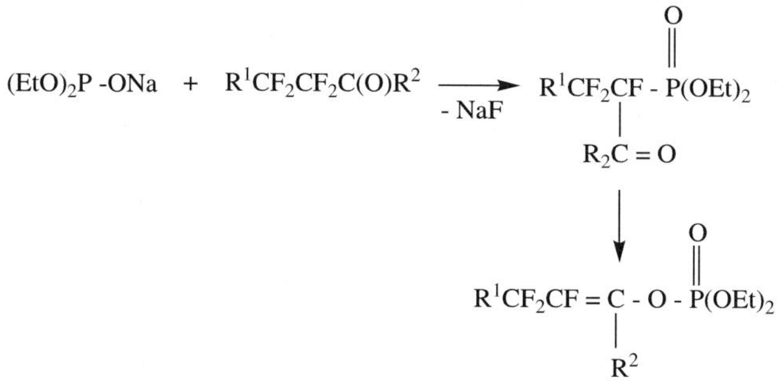

$$(EtO)_2P\text{-ONa} + R^1CF_2CF_2C(O)R^2 \xrightarrow[-NaF]{} \underset{\underset{R_2C=O}{|}}{R^1CF_2CF\text{-}P(OEt)_2\!=\!O}$$

$$\downarrow$$

$$R^1CF_2CF=\underset{\underset{R^2}{|}}{C}\text{-}O\text{-}P(OEt)_2\!=\!O$$

N-alkylpyridinium salts undergo dearomatization and phosphorylation by sodium dialkyl phosphites in the ortho- and para- positions with respect to the nitrogen atom [382,383].

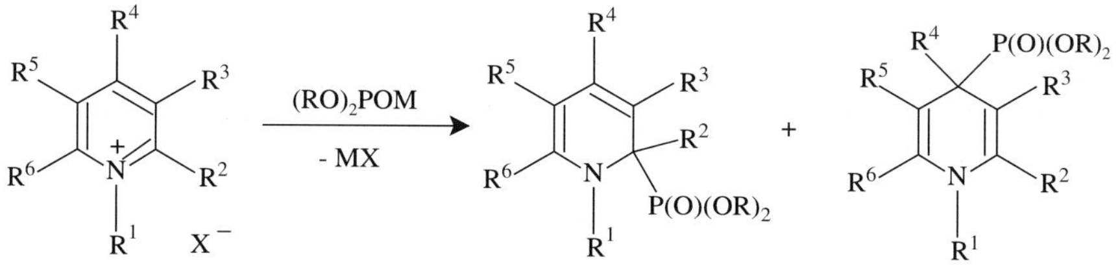

N-alkoxypyridinium salts, on the other hand, retain their aromatic structure under the above conditions via elimination of the alkoxy group [384,385].

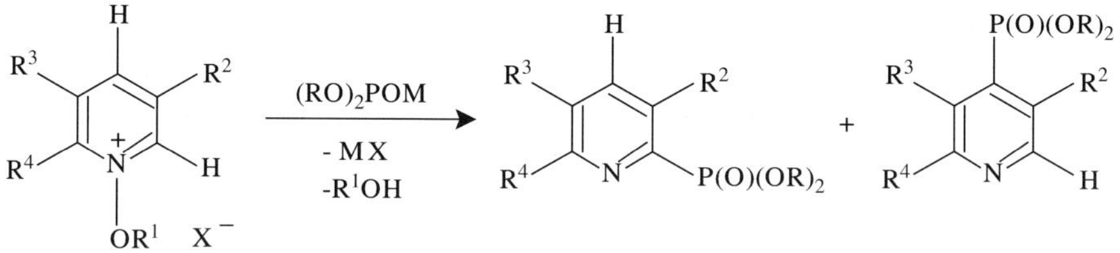

Heating the diethyl ester of 1,2-epoxy-2-phenyl-ethanephosphonic acid with equimolar quantities of dialkyl H-phosphonate and the sodium salt of dialkyl phosphite yields the diethyl ester of 2-phenyl-2-dialkylphosphonoethane phosphonic acid [386].

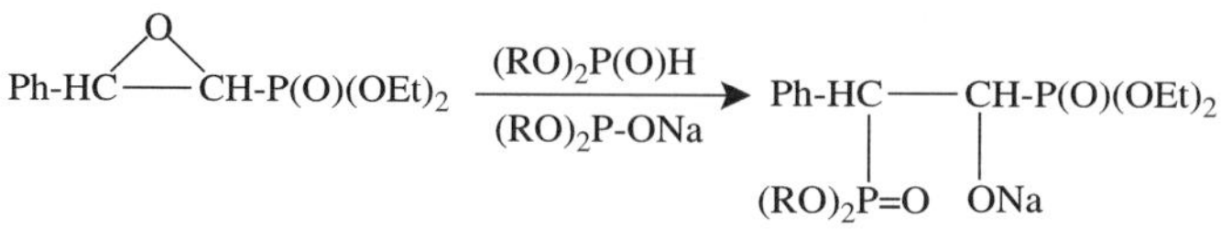

$$\underset{\text{Ph-HC}\overset{\overset{O}{\triangle}}{\text{——}}CH\text{-}P(O)(OEt)_2}{} \xrightarrow[(RO)_2P\text{-ONa}]{(RO)_2P(O)H} \underset{\underset{(RO)_2P=O}{|}\;\underset{ONa}{|}}{Ph\text{-}HC\text{——}CH\text{-}P(O)(OEt)_2}$$

4.6 Metal Salts of Dialkyl H-phosphonates

The diethyl ester of 2,3-epoxypropyl phosphonic acid reacts with equimolar quantities of dialkyl H-phosphonate and the corresponding sodium dialkyl phosphite, forming 1,2,3-tris-dialkylphosphono-propane. This reaction is suggested to take place according to the following scheme [386]:

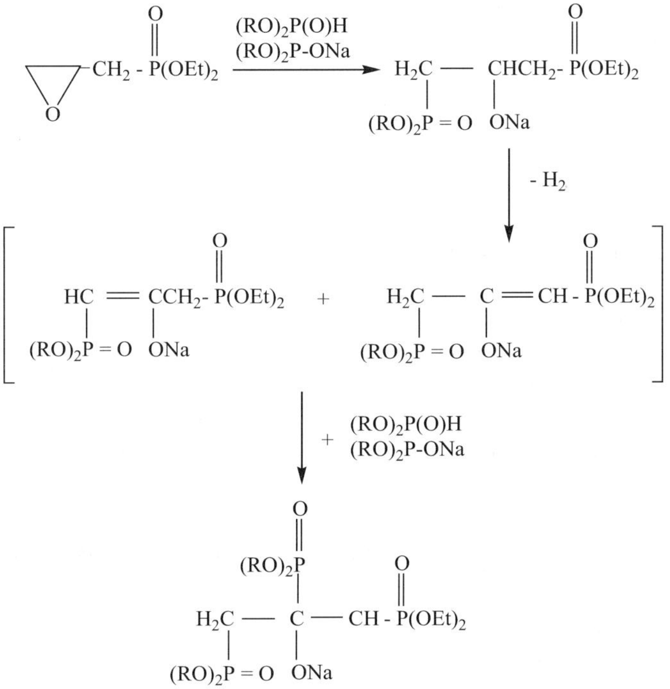

Under the above conditions, epichlorohydrin also yields 1,2,3-tris-dialkylphosphono-propane, presumably via the initial formation of the corresponding 2,3-epoxyphosphonates, which undergo the same transformations as described in the previous scheme [386].

The reaction of dialkyl H-phosphonates with epoxyketones yields two different products depending on the reaction conditions [387]. In the presence of the sodium salts of the corresponding dialkyl phosphite, an epoxy-containing Abramov's addition product is formed. In the presence of a Lewis acid, however, a nonregioselective ring-opening takes place.

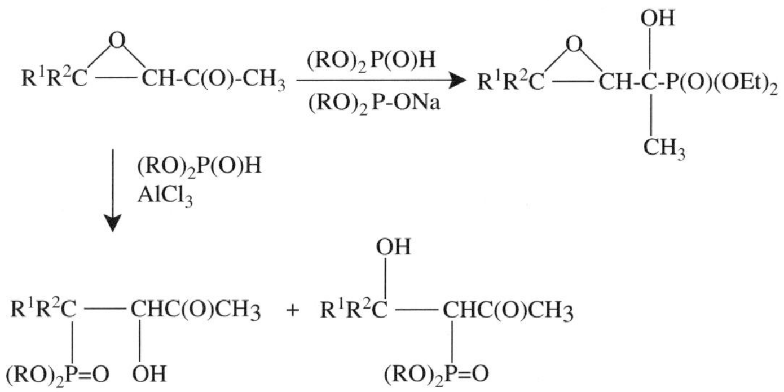

An approach for the preparation of combustion-resistant polystyrene involves its phosphorylation, which is achieved by the reaction of chloromethylated or bromomethylated polystyrene with sodium dialkyl phosphite [388].

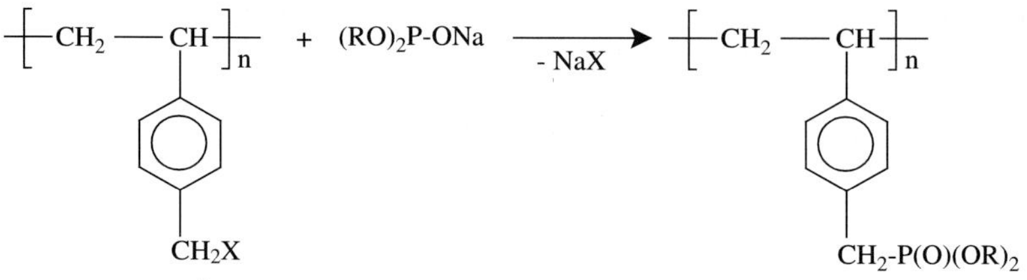

$$\text{\textendash}[CH_2\text{\textendash}CH]_n\text{\textendash}(C_6H_4)\text{\textendash}CH_2X + (RO)_2P\text{-}ONa \xrightarrow{-NaX} \text{\textendash}[CH_2\text{\textendash}CH]_n\text{\textendash}(C_6H_4)\text{\textendash}CH_2\text{-}P(O)(OR)_2$$

4.6.2 Phosphonate-type metal salts

These metal salts retain the phosphonate structure of the starting dialkyl H-phosphonates, which is characterized by a four-coordinated phosphorus atom in the O_3PH fragment. They can be prepared in one of the following methods:

(i) by dealkylation of dialkyl H-phosphonates by metal salts [389,390]

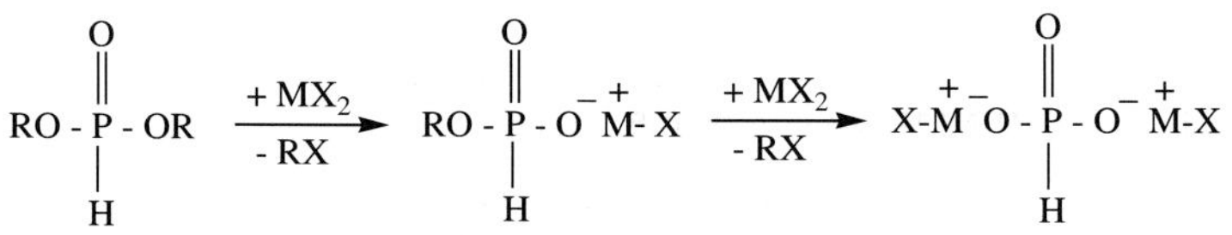

$$\text{RO-P(O)(H)-OR} \xrightarrow[-RX]{+MX_2} \text{RO-P(O)(H)-O}^- M^+\text{-}X \xrightarrow[-RX]{+MX_2} X\text{-}M^+ {}^-\text{O-P(O)(H)-O}^- M^+\text{-}X$$

(ii) by the reaction between dialkyl H-phosphonates and their sodium salts (Pelchowicz reaction) [372]

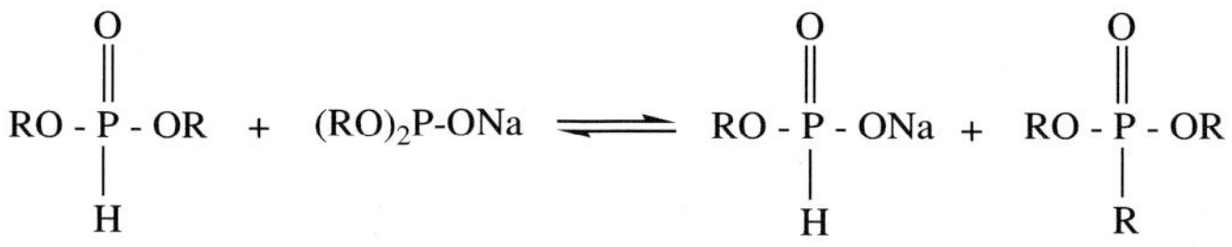

$$\text{RO-P(O)(H)-OR} + (RO)_2P\text{-}ONa \rightleftharpoons \text{RO-P(O)(H)-ONa} + \text{RO-P(O)(R)-OR}$$

(iii) or by alkali hydrolysis of dialkyl H-phosphonates

$$\text{RO-P(O)(H)-OR} \xrightarrow[-ROH]{+NaOH/H_2O} \text{RO-P(O)(H)-O}^-\text{Na}^+ \xrightarrow[-ROH]{+NaOH/H_2O} \text{Na}^+ {}^-\text{O-P(O)(H)-O}^-\text{Na}^+$$

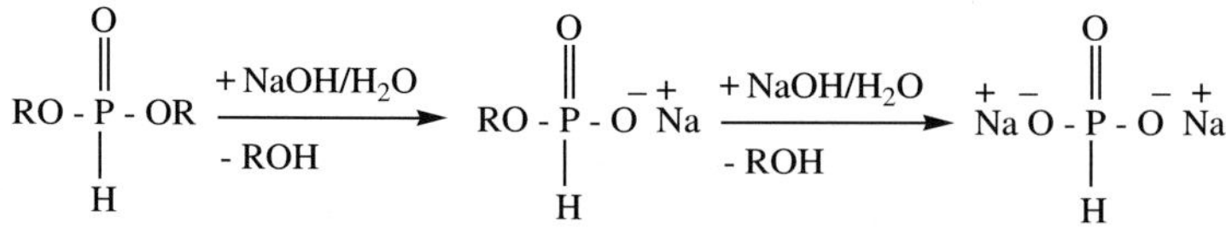

Although there are no X-ray structural data published yet for the above-type phosphonate salts, the structure of the closely related monolithium salt of H-phosphonic acid has been already reported [391]. In this structure, the tetrahedral environment of the phosphorus center is completed by three oxygens and one hydrogen atom. The P–H bond length is 1.17 Å and this hydrogen atom does not take part in a hydrogen-bond network. The specific NMR characteristics of these compounds are in most respects identical to those for the

4.6 Metal Salts of Dialkyl H-phosphonates

alkylammonium salts of the monoalkyl H-phosphonates discussed in Chapter 3, Section 3.8. The common features include decrease of the $^1J(PH)$ coupling constant of approximately 100 Hz as well as an up-field shift of the phosphorus signal in the ^{31}P NMR spectrum of the metal salts relative to the dialkyl H-phosphonate precursors. Tables 4.2 and 4.3 show ^1H and ^{31}P NMR data for some metal salts of the monoethyl ester of H-phosphonic acid. The $^{13}C\{H\}$ NMR spectrum of this compound consists of two signals at 18.72 and 63.18 ppm, corresponding to the CH_3 and CH_2 carbon atoms, respectively.

A useful comparison is also the ^{31}P NMR data for H-phosphonic acid and its mono- and disodium salts, summarized in Table 4.4.

The delocalized negative charge in the molecule of the sodium salts of H-phosphonic acid determines the decrease in the phosphorus chemical shift as well as in the $^1J(PH)$ coupling constants as compared to their values in the pure acid. These data show that this reduction depends on the degree of ionization of these compounds. Thus, the lowest $^1J(PH)$ value corresponds to the disodium salt of phosphonic acid.

Table 4.2

^1H NMR data for some metal salts of monoethyl H-phosphonate with a common formula $[C_2H_5OP(O)HO]^-M^+$

Metal	^1H NMR, chemical shift (ppm)	Reference
Na[a]	1.26 (t,3H,$^3J(HH)$ = 7.1 Hz, CH_3); 3.91 (dq,2H, $^3J(HH)$ = 7.2 Hz, $^2J(PH)$ = 8.2 Hz, CH_2); 7.21 (d, 1H, $^1J(PH)$ = 633.2 Hz, PH)	[392]
Zn[b]	1.26 (t,3H,$^3J(HH)$ = 7.1 Hz, CH_3); 3.93 (dq,2H, $^3J(HH)$ = 7.1 Hz, $^2J(PH)$ = 8.5 Hz, CH_2); 6.68 (d, 1H, $^1J(PH)$ = 663.1 Hz, PH)	[393]
Ca[b]	1.16 (t,3H,$^3J(HH)$ = 7.1 Hz, CH_3); 3.93 (dq,2H, $^3J(HH)$ = 7.2 Hz, $^2J(PH)$ = 8.5 Hz, CH_2); 6.64 (d, 1H, $^1J(PH)$ = 613.1 Hz, PH)	[393]

[a]in D_2O.
[b]in DMSO-d_6.

Table 4.3

^{31}P NMR data for some metal salts of monoethyl H-phosphonate with a common formula $[C_2H_5OP(O)HO]^-M^+$

Metal	^{31}P NMR, chemical shift (ppm)	Reference
Na[a]	4.91 (dt, $^1J(PH)$=596.0 Hz, $^3J(PH)$=11.6 Hz)	[392]
Zn[b]	4.84 (dt, $^1J(PH)$=663.0 Hz, $^3J(PH)$=8.1 Hz)	[393]
Ca[b]	0.28 (dt, $^1J(PH)$=608.9 Hz, $^3J(PH)$=8.9 Hz)	[393]

[a]in D_2O.
[b]in DMSO-d_6.

Table 4.4

^{31}P NMR spectra of H phosphonic acid and its sodium salts

Compound	Chemical shift (ppm)	^1J(PH) (Hz)	Reference
HPO$_3$Na$_2$	4.0	585	[394–396]
H$_3$PO$_3$	5.8; 4.0; 4.5; 3.2; 4.4; 3.2	700, 670	[395,396], [397–402]
H$_2$PO$_3$Na	4.0	620	[395,396,403]

4.7 COMPLEXES OF DIALKYL H-PHOSPHONATES

The dialkyl H-phosphonate molecule exhibits several different modes of coordination depending on the type of the corresponding metal or the Lewis acid [404]. A number of these complexes—especially in the case of the late transition metals—contain the tautomeric form of the starting dialkyl H-phosphonate, furnished presumably by a complexation-assisted equilibrium shift.

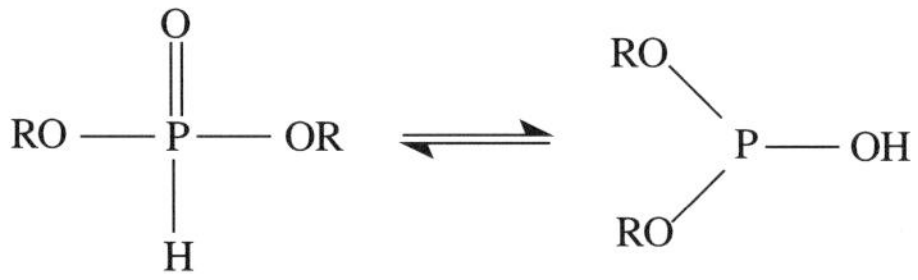

Taking into account both species in the above equilibrium as potential ligands, the following classification of their complexes has been developed [404]:

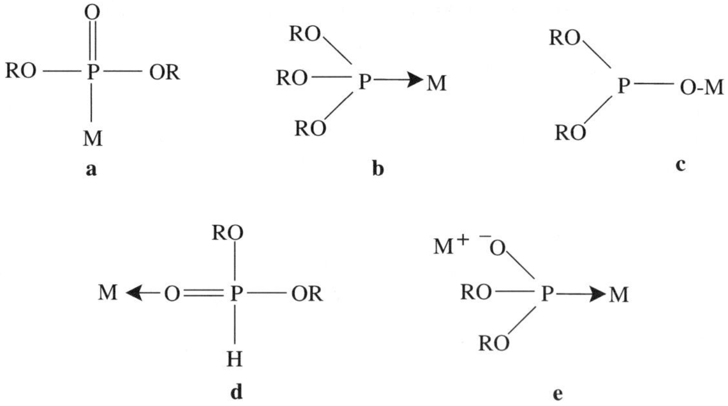

In a number of transition metal complexes, the coordination modes of dialkyl H-phosphonates correspond to **a**, **b**, or **e** in the above scheme. The coordination mode **c** is characteristic for silver, alkaline, and earth-alkaline metals. Coordination of type **d** is encountered in molecular complexes of dialkyl H-phosphonates with Lewis acids as well as in some nontransition and early transition metals as discussed further in the text.

4.7 Complexes of Dialkyl H-phosphonates

4.7.1 Molecular complexes with Lewis acids

Strong Lewis acids such as BF_3 [405] and BCl_3 [406] form with dialkyl H-phosphonates complexes of the following type:

$$X_3B \leftarrow O = \underset{\underset{H}{|}}{\overset{\overset{RO}{|}}{P}} - OR$$

In these complexes, the phosphoryl oxygen atom acts as a lone pair–electron donor. It has been established [405] that in mixtures containing excess free diethyl H-phosphonate and its complex with BF_3, a rapid exchange process on the ^1H NMR time scale takes place at room temperature.

$$\left[X_3B \leftarrow O = \underset{\underset{H}{|}}{\overset{\overset{RO}{|}}{P}} - OR \; + \; O = \underset{\underset{H}{|}}{\overset{\overset{RO}{|}}{P}} - OR \right]$$

As a result of this rapid exchange, intermediate values (between those of the free and the complexed phosphonate) have been observed for the $^1J(PH)$ coupling constants as well as for the chemical shifts of the P–H protons. These averaged values depend on the ratio between the free and complexed phosphonate species. Thus, a mixture containing 80% complexed and 20% free diethyl H-phosphonate exhibits chemical shift for the P–H proton at 7.0 ppm with $^1J(PH)=779$ Hz, whereas the corresponding values for diethyl H-phosphonate and its BF_3 complex are 6.65 ppm ($^1J(PH) = 700$ Hz) and 7.02 ppm ($^1J(PH) = 792$ Hz), respectively. This comparison also indicates that as a result of the complexation, the electron shielding on the P–H proton increases, which results in a downfield shift of its ^1H NMR signal, accompanied by an increase in $^1J(PH)$. Cooling down the above mixture to –48 °C results in a slowing down of the exchange process up to a point where separated signals for the P–H protons of both the free and the complexed diethyl H-phosphonate can be observed. The complexes of dialkyl H-phosphonates with boron trichloride are unstable and readily undergo dealkylation with elimination of alkyl chloride [406].

$$Cl_3B \leftarrow O = \underset{\underset{H}{|}}{\overset{\overset{RO}{|}}{P}} - OR \; \xrightarrow{- RCl} \; Cl_2BO - \underset{\underset{H}{|}}{\overset{\overset{O}{\|}}{P}} - OR$$

Dialkyl H-phosphonates undergo addition to 1,3,5-trinitrobenzene with the initial formation of the Jackson–Meizenheymer complex [407].

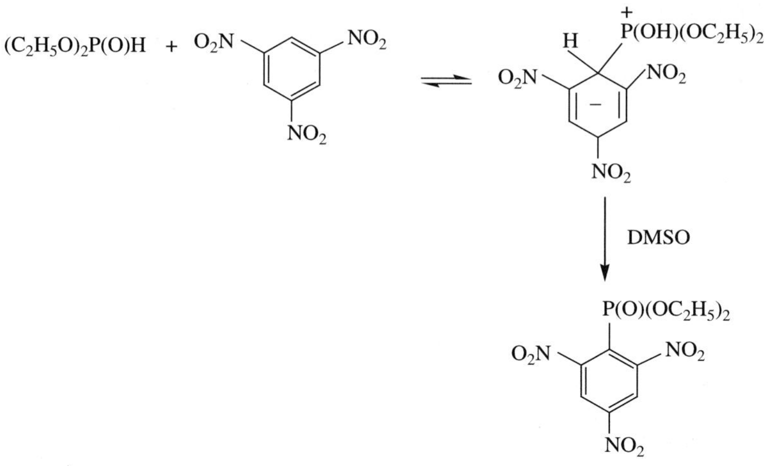

$(C_2H_5O)_2P(O)H$ + [1,3,5-trinitrobenzene] ⇌ [Meisenheimer complex with $P(OH)(OC_2H_5)_2$ substituent]

↓ DMSO

[product: 2,4,6-trinitrophenyl-$P(O)(OC_2H_5)_2$]

The formation of this initial complex has been attributed to the base-catalyzed shift in the phosphonate–phosphite equilibrium of these compounds to produce the reactive phosphorus (III) nucleophile. Trialkyl phosphites, which also easily form this type of complex [408–410], undergo an additional transformation to the corresponding phosphonate form and rearomatization of the system.

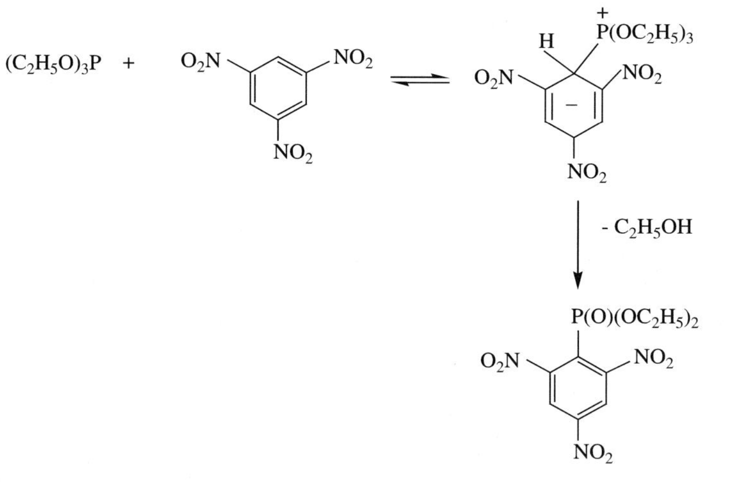

$(C_2H_5O)_3P$ + [1,3,5-trinitrobenzene] ⇌ [Meisenheimer complex with $P(OC_2H_5)_3$ substituent]

↓ $- C_2H_5OH$

[product: 2,4,6-trinitrophenyl-$P(O)(OC_2H_5)_2$]

4.7.2 Complexes with some main group and f-elements

Silver derivatives of dialkyl H-phosphonates have been prepared by several methods from silver acetate or silver oxide (1). IR spectroscopic studies of these compounds [411,412] indicate a phosphite-type structure $(RO)_2P–OAg$. Its assignment is based on the disappearance of the absorption bands corresponding to the P–H and P=O stretches in the IR spectra of these derivatives. In contrast to these results, the products of the interaction

4.7 Complexes of Dialkyl H-phosphonates

between mercury chloride $HgCl_2$ and dialkyl H-phosphonates [413] are bis(dialkylphosphonato) mercury (II) complexes. These contain a direct mercury–phosphorus bond and are presumably obtained via deprotonation of the dialkyl H-phosphonate.

$$2\ RO-\overset{\overset{O}{\|}}{\underset{H}{P}}-OR\ +\ HgCl_2\ \xrightarrow{-2HCl}\ \underset{RO}{\overset{RO}{>}}P\overset{O}{\underset{Hg}{<}}P\overset{O}{\underset{OR}{<}}\overset{RO}{}$$

The structure of the corresponding methyl derivative (R = Me) has been confirmed by X-ray single-crystal analysis [414]. In this crystal structure, the Hg–P bond distance is 2.411 Å with a P–Hg–P bond angle of 165.90°. The phosphoryl bond length of 1.46 Å is shorter, as expected, than the other two P–O bonds at 1.50 and 1.68 Å, respectively. The coordination sphere of the mercury atom is completed by two weakly coordinated phosphoryl oxygens at 2.54 Å, which results in a distorted tetrahedral environment around the metal center.

A number of studies have been devoted to the complexation of tin or alkyltin halides with dialkyl H-phosphonates [415–418]. The early investigations of Pudovik and coworkers [415,416], based primarily on IR spectroscopic studies, indicate that the complex formation is accompanied by a shift of the characteristic vP–H and vP–OC absorptions toward higher frequencies, corresponding to the shortening of these types of bonds. At the same time, there is a substantial lowering of the vP=O vibrational frequencies indicating elongated P=O bonds. These results correspond to the coordination of the phosphonate ligand through the phosphoryl oxygen.

$$Sn\leftarrow O=\underset{H}{\overset{RO}{P}}-OR$$

This suggestion has been further confirmed in a detailed IR and multinuclear ^1H, ^{31}P, and ^{119}Sn-NMR spectroscopic studies of the reaction products of tin and alkyltin halides with dialkyl H-phosphonates [417]. Common features in the ^1H NMR spectra of the studied complexes are the downfield shift of the signals for the P–H protons (compared to the free dialkyl H-phosphonates), accompanied by the increase of the $^1J(PH)$ coupling constants. The corresponding phosphorus signals of these compounds are also shifted downfield in the $^{31}P\{H\}$ NMR spectra. The sensitivity of the $^1J(PH)$ coupling constants towards complex formation has been utilized in order to calculate the formation constants for the complexation between diethyl and dimethyl phosphonates and Me_2SnCl_2 or Me_3SnCl [418]. A common feature of all studied tin halide complexes is their thermal instability with respect to dealkylation, leading to the formation of oligomeric phosphorus- and tin-containing products and evolution of alkyl halides [415–417]. It has been established that the thermal stability of the initially formed tin complexes increases with the number of the alkyl substituents at the tin center [417].

Diethyl H-phosphonate forms a σ-donor complex with uranyl nitrate $UO_2(NO_3)_2$ similar to those already described for BX_3 and organotin halides [419].

$$[UO_2(NO_3)_2\{(EtO)_2P(O)H\}_2]$$

The ^{1}H and ^{31}P{H} NMR data for this complex reveal the same characteristic patterns as in the previously described σ-complexes of dialkyl H-phosphonates. They are expressed in a downfield shift of the P–H signals, and an increase in ^1J(PH) coupling constant in the corresponding ^{1}H and ^{31}P{H}NMR spectra.

Lead acetate initially forms a 1:1 complex with dimethyl H-phosphonate, which slowly undergoes dealkylation, leading finally to the formation of lead phosphonate PbHPO$_3$ [420]. ^{1}H, ^{31}P{H}NMR, and IR spectral data of the initial 1:1 adduct clearly indicate the σ-donor character of the metal–ligand bond, realized apparently through the phosphoryl oxygen lone-pair donation as in the other cases in this section.

4.7.3 Complexes with transition metals

Dialkyl H-phosphonates exhibit a number of complexation modes with transition metals depending on their type. In complexes of dialkyl H-phosphonates with some early transition metals (Zr, Ti) in their higher oxidation states, the phosphonate ligands act as lone-pair σ-donors and are coordinated to the metal center through the phosphoryl oxygen atom. In contrast to this, the late transition metals (Ni, Pd, Pt, Co, Ir, Rh, and Ru) exhibit a well-expressed preference to the more 'soft' phosphorus donor of the phosphite tautomeric form. The late transition metal complexes of dialkyl H-phosphonates are additionally stabilized by the electron back donation from the electron-rich metal center to the empty 3d and 3σ* orbitals at the phosphorus donor.

Complexes containing P(V)-type ligands

Reaction of dialkyl H-phosphonates with ZrCl$_4$ yields, depending on the ratio of the reagents, complexes with different coordination numbers at the metal center [421]. These complexes contain chlorine- and phosphonate-type bridges that form a polymeric network. In the course of this reaction, partial dealkylation of the dialkyl H-phosphonates takes place, and results in the formation of Zr–O–P links. On the other hand, the phosphoryl oxygen forms additional donor–acceptor bonds with the zirconium atom.

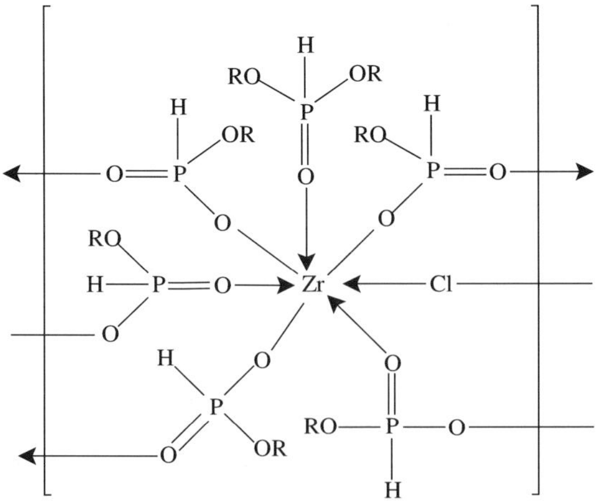

4.7 Complexes of Dialkyl H-phosphonates

An IR study of the complexes formed from TiCl$_4$ and dialkyl H-phosphonates indicates that at low temperatures, a σ-donor complex through the phosphoryl oxygen atom is formed [415]. Similar to the analogous complexes of tin halides, these titanium complexes are thermally unstable and easily undergo dealkylation with cleavage of alkyl halides even at low temperatures.

4.7.4 Complexes containing P(III)-type ligands

Platinum complexes

Troitzkaya and Grindberg were the first to report the formation of a 1:4 complex between platinum (II) and dialkyl H-phosphonates [422,423]. The anticipated structure contains two anionic and two neutral phosphonate ligands.

$$K_2PtCl_4 + 4(EtO)_2P(O)H \xrightarrow[-2\ HCl]{-2\ KCl} Pt[P(OEt)_2OH]_2[P(OEt)_2O]_2$$

Further IR and NMR studies by Pidcock and coworkers [424–426] indicated that these complexes are square-planar and have a symmetric structure with the four phosphorus atoms directly bound to the platinum.

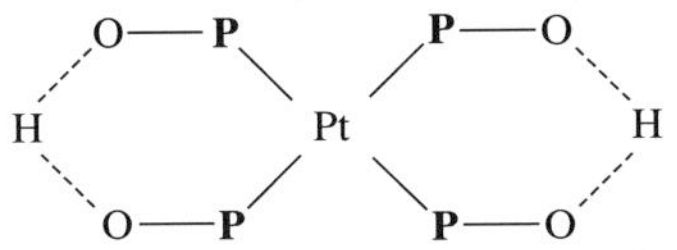

P = (RO)$_2$P   R = Ph, Me

All of the phosphorus atoms in the above complex are equivalent in the ^{31}P{H} NMR, which implies that a rapid proton exchange takes place in solution. The two neutral dialkyl phosphite molecules in these complexes can be easily displaced by 1,2-bis(diphenylphosphino) ethane [424].

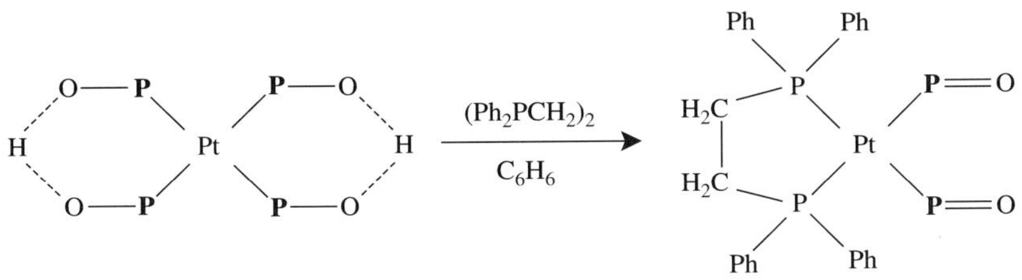

P = (RO)$_2$P   R = Ph, Me

On the other hand, the initial symmetrical Pt (II) complexes of dialkyl H-phosphonates can be titrated by base, whereby they behave as dibasic acids [427].

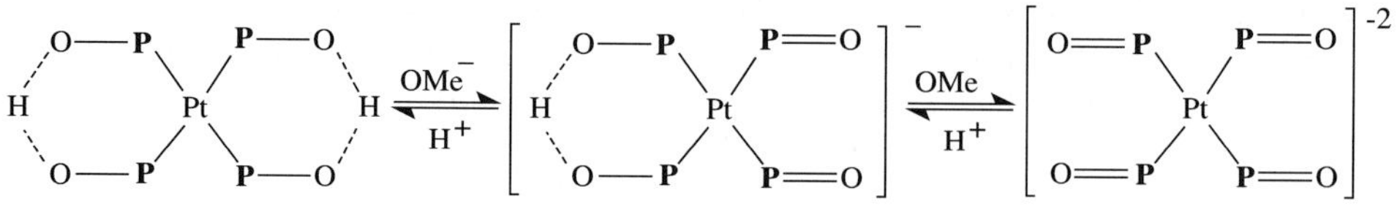

$$\text{P} = (\text{RO})_2\text{P}$$

Diphenyl and dimethyl H-phosphonates react with complexes of the type cis-[PtCl$_2$L$_2$] (L = R$_3$P or Et$_3$As) with liberation of one equivalent hydrogen chloride [426].

$$(\text{RO})_2\text{P(O)H} + \text{cis-[PtCl}_2\text{L}_2] \xrightarrow{-\text{HCl}} \text{trans-[PtCl\{(RO)}_2\text{PO\}L}_2]$$

Complexes of the type [PtCl{(MeO)$_2$PO}{(MeO)$_2$POH}L], where L = PEt$_3$ or PPh$_3$, can act as anionic ligands when treated with a base to remove the acidic proton. These anionic compounds chelate metal species as Cu(II), Co(II), UO$_2^{2+}$, and Th(IV) through the phosphoryl oxygens furnishing mixed metal species of the type {PtCl[(RO)$_2$PO]$_2$L}$_n$M, where n = 2 for M = Co, Cu, UO$_2$; n = 4 for M = Th [428]. Mixed metal species are also obtained by treatment of mononuclear precursor complexes [Pt{(RO)$_2$PO}$_2$L-L], where L-L = diphosphine or diarsine with perchlorates of divalent metals [428].

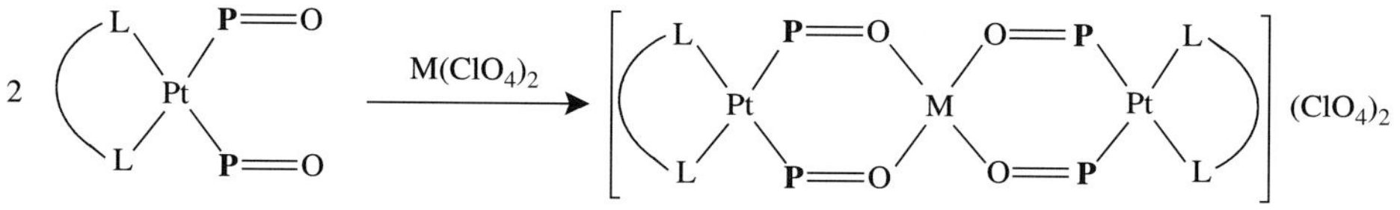

$$\text{P} = (\text{RO})_2\text{P} \quad \text{L—L} = (\text{Ph}_2\text{PCH}_2)_2 \text{ or } (\text{Ph}_2\text{AsCH}_2)_2$$
$$\text{M} = \text{Co, Cu, Ni, Zn}$$

Bridged platinum complexes of the type [Pt$_2$Cl$_4$L$_2$] (where L = Et$_3$As or pyridine) are converted by phosphonate diesters into mononuclear phosphonato complexes in the following anticipated reaction sequel [426]:

$$2(\text{RO})_2\text{P(O)H} + [\text{Pt}_2\text{Cl}_4\text{L}_2] \longrightarrow 2\,[\text{PtCl}_2\{(\text{RO})_2\text{POH}\}\text{L}]$$

$$[\text{PtCl}_2\{(\text{RO})_2\text{POH}\}\text{L}] + (\text{RO})_2\text{P(O)H} \xrightarrow{-\text{HCl}} [\text{PtCl}\{(\text{RO})_2\text{PO}\}\{(\text{RO})_2\text{POH}\}\text{L}]$$

Palladium complexes

Troitzkaya et al. reported the formation of the complexes [Pd{(EtO)$_2$PO}$_2$] and [PdCl{(EtO)$_2$PO}{(EtO)$_2$POH}]$_2$ from the reaction between K$_2$[PdCl$_4$] and diethyl H-phosphonate (or triethyl phosphite) in water [429]. Treatment of sodium tetrachloropalladate

4.7 Complexes of Dialkyl H-phosphonates

with diphenyl H-phosphonate/ethanol also furnishes two types of Pd(II) complexes for which the following structures have been assigned on the basis of ^{31}P-NMR data [426]:

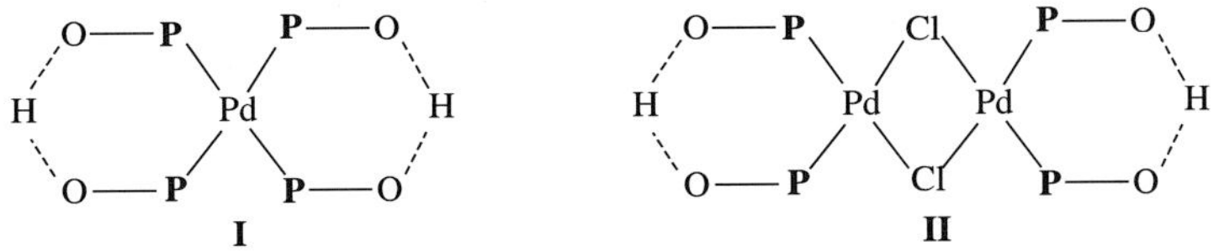

P = (RO)$_2$P

The above dichloro-bridged dimer reacts with triphenyl phosphine furnishing the mononuclear complex [PdCl{(PhO)$_2$PO}{(PhO)$_2$POH}(Ph$_3$P)]. Similar Pd (II) dimers with R = Me or Et have also been obtained by reaction of the π-allyl complexes [(2-MeC$_3$H$_4$)PdCl$_2$] [430] or [C$_3$H$_5$PdCl]$_2$ [431] with the corresponding dialkyl H-phosphonates. Treatment of one of the complexes in the above scheme (R = Me) with boron trifluoride and subsequently cyclopentadienyl thallium(I) leads to the formation of a mononuclear Pd(II) metallocycle [430].

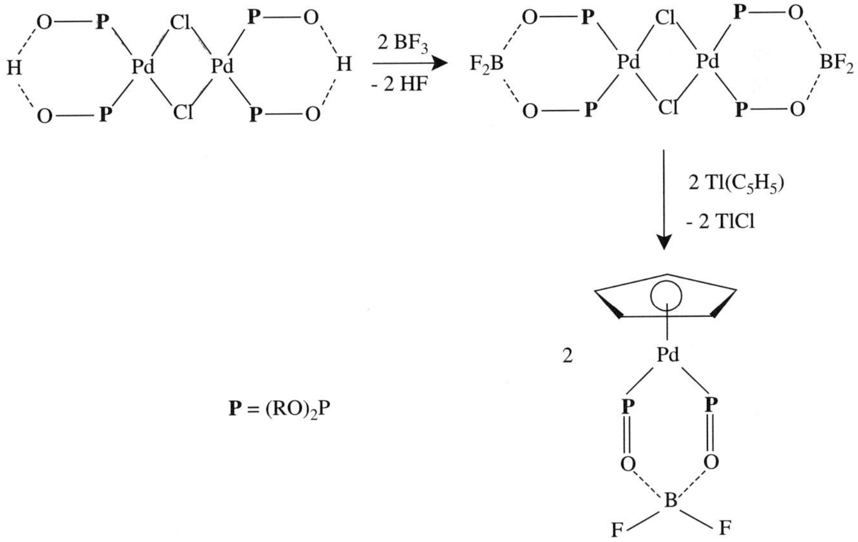

P = (RO)$_2$P

Pd (II) complexes similar to this metallocycle have also been prepared by reaction of the π-allyl-cyclopentadienyl-palladium precursor [(2-MeC$_3$H$_4$)Pd(C$_5$H$_5$)] with dialkyl H-phosphonates [430],

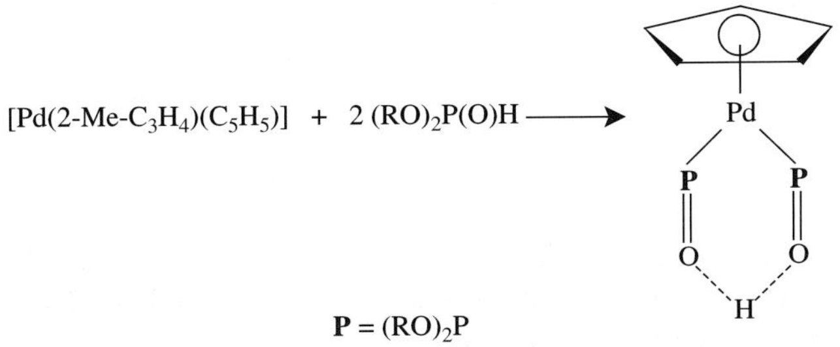

[Pd(2-Me-C$_3$H$_4$)(C$_5$H$_5$)] + 2 (RO)$_2$P(O)H ⟶

P = (RO)$_2$P

or by reaction of the dichloro-bridged Pd (II) dimers, described previously, with cyclopentadienyl thallium (I).

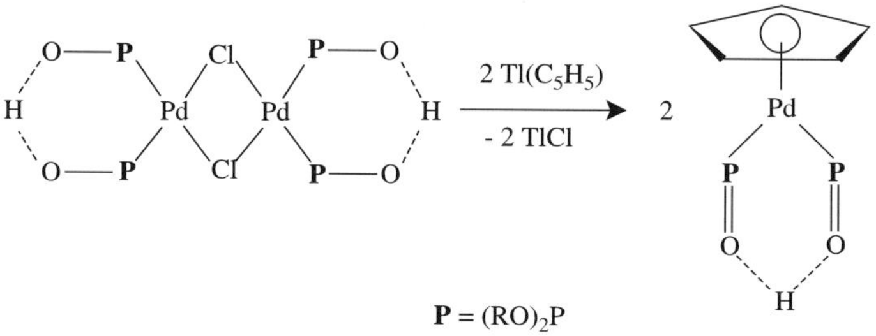

$$[\text{dichloro-bridged Pd dimer with P}=(RO)_2P\text{ ligands}] \xrightarrow[-2\text{ TlCl}]{2\text{ Tl}(C_5H_5)} 2\ [(Cp)Pd(P-O\cdots H\cdots O-P)]$$

P = (RO)$_2$P

When the reaction between [(2-MeC$_3$H$_4$)Pd(C$_5$H$_5$)] and dimethyl H-phosphonate is carried out at low temperatures and excess of phosphonate, the main reaction product is a π-(2-methyl)allyl-containing metallocycle [430].

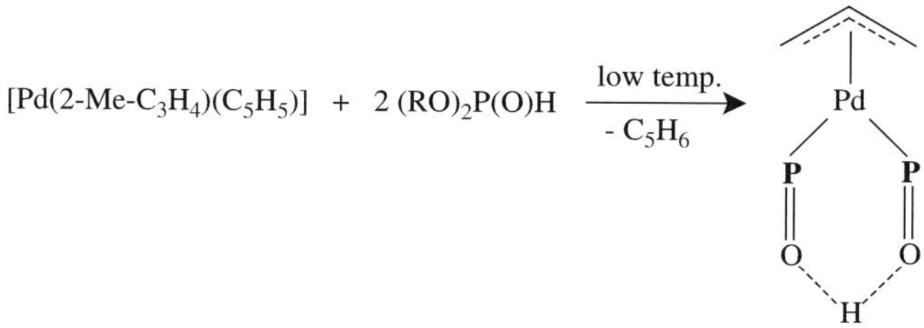

$$[\text{Pd}(2\text{-Me-}C_3H_4)(C_5H_5)] + 2\ (RO)_2P(O)H \xrightarrow[-C_5H_6]{\text{low temp.}} [(\text{2-Me-allyl})Pd(P-O\cdots H\cdots O-P)]$$

P = (RO)$_2$P

The labile proton in these metalocyclic complexes can be easily removed to form a chelate ligand from which a variety of polymetallic species have been prepared [362]:

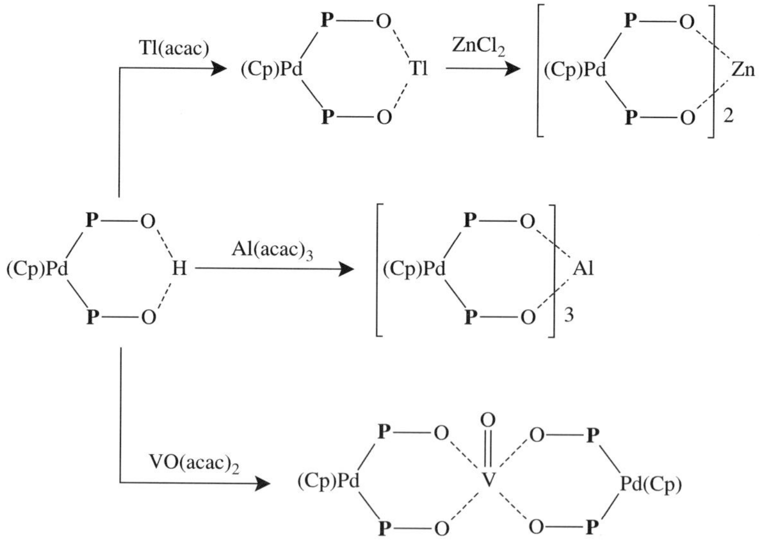

$$(Cp)Pd(P-O)_2H \xrightarrow{\text{Tl(acac)}} (Cp)Pd(P-O)_2Tl \xrightarrow{\text{ZnCl}_2} [(Cp)Pd(P-O)_2]_2Zn$$

$$(Cp)Pd(P-O)_2H \xrightarrow{\text{Al(acac)}_3} [(Cp)Pd(P-O)_2]_3Al$$

$$(Cp)Pd(P-O)_2H \xrightarrow{\text{VO(acac)}_2} (Cp)Pd(P-O)_2V(O)(O-P)_2Pd(Cp)$$

P = (RO)$_2$P   Cp = cyclopentadienyl   acac = acetylacetonato

4.7 Complexes of Dialkyl H-phosphonates

Nickel complexes

A number of cyclopentadienyl nickel(II) bis(phosphonato) complexes have been prepared by the reaction of bis-(cyclopentadienyl)nickel(II) with dialkyl H-phosphonates [432,433].

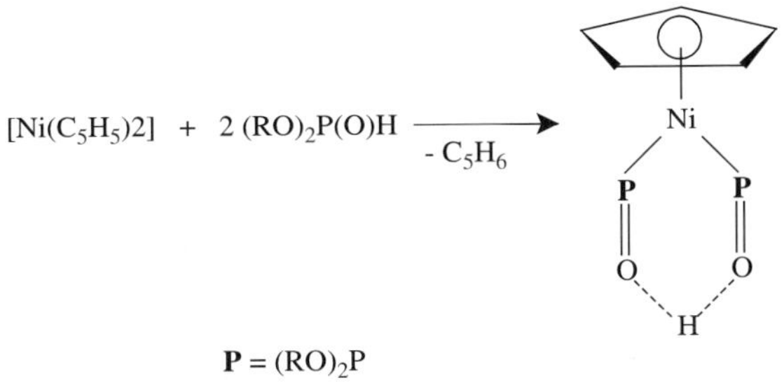

$$[Ni(C_5H_5)_2] + 2\,(RO)_2P(O)H \xrightarrow{-C_5H_6}$$

P = (RO)$_2$P

The acidic proton, in analogy to the analogous Pd(II) and Pt(II) complexes, can be easily removed by NH$_3$ or HBF$_4$ [434], metal halides [432–434], or acetylacetonates [433,434]:

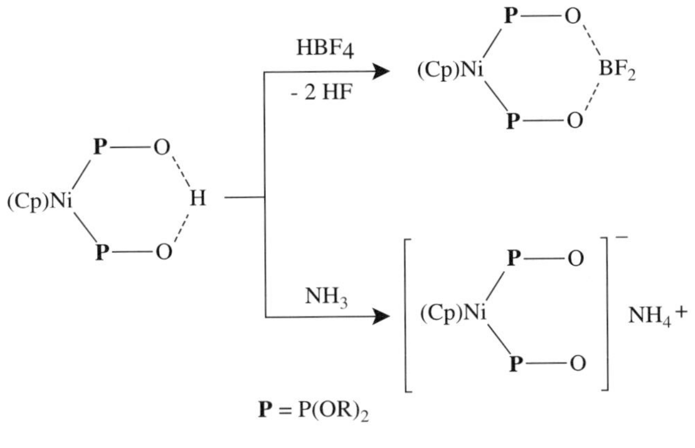

P = P(OR)$_2$

The ammonium salts formed by NH$_3$ react further with Zn (II) and Co (II) chlorides, furnishing trinuclear bimetallic complexes [434].

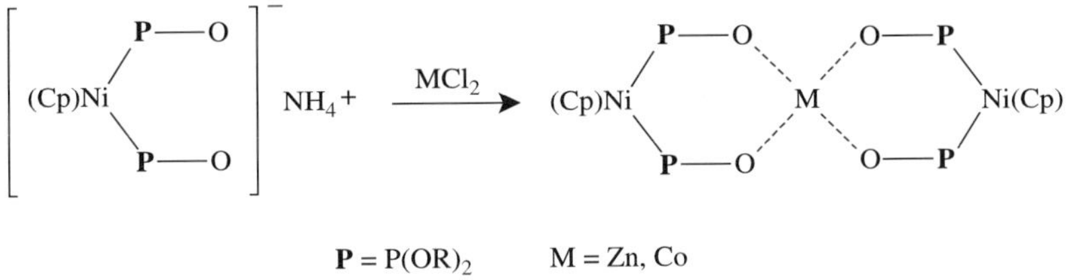

P = P(OR)$_2$ M = Zn, Co

Reaction of [(C$_5$H$_5$)Ni{OP(OR)$_2$}{H(O)P(OR)$_2$}] with some acetylacetonato complexes also results in the formation of several bimetallic species [433,434].

P = $(RO)_2P$ Cp = cyclopentadienyl acac = acetylacetonato

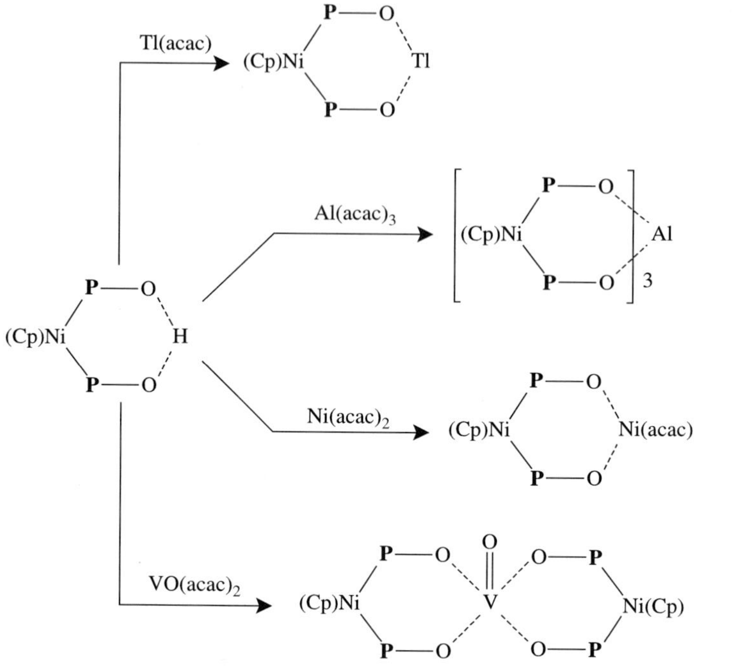

Cobalt complexes

Reaction of cobaltocene with dialkyl H-phosphonates results in trinuclear complexes in almost quantitative yields [435].

$$3\ Co(C_5H_5)_2 + 6\ (RO)_2P(O)H \longrightarrow [Co_3(C_5H_5)_2\{(RO)_2PO\}_6]$$

R = Et, Me

Crystal structure analysis of one of these complexes (R = Me) reveals that the three cobalt atoms are arranged linearly [436]. The two terminal ones are equivalent with each being coordinated to a cyclopentadienyl ring and the phosphorus atoms of three phosphonate groups. The middle cobalt atom is located at the center of symmetry of the metal cluster and is surrounded octahedrally by six oxygen atoms.

P = $P(OMe)_2$ Cp = cyclopentadienyl

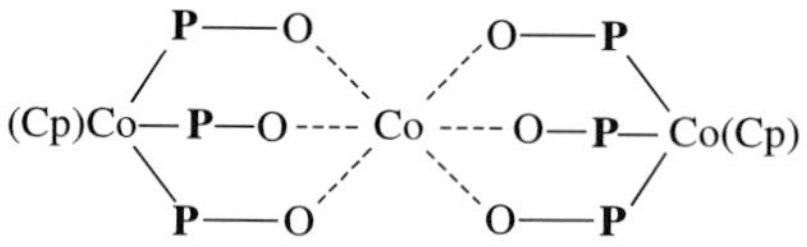

These trinuclear cobalt complexes can be protonated by HBF_4, whereby the following equilibrium is established [435]:

4.7 Complexes of Dialkyl H-phosphonates

$$[Co_3(C_5H_5)\{(RO)_2PO\}_6] \underset{-Co(BF_4)_2}{\overset{4\ HBF_4}{\rightleftharpoons}} 2\ [Co(C_5H_5)\{(RO)_2PO\}\{(RO)_2POH\}_2]\cdot BF_4$$

The so-formed mononuclear cobalt species has been used further as a tridentate ligand in the synthesis of a series of trinuclear heterometallic complexes of the following type [404,435]:

$$2\begin{bmatrix}(Cp)Co \begin{array}{c} P-OH \\ P-O \\ P-OH \end{array}\end{bmatrix} BF_4 \xrightarrow{M^{2+}} (Cp)Co \begin{array}{c} P-O \\ P-O \end{array} M \begin{array}{c} O-P \\ O-P \end{array} Co(Cp)$$

P = P(OMe)₂   Cp = cyclopentadienyl

The mononuclear cobalt complexes also undergo a capping reaction with an additional HBF₄ molecule forming a cage-like compound [437].

$$\begin{bmatrix}(Cp)Co \begin{array}{c} P-OH \\ P-O \\ P-OH \end{array}\end{bmatrix} BF_4 \xrightarrow[-3HF]{HBF_4} \begin{bmatrix}(Cp)Co \begin{array}{c} P-O \\ P-O \end{array} BF\end{bmatrix} BF_4$$

P = P(OMe)₂   Cp = cyclopentadienyl

More recently, the preparation of a series of octahedral Co(III) spin-crossover complexes with a common formula $[(C_5H_5)_2Co_3\{OP(OR)_2\}_6]$ has been reported [438]. Depending on steric and electronic effects on the substituents R, a systematic fine-tuning of the magnetic behavior can be achieved, whereby the magnetic properties at a given temperature can vary from high spin over spin equilibrium to completely diamagnetic species.

Iridium and rhodium complexes

Dimethyl H-phosphonate oxidatively adds to Ir (I) and Rh (I) compounds with the formation of hydrido-Ir(III) and hydrido-Rh(III) phosphonato complexes [439]. Treatment of chloro-bis(cyclooctadiene) iridium (I) $[IrCl(C_8H_{14})_2]_2$ with two equivalents of triphenyl phosphine and subsequent reaction with dimethyl H-phosphonate affords two products of oxidative addition **I** and **II**, which have not been spectroscopically distinguished:

```
        H                              H
   Cl   |   PPh₃                  Cl   |   PPh₃
     \  |  /                        \  |  /
       Ir                             Ir
     /  |  \                        /  |  \
  Ph₃P  |   P(OH)(OMe)₂         Ph₃P  |   P(O)(OR)₂
      P(O)(OR)₂                      P(OH)(OMe)₂

        I                              II
```

Complexes of the type [Ir(H)Cl{OP(OMe)$_2$}{P(OH)(Ome)$_2$}(AsPh$_3$)$_2$] and [M(H)Cl{OP(OMe)$_2$}(PMe$_2$Ph)$_3$], where M = Rh or Ir, have been similarly prepared from [MCl(C$_8$H$_{14}$)$_2$]$_2$, tertiary phosphine or arsine, and dimethyl H-phosphonate [369].

REFERENCES

1. M. Horiguchi, M. Kandatsu, *Nature*, **1959**, 184, 901.
2. M. V. Miceli, T. O. Henderson, T. C. Myers, *Science*, **1980**, 209, 1245.
3. M. I. Kabachnik, T. Y. Medved, N. M. Diatlyova, M. V. Rudomino, *Usp. Khim.*, **1974**, 43, 1554.
4. K. Praier, J. Rachon, *Z. Chem.*, **1975**, 15, 209.
5. D. Redmore. In: *Topics in Phosphorus Chemistry*, Vol. VIII (eds. M. Grayson, E. J. Griffith), InterScience, New York, **1976**, 515–585.
6. E. N. Rizkalla, *Rev. Inorg. Chem.*, **1983**, 5, 223.
7. J. Uziel, J. P. Genet, *Zh. Org. Khim.*, **1997**, 33, 1605.
8. (a) M. Kabachnik, T. Medved, *Dokl. Akad. Nauk SSSR*, **1952**, 83, 689; (b) M. Kabachnik, T. Medved, *Izv. Akad. Nauk SSSR, Otd. Khim. Nauk.*, **1953**, 868; (c) T. Medved, M. I. Kabachnik, *Izv. Akad. Nauk SSSR, Otd. Khim. Nauk.*, **1954**, 314.
9. E. K. Fields, *J. Am. Chem. Soc.*, **1952**, 74,1528; U.S. Pat. 2,635,112 (**1953**); C. A, 48, 7049 (**1954**).
10. C. Mannich, *Chem. Ber.*, **1932**, 65, 378.
11. K. A. Petrov, V. A. Chuzov, T. S. Erohina, *Zh. Obshch. Khim.*, **1975**, 45, 737.
12. (a) V. I. Galkin, E. R. Zvereva, A. A. Sobanov, I. V, Galkina, R. A. Cherkasov, *Zh. Obshch. Khim.*, **1993**, 63, 2224; (b) V. I. Galkin, E. R. Zvereva, I. V, Galkina, A. A. Sobanov, R. A. Cherkasov, *Zh. Obshch. Khim.*, **1998**, 68, 1453; I. V. Galkin, E. R. Zvereva, R. A. Cherkasov, *Zh. Obshch. Khim.*, **1998**, 68, 1457; (d) I. V. Galkina, A. A. V. Galkina, V. I Galkin, R. A. Cherkasov, *Zh. Obshch. Khim.*, **1998**, 68, 1469.
13. R. Gancarz, I. Gancarz, *Tetr. Lett.*, **1993**, 34, 145.
14. R. Gancarz, *Tetrahedron*, **1995**, 51, 10627.
15. C. Maury, J. Royer, H. P. Husson, *Tetr. Lett.*, **1992**, 33, 6127.
16. A. R. Sardarian, B. Kaboudin, *Tetr. Lett.*, **1997**, 38, 2543.
17. H. J. Ha, G. S. Nam, *Synth. Comm.*, **1992**, 22, 1143.
18. A. Heydari, A. Karimian, J. Ipaktschi, *Tetr. Lett.*, **1998**, 39, 6729.
19. A. B. Smith III, C. M. Taylor, S. I. Bencovic, R. Hirschmann, *Tetr. Lett.*, **1994**, 35, 6853.
20. Ch. Qian, T. Huang, *J. Org. Chem.*, **1998**, 63, 4125.
21. M. E. Chalmers, G. M. Kosolappoff, *J. Am. Chem. Soc.*, **1953**, 75, 5278.
22. U.S. Pat. 5,508,463 (**1996**).
23. U.S. Pat. 5, 824,809 (**1998**).
24. B. Boduszek, *Tetrahedron*, **1996**, 52, 12483.
25. W. F. Gilmore, H. McBride, *J. Am. Chem. Soc.*, **1972**, 94, 4361.
26. B. Bernet, E. Krawczyk, A. Vasella, *Helv. Chim. Acta*, **1985**, 68, 2299.
27. S. Laschat, H. Kunz, *Synthesis*, **1992**, 90.
28. A. Smith, K. M. Yager, C. Taylor, *J. Am. Chem. Soc.*, **1995**, 117, 10879.
29. C. Agami, F. Couty, N. Rabasso, *Tetr. Lett.*, **2002**, 43, 4632.
30. B. Boduszek, A. Halama, J. Zon, *Tetrahedron*, **1997**, 53, 11399.
31. V. Boev, U. Volanskii, *Khim. Farm. Zh.*, **1977**, 11, 39.
32. R. A. Cherkasov, V. I. Galkin, *Uspehi Khimii*, **1998**, 67, 952.
33. I. S. Antipin, I. I. Stoikov, E. M. Pinkahassik, N. A. Fitseva, I. Stibov, A. I. Konovalov, *Tetr. Lett.*, **1997**, 38, 5865.

34. Ru-Yu Chen, L. J. Mao, *Phosphorus, Sulfur, & Silicon*, **1994**, 89, 97.
35. G. Z. Akhmetova, P. A. Gurevich, V. V. Moskva, N. V. Kokhan, V. P. Arkhipov, *Zh. Obshch. Khim.*, **1997**, 67, 827.
36. D. Trendafilova-Gercheva, K. Troev, M. Georgieva, V. Vassileva, *Angew. Makromol. Chem.*, **1992**, 199, 137.
37. J. P. Finet, C. Frejaville, R. Lauricella, F. Le Moigne, P. Stipa, P. Tordo, *Phosphorus, Sulfur, & Silicon*, **1993**, 81, 17.
38. B. Boduszek, *Phosphorus, Sulfur, & Silicon*, **1995**, 104, 63.
39. L. Cottier, G. Descotes, J. Lewkowski, R. Skowronski, *Phosphorus, Sulfur,& Silicon*, **1996**, 116, 93.
40. X. J. Mu, M. Y. Lei, J. P. Zou, W. Zhang, *Tetr. Lett.*, **2006**, 47, 1125.
41. V. Yu. Pavlov, M. M. Kabachnik, E. V. Zobnina, V. P. Timofeev, I. O. Konstasntinov, B. G. Kimel, G. V. Ponomarev, I. P. Beletskaya, *Synlett*, **2003**, 2193.
42. O. A. Aimakov, K. B. Erzhanov, T. A. Mastryukova, *Izv. Akad. Nauk, Ser. Khim.*, **1999**, 48, 1815.
43. G. L. Matevosyan, P. M. Zavlin, A. V. Zhuravlev, *Zh. Obshch. Khim.*, **1980**, 50, 1506.
44. G. L. Matevosyan, P. M. Zavlin, *Zh. Obshch. Khim.*, **1998**, 68, 1536.
45. V. Pak, N. Kozlov, I. Balikova, G. Gartman, *Zh. Obshch. Khim.*, **1974**, 46, 497.
46. S. Failla, P. Finocchiaro, *Phosphorus, Sulfur, & Silicon*, **1995**, 107, 79.
47. S. Failla, P. Finocchiaro, M. Latronico, M. Libertini, *Phosphorus, Sulfur, & Silicon*, **1994**, 88, 185.
48. S. Failla, P. Finocchiaro, M. Latronico, *Phosphorus, Sulfur, & Silicon*, **1995**, 101, 261.
49. S. Failla, P. Finocchiaro, G. Haegele, V. Kalchenko, *Phosphorus, Sulfur, & Silicon*, **1997**, 128, 63.
50. S. Failla, P. Finocchiaro, G. Haegele, R. Rapisardi, *Phosphorus, Sulfur, & Silicon*, **1993**, 82, 79.
51. G. Gartman, V. Pak, N. Kozlov, *Zh. Obshch. Khim.*, **1979**, 49, 2375.
52. A. N. Pudovik, M. Kochemkina, *Izv. Akad. Nauk SSSR, Otd. Khim. Nauki.*, **1952**, 940.
53. I. Levashov, N. Kozlov, B. Pak, *Zh. Obshch. Khim.*, **1974**, 44,1112.
54. J. Lukszo, J. Kowalik, P. Mastalerz, *Chem. Lett.*, **1978**, 1103.
55. (a) D. Redmore, *J. Org. Chem.*, **1978**, 43, 992; (b) D. Redmore, *Chem. Rev.*, **1971**, 71, 315.
56. D. W. Boykin, A. Kumar, G. Yiao, W. D. Wilson, B. C. Bender, D. R. McCurdy, J. E. Hall, R. R. Tidwell, *J. Med. Chem.*, **1998**, 41, 124.
57. I. Kraicheva, P. Finocchiaro, S. Faila, *Phosphorus, Sulfur, & Silicon*, **2002**, 177, 2915.
58. S. Faila, P. Finocchiaro, G. A. Consiglio, *Heteroatom Chem.*, **2000**, 11, 493.
59. I. Kraicheva, *Phosphorus, Sulfur, & Silicon*, **2003**, 178, 191.
60. G. B. Bowden, R. Roberts, D. S. Alberts, Y. M. Peng, D. Garcia, *Cancer Res.*, **1985**, 45, 4915.
61. U. Herrmann, B. Tummler, G. Maass, P. K. T. Mew, F. Vogtle, *Biochemistry*, **1984**, 23, 4059.
62. B. M. Krasovitsky, B. M. Bolotin, *Organicheskie Kyuminofory*, Khimiya, Moscow, **1984**, 30.
63. K. Moedritzer, R. R. Irani, *J. Org. Chem.*, **1966**, 31, 1603.
64. K. H. Worms, K. Wollmann, Z. *Anorg. Allg. Chem.*, **1971**, 381, 260.
65. B. Dhawan, D. Redmore, *Phosphorus, Sulfur, & Silicon*, **1990**, 48, 41.
66. (a) M. Kabachnik, T. Medved, *Dokl. Akad. Nauk SSSR*, **1952**, 83, 689; (b) M. Kabachnik, T. Medved, *Izv. Akad. Nauk SSSR, Otd. Khim. Nauk*, **1953**, 868; (c) T. Medved, M. I. Kabachnik, *Izv. Akad. Nauk SSSR, Otd. Khim. Nauk*, **1954**, 314; (d) E. K. Fields, *J. Am. Chem. Soc.*, **1952**, 74, 1528.
67. G. L. Matevosyan, I. G. Chezlov, I. Yu. Matyushichev, P. M. Zavlin, *Zh. Obshch. Khim.*, **1986**, 56, 1426.
68. A. F. Grapov, A. C. Remizov, V. K. Promenekov, *Zh. Obshch. Khim.*, **1980**, 50, 1207.
69. C. M. McLeod, G. M. Robinson, *J. Chem. Soc.*, **1921**, 119, 1472.
70. H. Gross, B. Costisella, Th. Gnauk, L. Brennecke, *J. Prakt. Chem.*, **1976**, 318, 116.
71. U.S. Pat. 4,060,571 (**1977**).

72. V. Orlovskii, B. Vovsi, *Zh. Obshch. Khim.*, **1976**, 46, 297.
73. U.S. Pat. 4,005,160 (**1977**).
74. Yu. G. Safina, G. Sh. Malkova, R. A. Cherkasov, *Zh. Obshch. Khim.*, **1991**, 61, 620.
75. Yu. G. Safina, G. Sh. Malkova, R. A. Cherkasov, V. V. Ovchinnikov, *Zh. Obshch. Khim.*, **1990**, 60, 221.
76. (a) R. W. Ratcliffe, B. G. Christensen, *Tetr. Lett.*, **1973**, 46, 4645; (b) M. Shiozaki, H. Masuko, *Heterocycles*, **1984**, 22, 1727.
77. H. Groger, Y. Saida, H. Sasai, K. Yamaguchi, J. Martens, M. Schibasaki, *J. Am. Chem. Soc.*, **1998**, 120, 3089.
78. P. G. Baraldi, M. Guarneri, F. Moroder, G. P. Pollini, D. Somoni, *Synthesis*, **1982**, 653.
79. R. Burpitt, W. Goodlett, *J. Org. Chem.*, **1965**, 30, 4307.
80. Belgium pat. 774,349 (**1972**).
81. U.S. Pat. 3,160,632 (**1964**).
82. U.S. Pat. 4,008,296 (**1977**).
83. DDR Pat. 149,929 (**1980**).
84. U.S. Pat. 4,237,065 (**1980**); Hungary Pat. 8015160 (**1980**).
85. U.S. Pat. 4,400,330 (**1983**).
86. U.S. Pat. 4,439,373 (**1984**).
87. K. Troev, *Phosphorus, Sulfur, & Silicon*, **1997**, 127, 167.
88. K. Troev, G. Haegele, K. Kreidler, R. Olschner, C. Verwey, D. M. Roundhill, *Phosphorus, Sulfur, & Silicon*, **1999**, 148, 161.
89. K. Troev, Sh. Cremer, G. Haegele, *Heteroatom Chem.*, **1999**, 10, 627.
90. K. Troev, G. Haegele, I. Ivanov, A. Kril, *Heteroatom Chem.*, submitted 2005.
91. I. Ivanov, I. Sainova, E. Shikova, K. Troev, *Exp. Patalogie*, **2000**, 5,43.
92. E. Najdenova, A. Vassilev, Yu. Popova, K. Troev, *Heteroatom Chem.*, **2003**, 14, 229.
93. A. Deron, R. Gancarz, I. Gancarz, A. Halama, L. Kuzma, T. Rychewski, J. Zon, *Phosphorus, Sulfur, & Silicon*, **1999**, 437, 144–145.
94. H. Hariharan, R. J. Motekajtis, A. E. Martell, *J. Org. Chem.*, **1975**, 40, 470.
95. V. Kukhar, H. Hudson, *Aminophosphonic and Aminophosphinic Acids*, Wiley, New York, **2000**, 173–203.
96. W. F. Gilmore, H. McBride, *J. Pharm. Sci.*, **1972**, 63, 1087.
97. A. B. Smith III, C. M. Taylor, S. J. Benkovic, R. Hirschmann, *Tetr. Lett.*, **1994**, 35, 6853.
98. J. A. Steere, P. B. Sampson, J. F. Honek, *Bioorg. & Med. Chem. Lett.*, **2002**, 12, 457.
99. F. R. Atherton, C. H. Hassall, R. W. Lambert, *J. Med. Chem.*, **1986**, 29, 29.
100. (a) W. T. Lowther, Y. Zhang, P. B. Sampson, J. F. Honek, B. W. Matthews, *Biochemistry*, **1999**, 38, 14810; (b) B. Lejczak, P. Kafarski, J. Zygmunt, *Biochemistry*, **1989**, 28, 3549.
101. M. Bray, J. Driscoll, J. W. Huggins, *Antiviral Res.*, **2000**, 45, 135.
102. A. J. Bitoni, J. Baumann, T. Varvi, J. R. McCarthy, P. P. McCann, *Biochem. Pharmacol.*, **1990**, 40, 601.
103. J. A. Wolos, K. A. Frondorf, R. E. Esser, *J. Immunol.*, **1993**, 151, 526.
104. (a) Y. Saso, E. M. Conner, B. R. Teedarden, C. S. Yuan, *Pharmacology*, **2001**, 296, 106; (b) J. A. Wolos, K. A. Frondorf, G. F. Davies, E. T. Jarvi, F. R. McCarthy, T. L. Bowlin, *J. Immunol.*, **1993**, 150, 3264.
105. (a) H. Fliesch, R. G. Russell, F. Straumann, *Nature*, **1966**, 212, 901; (b) M. D. Francis, R. G. Russell, H. Fliesch, *Science*, **1969**, 165, 1264; (c) H. Fliesch, R. G. Russell, M. D. Francis, *Science*, **1969**, 165, 1262.
106. (a) U.S. Pat. 3,988,443 (**1976**); (b) U.S. Pat. WO 92/00721 (**1992**).
107. M. Nakamura, T. Ando, M. Abe, K. Kumagai, Y. Endo, *Br. J. Pharmacol.*, **1996**, 119, 205.
108. J. G. Slatter, K. L. Feenstra, M. J. Hauer, D. A. Kloosterman, A. H. Parton, P. E. Sandres, G. Scott, W. Speed, *Drug Metab. Dispos.*, **1996**, 24, 65.

109. J. P. Rasanen, E. Pohjala, H. Nikander, T. A. Pakkanen, *J. Phys. Chem.*, **1996**, 100, 8230.
110. C. Dufau, G. Sturtz, *Phosphorus, Sulfur, Silicon Relat. Elem.*, **1992**, 69, 93.
111. J. H. Lin, *Bone*, **1996**, 18, 75.
112. R. T. Chlebowski, A. McTiernan, *J. Clin. Oncol.*, **1999**, 17, 130.
113. M. Mimura, M. Hayashida, K. Nomiyama, S. Ikegami, Y. Iida, M. Tamura, Y. Hiyama, Y. Ohishi, *Chem. Pharm. Bull.*, **1993**, 41, 1971.
114. H. Fliesh. In: *Handbook of Experimental Pharmacology* (ed. P. F. Baker), Springer, Berlin, **1988**, 83, 455.
115. F. Cheng, X. Yang, C. Fan, H. Zhu, *Tetrahedron*, **2001**, 57, 7331.
116. R. A. Nugent, M. Murphy, S. T. Schlachter, C. J. Dunn, R. J. Smith, N. D. Staite, L. A. Galinet, S. K. Shields, D. G. Aspar, K. A. Richard, N. A. Rohloff, *J. Med. Chem.*, **1993**, 36, 134.
117. Eur. Pat. 100,718 (**1983**); C. A. 100, 192078j (**1984**).
118. H. J. Cristau, C. Brahic, J. L. Pirat, *Tetrahedron*, **2001**, 57, 9149.
119. D. Kantoci, J. K. Denike, W. J. Wechter, *Synth. Commun.*, **1996**, 26, 2037.
120. U.S. Pat. WO 96/31124 (**1996**).
121. D. Q. Qian, X. D. Shi, R. Z. Cao, L. Z. Liu, *Heteroatom Chem.*, **1999**, 10, 271.
122. J. Ohanessian, D. Avenel, D. El. Manouni, M. Benramdane, *Phosphorus, Sulfur, Silicon Relat. Elem.*, **1997**, 129, 99.
123. L. Lemee, M. Gulea, M. Saquet, S. Masson, N. Collignon, *Heteroatom Chem.*, **1999**, 10, 281.
124. D. C. Stepinski, D. W. Nelson, P. R. Zalupski, A. W. Herlinger, *Tetrahedron*, **2001**, 57, 8637.
125. G. A. Rodan, H. A. Fleisch, *J. Clin. Invest.*, **1996**, 97, 2692.
126. J. R. Green, *Med. Klin.*, **2000**, 95, Suppl. II, 23.
127. S. P. Luckman, D. E. Hughes, F. P. Coxon, *J. Bone Miner. Res.*, **1998**, 13, 581.
128. C. Fernandes, R. S. Leite, F. M. Lancas, *Quim. Nova*, **2005**, 28, 247.
129. H. Gross, B. Costisella, L. Haase, *J. Prakt. Chem.*, **1969**, 311, 577.
130. H. Gross, B. Costisella, Th. Gnauk, L. Brennecke, *J. Prakt. Chem.*, **1976**, 318, 116.
131. (a) Jpn. Pat. 79,135,724 (**1978**); C. A. 92, 146905v (**1980**); (b) Jpn. Pat. 79,144,383 (**1979**); C. A. 93, 26547 (**1980**); (c) V. I. Krutikov, A. V. Erkin, P. A. Pautov, M. M. Zolotukhina, *Zh. Obshch. Khim.*, **2003**, 73, 205.
132. H. Wada, Q. Fernando, *Anal. Chem.*, **1971**, 43, 751.
133. Von B. Blaser, K. H. Worms, H. G. Gremscheid, K. Wollmann, *Z. Anorg. Allg. Chem.*, **1971**, 381, 247.
134. (a) Ger. Offen DE 3,016,289 (**1981**); C. A. 96, 52503t (**1985**); (b) U.S. Pat. 4,267,108 (**1981**); (c) U.S. Pat. 4,304,734 (**1981**); (d) U.S. Pat. 4,327,039 (**1982**); (e) U.S. Pat. 4,407,761 (**1983**); (f) U.K. Pat. 2,166,741 A (**1986**);
135. G. R. Kieczykowski, R. B. Jobson, D. G. Melillo, D. F. Reinhold, V. J. Grenada, I. Shinkai, *J. Org. Chem.*, **1995**, 60, 8310.
136. (a) U.S. Pat. 4,927,077 (**1990**); (b) U.S. Pat. 5,019,651 (**1991**).
137. U.S. Pat. 5,908,959 (**1999**).
138. U.S. Pat. 2004/0043967 A1 (**2004**).
139. U.S. Pat. 4,608,368 (**1986**).
140. Y. Xie, H. Ding, L. Qian, X. Yan, C. Yang, Y. Xie, *Biorg. Med. Chem. Lett.*, **2005**, 15, 3267.
141. (a) D. M. Mizrahi, T. Wagner, Y. Segall, *Phosphorus, Sulfur, & Silicon*, **2001**, 173, 1; (b) WO Pat. 96/40156 (**1996**).
142. D. V. Griffiths, J. M. Hughes, J. M. Brown, J. C. Caesar, S. P. Swetnam, S. A. Cumming, J. D. Kelly, *Tetrahedron*, **1997**, 53, 17815.
143. J. M. Sanders, Y. Song, J. M. W. Chan, Y. Zhang, S. Jennings, T. Kosztowski, S. Odeh, R. Flessner, C. Schwerdtfeger, E. Kotsikorou, G. A. Meints, A. O. Gomez, D. Gonzalez-Pacanowska, A. M. Raker, H. Wang, E. R. van Beek, S. E. Papapoulos, C. T. Morita, E. Oldfield, *J. Med. Chem.*, **2005**, 48, 2957.

144. World Pat. WO03097655 (**2003**).
145. L. Lazzarato, B. Rolando, M. L. Lolli, G. C. Tron, R. Fruttero, A. Gasco, G. Deleide, H. L. Guenther, *J. Med. Chem.*, **2005**, 48, 1322.
146. R. M. Palmer, A. G. Ferrige, S. Moncada, *Nature*, **1987**, 327, 524.
147. S. Moncada, A. Higgs, *N. Engl. J. Med.*, **1993**, 329, 2002.
148. T. Yokomatsu, Y. Yoshida, N. Nakabayashi, S. Shibuya, *J. Org. Chem.*, **1994**, 59, 7562.
149. (a) L. B. Han, C. Q. Zhao, M. Tanaka, *J. Org. Chem.*, **2001**, 66, 5929; (b) L. B. Han, C.Q. Zhao, S. Y. Onozawa, M. Goto, M. Tanaka, *J. Am. Chem. Soc.*, **2002**, 124, 3842.
150. B. Burgada, A. Mohri, *Phosphorus, Sulfur Silicon Relat. Elem.*, **1982**, 13, 85.
151. A. Allen Jr., D. R. Manke, W. Lin, *Tetr. Lett.*, **2000**, 41, 151.
152. S. Takahashi, Y. Kuroyama, K. Sonogashira, N. Hagihara, *Synthesis*, **1980**, 627.
153. L. B. Han, R. Hua, M. Tanaka, *Angew. Chem., Int. Ed. Engl.*, **1998**, 37, 94.
154. E. Balaraman, K. Swamy, *Synthesis*, **2004**, 18, 3037.
155. V. Orlovskii, B. Vovsi, *Zh. Obshch. Khim.*, **1976**, 46, 297.
156. Von K. H. Worms, H. Blum, *Z. anorg. allg. Chem.*, **1979**, 457, 209.
157. S. H. Szajnman, E. L. Ravaschino, R. Docampo, J. B. Rodriguez, *Bioorg. Med. Chem. Lett.*, **2005**, 15, 4685.
158. Brit. Pat. 1,357,901 (**1974**).
159. Y. Du, K. Y. Jung, D. F. Wiemer, *Tetr. Lett.*, **2002**, 43, 8665.
160. A. El-Mabhouh, Ch. Angelov, A. McEwan, G. Jia, J. Mercer, *Cancer Biotherapy Radiopharm.*, **2004**, 19, 627.
161. U.K. Pat. 2,118,042 A (**1983**).
162. D. W. Hutchinson, D. M. Thornton, *Synthesis*, **1990**, 135.
163. C. Yuan, C. Li, Y. Ding, *Synthesis*, **1991**, 854.
164. H. Couthon, J. P. Gourves, J. Guervenou, B. Corbel, G. Sturz, *Synth. Commun.*, **1999**, 29, 4251.
165. U.S. Pat. 3,686,290 (**1972**).
166. C. R. Degenhardt, D. C. Burdsall, *J. Org. Chem.*, **1986**, 51, 3488.
167. U.S. Pat. 3,062,792 (**1962**); 3,576,793 (**1971**).
168. J. A. Matthews, L. J. Kricka, *Anal. Biochem.*, **1988**, 169, 1.
169. P. C. Zamecnik, M. L. Stephenson, *Proc. Natl. Acad. Sci. USA*, **1978**, 75, 280.
170. R. Teoule, *Nucleosides Nucleotides*, **1991**, 10, 129.
171. R. J. Johnes, N. Bischofberger, *Antiviral Res.*, **1995**, 27, 1.
172. N. S. Corby, G. W. Kenner, A. R. Todd, *J. Chem. Soc.*, **1952**, 3669.
173. M. Sekine, T. Hata, *Tetr. Lett.*, **1975**, 21, 1711.
174. N. Bhongle, U.S. Pat. 5,639,875 (**1997**).
175. M. Sekine, H. Mori, T. Hata, *Tetr. Lett.*, **1979**, 13, 1145.
176. D. B. Denney, D. Z. Denney, P. Hammond, L. t. Lui, Y. P. Wang, *J. Org. Chem.*, **1983**, 48, 2159.
177. O. Sakatsume, H. Yamane, H. Takaku, N. Yamamoto, *Tetr. Lett.*, **1989**, 30, 6375.
178. H. Takaku, S. Yamakage, O. Sakatsume, M. Ohtsuki, *Chem. Lett.*, **1988**, 1675.
179. B. C. Froehler, M. D. Matteucci, U.S. Pat. 5,548,076 (**1996**).
180. M. P. Reddy, N. B. Hanna, F. Faroogui, U.S. Pat. 5,726,301 (**1998**).
181. J. Jankowska, M. Sobkowski, J. Stawinski, A. Kraszewski, *Tetr. Lett.*, **1994**, 35, 3355.
182. (a) J. Cieslak, M. Sobkowski, A. Kraszewski, *Tetr. Lett.*, **1996**, 37, 4561; (b) I. Kers, J. Stawinski, A. Kraszewski, *Tetrahedron*, **1999**, 55, 11579; (c) M. Sobkowski, A. Kraszewski, J. Stawinski, *Nucleotides Nucleotides*, **1998**, 17, 253; (d) J. Cieslak, M. Sobkowski, J. Jankowska, M. Wenska, M. Szymczak, B. Imiolczyk, I. Zagorowska, D. Shugar, J. Stawinski, A. Kraszewski, *Acta Biochim. Pol.*, **2001**, 48, 429.
183. J. Cieslak, M. Szymczak, M. Wenska, J. Stawinski, A. Kraszewski, *J. Chem. Soc., Perkin Trans.1*, **1999**, 3327.

184. N. B. Tarusova, A. A. Khorlin, A. A. Krayevsky, M. N. Korneyeva, D. N. Nossik, I. V. Kruglov, G. A. Galegov, R. Sh. Bibilashvili, *Mol. Biol.* (Moskow), **1989**, 23, 1716.
185. N. B. Tarusova, M. Kukhanova, A. A. Krayevsky, E. K. Karamov, V. Lukashov, G. B. Kornilayeva, M. A. Rodina, G. A. Galegov, *Nucleosides Nucleotides*, **1991**, 10, 351.
186. A. A. Khorlin, N. B. Tarusova, N. B. Dyatkina, A. A. Kraevsky, R. Sh. Bibilashvili, G. A. Galegov, M. N. Korneeva, D. N. Nosik, S. N. Maiorova, V. M. Shobukhov, V. M. Zhdanov, Eur. Pat. 354426 (**1988**).
187. C. Meier, *Angew. Chem.*, **1993**, 105, 1854.
188. C. Meier, L. Habel, W. Laux, E. De Cletcq, Jan Balzarini, *Nucleosides Nucleotides*, **1995**, 14, 759.
189. R. H. Hall, A. Todd, R. F. Webb, *J. Chem. Soc.*, **1957**, 3291.
190. B. C. Froehler, M. D. Matteucci, *Tetrahedron Lett.*, **1986**, 27, 469.
191. J. Cieslak, J. Jankowska, M. Sobkowski, A. Kers, I. Kers, J. Stawinski, A. Kraszewski, *Collect. Symp. Ser.*, **1999**, 2, 63.
192. J. Cieslak, J. Jankowska, A. Kers, I. Kers, A. Sobkowska, M. Sobkowski, J. Stawinski, A. Kraszewski, *Collect. Czech. Chem. Commun.*, **1996**, 61 (Special issue), S242.
193. M. Sobkowski, M. Wenska, A. Kraszewski, J. Stawinski, *Nucleosides, Nucleotides, Nucleic Acids*, **2000**, 19, 1487.
194. F. Eckstain, G. Gish, *TIBS*, **1989**, 97.
195. G. Zon, *Pharm. Res.*, **1988**, 5, 539.
196. E. Uhlmann, A. Peyman, *Chem. Rev.*, **1990**, 90, 543.
197. A. Sobkowska, M. Sobkowski, J. Cieslak, A. Kraszewski, I. Kers, J. Stawinski, *J. Org. Chem.*, **1997**, 62, 4791.
198. C. A. Stein, J. S. Cohen. In: *Oligodeoxynucleotides – Antisense Inhibitors of GeneExpression* (ed. J. S. Cohen), The Macmillan Press Ltd, New York, **1989**, 79.
199. W. S. Marshall, M. H. Caruthers, *Science*, **1993**, 259, 1564.
200. J. Stawinski, M. Thein, *Tetrahedron Lett.*, **1992**, 33, 3189.
201. J. Stawinski, R. Stromberg, R. Zain, *Tetrahedron Lett.*, **1992**, 33, 3185.
202. J. Nielsen, W. K. D. Brill, M. H. Caruthers, *Tetrahedron Lett.*, **1988**, 29, 2911.
203. J. Stawinski, M. Thelin, R. Zain, *Tetrahedron Lett.*, **1989**, 30, 2157.
204. J. Stawinski, M. Thelin, E. Westman, R. Zain, *J. Org. Chem.*, **1990**, 55, 3503.
205. R. Zain, R. Stromberg, J. Stawinski, *J. Org. Chem.*, **1995**, 60, 8241.
206. G. M. Porritt, C. B. Reese, *Tetrahedron Lett.*, **1989**, 30, 4713.
207. J. Cieslak, J. Jankowska, J. Stawinski, A. Kraszewski, *J. Org. Chem.*, **2000**, 65, 7049.
208. J. Cieslak, J. Jankowska, M. Sobkowski, M. Wenska, J. Stawinski, A. Kraszewski, *J. Chem. Soc., Perkin Trans. 1*, **2002**, 31.
209. J. Jankowska, A. Sobkowska, J. Cieslak, M. Sobkowski, A. Kraszewski, J. Stawinski, D. Shugar, *J. Org. Chem.*, **1998**, 63, 8150.
210. F. Eckstein, *Angew. Chem., Int. Ed. Engl.*, **1983**, 22, 423.
211. F. Eckstein, *Annu. Rev. Biochem.*, **1985**, 54, 367.
212. F. Eckstein, G. Gish, *TIBS*, **1989**, 97.
213. F. Eckstein, *Tetrahedron Lett.*, **1967**, 3495.
214. P. M. Burgers, F. Eckstein, *Tetrahedron Lett.*, **1978**, 3835.
215. P. M. Burgers, F. Eckstein, *Biochemistry*, **1979**, 18, 592.
216. J. F. Marleir, S. J. Bencovic, *Tetrahedron Lett.*, **1980**, 21, 1121.
217. H. Almer, R. Stromberg, *Tetrahedron Lett.*, **1991**, 30, 3723.
218. F. Seela, U.J. Kretschmer, *J. Org. Chem.*, **1991**, 56, 3861.
219. H. Almer, J. Stawinski, R. Stromberg, M. Thelin, *J. Org. Chem.*, **1992**, 57, 6163.
220. M. Sekine, T. Hata, *Tetrahedron Lett.*, **1975**, 1711.

221. M. Sekine, H. Mori, T. Hata, *Tetrahedron Lett.*, **1979**, 1145.
222. P. J. Garegg, I. Lindh, T. Regberg, J. Stawinski, R. Stomberg, *Chem.cr.*, **1985**, 25, 280.
223. P. J. Garegg, I. Lindh, T. Regberg, J. Stawinski, R. Stomberg, *Chem.cr.*, **1986**, 25, 59.
224. P. J. Garegg, I. Lindh, T. Regberg, J. Stawinski, R. Stomberg, C. Henrichson, *Tetrahedron Lett.*, **1986**, 27, 4051.
225. B. C. Froehler, P. C. Ng, M. D. Matteucci, *Nucleic Acids Res.*, **1986**, 14, 5399.
226. P. J. Garegg, I. Lindh, T. Regberg, J. Stawinski, R. Stomberg, C. Henrichson, *Tetrahedron Lett.*, **1986**, 27, 4055.
227. A. Andrus, J. M. Efcavitch, L. J. McBride, B. Giusti, *Tetrahedron Lett.*, **1988**, 29, 861
228. S. Agrawal, P. C. Zamechnick, U.S. Pat. 5,149,798 (**1992**).
229. F. Chow, T. Kempe, G. Palm, *Nucl. Acids Res.*, **1981**, 9, 2807.
230. T. Tanaka, S. Tamatsukuri, M. Ikehara, *Nucl. Acid Res.*, **1987**, 15, 7253.
231. R. T. Jiang, Y. J. Shyy, M. D. Tsai, *Biochemistry*, **1984**, 23, 1661.
232. I. Vasilenko, B. De Kruijff, A. Verkleij, *J. Biochim. Biophys. Acta*, **1982**, 685, 144.
233. M. Jett, C. R. Alving, *Methods Enzymol.*, **1987**, 141, 459.
234. R. L. Juliano. In: *Liposomes: from Physical Structure to Therapeutical Application;* (ed. C. C. Knight,), Elsevier, Amsterdam, **1981**, 391.
235. K. Satouchi, R. N. Pinckard, L. M. McMagnus, D. J. Hanshan, *J. Biol. Chem.*, **1981**, 26, 405.
236. I. Lindt, J. Stawinski, *J. Org. Chem.*, **1989**, 54, 1338.
237. R. L. McConnell, H. W. Coover, *J. Org. Chem.*, **1959**, 24, 630.
238. Q. Xiao, J. Sun, Y. Ju, Yu-fen Zhao, Y. Cui, *Tetrahedron Lett.*, **2002**, 43, 54.
239. S. L. Beaucage, R. P. Iyer, *Tetrahedron*, **1993**, 49, 129.
240. V. A. Korshun, Yu. A. Berlin, *Bioorg. Khim.*, **1994**, 20, 565.
241. D. Hendlin, E. O. Stapley, M. Jackson, H. Wallick, A. K. Miller, F. J. Wolf, T. W. Miller, L. Chaier, F. M. Kahan, E. L. Foltz, H. B. Woodruff, J. M. Mata, S. Hernandez, S. Mochales, *Science*, **1969**, 166, 122.
242. F. M. Kahan, J. S. Kahan, P. J. Cassidy, H. Kropp, *Ann. N. Y. Acad. Sci.*, **1974**, 235, 364.
243. Fr. Pat. 2,596,393 (**1988**), C.A. 108, 94780c (**1988**).
244. Eur. Pat. Appl. EP 78613 (**1998**), C. A. 99, 140148c (**1981**).
245. U.S. Pat. 3,658,668 (**1997**), C. A. 77, 152356y (**1999**).
246. U.S. Pat. 4,693,742 (**1997**), C. A. 108, 6183g (**1999**).
247. U.S. Pat. 4,632,973 (**1998**), C. A. 106, 197379s (**1998**).
248. U.S. Pat. 4,654,277 (**1997**), C. A. 107, 60892e (**1997**).
249. U.S. Pat. 4,692,355 (**1997**), C. A. 107, 238808x (**1997**).
250. D. Redmore, *Chem. Rev.* **1971**, 71, 315.
251. S. Inokawa, H. Yamamoto, *Phosphorus Sulfur*, **1983**, 16, 79.
252. B. Iorga, F. Eymery, Ph. Savignac, *Synthesis*, **1999**, 207.
253. (a) B. A. Arbuzov, V. S. Vinogradova, N. A. Polezhaeva, *Dokl. Acad. Nauk SSSR*, **1956**, 111, 107; (b) G. Sturtz, *Bull. Soc. Chim. Fr.*, **1964**, 2333.
254. B. Springs, P. Haake, *J. Org. Chem.*, **1976**, 41, 1165.
255. F. T-Boullet, A. Foucaud, *Tetrahedron Lett.*, **1980**, 21, 2161.
256. (a) S. Inokawa, Y. Kawata, K. Yamamoto, H. Kawamoto, H. Yamamoto, K. Takagi, M. Yamashita, *Carbohydr. Res.*, **1981**, 88, 341; (b) S. Kasino, S. Inokawa, M. Haisa, N. Yasuoka, M. Kakudo, *Acta Crystallogr., Sect. B*, **1981**, 37, 1572; (c) T. Hanaya, K. Yasuda, K. Yamamoto, H. Yamamoto, *Bull. Chem. Soc. Jpn.*, **1993**, 66, 2315.
257. Z. I. Glebova, O. V. Eryuzheva, Yu. A. Zhdanov, *Zh. Obshch. Khim.*, **1993**, 63, 1677.
258. K. Kossev, K. Troev, D. Max Roundhill, *Phosphorus, Sulfur, Silicon*, **1993**, 83, 1.
259. M. S. Newmann, B. Magerlein, *J. Org. React.*, **1949**, 5, 413.
260. V. F. Martinov, V. E. Tomofeev, *Zh. Obshch. Khim.*, **1962**, 32, 3449.

261. P. Coutrot, P. Savignac, *Synthesis*, **1978**, 34.
262. K. Lee, W. S. Shin, D. Y. Oh, *Synth. Commun.*, **1992**, 22, 649.
263. DE Pat. 1,924,135 (**1965**); C. A. 72, 43870n (**1970**).
264. E. J. Glamkowski, G. Gal, R. Purick, A. J. Davidson, M. Sletzinger, *J. Org. Chem.*, **1970**, 35, 3510.
265. A. N. Pudovik, M. G. Zimin, A. A. Sobanov, *Zh. Obshch. Khim.*, **1973**, 45, 1232.
266. R. H. Churi, C. E. Griffin, *J. Am. Chem. Soc.*, **1966**, 88, 1824.
267. E. N. Ryazantsev, D. A. Ponomarev, V. A. Al'bitskaya, *Zh. Obshch. Khim.*, **1987**, 57, 2300.
268. C. E. Griffin, S. K. Kundu, *J. Org. Chem.*, **1969**, 34, 1532.
269. M. Baboulene, G. Strutz, *Synthesis*, **1978**, 456.
270. J. J. Eisch, J. E. Galle, *J. Organomet. Chem.*, **1976**, C10, 121; *Ibid.*, **1988**, 341, 293.
271. S. Inokawa, H. Yamamoto, *Phosphorus & Sulfur*, **1983**, 16, 79.
272. S. Kadiyala, H. Lo, M. S. Ponticello, K. W. Leong. In: *Biomedical Applications of Synthetic Biodegradable Polymers* (ed. J. O. Hollunger), CRC Press: Baca Raton, FL, **1997**.
273. H. Mao, I. Kadiyala, K. W. Leong, Z. Zhao, W. Dang. In: *Encyclopedia of Controlled Drug Delivery*, Vol. 1 (ed. E. Manthiowitz), Wiley-Interscience, New York, **1999**, 45–60.
274. S. Wang, A. C. Wan, X. Xu, S. Gao, H. Q. Mao, K. W. Leong, H. Yu, *Biomaterials*, **2001**, 22, 1157.
275. J. Wen, R. X. Zhuo, *Macromol. Rapid Commun.*, **1998**, 19, 641.
276. M. V. Chaubal, B. Wang, G. Su, Z. Zhao, *J. Appl. Polym. Sci.*, **2003**, 90, 4021.
277. H. Coates, Brit. Pat. 796,446 (**1958**); C. A. 55, 17799 f (**1958**); U.S. Pat. 2,963,451 (**1960**).
278. (a) K. Petrov, E. Nifantiev, N. Novoselov, *Vysokomol. Soedin.*, **1962**, 4, 1214; (b) K. Petrov, E. Nifantiev, *Vysokomol. Soedin.*, **1964**, 9, 1545.
279. W. Vogt, S. Balasubramanian, *Makromol. Chem.*, **1973**, 163, 89.
280. K. Troev, *Doctoral Thesis*, Institute of Polymers, Bulgarian Academy of Sciences, Sofia, **1985**.
281. J. Pretula, S. Penczek, *Makromol. Chem., Rapid Commun.*, **1988**, 9, 731.
282. J. Pretula, S. Penczek, *Makromol. Chem.*, **1990**, 191, 671.
283. S. Penczek, J. Pretula, *Macromolecules*, **1993**, 26, 2228.
284. R. Tzevi, G. Todorova, K. Kossev, K. Troev, E. Georgiev, D. M. Roundhill, *Makromol. Chem.*, **1993**, 194, 3261.
285. J. A. Pople, W. C. Schneider, H. J. Berstein, *High-Resolution Nuclear Magnetic Resonance*, McGraw-Hill Book Company, NY, **1959**, Chapter 19.
286. J. Pretula, K. Kaluzynski, R. Szymanski, S. Penczek, *Macromolecules*, **1997**, 30, 8172.
287. K. E. Branham, J, W. Mays, G. M. Gray, P. C. Bharara, H. Byrd, R. Bittenger, B. Farmer, *Polymer*, **2000**, 41, 3371.
288. R. D. Myrex, B. Farmer, G. M. Gray, Y. J. Wright, J. Dees, P. C. Bharara, H. Byrd, K. E. Branham, *Eur. Polym. J.*, **2003**, 39, 1105.
289. S. Penczek, T. Biela, P. Klosinski, G. Lapienis, *Makromol. Chem., Macromol. Symp.*, **1986**, 6, 123.
290. E. Bezdushna, H. Ritter, K. Troev, *Macromol. Rapid Comm.*, **2005**, 26, 471.
291. K. Takemoto, H. Tahara, A. Yamada, Y. Inaki, N. Ueda, *Makromol. Chem.*, **1973**, 169, 327.
292. T. Biela, S. Penczek, S. Slomkowski, O. Vogl, *Makromol. Chem., Rapid Commun.*, **1982**, 3, 167.
293. (a) P. A. Lambert, I. C. Hancock, J. Baddiley, *Biochem J.*, **1975**, 149, 519; (b) J. E. Heckels, P. A. Lambert, J. Baddiley, *Biochem. J.*, **1977**, 162, 359.
294. P. Klosinski, S. Penczek, *Macromolecules*, **1983**, 16, 316.
295. B. Gorjanova, Iv. Gitsov, K. Troev—unpublished data.
296. K. Kaluzynski, J. Libiszowski, S. Penczek, *Macromolecules*, **1976**, 9, 365.
297. J. Baran, S. Penczek, *Macromolecules*, **1995**, 28, 5167.
298. J. C. Brosse, D. Derouet, L. Fontaine, S. Chairatanathvorn, *Makromol. Chem.*, **1989**, 190, 2339.

299. L. Elling, M. R. Kula, *Biotechnol. Appl. Chem.*, **1991**, 13, 354.
300. R. Tzevi, P. Novakov, K. Troev, D. M. Roundhill, *J. Polym. Sci.:Part A: Polym. Chem.*, **1997**, 35, 625.
301. D. Nagai, H. Kuramoto, A. Sudo, F. Sanda, T. Endo, *Macromolecules*, **2002**, 35, 6149.
302. K. Kossev, A. Vassilev, Yu. Popova, I. Ivanov, K. Troev, *Polymer*, **2003**, 44, 1987.
303. H. Ringsdorf, *J. Polym. Sci., Polym. Symp.*, **1975**, 51, 135.
304. R.M. Ottenbrite, W. Regelson, A. Kaplan, R. Carchman, P. Morahan, A. Munson. In: *Polymeric Drugs* (eds. L. G. Donaruma, O. Vogl), Academic Press, New York, **1978**.
305. K. Uhrich, *Trends in Polym. Sci.*, **1997**, 5, 388.
306. R. Duncan, *Anti-Cancer Drugs*, **1992**, 3, 175.
307. R. Georgieva, R. Tsevi, K. Kossev, R. Kusheva, M. Balgjiska, R. Petrova, V. Tenchova, I. Gitsov, K. Troev, *J. Med. Chem.*, **2002**, 45, 5797.
308. P. Senter, H. P. Svensson, G. J. Schreiber, J. L. Rodriguez, V. M. Vrudhula, *Bioconjugate Chem.*, **1995**, 64, 389.
309. E. H. Schacht, *Controlled Drug Delivery*, **1983**, 1, 149 (Chapter 6).
310. *Controlled Release Pesticides* (ed. H. B. Scher), ACS Symposium Series, No 53, American Chemical Society, Washington, DC, **1977**.
311. R. B. Greenwald, A. Pendri, C. Conover, C. Gilbert, R. Yang, J. Xia, *J. Med. Chem.*, **1996**, 39, 1938.
312. C. Conover, A. Pendri, C. Lee, C. W. Gilbert, K. L. Shum, R. B. Greenwald, *Anticancer Res.*, **1997**, 17, 3361.
313. R. B. Greenwald, A. Pendri, C. D. Conover, C. Lee, Y. H. Choe, C. Gilbert, A. Martinez, *Biorg. Med. Chem.*, **1998**, 65, 551.
314. R. Greenwald, C. W. Gilbert, A. Pendri, C. D. Conover, J. Xia, A. Martinez, *J. Med. Chem.*, **1996**, 39, 424.
315. C. Li, D. F. Yu, T. Inoue, D. J. Yang, L. Milas, N. Hunter, *Anticancer Drugs*, **1996**, 7, 642.
316. A. Nathan, S. Zalipsky, S. I. Ertel, S. N. Agathas, M. L. Yarmush, J. Kohn, *Bioconjugate Chem.*, **1993**, 4, 54.
317. A. Nathan, S. Zalipsky, J. Kohn, *J. Bioact, Compat. Polym.*, **1994**, 9, 239.
318. M. Pechar, J. Stroalm, K. Ulbrich, E. Schacht, *Macromol. Chem. Phys.*, **1997**, 198, 1009.
319. M. Pechar, J. Stroalm, K. Ulbrich, *Collect. Czech. Chem. Commun.*, **1995**, 60, 1765.
320. V. Alkanov, A. Kabanov, *Expert Opin. Invest. Drugs*, **1998**, 79, 1453.
321. K. Katoaka. In: *Controlled Drug Delivery: Challenges& Strategies* (ed. K. Park), ACS, Washington DC, **1997**, 49–71.
322. K. Kataoka, G. S. Kwon, M. Yokoyama, T. Okano, Y. Sakurai, *J. Controlled Release*, **1993**, 24, 119.
323. T. Inoue, G. Chen, K. Nakamae, A. S. Hoffman, *J. Controlled Release*, **1998**, 51, 221.
324. G. Kwon, T. Okano, *Adv. Drug Deliv. Rev.*, **1996**, 21, 107.
325. J. C. Brosse, L. Fontaine, D. Derouet, S. Chairatanathvorn, *Makromol. Chem.*, **1989**, 190, 2329.
326. Z. M. Bacq, G. Dechamps, P. Fischer, A. Herve, H. Lebihan, J. Lecomte, M. Pirotte, P. Rayet, *Science*, **1953**, 117, 633.
327. G. Baldini , *L. Br. J. Radiol.*, **1957,** 30, 271.
328. J. Durkovsky, E. Siracka-Vasela, *Neoplasma*, **1958**, 5, 417.
329. H. Monig, O. Messershmidt, *Radiation Exposure and Occupational Risks*, Springer, Heidelberg, **1990**, 97.
330. I. Nikolov, V. Rogozkin, T. Pantev, K. Chertkov, E. Dikovenko, S. Paridova, *Strahlenther Oncol.*, **1986**, 162, 200.
331. T. Pantev, R. Georgieva, S. Topalova, *Strahlenther. Oncol.*, **1991**, 167, 422.
332. Z. M. Bacq, P. Alexander, *Oxygen in the Animal Organizm*, Pregamon Press, Oxford, **1964**, 509–536.
333. R. J. Jones, N. Bischofberger, *Antivir. Res.*, **1995**, 27, 1.

334. A. A. Krayevsky, N. B. Tarussova, Q. Y Zhu, P. Vidal, T. C. Chou, P. Baron, B. Polsky, X. J Jiang, J. Matulic-Adamic, I. Rosenberg, K. Watanabe, *Nucleosides Nucleotides*, **1992**, 11, 177.
335. (a) R. Yarchoan, H. Mitsuya, C. E. Myers, S. Broder, *New England J. Med.* **1989**, 232, 726; (b) K. Parang, L. I. Wiebe, E. E. Knaus, *Curr. Med. Chem.*, **2000**, 7, 995.
336. (a) N. B. Tarassova, A. A. Khorlin, A. A. Krayevsky, N. N. Korneyeva, D. N. Nosik, I. V. Kruglov, G. A. Galegov, R. Sh. Beabealashvilli, *Mol. Biol.* (Russian) **1989**, 23, 1716; (b) N. B. Tarassova, M. K. Kuhanova, A. A. Krayevsky, E. K. Karamov, V. V. Lukashov, G. B. Kornilayeva, M. A. Rodina, G. A. Galegov, *Nucleosides Nucleotides*, **1991**, 10, 351; (c) C. Meir. L. Habel, W. Laux, E. De Clercq, J. Balzarini, *Nucleosides Nucleotides*, **1995**, 14, 759; (d) Q. Xiao, J. Sun, Y. Ju, Yu-Fen Zhao, Yu-xin Cui, *Tetr. Lett.*, **2002**, 30, 5281.
337. J. Machado, H. Salomon, M. Oliveira, Ch. Tsoukas, A. A. Krayevsky, M. A. Wainberg, *Nucleosides Nucleotides*, **1999**, 18, 901.
338. K. Troev, I. Gitsov, I. Ivanov, *Biocong. Chem.*, submitted, **2005**.
339. (a)S. Aaronson, G. Torado *J. Cell Physiol.* **1968**, 72, 141; (b) E. Borenfreund, J. Puerner *Toxicol. Lett.*, **1985**, 24, 119; (c) U. Hackenberg, H. Bartling *Arch. Exp. Pharmakol.*, **1959**, 235, 437; (d) T. Pick *J. Exp. Med.* **1958**, 108, 945; (d) M. Liebsch, H. Spiclmann. In: *Methods in Molecular Biology, Vol. 43, In vitro Toxicity Testing Protocols* (eds. S. O'Hare, C. K. Auerwill), Humana Press, Totowa, NJ, **1995**, 177–187.
340. (a) J. Wang, H. Q Mao, K. W. Leong, *J. Am. Chem. Soc.*, **2001**, 123, 9480; (b) J. Wang, S.W. Huang, P. C. Zhang, H. Q. Mao, K. W. Leong, *Intern. J. Pharmac.*, **2003**, 265, 75.
341. J. Wang, P. C. Zhang, H. F. Lu, N. Ma, S. Wang, H. Q. Mao, K. W. Leong, *J. Controlled Release*, **2002**, 83, 157.
342. S. W. Kim, Y. H. Bac, T. Okano, *Pharmacol. Res.*, **1992**, 81, 85.
343. O. Wichtrle, D. Lim, *Nature*, **1960**, 185, 177.
344. M. Richard, B. I. Dahiyat, D. M. Arm, P. R. Brown, K. W. Leong, *J. Biomed. Mater. Res.*, **1991**, 25, 1151.
345. K. W. Leong, H. Q. Mao, R. X. Zhuo, *Chim. J. Polym.*, **1995**, 13, 289.
346. B. I. Dahiyat, E. Hostin, E. M. Posadas, K. W. Leong, *J. Biomater. Sci., Polymer Edition*, **1993**, 4, 529.
347. J. Wang, R. Zhuo, *Eur.Polym. J.*, **1999**, 35, 491.
348. L. W. Daasch, *J. Am. Chem. Soc.*, **1958**, 80, 5301.
349. K. Moedritzer, *J. Inorg. Nucl. Chem.*, **1961**, 22, 19.
350. D. E. C. Corbridge, *Phosphorus. An Outline of its Chemistry, Biochemistry and Technology*, Elsevier, Amsterdam, **1990**, 353.
351. E. M. Georgiev, J. Kaneti, K. Troev, D. M. Roundhill, unpublished data.
352. A. Michaelis, T. Becker, *Chem. Ber.*, **1897**, 30, 1003.
353. A. F. Grapov, *The Chemistry and Methods for Investigation of the Organic Compounds*, Mir, Moskow, **1966**, 15, 41–231.
354. R. Harvey, E. de Sombre, *Topics in Organophosphorus Chemistry*, Wiley, New York, **1964**, 1, 61.
355. C. Benezra, J. Bravet, *Can. J. Chem.*, **1975**, 53, 474.
356. L. P. Reiff, H. S. Aaron, *J. Am. Chem. Soc.*, **1970**, 92, 5275.
357. W. B. Farnham, R. K. Murray, K. Mislow, *Am. Chem. Soc.*, **1970**, 92, 5809.
358. W. B. Farnham, R. K. Murray, K. Mislow, *J. Chem. Soc., Chem. Commun.*, **1971**, 605.
359. G. R. Van den Berg, D. H. Platenbury, H. P. Benschop, *Chem. Soc., Chem. Commun.*, **1971**, 606.
360. J. Michalsky, Z. Skrzypzynsky, *J. Organom. Chem.*, **1975**, 97, C31.
361. A. N. Pudovik, B. A. Arbuzov, *Izw. Akad. Nauk SSSR, Otd. Khim. Nauk*, **1949**, 522.
362. F. Hammerschmiedt, E. Zbiral, *Liebig's Ann. Chem.*, **1979**, 492.
363. I. N. Azerbayev, N. N. Godovikov, N. B. Abdullaev, B. D. Abjurov, *Zh. Obshch. Khim.*, **1978**, 48, 1271.

364. M. Makosza, K. Wojciechowski, *Bull. Pol. Acad. Sci., Chem.*, **1984**, 32, 175.
365. A. Zwierzak, *Synthesis*, **1975**, 507.
366. N. Bondarenko, M. Rudomino, E. N. Tsvetkov, *Zh. Obshch. Khim.*, **1990**, 60, 1196.
367. N. A. Bondarenko, A. M. Yurkevich, A. N. Bovin, E. N. Tsvetkov, *Izw. Akad. Nauk SSR, Otd. Khim. Nauk*, **1990**, 1387.
368. T. Hirao, T. Masgunaga, N. Yamada, Y. Ohshiro, T. Agawa, *Bull. Chem. Soc. Jpn.*, **1982**, 55, 909.
369. A. Osuka, N. Ohmasa, Y. Yoshida, H. Suzuky, *Synthesis*, **1983**, 69.
370. G. M. Blackburn, G. E. Taylor, *J. Organomet. Chem.*, **1988**, 348, 55.
371. U.S. Pat. 4,330,486 (**1980**). C. A. 97, 145014v (**1980**).
372. Z. Pelchowicz, S. Brukson, E. Bergmann, *J. Chem. Soc.*, **1961**, 4348.
373. A. N. Pudovik, R. Tarasova, *Zh. Obshch. Khim.*, **1964**, 34, 1151.
374. W. Stec, A. Zwierzak, J. Michalski, *Bull. Pol. Acad. Sci. Chim.*, **1970**, 18, 23.
375. W. Daniel, L. Cirisley, *J. Org. Chem.*, **1961**, 26, 2544.
376. M. Zimin, T. Dvoinishnikova, I. Konovalov, A. N. Pudovik, *Zh. Obshch. Khim.*, **1978**, 48, 2790.
377. M. Zimin, A. Burilov, A. N. Pudovik, *Zh. Obshch. Khim.*, **1980**, 50, 751.
378. A. N. Pudovik, M. Zimin, I. Konovalova, V. Pozidaev, L. Vinogradova, *Zh. Obshch. Khim.*, **1975**, 45, 30.
379. A. N. Pudovik, I. Konovalova, M. Zimin, T. Dvoinishnikova, L. Vinogradov, *Zh. Obshch. Khim.*, **1977**, 47, 1698.
380. T. Minami, N. Matsuzaki, Y. Ohshiro, T. Agawa, *J. Chem. Soc., Perkin Trans. 1*, **1980**, 1731.
381. Y. Okada, M. Kuroboshi, T. Ishihara, *First International Conference on Heteroatom Chemistry*, Kobe, Japan, **1987**, 98.
382. U.S. Pat. 3,694,144 (**1972**).
383. U.S. Pat. 3,830,815 (**1974**).
384. U.S. Pat. 3,775,057 (**1973**).
385. U.S. Pat. 3,786,055 (**1974**).
386. A. N. Pudovik, M. Zimin, A. Sobanov, B. Kamaldzanova, *Zh. Obshch. Khim.*, **1975**, 45, 2403.
387. A. N. Pudovik, M. Zimin, A. Sobanov, *Zh. Obshch. Khim.*, **1975**, 45, 1232.
388. U.S. Publ. Pat. Appl. 530174 (**1976**); C. A. 84, 151476v (**1976**).
389. V. Orlovski, B. Vovsi, A. Mishkevich, *Zh. Obshch. Khim.*, **1973**, 42, 1930.
390. K. Troev, G. Borisov, *Phosphorus & Sulfur*, **1983**, 17, 257.
391. E. Philippot, O. Lindquist, *Acta Chem. Scand.*, **1970**, 24, 2803.
392. K. Troev, E. M. G. Kirilov, D. M. Roundhill, *Bull. Chem. Soc. Jpn.*, **1990**, 63, 1284.
393. K. Troev, D. M. Roundhill, *Phosphorus & Sulfur*, **1988**, 36, 189.
394. M. Crutchfield, C. Dungan, J. Letcher, V. Mark, J. van Wazer, *Topics in Phosphorus Chemistry*, **1967**, Vol. 5.
395. J. van Wazer, C. Callis, J. Shoolery, *J. Am. Chem. Soc.*, **1956**, 78, 5715.
396. R. Jones, A. Katritzky, *Angew. Chem.*, **1962**, 74, 60.
397. C. Callis, J. van Wazer, J. Shoolery, *Annal. Chem.*, **1956**, 28, 269.
398. H. Gutowsky, D. McCall, C. Slichter, *J. Chem. Phys.*, **1953**, 21, 279.
399. W. Quinn, R. Brown, *J. Chem. Phys.*, **1953**, 21, 1605.
400. H. Gutowsky, D. McCall, *J. Chem. Phys.*, **1954**, 22, 162.
401. R. Mesmer, R. Carroll, *J. Am. Chem. Soc.*, **1966**, 88, 1381.
402. J. Riess, J. van Wazer, *Inorg. Chem.*, **1966**, 5, 178.
403. W. Knight, *Phys. Rev.*, **1949**, 76, 1259.

404. W. Klaui, K. Dehnicke, *Chem. Ber.*, **1978**, 111, 451.
405. L. Elegant, R. Wolf, M. Aztaro, *Bull. Chim. Soc. France*, **1969**, 12, 4269.
406. R. Bedell, M. Frazer, M. Gerrard, *J. Chem. Soc.*, **1967**, 4037.
407. P. Onisko, Y. Gololobov, *Zh. Obshch. Khim.*, **1979**, 49, 39.
408. Y. Gololobov, P. Onisko, V. Fadeev, N. Tasur, *Zh. Obshch. Khim.*, **1975**, 45, 1391.
409. Y. Gololobov, P. Onisko, B. Prokoinenko, *Dokl. Akad. Nauk SSSR*, **1977**, 237, 105.
410. P. Onisko, Y. Gololobov, *Zh. Obshch. Khim.*, **1977**, 47, 2480.
411. L. W. Daasch, *J. Am. Chem. Soc.*, **1958**, 80, 5301.
412. T. D. Smith, *J. Inorg. Nucl. Chem.*, **1960**, 15, 95.
413. R. B. Fox, D. L. Vanezky, *J. Am. Chem. Soc.*, **1956**, 78, 1664.
414. G. G. Mather, A. Pidcock, *J. Chem. Soc., Dalton. Trans.*, **1973**, 560.
415. A. A. Muratova, E. G. Yarkova, A. P. Pribylova, A. N. Pudovik, *Zh. Obshch. Khim.*, **1968**, 38, 2772.
416. A. N. Pudovik, A. A. Muratova, R. N. Safiulina, E. G. Yarkova. V. P. Plekhov, *Zh. Obshch. Khim.*, **1975**, 45, 520.
417. P. N. Nagar, *J. Indian Chem. Soc.*, **1990**, 67, 703.
418. R. C. Reiter, N. Takahashi, R. K. Bunting, G. W. Caldwell, *Inorg. Chim. Acta.*, **1982**, 64, L45.
419. J. L. Burdet, L. L. Burger, *Can. J. Chem.*, **1966**, 44, 111.
420. P. G. Harrison, M. A. Healy, *Inorg. Chim. Acta.*, **1983**, 80, 279.
421. D. Pun, A. Parnash, *Indian J. Chem.*, **1975**, 13, 384.
422. A. A. Grinberg, A. D. Troitzkaya, *Zh. Obshch. Khim.*, **1944**, 4, 178.
423. I. N. Itskovich, A. D. Troitzkaya, *Trudy Kazan. Khim. Teknol. Inst.*, **1953**, 18, 59.
424. A. Pidcock, C. R. Waterhouse, *Inorg. Nucl. Chem. Lett.*, **1967**, 3, 487.
425. A. Pidcock, C. R. Waterhouse, *J. Chem. Soc. (A)*, **1970**, 2080.
426. F. H. Allen, A. Pidcock, C. R. Waterhouse, *J. Chem. Soc. (A)*, **1970**, 2087.
427. R. P. Sperline, M. K. Dickson, D. M. Roundhill, *J. Chem. Soc., Chem Commun.*, **1977**, 62.
428. R. P. Sperline, D. M. Roundhill, *Inorg. Chem.*, **1977**, 16, 2612.
429. G. A. Levishina, A. D. Troitzkaya, R. R. Shagidullin, *Zh. Neorg. Khim.*, **1966**, 11, 985.
430. H. Werner, T. N. Khac, *Z. Anorg. Allg. Chem.*, **1981**, 479, 134.
431. T. S. Kukharieva, I. D. Rozhdestvenskaya, E. E. Nifantiev, *Koord. Khim.*, **1977**, 241.
432. H. Werner, T. N. Khac, *Inorg. Chim. Acta*, **1979**, 30, L347.
433. H. Werner, T. N. Khac, *Z. Anorg. Allg. Chem.*, **1981**, 475, 241.
434. H. Werner, T. Khac, *Angew. Chem. Int. Ed.*, **1977**, 16, 324.
435. W. Klaui, H. Werner, *Angew. Chem. Int. Ed.*, **1976**, 15, 172.
436. V. Harder, E. Dubber, H. Werner, *J. Organom. Chem.*, **1974**, 71, 427.
437. W. Klaui, H. Werner, G. Hutter, *Chem. Ber.*, **1977**, 110, 2283.
438. W. Klaui, W. Eberspach, P. Gutlich, *Inorg. Chem.*, **1987**, 26, 3977.
439. M. A. Bennett, T. R. B. Mitchell, *J. Organom. Chem.*, **1974**, 70, C30.

APPENDIX

General procedure of Kabachnik–Fields reaction (ref. *J. Am. Chem. Soc.*, **1952**, 74, 1528).

The aldehyde or ketone was added to a stirred mixture of equimolar quantities of secondary amine and H-phosphonate diesters at a rapid rate. The temperature was kept below

85 °C. Fifteen minutes after the addition was ended, the mixture was cooled, dried over anhydrous sodium sulfate, filtered, and distilled *in vacuo*.

General procedure for the preparation of aminophosphonic acids from dimethyl H-phosphonate, aldehydes, and amines in the presence of lithium perchlorate/diethylether catalyst [LPDE] (*ref. Tetr. Lett.*, **1998**, 39, 6729).

To a mixture of aldehyde (2 mmol) in 5 M LPDE (4 mL), amine (2.2 mmol) was added at room temperature. The mixture was stirred for 2 min and dimethyl H-phosphonate (2 mmol) was added. The mixture was stirred for 8 min, then water was added and the product was extracted with CH_2Cl_2. The organic phase was collected, dried (Na_2SO_4), and evaporated to afford the crude product. The product was purified by recrystallization.

General procedure for preparation of α-aminophosphonates under microwave irradiation (*ref. Tetr. Lett.*, **2006**, 47, 1125).

Benzaldehyde (1 mmol), aniline (1 mmol), and dimethyl H-phosphonate (2 mL) were added into a 25-mL three-necked flask. The mixture was then heated at 80 °C for 2 min at 180 W in a multimood microwave oven. The reaction mixture was diluted with water and extracted with CH_2Cl_2 (20 mL). The organic layer was washed with H_2O (3 × 10 mL), dried over anhydrous Na_2SO_4, filtered, concentrated, and the crude product was purified by silica gel flash column chromatography eluted with 2:1 petroleum ether/acetone to give pure α-aminophosphonate in 98% yield, with melting point at 90–91 °C.

General procedure for the preparation of aminophosphonic acids from phosphorous acid, formaldehyde and aminoalcohols (*ref. Phosphorous, Sulfur, and Silicon*, **1990**, 48, 41).

A mixture of an aminoalcohol (0.2 mol), water (100 mL), phosphorous acid (0.4 mol), and concentrated HCl (100 mL) was heated to reflux. Formaldehyde (66.6 mL of 36% aqueous, 0.8 mol) was added dropwise over a period of 2 h. The reaction mixture was refluxed for an additional 3 h and the solvent was removed on a rotary evaporator. Water (200 mL × 3) was added and the solution was reevaporated to dryness to get the crude product as a colorless syrup.

General procedure for the preparation of aminophosphonic acids from phosphorous acid (*ref. J. Org. Chem.*, **1978**, 43 (5) 992)

A mixture of imine (0.2 mol) and phosphorous acid (0.2 mol) was stirred with a mechanical stirrer and slowly heated to 75–80 °C, whereupon the reactants gave a homogeneous liquid. Further heating to 100–120 °C brought about a vigorous reaction resulting in a significant viscosity increase and an internal temperature of 140–160 °C. The source of heat was removed and water (100 mL) was added as the temperature reached 95–100 °C. The crude α-aminophosphonic acid is purified by crystallization or by ion-exchange chromatography.

General procedure for the synthesis of N-phosphonomethyl glycine (*ref. U. S. Pat. 4,237,065*, **1980**).

In a reaction vessel, methanol (400 mL), paraformaldehyde (45.0 g, 1.5 mol), and a catalytic amount of triethylamine (5 mL) are mixed. The mixture is heated at reflux until complete dissolution is attained. Glycine (75.0 g, 1 mol) followed by triethylamine (102 g, 1 mol) is then added to the solution. The temperature of the mixture is maintained at 60–70 °C until complete dissolution is attained, and then diethyl H-phosphonate (91 g, 0.66 mol) is added over a period of 10 min. The reaction mixture is stirred for 1.5 h at 65–72 °C. During the cooling of the reaction mixture, sodium hydroxide (250 g) in the form of a 40% aqueous solution is added and the reaction mixture is then refluxed for another 1.5 h. After this time, the volatile organic materials are distilled and the remaining solution is acidified to a pH of 1.5 by addition of concentrated hydrochloric acid. On cooling to 15 °C, the pure N-phosphonomethyl glycine precipitates and is collected by filtration (71 g, 64%).

General procedure for the preparation of bisphosphonates from carboxylic acid (ref. J. Org. Chem., 1995, 60, 8311).
Preparation of (4-amino-1-hydroxybutylidene)bisphosphonic acid monosodium salt
A 250-mL flask was fitted with a mechanical stirrer, a thermocouple, an addition funnel, and a reflux condenser through which was circulated −10 °C brine and connected to a caustic scrubber. The system was flushed with nitrogen and charged with 4-aminobutyric acid (20 g, 0.19 mol), H-phosphonic acid (16 g, 0.19 mol), and methanesulfonic acid (80 mL). The mixture was heated to 65 °C, PCl_3 (35 nL, 0.40 mol) added over 20 min, and the mixture maintained at 65 °C overnight (generally 16–20 h). The clear, colorless solution was cooled to 25 °C and quenched into 0–5 °C water (200 mL) with vigorous stirring. The reaction flask was rinsed with an additional 100 mL of water and combined solution refluxed for 5 h. The solution was cooled to 20 °C, the pH adjusted to 4.3 with 50% NaOH (ca. 80 mL), and the resulting suspension aged for 2 h at 0–5 °C. The product was collected by filtration, washed with cold water (2 × 50 mL) and 95% ethanol (100 mL), and dried (air drying then *in vacuo* at 40 °C) yielding 56.4 g (89% yield) of the monosodium salt trihydrate.

Preparation of (4-amino-1-hydroxybutylidene)bisphosphonic acid
To obtain the free acid, the above procedure is followed and the pH is adjusted to 1.8 instead of 4.3. Isolation and drying is identical, providing 44.0 g (85% yield) of white crystalline of (4-amino-1-hydroxybutylidene)bisphosphonic acid as a monohydrate.

General procedure for the preparation of 6-amino-1-hydroxyhexylidene diphosphonic acid (ref. U.S. Pat. 4,304,734, 1981).
A mixture of 13 g of ε-aminocaproic acid and 12.7 g of H-phosphonic acid (H_3PO_3) in 100 mL of chlorobenzene was heated under stirring to 100 °C, and 14 mL of phosphorus trichloride was added dropwise to it within a period of 30 min. The solution was then heated under stirring for three more hours; insoluble solid matter separated during this time. The solvent was poured off after cooling, the residue was boiled with 60 mL of water for about 30 min. and subjected to hot filtration with activated charcoal through a layer of supercel. Activated charcoal and supercel were washed with hot water, and the united solids were concentrated at 40 °C in vacuum. Separated crystals were filtered off, and a further fraction was obtained from the mother liquors by adding methanol in excess. Yield was 15 g, 55%.

Using the same procedure, 1-hydroxy-2-amino-2-methylpropane-1,1-diphosphonic acid and 2′-pyrrolidine-1-hydroxymethane-1,1-diphosphonic acid were obtained (*ref. U.S. Pat. 4,267,108,* **1981**).

General procedure for the preparation of bisphosphonates from lactams (*ref. Tetrahedron Lett.,* **2002**, 43, 8665).

A solution of *n*-octyl pyrrolidinone (211 mg, 1.07 mmol, 1.0 equiv.) in anhydrous diethyl ether (2 mL) was added dropwise via cannula to a stirred solution of LDA [2.2 equiv., prepared *in situ* from diisopropylamine (0.32 mL, 2.36 mmol) and *n*-butyllithium in hexanes (2.50 M, 0.94 mL, 2.36 mmol)] in anhydrous diethyl ether (5mL) at −78 °C. After 60 min, HMPA (0.42 mL, 2.46 mmol) and diethyl phosphorochloridite (0.36 mL, 2.46 mmol) were added in turn, and the resulting mixture was allowed to warm to O °C over the period of 2 h. The reaction was quenched by slow addition of hydrogen peroxide (30%, 1.21 mL, 10 equiv.), and the resulting mixture was stirred vigorously at O °C for 10 min. The organic phase was separated, washed with brine, and dried ($MgSO_4$). After concentration *in vacuo*, final purification by column chromatography on silica gel (from EtOAc to 8:2 EtOAc:CH_3OH) afforded bisphosphonate (470 mg, 94%) as a clear oil.

General procedure for the preparation of bisphosphonates from nitriles (*ref. Biorganic & Med. Chem. Lett.,* 2005, 15, 4685; *U.S. Pat. 5,990,098,* **1999**).

A mixture of H_3PO_3 (4.1 g, 50 mmol) and acetonitrile (1.0 mL, 20 mmol) was stirred at 130 °C for 12 h under argon atmosphere. Methanol (10 mL) was added and the solid was filtered and crystallized from water–methanol to afford 2.10 g (51% yield).

General procedure for the synthesis of nucleoside H-phosphonates (*ref. J. Org. Chem.,* **1998**, 63, 8150).

A suitable protected nucleoside (3′-O-benzoylthymidine, 3′-O-benzoyl-2-deoxyuridine, 3′-O-benzoyl-5-fluoro-2′-deoxyuridine, etc.) (1 mmol) was rendered anhydrous by repeated evaporation of added pyridine (2 × 10 mL), dissolved in the same solvent (10 mL) and treated with diphenyl H-phosphonate (3 molar equiv, 15 min). The reaction was quenched by the addition of water–triethylamine (1:1, v/v, 4 mL) and left for 15 min. The mixture was concentrated and the residue was partitioned between water (10 mL) and CH_2Cl_2 (10 mL). The aqueous layer was extracted with CH_2Cl_2 (2 × 10 mL) and concentrated and the residue, after being dissolved in a minimum volume of MeOH, was applied on a silica gel column equilibrated with CH_2Cl_2–MeOH (95:5, v/v). The column was washed with CH_2Cl_2 containing Et_3N (95:5, v/v) and the unprotected nucleoside H-phosphonate was eluted from the gel with MeOH. Fractions containing pure products were collected and evaporated to dryness to yield colorless, hygroscopic foams. ^{31}P NMR data: solvent D_2O, δ (ppm) = 6.54 (dt, $^1J(P,H)$ = 639.6 Hz, $^3J(P,H)$ = 6.5 Hz).

General procedure for the preparation of 3′-azido-2′, 3′-dideoxythymidine-5′-hydrogen phosphonates (*ref. Mol. Biology,* **1989**, 23, 1716, *EP* 354246, **1988**).

To a solution of imidazol (0.50 g, 7.36 mmol) in anhydrous acetonitrile (15 ml) at 0 °C, phosphorus trichloride (0.19 ml, 2.17 mmol) and Et_3N (1.05 ml, 7.54 mmol) was added dropwise. After 15 min at 0 °C, a solution of 3′-azido-2′, 3′-dideoxythymidine (134 mg, 0.5 mmol) in acetonitrile (10 ml) was added dropwise to the reaction mixture for 30 min, keeping the reaction temperature at around 5 °C. The reaction mixture was stirred for an additional 2 h at 20 °C and the solvent was removed on a rotary evaporator.

General procedure for the synthesis of poly(alkylene H-phosphonate)s
Conventional heating (ref. Makromol. Chem., **1990**, *191, 671).*

Condensation was carried out either in bulk or in 1,2-dichlorobenzene as solvent in a round-bottom flask, fitted with a magnetic stirrer, thermometer, and condenser, attached to the vacuum line. Usually, dialkyl H-phosphonate (dimethyl H-phosphonate) in 20% excess over stoichiometry was used. First, the temperature was raised to 120 °C, and the methanol formed was removed under normal pressure. When the amount of methanol was equal to 90 % of the theoretical amount to be removed, the temperature was slowly increased and a vacuum was applied. The temperature should not exceed 160 °C when hydroxyl end-groups are still detectable, until methyl alcohol is distilled off and dimethyl H-phosphonate is still present in the reaction mixture. When methanol evolution ceased, for the polycondensation in bulk, temperature was slowly (approximately after 1–6 h) raised to 160 °C, and at the final stage, to 180 °C. Vacuum was applied, first at 160 °C (330–660 mbar and then, at the final stage, down to a few mbar). When the condensation was completed, the product was at 200 °C; a highly viscous, colorless mass, solidifying into an amorphous, elastic or crystalline product, depending on the starting diol.

^{1}H NMR (CDCl$_3$): δ (ppm) 3.70 (t, ^{3}J(H,H) = 4.6 Hz –OCH_2CH_2O–), 3.78 (d, ^{3}J(P,H) =12 Hz –POCH_3), 4.15–4.28 (m, P–OCH_2–), 6.80 (d, ^{1}J(P,H) = 690.4 Hz –HO–P–H end group), 6.88 (d, ^{1}J(P,H) = 709.1 Hz –CH$_3$O–P–H end group), 6.95 (d, ^{1}J(P,H) = 716.6 Hz –P–H repeating unit); ^{13}C{H} NMR (CDCl$_3$): δ (ppm) 54.8 (d, ^{2}J(P,C) = 5.6 Hz –POCH_3), 65.0 (d, ^{2}J(P,H) = 7.3Hz –P–OCH_2–), 70.4 (d, ^{3}J(P,C) = 4.9 Hz –POCH_2CH_2–), 70.9 (s, OCH_2CH_2O– repeating unit of PEG); ^{31}P NMR (CDCl$_3$): δ (ppm) 11.23 (dsex., ^{1}J(P,H) = 709.6 Hz, ^{3}J(P,H) = 11.7 Hz –CH$_3$O–P(H)OCH$_2$ end group), 10.43 (dquintets, 1J(P,H) = 716.7 Hz, ^{3}J(P,H) = 9.9 Hz, P–H repeating unit), 8.25 (dt, ^{1}J(P,H) = 695.5 Hz, ^{3}J(P,H) = 10.3 Hz –HO–P–H end group).

Microwave irradiation (ref. Macromol. Rapid Commun., **2005**, *26, 471).*

A monomodal microwave (CEM-Discover) operating at a maximum power of 300 W was used. Dimethyl H-phosphonate 5.58 g (0.0525 mol) and poly(ethylene glycol) 20 g (0.05 mol) were placed in a round-bottom flask, equipped with a condenser, and subjected to microwave irradiation for 55 min, using microwave power in a range between 30 and 40 W. The reaction was carried out under nitrogen atmosphere. During the process, the temperature varied between 140 and 190 °C. For the first 10 min, the reaction was carried out at 140 °C and for the next 15 min, at 150 °C. After that, for the next 15 min, the temperature was increased to 165 °C, for the next 15 min to 180 °C, and for the last 5 min, to 190 °C. The temperature in the microwave system was controlled outside the vessel by use of an IR sensor. During the last 10 min, the reaction was performed under vacuum.

^{1}H NMR (CDCl$_3$): δ (ppm) 3.59–3.68 (m, 4H, OCH$_2$CH$_2$); 3.68 (d, ^{3}J(P,H) = 11.8 Hz, 3H, POCH$_3$); 4.09–4.16 (m, 4H, POCH$_2$CH$_2$); 6.85 (d, 1H, ^{1}J(P,H) = 716.5 Hz, P–H); 6.76 (d, 1H, ^{1}J(P,H) = 709.0 Hz, P–H); 6.74 (d, 1H, ^{1}J(P,H) = 698.4 Hz, P–H).
^{13}C{H} NMR (CDCl$_3$): δ (ppm) 52.8 (d, ^{2}J(P,C) = 12.3 Hz, POCH$_3$); 65.03 (d, ^{2}J(P,C) = 6.1 Hz, POCH$_2$); 68.50 (d, ^{3}J(P,C) = 5.8 Hz, POCH$_2$CH$_2$); 70.22 (CH$_2$OCH$_2$CH$_2$).
^{31}P NMR (CDCl$_3$); δ (ppm) 10.46 (dquintets,^{1}J(P,H) = 716.1 Hz, ^{3}J(P,H) = 9.8 Hz, phosphorus atom in the repeating unit); 11.18 (dsextets, ^{1}J(P,H) = 709.0 Hz, ^{3}J(P,H) = 10.5 Hz, phosphorus atom in the end group, bonded with OCH$_3$ and OCH$_2$ groups); 8.31 (dt, ^{1}J(P,H) = 697.3 Hz, ^{3}J(P,H) = 9,8 Hz, phosphorus atom bonded with –OH and –OCH$_2$– groups).

General procedure for the synthesis of poly(alkylene phosphate)s (Atherton–Todd reaction conditions) (ref. Mackromol. Chem., **1993***, 194, 3261).*

Dichloromethane (9 mL), carbontetrachloride (23.5 mL), triethylamine (0.94 g), and methanol (0.72 mL) were placed in a three-necked flask equipped with a magnetic stirrer, thermometer, reflux condenser, and a dropping funnel. A solution of poly(alkylene H-phosphonate) (2.3 g, 0.0036 mol, 646 molecular weight of the repeating unit) in dichloromethane (13.5 mL) was added dropwise at ambient temperature under continuous stirring. The reaction mixture was allowed to stand at room temperature for 24 h. After filtration of the precipitated triethylamine hydrochloride, the filtrate was concentrated and the poly(alkylene phosphate) was precipitated by addition of diethyl ether. Poly(alkylene phosphate) was purified by dissolution in *N,N*-dimethylformamide and precipitation in diethyl ether. The isolated product was dried at 30–40 °C under reduced pressure (1 mmHg). Yield was 2.4 g, 100%.
^{1}H NMR (CDCl$_3$): α (ppm) 1.36 (t, ^{4}J$_{(H,H)}$ = 7.1 Hz, CH$_3$CH$_2$OP), 3.48–3.61 (m, –OCH$_2$CH$_2$–), 3.76 (d, ^{3}J$_{(P,H)}$ = 11.6 Hz, POCH$_3$). ^{13}C{H} NMR (CDCl$_3$): α(ppm) 14.9 (d, ^{3}J$_{(P,C)}$ = 4.8 Hz, CH$_3$); 54.6 (d, ^{2}J$_{(P,C)}$ = 5.6 Hz, POCH$_3$); 60.8 (d, ^{2}J$_{(P,C)}$ = 2.5 Hz, POCH$_2$); 63.8 (CH$_2$OCH$_2$). ^{31}P NMR (CDCl$_3$): α (ppm): 0.57 (octet for *P* in the end group, (^{3}J$_{(H,H)}$ = 8.14 Hz, POCH$_3$ + POCH$_2$CH$_3$); 1.28 (octet, for *P* in the repeating group ^{3}J$_{(P,H)}$ = 10.05 Hz).

General procedure for the immobilization of cysteamine aminohydrochlorides onto poly(alkylene phosphate)s (ref. J. Med. Chem. **2002***, 45, 5797).*
Immobilization of Cy hydrochloride

Poly(oxyethylene phosphate) (5.2 g, 7.7 mmol of repeating units, molecular weight of the repeating unit is 676) and Cy hydrochloride (0.85 g, 7.5 mmol) were mixed in a two-necked flask fitted with a magnetic stirrer, reflux condenser, and a thermometer. The reaction was carried out at 110 °C and was stopped after the evolution of methylchloride ceased. The crude product was dissolved in chloroform and precipitated by addition of diethyl ether and was dried under vacuum at 30–40 °C. Yield was 5.6 g (95.4%).
^{1}H NMR (CDCl$_3$), α (ppm): 1.36 (t, ^{3}J$_{(H,H)}$ = 7.1 Hz, CH$_3$CH$_2$OP); 3.48–3.61 (m, –OCH$_2$CH$_2$); 3.76 (d, ^{3}J$_{(P,H)}$ = 11.6 Hz, POCH$_3$); 4.11–4.14 (m, POCH$_2$CH$_3$); 4.65 (s, –SH); 8.19 (s, –NH$_3$).^{13}C{H} NMR (CDCl$_3$), α (ppm) 14.9 (d, ^{3}J$_{(P,C)}$ = 4.8 Hz, CH$_3$); 39.3 (CH$_2$SH); 43.1 (NCH$_2$); 54.6 (d, ^{2}J$_{(P,C)}$ = 5.6 Hz, POCH$_3$); 60.8 (d, ^{2}J$_{(P,C)}$ = 2.5 Hz,

POCH_2); 63.8 (CH_2OCH_2). ^{31}P NMR (CDCl$_3$): α (ppm) 0.32 (PO⁻), 1.34 (octet, $^3J_{(P,H)}$ = 10.05 Hz, $^3J_{(H,H)}$ = 8.14 Hz, POCH_3 + POCH_2CH_3).

General procedure for the immobilization of 3'-azido-2,3'-dideoxythymidine onto poly(oxyethylene H-phosphonate)s

Dichloroethane (5 mL), carbon tetrachloride (15 mL), triethylamine (2 mL), acetonitrile (7 mL), and 3'-azido-2,3'-dideoxythymidine (250 mg, 0.94 mmol) were placed in a three-necked flask equipped with magnetic stirrer, reflux condenser, dropping funnel, and inert-gas outlet. A solution of poly(oxyethylene H-phosphonate) (232 mg, 0.94 mmol of repeating units) in dichloroethane (5 mL) was added dropwise at ambient temperature under continuous stirring. The reaction was allowed to proceed for 24 h. After filtration of the precipitated triethylamine hydrochloride, the filtrate was concentrated and the polymer conjugate was precipitated by addition of diethyl ether. The isolated product was dried at 35–40 °C under reduced pressure (10 mbar). The elemental analysis of chlorine showed a trace of Cl$_2$. Yield was 485 mg (100%).

^{1}H NMR (CD$_3$OD): δ (ppm) 1.88 (s, 3H, CH_3-5), 2.44 (t, 2H, $^3J(H,H)$ = 6.3Hz, H-2'), 3.70 (t, $^3J(H,H)$ = 4.6 Hz –OCH_2CH_2O–), 3.78 (d, $^3J(P,H)$ =12 Hz –POCH_3), 3.66–3.92 (m, P–OCH_2–, H-5'), 4.37(q, 1H, $^3J(H,H)$ = 6.6Hz, H-3'), 6.17 (t, 1H, $^3J(H,H)$ = 6.4 Hz, H-1', H-4'), 7.81 (s, 1H, H-6), 11.15 (br, H-3). ^{13}C{H} NMR (CD$_3$OD): δ (ppm) 12.9 (s, CH$_3$-5), 38.6 (s, C-2'), 54.8 (d, $^2J(P,C)$ = 5.6 Hz –POCH$_3$), 62.2 (s, C-3'), 62.9 (s, C-5'), 66.9 (d, $^2J(P,H)$ = 7.3Hz –P–OCH$_2$–), 71.3 (d, $^3J(P,C)$ = 4.9 Hz –POCH$_2$$CH_2$–), 71.9 (s, O$CH_2$$CH_2$O– repeating unit of PEG), 86.5 (d, 3J(P,C) = 8.8 Hz, C-4'), 86.5 (s, C-1' overlapped with C-4'), 112.0 (s, C-5), 138.5 (s, C-6), 152.7 (s, C-2),166.7 (s, C-4); ^{31}P{H} NMR(CD$_3$OD): δ (ppm) 1.02.

– 5 –

Application of H-Phosphonate Diesters and Their Derivatives

H-phosphonates and their immediate derivatives have numerous applications in different practical areas including agriculture, industry, medicine, etc. This is largely due to the versatility of these compounds, which allows for their use as phosphorylation agents for a variety of substrates. This chapter summarizes some of the most important practical applications of the H-phosphonates and their derivatives, as well as some new promising developments with application potential.

5.1 PHYSIOLOGICALLY ACTIVE SUBSTANCES

Organophosphorus compounds, many of which are derivatives of dialkyl H-phosphonates, represent one of the largest classes of pesticides currently in use. They are mainly applied as insecticides, fungicides, herbicides, and bactericides, and also as plant-growth regulators [1,2].

The ferments acetylcholinesterase and cholinesterase are the target of the organophosphorus compounds *in vivo*. The suppressing effect of some organophosphorus derivatives on the function of these two ferments results in the hindering of the nerve signals transferring processes in the organism. It is assumed that these effects are due to the fact that most physiologically active organophosphorus compounds attack the M– and H– cholinoacceptors [3,4].

The main reactions leading to the deactivation of practically all organophosphorus compounds are hydrolysis and oxidation. These reactions take place in the atmosphere, water, and soil as well as within most biological systems. Under the action of the microorganisms living in plants and animals as well as some proteins (hydrolases), these processes are relatively fast [5–7]. Organophosphorus compounds having multiple ester bonds in their molecules play a major role in biochemical processes. This is one of the main reasons why the diesters of phosphonic acid and their derivatives have numerous applications as physiologically active substances.

5.1.1 Fungicides and bactericides

It has been established that silicon-containing phosphonic acid diesters having the general formula have fungicidal and bactericidal properties [8]. Fungicide products developed on

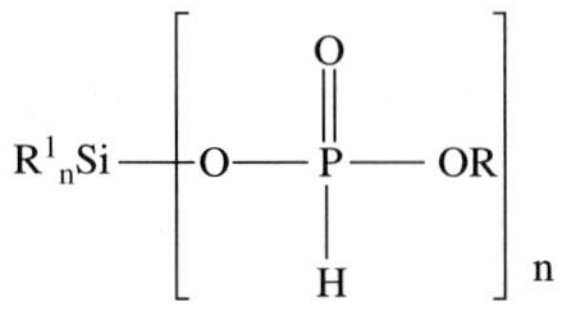

$$\left[R^1_n Si - O - \underset{\underset{H}{|}}{\overset{\overset{O}{\|}}{P}} - OR \right]_n$$

n = 1–4; R^1 = H, Alkyl or *iso*-Alkyl (C_1 to C_8), Ph, Ar

their basis possess low toxicity. An additional advantage is the fact that these products do not exhibit phytotoxic activity due to their decomposition to nontoxic substances, and hence are not considered as pollutants. The working concentration of these active substances ranges from 0.01 to 2%.

Some alkali- and transition-metal derivatives of dialkyl H-phosphonates of the type are

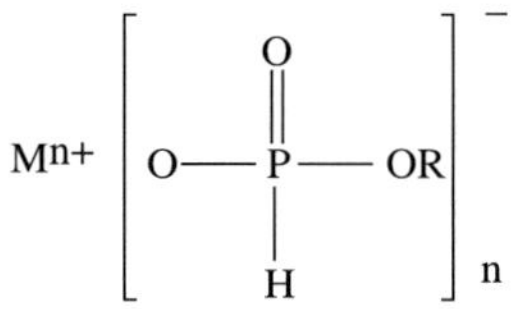

$$M^{n+} \left[O - \underset{\underset{H}{|}}{\overset{\overset{O}{\|}}{P}} - OR \right]^{-}_n$$

R = H or alkyl (C_1 to C_8)

used as plant preservatives against bacterial diseases [9]. Amino derivatives of dialkyl phosphonates constitute another large family of compounds having bactericidal properties. The low toxicity of the aminoalkylphosphonic acids determines their application as antimicrobial preparations for warm-blooded animals [10].

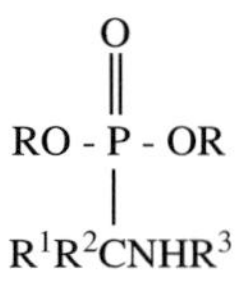

$$RO - \underset{\underset{R^1 R^2 CNHR^3}{|}}{\overset{\overset{O}{\|}}{P}} - OR$$

R = Me, Et, Pr, i-Pr; R^1, R^2 = H, Me or Ph; R^3 = Me, n-Bu

Ferrocenylaminomethyl-phosphonates having a general formula shown below [11] also exhibit antimicrobial activity.

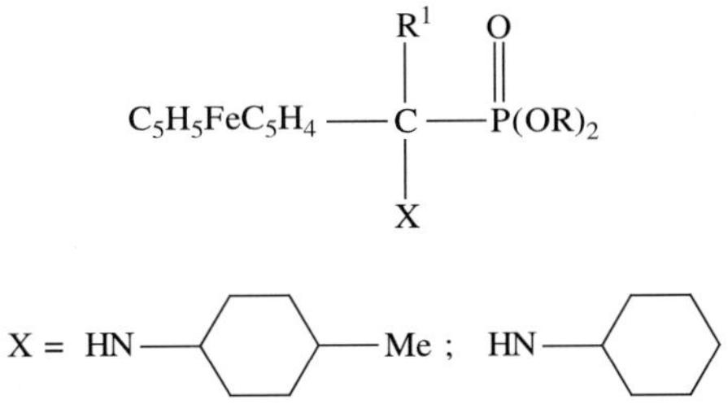

5.1 Physiologically Active Substances

$$C_5H_5FeC_5H_4 - \underset{\underset{X}{|}}{\overset{\overset{R^1}{|}}{C}} - \overset{\overset{O}{\|}}{P}(OR)_2$$

X = HN—⟨C6H4⟩—Me ; HN—⟨C6H5⟩

The aminophosphonates formed from aromatic amines have the highest bioactivity. Bioactivity tests carried out with this family of compounds indicate that the increase in the number of carbon atoms of the alkyl group R^1 leads to reduction of the preparations' efficiency. The degree of branching of the residue does not influence the biological activity of these compounds. The introduction of the ferrocene fragment in the above molecule, on the other hand, leads to a substantial increase of the antimicrobial activity.

Internal salts of H-phosphonic acid diesters with the general formula given below possess plant preservative properties against pathogenesis caused by fungicides [12].

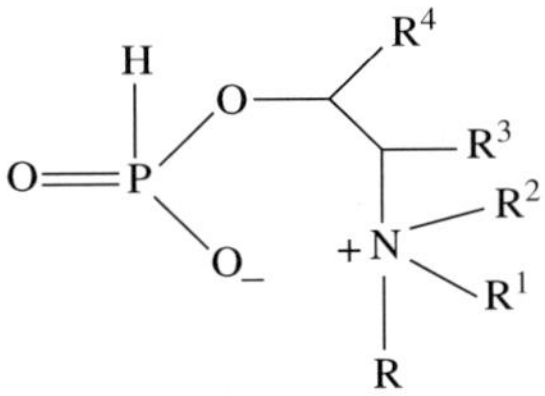

[Structure: H-phosphonate internal salt with O=P(H)(O⁻)(O–CR⁴–CR³R²–⁺NR R¹)]

Another derivative of a similar type—isothioronium salts of monoalkyl H-phosphonate—are also used as fungicides [13].

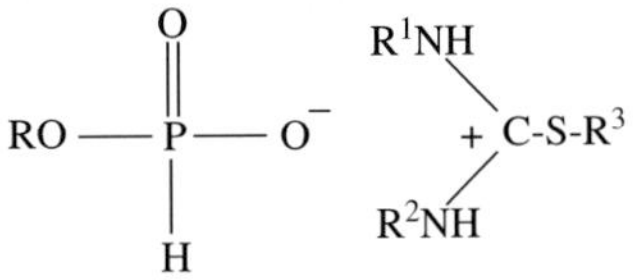

[Structure: RO–P(=O)(H)–O⁻ · ⁺(R¹NH)(R²NH)C–S–R³]

Phosphoramides having an aziridine functionality in their molecules are used as cytostatics, mutagenes, and sexual sterilizers for harmful insects [14].

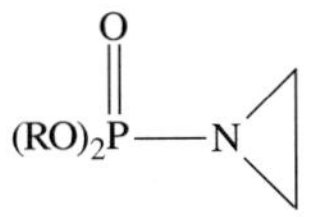

$$(RO)_2\overset{\overset{O}{\|}}{P} - N\triangleleft$$

5.1.2 Herbicides

The problems associated with undesirable plant species result mainly in reduced food-crop production. One approach to weed control is the application of herbicides. The primary phosphonate herbicide is glyphosate-N-(phosphonomethyl) glycine **1**. Herbicidal activity of N-(phosphonomethyl) glycine is reported by D. D. Baird et al. [15]. This aminophosphonic acid is a broad-spectrum herbicide having little or no residual effect. This herbicide is nonselective and is effective against johnson grass, quack grass, and other annual and biennial species of grass, sedges, and broad-leaved weeds. It is primarily used as a post-emergence spray on the foliage of the vegetation to be controlled. Annual species will normally be controlled by 0.3 to 1.0 lb of active ingredient per acre and perennials will be controlled by 1.0 to 4.0 lb of active ingredient per acre [16].

$$\text{(HO)}_2\text{P(=O)-CH}_2\text{-NH-CH}_2\text{-COOH}$$
1

Although the free compound glyphosate has relatively low solubility in water, the monoisopropylammonium, sodium, and ammonium salts of N-(phosphonomethyl) glycine introduced by Monsanto Co [17] and the trimethylsulfonium salt introduced by ICI [18] are water-soluble. Glyphosate is rapidly translocated in the phloem of both annual and perennial weeds to meristematic areas such as shoot and rhizome apices and rhizome nodes [19–22]. The efficacy of a phloem-translocated herbicide is related to plant growth and metabolically available herbicide. Humidity and temperature are important in the absorption and translocation [23], and light will affect translocation [24]. Glyphosate is rapidly metabolized by microbial action in soil and/or surface water to give harmless products. The only major matabolite is aminomethylphosphonic acid **2** [25], which quickly degrades further to carbon dioxide, ammonia, and inorganic phosphate [25,26]. Glyoxalate **3** may also be formed in the initial enzymic cleavage of the glyphosate, but this also degrades readily to carbon dioxide.

$$(\text{HO})_2\text{P(=O)-CH}_2\text{NH-CH}_2\text{-COOH} \xrightarrow{\text{microbial degradation}} (\text{HO})_2\text{P(=O)-CH}_2\text{NH}_2 + \text{OHC-COOH}$$
1   **2**   **3**

Other minor metabolites that have in some cases been identified in trace quantities (chromatographically), after degradation in soil, are N-methylaminomethyphosphonic acid **4**, glycine **5**, N,N-dimethylaminomethylphosphonic acid **6**, and hydroxymethylphosphonic acid **7** [25].

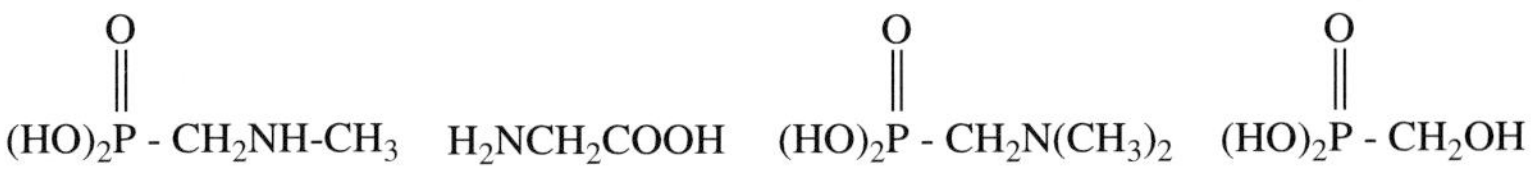

5.1 Physiologically Active Substances

$$(HO)_2P(=O)-CH_2NH-CH_3 \quad H_2NCH_2COOH \quad (HO)_2P(=O)-CH_2N(CH_3)_2 \quad (HO)_2P(=O)-CH_2OH$$

      4   **5**   **6**   **7**

In plants, glyphosate is metabolized slowly, mainly to aminomethyphosphonic acid, and it is excreted rapidly from mammals without change [27].

The mechanism of the action of glyphosate is thought to be associated with the metabolism of aromatic amino acids. *N*-(phosphonomethyl) glycine is a highly effective herbicide because of its potent and specific inhibition of 5-enolpyruvoylshikimate 3-phosphate synthase (EPSPS) [28]. This enzyme, which is found in plants and microorganisms but not in other forms of life, catalyzes the penultimate step of the shikimic acid pathway [29], that is, the interaction of shikimate-3-phosphate **8** with phosphenolpyruvate **9** to give 5-enolpyruvoylshikimate 3-phosphate **11** and inorganic phosphate (Pi)[30,31–33].

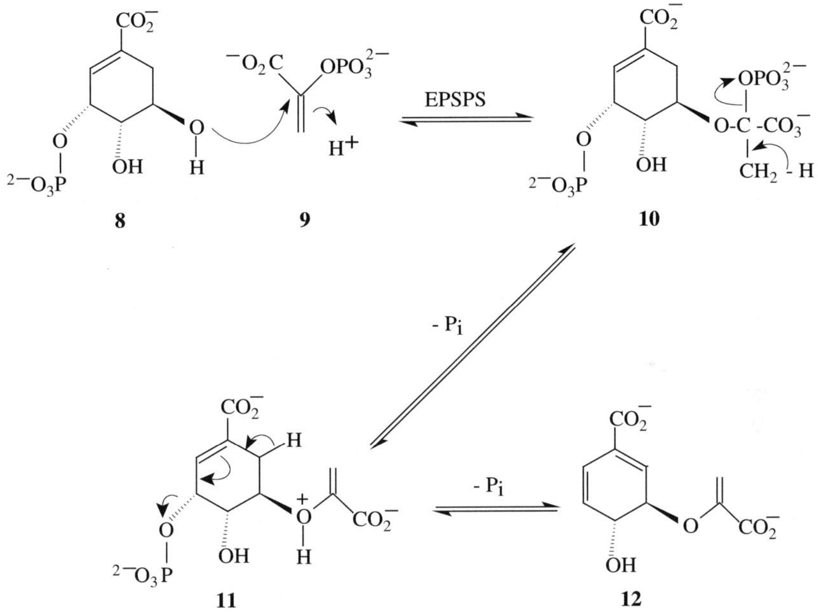

Further elimination of inorganic phosphate from 5-enolpyruvoylshikimate 3-phosphate in the final step of the shikimic acid pathway yields chorismate **12**, a key intermediate in the biosynthesis of numerous aromatic plant metabolites including the aromatic acid phenylalanine, tyrosine, and tryptophan.

Glyphosate acts as a competetive inhibitor to phosphenolpyruvate **9** [34] and forms a stable ternary complex by synergistic binding with the enzyme and shikimate-3-phosphate **8** [35]. It was found that it is the glyphosate dianion **13** that is present in this ternary complex [36]. This anion is considered to be the species associated with the herbicidal activity of glyphosate in plants [30]. It is assumed that the dianionic form **13** could mimic the protonated form **14** of phosphonoenol pyruvate **9**, which is involved in the formation of the tetrahedral intermediate **10**. However, this hypothesis is questionable [30].

$$\underset{13}{\overset{-O_2C\quad PO_3^{2-}}{\underset{H}{\overset{|}{H_2C}}\underset{\overset{|}{H}}{\overset{|}{\underset{N}{\overset{+}{\diagdown}}}}\overset{|}{\underset{}{CH_2}}}} \qquad \underset{14}{\overset{PO_3^{2-}}{\underset{CH_3}{\overset{|}{-O_2C\diagdown\overset{+}{\underset{|}{O}}\diagup}\atop\overset{|}{CH}}}}$$

The first visible effect following application of glyphosate is chlorosis [37–39]. Chlorophyll accumulation has been found to be reduced in corn shoots and etiolated barley grown in the laboratory and in field-grown soybeans, indicating that interference with greening may be important in the mechanism of phytotoxcicity. In addition, there is glyphosate-promoted chlorophyl degradation in light but not in the dark in tobacco leaf discs [40].

The phytotoxicity of glyphosate in soil is very low and may be due either to rapid degradation to aminomethylphosphonic acid or to a high degree of soil binding [41].

The acute oral LD_{50} of glyphosate is 4900 mg/kg in the rat and 1600 mg/kg in the mouse, while the acute intraperitoneal LD_{50} is 240 and 130 mg/kg in the rat and mouse, respectively [42]. Recently, the followng values were reported: the acute oral LD_{50} for rat is greater than 5000 mg/kg for the monoisopropylammonium salt of glyphosate [27].

5.1.3 Plant-growth regulators

Growth regulators, together with the fertilizers and plant protection preparations, play an important role in modern intensive agriculture. They are used for inducing flowering, controlling vine formation (pruning), thinning fruit, and retarding or accelerating ripening.

Glyphosine (*N,N*-bis(phosphonomethyl) glycine) is an analogue of glyphosate but is used to accelerate ripening and increase the level of sucrose in sugar cane, *Saccharum officinarum*.

$$\left[(HO)_2\overset{\overset{O}{\|}}{P}-CH_2\right]_2 N-CH_2-COOH$$

The visible effects of its application are cessation of growth, chlorosis, and dessication [16,43]. At relatively high doses, glyphosine has effects that are similar to those of glyphosate. However, the effects are quantitatively smaller. Potential metabolities of glyphosine (glyphosate, aminomethylphosphonic acid, sarcosine, and glycine) reduced growth but had little effect on reducing dry weight accumulation or on stimulating phenylalanine ammonia-lyase [43]. This indicates that the conversion of glyphosine to glyphosate

5.1 Physiologically Active Substances

is not rapid in intact seedlings, or that the glyphosate is rapidly metabolized with concomitant decrease in concentrations. The rapid translocation of glyphosate has allowed widespread utilization of new methods of application so that only weeds are treated and there is little or no contact of herbicide with low-growing food crops.

Alkylammonium salts of monoalkyl H-phosphonic acids having the following general formula are used as plant growth regulators [44,45]:

$$\text{RO}-\overset{\overset{\displaystyle O}{\|}}{\underset{\displaystyle H}{P}}-O^-\ \ ^+NHR^1R^2$$

where R = alkyl group containing C_1 to C_4; R^1 = alkoxy group containing C_1 to C_4; alkyl group containing C_1 to C_4 or C_6 to C_{20}; R^2 = H or alkyl group containing C_9 to C_{15}.

Continuous treatment of the crops with plant growth regulators of the same type causes a resistance to them, and the application of larger quantities of regulators may be dangerous to human health. One way to enhance the action of bioactive substances is via their immobilization on polymeric supports. This approach has the following advantages: (a) reduction of the toxicity of the active substance; and (b) prolongation of its action. Immobilization of the physiologically active substance on a polymer-carrier leads to considerable changes in their pharmacokinetics [46]. The reduced ability of polymers and other high molecular–weight compounds to penetrate through cellular membranes and other biological barriers causes limited diffusion of the bioactive compounds into the organism. At the same time, their ability to interact with the substrates and receptors does not change significantly.

The copolymer of vinylpyrrolidone with vinylamine shows herbicidal activity when phosphorylated with dimethyl H-phosphonate via alkylation [47].

The problem of environmental protection is becoming increasingly significant all around the world. One of the pollution sources is the chemical preparations used in agriculture. For that reason, questions related to these chemicals' biodegradation, the type of degradation products, and their influence on the human organism determine the application range of each individual preparation. The stability of the phosphonic acid esters with respect to biodegradation is determined largely by the type of the alkoxy groups. Methyl esters are the most unstable and their biodegradation takes place considerably faster than that of their higher homologues. This fact is determined by the high alkylation ability and hydrolytic instability of dimethyl phosphonates and their derivatives.

α-Hydroxyalkylphosphonates with general formula given below—wherein X and Y are independently selected from the group consisting of lower alkyl, halo lower alkyl, phenyl,

$$\begin{array}{c} \text{H} \quad \text{H} \quad \text{OH} \quad \text{O} \\ | \quad \ | \quad \ | \quad \ \| \\ \text{R—C—C—C—P—OX} \\ | \quad \ | \quad \ | \quad \ | \\ \text{R}^3 \ \ \text{R}^2 \ \ \text{R}^1 \ \ \text{OY} \end{array}$$

or benzyl; R^1, R^2, and R^3 are selected from the group consisting of hydrogen, lower alkyl, or phenyl; R is a group consisting of oxygen or sulfur—are used as herbicides [48].

5.1.4 Insecticides

Dimethyl-2,2,2-trichloro-1-hydroxyethylphosphonate is widely used as an insecticide, known under the trade name 'Chlorofos'.

$$\begin{array}{c} \text{O} \\ \| \\ (CH_3O)_2P\text{- CH- }CCl_3 \\ | \\ \text{OH} \end{array}$$

5.1.5 Drugs suppressing the growth of cancer, tumor, virus, or parasites

It has been demonstrated that the N-(phosphonomethyl) glycine is effective in inhibiting test-tube growth of *Plasmodium falciparum*, the parasite that causes malaria [49]. It has the same effect on related types of single-celled parasites such as *Toxoplasma* and *Cryptosporidium* that cause opportunistic infections in AIDS patiens [50].
It is known that aminophosphonic acids and peptides penetrate 4 to 5 times more easily through the membranes of cancer than those of normal cells [51]. The Procter & Gamble Company has found that the N-(phosphonomethyl) glycines are especially effective in suppressing the growth of cancer, tumors, viruses, or bacteria [52]. A pharmaceutical composition for the treatment of mammals, warm-blooded animals, and humans—comprising a pharmaceutical carrier and an effective amount of a chemotherapeutic agent and anticancer compound selected from the group consisting of N-(phosphonomethyl) glycine derivatives of the formula obtained according the method described in U. S. patent 3,799,758 [16]—is given below.

$$\begin{array}{c} \quad\quad\quad\quad \text{R} \quad\quad\quad \text{O} \\ \quad\quad\quad\quad | \quad\quad\quad\quad \| \\ \text{X - C - CH}_2\text{ - N - CH}_2\text{ - P - OY} \\ \quad \| \quad\quad\quad\quad\quad\quad\quad\quad | \\ \quad \text{O} \quad\quad\quad\quad\quad\quad\quad \text{OZ} \end{array}$$

5.1 Physiologically Active Substances

X is selected from the group consisting of hydroxyl, alkoxy or chloroxy up to 12 carbon atoms; lower alkenoxy, cyclohexyloxy, morpholino, pyrrolidinyl, piperidino, and NHR¹; Y and Z each independently selected from hydrogen and lower alkyl; R is selected from the group consisting of hydrogen, formyl, acetyl, benzoyl, nitribenzoyl, and chlorinated benzoyl; R¹ is selected from group consisting of hydrogen, lower alkyl, and lower alkenyl, cyclohexyl, penalkyl, of up to 8 carbon atoms, phenyl, chlorinated phenyl, and anisyl; certain salts of these compounds, selected from the group consisting of group I and II metals having an atomic number of up to 30, hydrochloride, acetate, salicylate, pyridine, ammonium, lower aliphatic hydrocarbon amine, lower alkanol amine, and aniline.

These compositions can be used to inhibit the growth of cancers and other tumors in humans or animals by administration of an effective amount of the N-(phosphonomethyl) glycine derivatives either orally, rectally, topically, or parenterally. Many types of chemotherapeutic agents have been shown to be effective against cancers and tumor cells, but not all types of cancers and tumors respond to these agents. Unfortunately, many of these agents destroy normal cells. It is believed that the phosphonoglycine derivatives in combination with chemotherapeutic agents can suppress and reduce the growth of cancer cells, including leukemia, without affecting normal cells. The use of these N-(phosphonomethyl) glycines in combination with other chemotherapeutic agents that are effective in destroying the tumor is a novel method of treatment.

It was shown that aminophosphonates with the general formula given below inhibit the growth of the leukemia L_{1210} cell *in vitro* [53].

$$\text{Ph}-CH_2NH-CH(\text{C}_6H_4-Y)-P(=O)(OR)_2$$

5.1.6 Anti-HIV prodrugs

Nucleoside analogues are widely used as antiviral agents in the treatment in AIDS and AIDS-related complex. The only clinical agent approved in the United States for the treatment of AIDS is 3′-azido-3′-deoxythymidine (AZT) [54,55]. The molecular mechanism of action for this nucleoside includes conversion into its corresponding 5′-monophosphate by the action of cellular nucleoside kinase, followed by stepwise phosphorylation catalyzed by cellular nucleoside kinase to the corresponding 5′-triphosphate. These inhibit proviral DNA synthesis [55–57], catalyzed by HIV reverse transcriptase (RT), and incorporation to the 3′ end of the growing DNA chain [55,58].

The AZT-5′-H-phosphonate **1** is one of the most significant compounds that is less toxic than AZT (CC50 values of 2.5 mM and 210 μM for AZT-5′-H-phosphonate and ATZ, respectively). The overall selectivity (SI) for AZT-5′-H-phosphonate (CC50/IC50) is superior to that of AZT [59], and AZT-5′-H-phosphonate is currently in Phase I of clinical trials [60]. Unlike the highly acidic nucleoside 5′-phosphate **2** that cannot enter the cell, 5′-AZT H-phosphonate may penetrate the cell membrane due to its weakly acidic undissociated nature. In this connection, 5′-O-hydrogen phospholipids were synthesized in order to increase transmembrane transport characteristic, and antiviral effects, and to investigate the intracellular metabolism of AZT-5′-H-phosphonate [61]. Grosselin *et al.* [62] showed that AZT-5′-H-phosphonate was rapidly metabolized to AZT in cell-culture medium.

A number of H-phosphonate derivatives of 3′-azido-2′,3′-dideoxynucleosides with adenine, guanine, and cytosine [59,63–65] exhibit antiviral acivities with selectivity similar to or higher than that of the corresponding nucleoside.

Unfortunately, charged phosphorylated or phosphonylated nucleosides are unable to penetrate the cell membranes or the blood–brain barrier because of their low lipophilicity. Meier *et al.* [66] have described the synthesis of a new prodrug system for antiviral nucleosides AZT and ddT based on α-hydroxybenzylphosphonates **1**, which present uncharged prodrugs of 5′-nucleoside H-phosphonates and 5′-nucleoside monophosphates. All compounds **1** exhibited pronounced activity against HIV-1- and HIV-2-infected cells without toxicity. Compounds 3 can be considerd as potential prodrugs of the 3′-azido-3′-deoxythimidine and 2′,3′-deoxythimidine, their H-phosphonate monoesters as well as their monophosphates.

5.1 Physiologically Active Substances

5.1.7 Antiresorption drugs, carriers of radioactive metals

Osteoporosis is a common skeletal disease characterized by gradual loss of bone mass and disrupted bone architecture as a result of an imbalance between the bone resorption activity of osteoclasts (bone-eroding cells) and the bone formation activity of osteoblasts (bone-building cells) [67]. Bisphosphonates are a family of drugs used to prevent and treat osteoporosis. Bisphosphonates like pyrophosphate have high affinity for bone mineral and were found to prevent calcification both *in vitro* and *in vivo*, and to inhibit the dissolution of hydroxyapatite crystals [68]. Nitrogen-containing bisphosphonates such as pamidronate (Aredia, **1**), aledronate (Fosamax, **2**), risendronate (Actonel, **3**), and zolendronate (Zometa, **4**), shown below, represent an important class of drugs currently used to treat osteoporosis, Paget's disease, and hypercalcemia due to malignancy [69–72]. These compounds function primarily by adsorbing to the bone and inhibiting the enzyme farnesyl diphosphate synthase in osteoclasts [73–78], resulting in decreased levels of protein prenylation [79–81]. ^{153}Sm

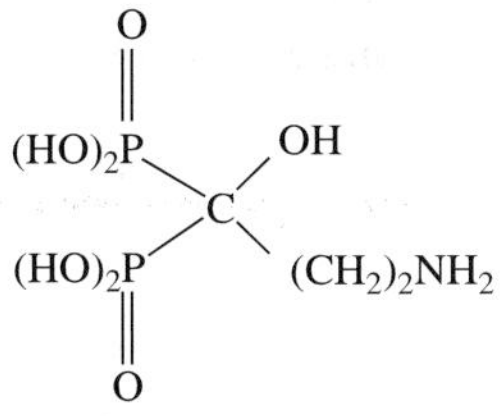

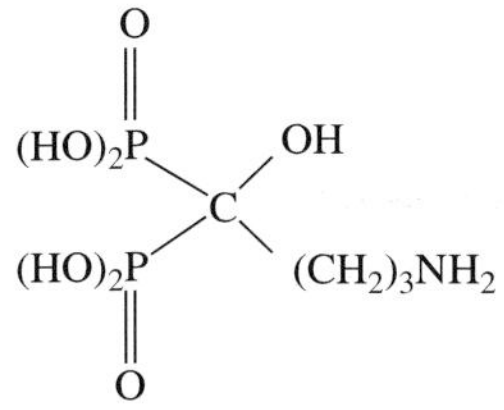

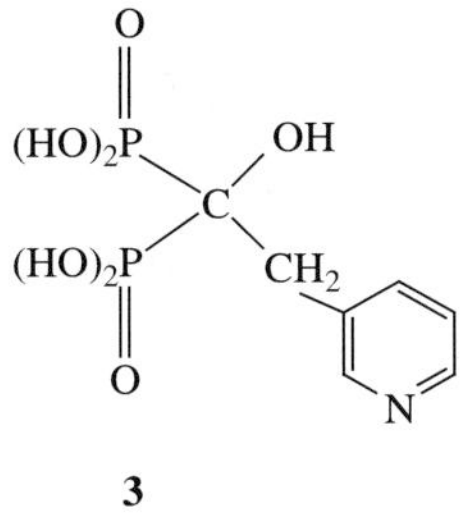

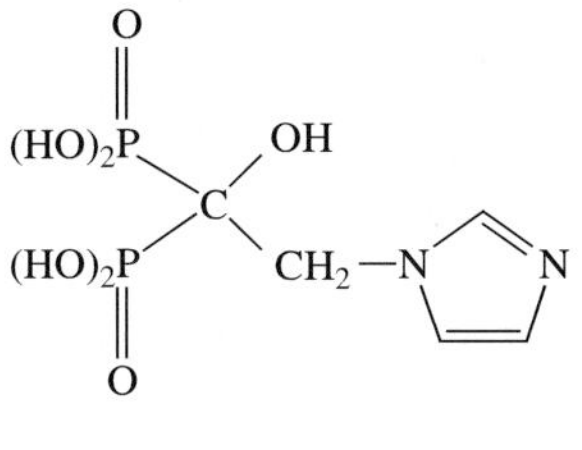

complexes with pamidronate **1**, aledronate **2** were described [82]. The results of experimental hydroxyapatite binding and skeletal uptake of ^{153}Sm-bisphosphonate complexes indicate that ^{153}Sm-alendronate is preferred comparatively to ^{153}Sm-pamidronate and ^{153}Sm-neridronate. This is in agreement with the calculated bisphosphonates–calcium interaction energy being significantly larger for alendronate than for the remaining bisphosphonates, as revealed by the shorter distance between the amino group and the nearest hydroxyapatite calcium atom observed in the alendronate–hydroxyapatite complex. These results revealed that bisphosphonates can be used as carriers of radioactive metals.

It has been found that bisphosphonic acids of the general formula given below—in which R^1 is fluorine or a linear or branched alkyl residue containing from 1 to 5 carbon

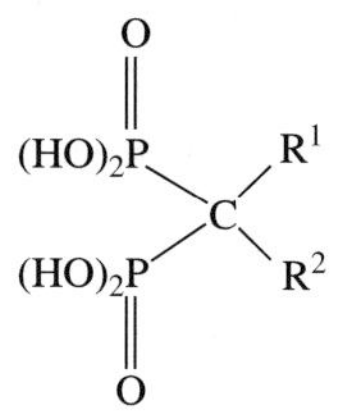

$$(HO)_2P(O)-C(R^1)(R^2)-P(O)(OH)_2$$

atoms that may optionally be substituted by one or more substituents such as amino groups and/or fluorine atoms, and R_2 is hydroxy or fluorine, and their salts with alkali metals, organic bases, and basic aminoacids—are very suitable for the treatment of urolithiasis and as inhibitors of the bone reabsorption. They exhibit high activity that is not accompanied by side effects with respect to pyrophosphate [83].

Pyridinium-1-yl-hydroxy-bisphosphonates with the general formula given below have been found to have high activity in bone resorption [84].

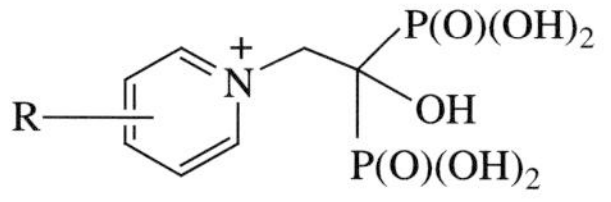

$$R-\text{[Pyridinium]}-N^+-CH_2-C(OH)(P(O)(OH)_2)_2$$

5.2 POLYMER ADDITIVES

The properties of polymers are hardly affected by the incorporation of phosphorus. Diesters of H-phosphonic acid and their immediate derivatives have found a number of applications in polymer synthesis such as flame retardants, antioxidants, heat and light stabilizers, catalysts, degrading agents, and alkylating agents. They are also used as corrosion inhibitors, scale inhibitors, and lubricants (antiwear and load-carrying additives).

5.2.1 Flame retardants and antioxidants

Most organic materials are thermodynamically unstable with respect to their combustion products. In an oxygen-containing atmosphere, they are, therefore, susceptible to flaming combustion if ignited. In recent years, interest in flame retardation has increased dramatically because of the large-scale introduction of synthetic macromolecular materials that are often more flammable than the original materials they replace. Flame retardant additives work by breaking one of the links that produce and support combustion—heat, fuel, and air. They may quench a flame by depriving it of the oxygen or may absorb heat and produce water, so reducing the temperature. Organophosphorus flame retardants work most effectively in the solid or condensed phase by inhibiting ignition and promoting char formation. The char acts as a mechanical barrier, so that the physical compactness of the material is preserved by preventing the access of oxygen to the material. Thus, the char at

5.2 Polymer Additives 265

least partially eliminates the exothermic oxidation reactions. The char also makes the evolution of gaseous and often toxic combustion products more difficult [85,86]. Reduction of smoke obscuration and corrosivity favor phosphorus-based flame retardants. Many of the current flame retardants are not suitable for use because they are too volatile, or not sufficiently thermally stable. New flame retardants have to demonstrate more permanence, greater thermal stability, and greater compatibility with polymers.

Poly(ethylene terephthalate)

Although the presence of two alkoxy groups in dialkyl H-phosphonates gives them the potential to be used as comonomers, their direct application to the synthesis of phosphorus-containing poly(ethylene terephthalate) is not feasible despite claims of some authors [87]. This is due to the relatively high temperatures (260–280 °C) employed in the polycondensation of dimethyl terephthalate with ethylene glycol, at which most dialkyl H-phosphonates are thermally unstable [88]. It is shown that dimethyl H-phosphonate suppresses the transesterification process between dimethyl terephthalate and ethylene glycol and accelerates the side reactions—ethylene glycol dehydration [89]. This problem may be avoided by using the sodium salts of dialkyl phosphites $(RO)_2P-ONa$ [90–92] or the disodium salt of 1,2-bis(methoxycarbonyl)ethane phosphonic acid [93,94] as comonomer.

Poly(ethylene terephthalate) with the structure given below was obtained when the sodium salt of diethyl phosphite was used as a comonomer [91,92]. The molecular weights of the resin was from 25,000 to 27,000 Da. Introduction of up to 2% P does not have any detrimental effect on the mechanical characteristics of the resin. The oxygen index (OI) increases from 19% O_2 for the nonmodified resin to 22.8% O_2 for the modified resin with 1.0% P. The coke residue increases from 8 to 28%. These results indicate that the introduction of phosphorus changes the mechanism of the polymer decomposition and catalyzes the coke formation.

When the disodium salt of 1,2-bis(methoxycarbonyl)ethane phosphonic acid is applied, poly(ethylene terephthalate) with the structure given below was obtained [93,94]. It was found that the polycondensation stage proceeds normally at 1–9% (by weight) of the

$$\left[\begin{array}{c} O\text{-}C \end{array} \!\!-\!\!\bigcirc\!\!-\!\! \begin{array}{c} C \\ \| \\ O \end{array} \!\text{-}OCH_2CH_2O\underset{O}{\overset{\|}{C}}CH_2\!-\!CH\!-\!\underset{O}{\overset{\|}{C}} \right]_x \!\!\!\begin{array}{c} \\ P=O \\ / \quad \backslash \\ NaO \quad ONa \end{array}\!\!\! \left[OCH_2CH_2O\underset{O}{\overset{\|}{C}}\!\!-\!\!\bigcirc\!\!-\!\!\underset{O}{\overset{\|}{C}}\!\!- \right]_y$$

modifier. Resin with molecular weight ranging between 22,000 and 30,000 Da was synthesized. The presence of the disodium salt of 1,2-bis(methoxycarbonyl) ethane phosphonic acid during the synthesis results in a lower concentration of carboxylic groups. The concentration decreases from 58.6×10^{-6} mg eqv./g for the nonmodified resin to 15.1×10^{-6} mg eqv./g for the one modified with 0.85% P. This result implies that the modifier acts also as an antioxidant. The oxygen index of the resin modified with 1.45% P increases to 24.8% O_2. The results obtained show that this modifier is more effective than the sodium salt of diethyl phosphite discussed above. The coke residue increases to 52%. The phosphorus- and metal-containing poly(ethylene terephthalate) thus synthesized has good characteristics and exhibits improved resistance to combustion and good thermal stability.

Phosphorus and metal-containing oligomers having the following general formula were used as modifiers of poly(ethylene terephthalate) [95–97]:

$$R\!\!-\!\!\left[O\!-\!\underset{H}{\overset{\overset{O}{\|}}{P}}\!-\!O\!-\!M \right]_n\!\!-\!\!X$$

M = Cd;   X = H_3CCOO

M = Zn;   X = Cl

M = Ca;   X = Cl

The modified resin containing 0.4% phosphorus and 1.4% cadmium [95] has the following characteristics [93]: $\eta_{rel} = 1.392$; [COOH] = 40.9×10^{-6} mg.eqv/g; m.p. = 251 °C. The concentration of carboxylic groups in the modified resin is lower compared to that of the nonmodified resin, which indicates that the modifier acts as an antioxidant. The combustion resistance of the modified polymer is also considerably improved. The time necessary for combustion increases from 181 sec for the nonmodified resin up to 630 sec for the resin having 1.5% phosphorus and 3.5% cadmium content. Similarly, the resin modified with the zinc-containing oligomers exhibits improved combustion resistance [94]. A resin having 0.6% phosphorus and 1.25% zinc content has $\eta_{rel} = 1.407$; [COOH] = 39.9×10^{-6} mg.eqv/g; m.p. = 255.8 °C. The content of carboxylic groups is also lowered, indicating an effect of thermal stabilization. The most substantial decrease in the carboxylic-group content has been determined for the resin modified with calcium-containing oligomers [97]. The modified resin, containing 1.0% phosphorus and 1.2% Ca has the following characteristics: $\eta_{rel} = 1.315$; [COOH] = 18.2×10^{-6} mg.eqv/g;

5.2 Polymer Additives

m.p. = 261.5 °C. In this resin, the carboxylic-group content is twice as low as the one determined for the nonmodified resins, which is an indication of a strongly expressed effect of thermal stabilization.

The calcium salt of the monoethyl ester of H-phosphonic acid was used a modifier of poly(ethylene terephthalate) [98]. A resin modified with 0.4% P and 0.25% Ca has η_{rel} =

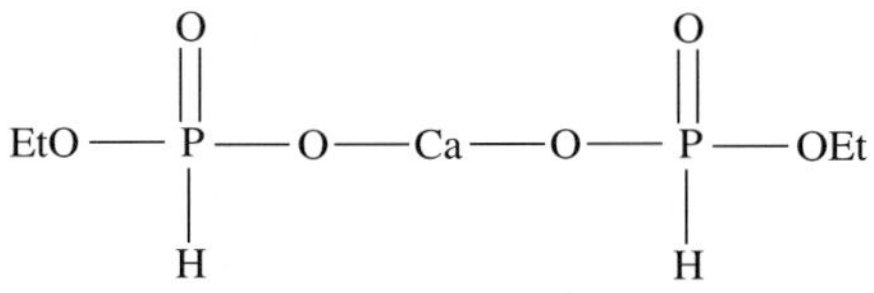

$$EtO-\underset{H}{\underset{|}{P}}(=O)-O-Ca-O-\underset{H}{\underset{|}{P}}(=O)-OEt$$

1.384; [COOH] = 54.27 × 10^{-6} mg.eqv/g; m.p. = 241.5 °C, and improved combustion resistance.

A phosphorus-containing modifier, prepared by addition of diethyl H-phosphonate to 2,6-octadienecarboxylic acid, is used as a comonomer for PET, which increases its viscosity [99].

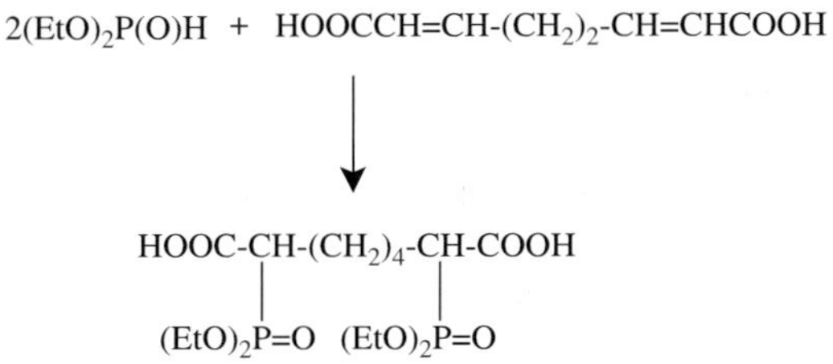

$$2(EtO)_2P(O)H + HOOCCH=CH-(CH_2)_2-CH=CHCOOH$$
$$\downarrow$$
$$HOOC-CH-(CH_2)_4-CH-COOH$$
$$\qquad\;\; | \qquad\qquad\quad\; |$$
$$\;(EtO)_2P=O \;\; (EtO)_2P=O$$

Polyurethanes

Attempts for the direct use of dialkyl H-phosphonates as additives for rigid polyurethane foams have so far been unsuccessful. This is probably due to the reaction of H-phosphonates with the amine catalysts used for this process, which results in its deactivation.

Diethyl-bis(2-hydroxyethyl)aminophosphonate, prepared from diethanolamine, formaldehyde, and diethyl H-phosphonate, known under the trade name 'Fyrol 6' (Stauffer Chemical Company) is used as a reactive flame retardant for rigid polyurethane foams on an industrial scale [100].

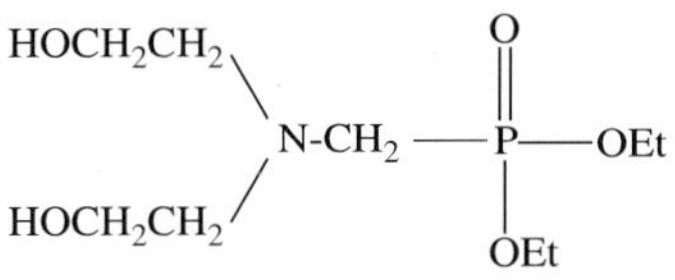

$$\begin{array}{c} HOCH_2CH_2 \\ \diagdown \\ \quad N-CH_2-\underset{OEt}{\underset{|}{P}}(=O)-OEt \\ \diagup \\ HOCH_2CH_2 \end{array}$$

Introduction of up to 30 parts by weight of either of these two modifiers into the composition of the rigid polyurethane foams leads to an improvement of the combustion resistance and increase in the temperature of softening for these materials.

The Kabachnik–Fields condensation product of cyclohexanone, ethanolamine, and dialkyl H-phosphonate [101] is also used as an effective flame retardant for rigid polyurethane foams and epoxy resins.

$$\text{cyclohexanone} + H_2N\text{-}CH_2CH_2OH + (RO)_2P(O)H \longrightarrow \underset{\text{NHCH}_2\text{CH}_2\text{OH}}{\overset{P(O)(OR)_2}{\text{cyclohexyl}}}$$

The reaction product of dimethyl H-phosphonate with diethanolamine and formaldehyde, the dimethyl analog of 'Fyrol' 6, is also applied as a flame retardant for rigid polyurethane foams [102].

$$\begin{array}{c} HOCH_2CH_2 \\ \diagdown \\ HOCH_2CH_2 N\text{-}CH_2\text{---}\underset{\underset{OMe}{|}}{\overset{\overset{O}{\|}}{P}}\text{---}OMe \\ \diagup \end{array}$$

Polyurethane foams having high combustion resistance have been prepared from phosphorus-containing polyols, used as hydroxyl-containing components [103].

$$R^1\text{-}X + (RO)_2P\text{-}ONa \longrightarrow (RO)_2P(O)R^1$$

where R^1 is a polyester alcohol, polyether alcohol, or polyester–polyether alcohol, and X is chlorine or bromine.

Phosphoramide, obtained by the Atherton–Todd reaction, is also applied as a flame retardant for rigid polyurethane foams [104].

$$(ClCH_2CH_2O)_2\overset{\overset{O}{\|}}{P}\text{-}H + H_2N\text{-}\text{Ph} \xrightarrow[-Et_3N\,HCl]{Et_3N\ +CCl_4} (ClCH_2CH_2O)_2\overset{\overset{O}{\|}}{P}\text{---}NH\text{-}\text{Ph}$$

Rigid polyurethane foams have been synthesized from industrial raw materials and phosphorus-containing flame retardants, obtained by the interaction of dialkyl H-phosphonates with amino alcohols. The phosphorus-containing compounds **1** and **2** are used as modifiers [105]. During polyurethane synthesis, both modifiers speed up the times for foaming, gel formation, and surface drying, and decrease the time for foam growth.

$$\underset{\underset{H}{|}}{\overset{\overset{O}{\|}}{CH_3O\text{-}P\text{-}O^-}} \underset{\overset{\diagup\ \diagdown}{CH_3\ \ CH_3}}{\overset{CH_3}{\overset{|}{{}^+N\text{-}CH_2CH_2OH}}} \qquad \underset{\underset{H}{|}}{\overset{\overset{O}{\|}}{CH_3O\text{-}P\text{-}O^-}} \underset{\overset{\diagup\ \diagdown}{CH_3\ \ CH_3}}{\overset{CH_2CH_2OH}{\overset{|}{{}^+N\text{-}CH_2CH_2OH}}}$$

1   2

5.2 Polymer Additives

For a low concentration of modifiers (5% by weight), rigid polyurethane foams have high fire resistance and satisfactorily retain their physical properties.

Polystyrene

An approach for the preparation of combustion-resistant polystyrene involves its

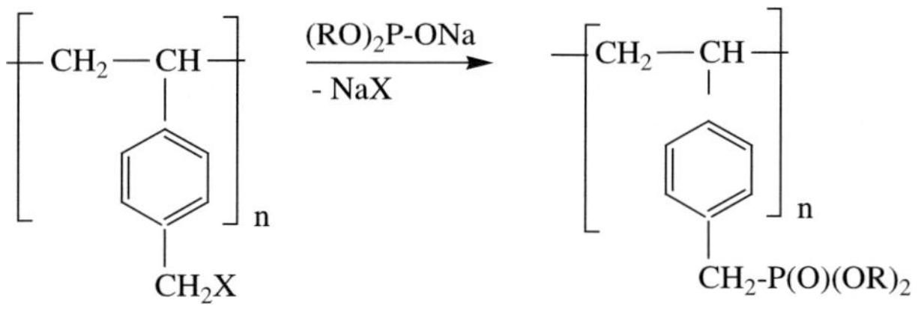

$$\left[-CH_2-CH(-C_6H_4-CH_2X)-\right]_n \xrightarrow[-NaX]{(RO)_2P\text{-}ONa} \left[-CH_2-CH(-C_6H_4-CH_2\text{-}P(O)(OR)_2)-\right]_n$$

phosphorylation, which is achieved by the reaction of chloromethylated or bromomethylated polystyrene with sodium dialkyl phosphite [106].

Polyacroleine

In polyacroleine, phosphorus is introduced directly by dialkyl H-phosphonates via the Abramov reaction [107].

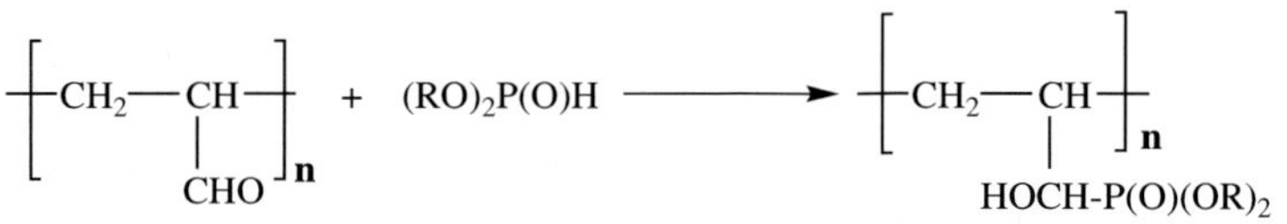

$$\left[-CH_2-CH(CHO)-\right]_n + (RO)_2P(O)H \longrightarrow \left[-CH_2-CH(HOCH\text{-}P(O)(OR)_2)-\right]_n$$

The phosphorus-containing polyacroleine obtained in this way is used as an additive to paints, polyurethanes, and epoxy resins in order to increase their combustion resistance.

The copolymer of acroleine with acrylonitrile is phosphorylated by the same method as polyacroleine [108].

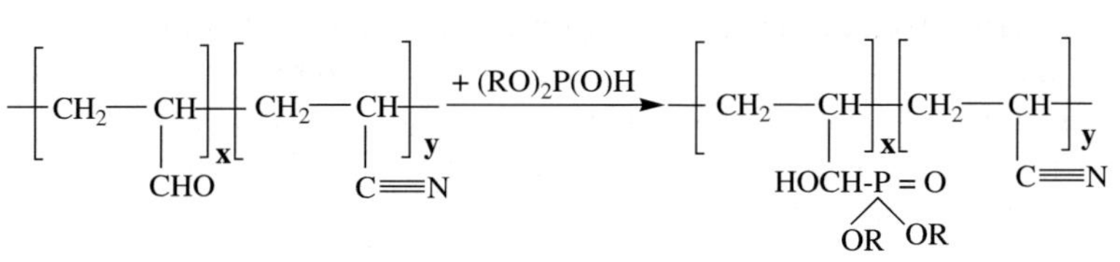

$$\left[-CH_2-CH(CHO)-\right]_x\left[-CH_2-CH(C\equiv N)-\right]_y \xrightarrow{+(RO)_2P(O)H} \left[-CH_2-CH(HOCH\text{-}P(=O)(OR)(OR))-\right]_x\left[-CH_2-CH(C\equiv N)-\right]_y$$

Rubber

Addition of dialkyl phosphonates to india-rubbers increases the combustion resistance of these materials [109].

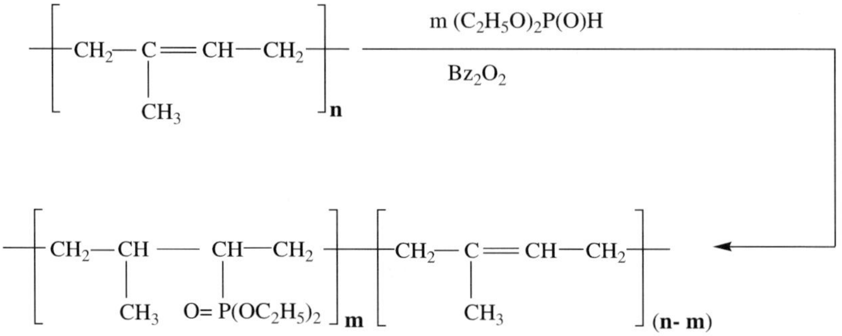

$$\left[-CH_2-\underset{\underset{CH_3}{|}}{C}=CH-CH_2-\right]_n \xrightarrow[Bz_2O_2]{m\ (C_2H_5O)_2P(O)H}$$

$$\left[-CH_2-\underset{\underset{CH_3}{|}}{CH}-\underset{\underset{O=P(OC_2H_5)_2}{|}}{CH}-CH_2-\right]_m \left[-CH_2-\underset{\underset{CH_3}{|}}{C}=CH-CH_2-\right]_{(n-m)}$$

These structures are formed as a result of anti-Markovnikov addition of dialkyl phosphonates to the allyl moiety in the presence of radical initiators.

Poly(ethylene)

Phosphorus-containing poly(ethylene) having a high combustion resistance can be obtained via irradiation of poly(ethylene) in the presence of dialkyl H-phosphonate [110].

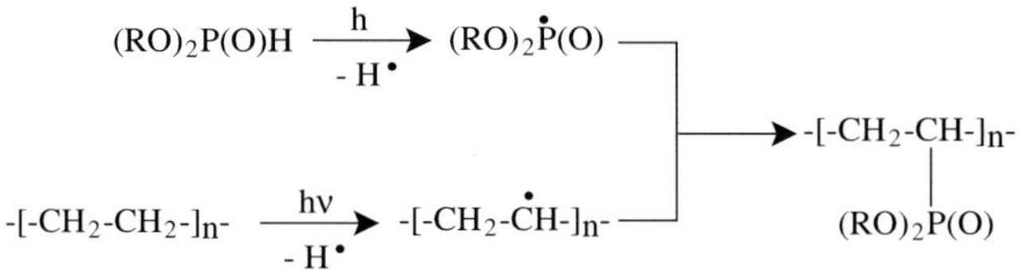

$$(RO)_2P(O)H \xrightarrow[-H^\bullet]{h\nu} (RO)_2\dot{P}(O)$$

$$-[-CH_2-CH_2-]_n- \xrightarrow[-H^\bullet]{h\nu} -[-CH_2-\dot{C}H-]_n-$$

$$\longrightarrow -[-CH_2-\underset{\underset{(RO)_2P(O)}{|}}{CH}-]_n-$$

Unsaturated polyesters

Unsaturated polyesters obtained from adipic acid and *cis*-2-butene-1,4-diol are phosphorylated by treatment with diethyl H-phosphonate in the presence of dibenzoyl peroxide or UV irradiation [111].

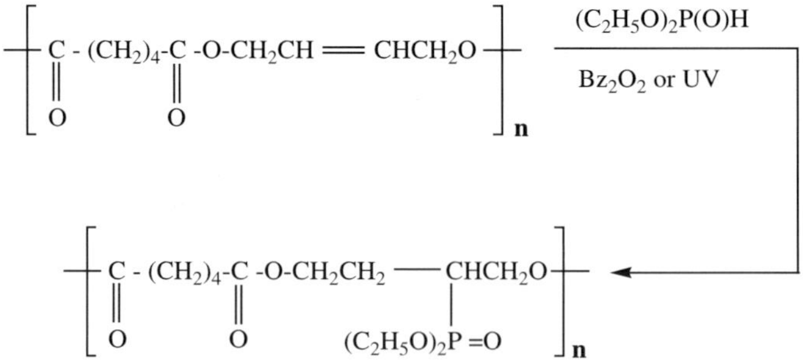

$$\left[-\underset{\underset{O}{\|}}{C}-(CH_2)_4-\underset{\underset{O}{\|}}{C}-O-CH_2CH=CHCH_2O-\right]_n \xrightarrow[Bz_2O_2\ or\ UV]{(C_2H_5O)_2P(O)H}$$

$$\left[-\underset{\underset{O}{\|}}{C}-(CH_2)_4-\underset{\underset{O}{\|}}{C}-O-CH_2CH_2-\underset{\underset{(C_2H_5O)_2P=O}{|}}{CHCH_2O}-\right]_n$$

The resulting resin exhibits higher combustion resistance and higher sorption ability toward Cu^{2+} ions.

5.3 Degrading and Alkylating Agents of Polymers

Epoxide resin

Bis(aminophosphonate)s obtained via addition of dialkyl H-phosphonate to Schiff

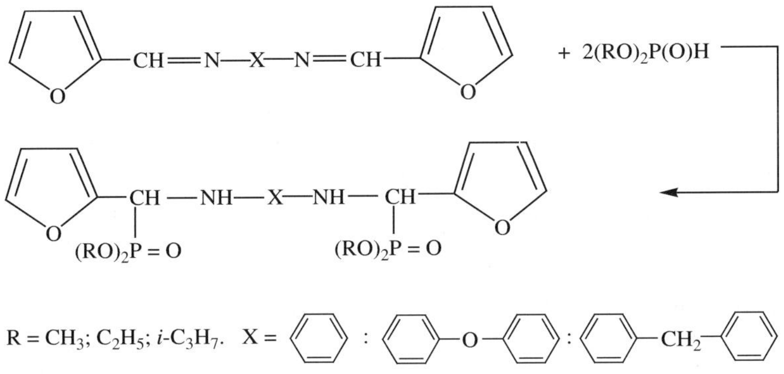

$$\text{furan-CH=N-X-N=CH-furan} + 2(RO)_2P(O)H \longrightarrow$$

$$\text{furan-CH(-(RO)}_2\text{P=O)-NH-X-NH-CH(-(RO)}_2\text{P=O)-furan}$$

R = CH$_3$; C$_2$H$_5$; i-C$_3$H$_7$.   X = phenyl : phenyl-O-phenyl : phenyl-CH$_2$-phenyl

bases are used as reactive or nonreactive additives for improving the resistance to combustion of polymers. Due to the presence of C=C double bonds, they polymerize and copolymerize in the presence of *p*-toluenesulfonic acid to give crosslinked polymers. The addition to epoxide resin resulted in increasing the combustion resistance [112].

5.3 DEGRADING AND ALKYLATING AGENTS OF POLYMERS

The observation of an exchange reaction between the alkoxy groups of dialkyl H-phosphonates represents an important scientific result for polymer chemistry.

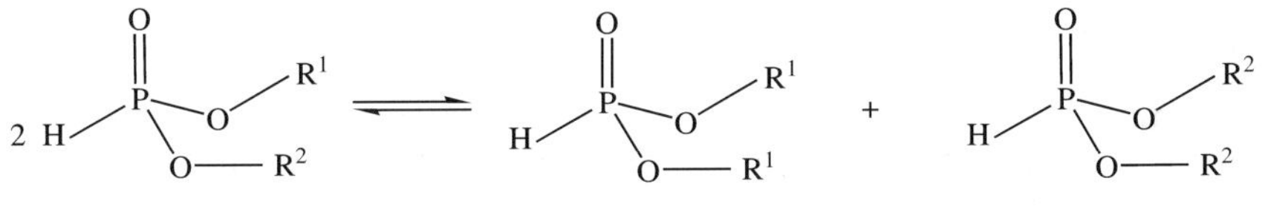

$$2\ H-P(=O)(O-R^1)(O-R^2) \rightleftharpoons H-P(=O)(O-R^1)(O-R^1) + H-P(=O)(O-R^2)(O-R^2)$$

This observation gives grounds for the assumption that diesters of H-phosphonic acid can participate in exchange reactions with urethane, carbonate, and amide groups.

Polyurethanes

The world consumption of polyurethanes (PUs) is about 7 million tons per year (2004). The furniture industry, automotive manufacture, building industry, and technical insulation are the major consumers of PUs. It has been established that approximately 10% of PUs convert into wastes during their processing. Therefore, the development of concept for the chemical recycling of PUs is of great importance. Recycling polymer waste materials and used polymers is one way to conserve natural resources and reduce environmental stress.

Microporous polyurethane elastomers, flexible polyurethanes, and rigid polyurethanes have been liquefied by dimethyl or diethyl H-phosphonates [113–116]. On heating a mixture of microporous polyurethane elastomer or flexible polyurethane and dimethyl or diethyl H-phosphonate at 160 °C, the polyurethane is liquefied (Scheme 5.1). Direct evidence for the exchange reaction of the methoxy group of dimethyl H-phosphonate and the urethane group is the signal of the reaction product at 11.09 ppm in the ^{31}P NMR spectrum, which represents a doublet of sextets with $^3J(P,H) = 12.03$ Hz and $^1J(P,H) = 708.1$ Hz. This signal can be assigned to the phosphorus atom bounded with CH$_3$O and OCH$_2$ groups—product **1** (R = CH$_3$O) (Scheme 5.1). The formation of POCH$_2$– groups from the chemical degradation of polyurethanes with dimethyl H-phosphonate is proved by the existence of a signal of the reaction product in the ^{1}H NMR spectrum in the region 4.21–4.33 ppm, which appears as a multiplet. In the ^{13}C{H} NMR spectrum, the POCH$_2$– carbon atom appears as a doublet at 63.82 ppm with $^2J(P,C) = 6.2$ Hz. It was shown that the heating of microporous or flexible polyurethanes with dimethyl or with diethyl H-phosphonates resulted in the alkylation of the urethane group. The structure of the reaction product is confirmed by ^{1}H and ^{31}P NMR spectroscopy. In the ^{1}H NMR spectrum (R = CH$_3$), the signal for the P–H proton appears as a doublet at 6.77 ppm with $^1J(P,H) = 653.0$ Hz. The signal at 7.28 ppm in the ^{31}P NMR spectrum presents a doublet of quartets with $^3J(P,H) = 12.03$ Hz and $^1J(P,H) = 657.4$ Hz. These data are characteristic for the phosphorus atom bounded with the negatively charged oxygen atom and can be assigned to the phosphorus atom of the product **2** (Scheme 5.1). In the ^{31}P NMR spectrum of the reaction mixture, there are two signals at 3.55 ppm and 2.03 ppm, which appear as an octet with $^3J(P,H) = 11.10$ Hz and as septet with $^3J(P,H) = 11.10$ Hz, respectively. The data from the ^{31}P NMR studies revealed that these phosphorus atoms are not bonded to a proton. These signals can be assigned to the phosphorus atoms of the products 3 and 4 (Scheme 5.1). The

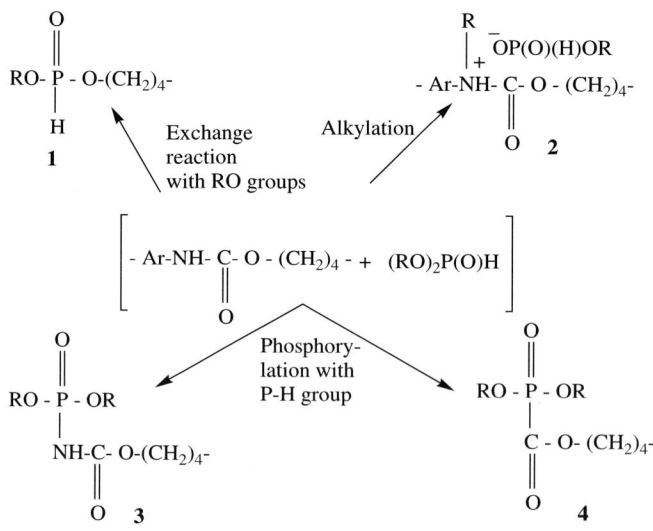

Scheme 5.1 Chemical degradation of microporous polyurethane elastomer with diesters of the H-phosphonic acid.

5.3 Degrading and Alkylating Agents of Polymers

content of these products is 9.3% and 6.9%, respectively. The degraded products were used as modifiers of polyurethanes [117] and polypropylene [118].

Polycarbonate

The development of a new synthetic route for synthesizing phosphorus-containing oligomers is an important direction in phosphorus chemistry, as these oligomers are widely used for modification of polymers.

Treatment of poly[2,2-bis(4-hydroxyphenyl)propanecarbonate]-polycarbonate (PC) with dimethyl or diethyl H-phosphonates at 160 °C yielded phosphorus-containing oligocarbonates (Scheme 5.2) [119]. Heating PC with dimethyl or diethyl H-phosphonates at 160 °C results in a decrease of the molecular weight of PC from 41,300 to 4600 Da and 4956 Da, respectively. The ^{31}P NMR spectrum of the reaction product shows a signal at 8.78 ppm, which represents a doublet of triplets with $^3J(P,H) = 9.0$ Hz and $^1J(P,H) = 698.9$ Hz. This signal can be assigned to the phosphorus atom in product **1** (R = C_2H_5), which is formed as a result of the exchange reaction between the ethoxy group of diethyl H-phosphonate and the carbonate group. On the other hand, the triplet at 1.34 ppm with $^3J(H,H) = 7.0$ Hz in the ^{1}H NMR spectrum and signal at 65.5 ppm in the ^{13}C{H} NMR spectrum confirm the formation of the product **2** (R = C_2H_5). In the ^{31}P{H} NMR

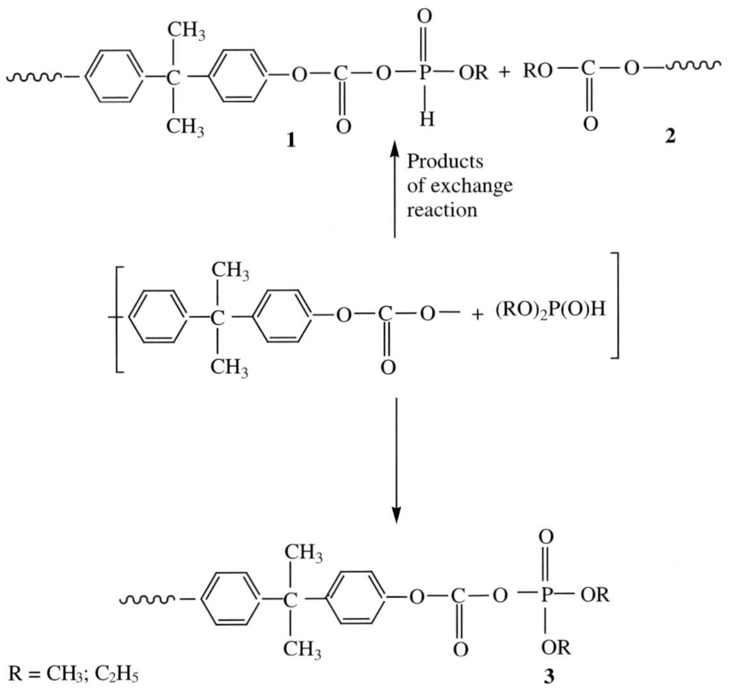

Scheme 5.2 Chemical degradation of polycarbonate with diesters of the H-phosphonic acid

spectrum of the reaction mixture, there is a signal at 3.02 ppm, which is septet with $^3J(P,H) = 12.01$ Hz. Obviously, this phosphorus atom is not bonded with a proton. These data can be assigned to the phosphorus atom of the product **3**. The content is 3.1%. Gel permeation chromatography (GPC) analyses show that the molecular weight decreases with increasing reaction time. The molecular weight strongly decreases for the first 5 h reaching a molecular weight near to the final value—4956 Da. The presence of phosphorus in the oligocarbonates improves their thermal stability. Thermogravimetric analysis (TGA) shows that the char yield at 850 °C is higher (R= CH_3—28.2%, R = C_2H_5—23.6%) compared to the polycarbonate (12.8%). Obviously, phosphorus promotes char formation.

Polyamide

The heating of polyamide-6 with the diesters of the H-phosphonic acid $(RO)_2P(O)H$ (where R could be $-CH_3$, $-C_2H_5$, or $-C_6H_5$) at elevated temperatures (160–180 °C) substantially decreases the molecular weight of the starting polymer [120]. In the ^{31}P NMR spectrum, the signal at 11.60 ppm appears as a doublet of septets with $^1J(P,H) = 700$ Hz and $^3J(P,H) = 12.1$ Hz, which are typical for the phosphorus atom of dimethyl H-phosphonate (its intensity strongly decreases after a vacuum distillation); the signal at 8.22 ppm appears as a doublet of quartets with $^1J(P,H) = 672.8$ Hz and $^3J(P,H) = 12.4$ Hz. This spectral signature is typical for the phosphorus atom in product **2** (Scheme 5.3). The presence of the negative charge on the oxygen atom bonded to the phosphorus determines a decrease of the $^1J(P,H)$ values from 700 Hz for the starting dimethyl H-phosphonate to 672.8 Hz for product **2** [121]. This product is formed as a result of alkylation of the amido group by dimethyl H-phosphonate. The signal at 4.79 ppm in the ^{31}P NMR spectrum appears as a doublet with $^1J(P,H) = 666.2$ Hz and can be assigned to the phosphorus atom of the phosphonic acid, which forms as a result of the hydrolysis of dimethyl H-phosphonate in the reaction conditions. The remaining two signals at 3.31 ppm and at 2.11 ppm appear in the corresponding ^{31}P NMR spectrum as an octet with $^3J(P,H) = 11.0$ Hz and a septet with $^3J(P,H) = 11.0$ Hz, respectively, and the signals can be assigned to the phosphorus atoms of products **3** and **4**, respectively. The content of these products is 9.4% and 3.4%, respectively. The ^{31}P NMR data suggest that in these products, the phosphorus atoms are not bonded to a hydrogen atom, that is, the P-H group has reacted with the polyamide.

The experimental data revealed that diethyl H-phosphonate reduces the degree of the alkylation reaction and favors the exchange reaction. The signal at 12.38 ppm, which is a doublet of quintets with $^3J(P,H) = 7.2$ Hz and $^1J(P,H) = 719$ Hz, can be assigned to the phosphorus atom in product **1** (R = C_2H_5) (Scheme 5.3).

The doublet in the $^{13}C\{H\}$ NMR spectrum at 63.17 ppm with $^2J(P,C) = 6.9$ Hz could originate from the PO*C*H$_2$ carbon atom of **1** and the one at 39.18 ppm with $^2J(P,C) = 6.4$ Hz can be assigned to the PNH*C*H$_2$ carbon atom. Phosphorus atoms with such substituents can be obtained as a result of the exchange reaction between the ethoxy group in diethyl H-phosphonate and an amide group.

5.4 Heat, Light, and UV Stabilizers

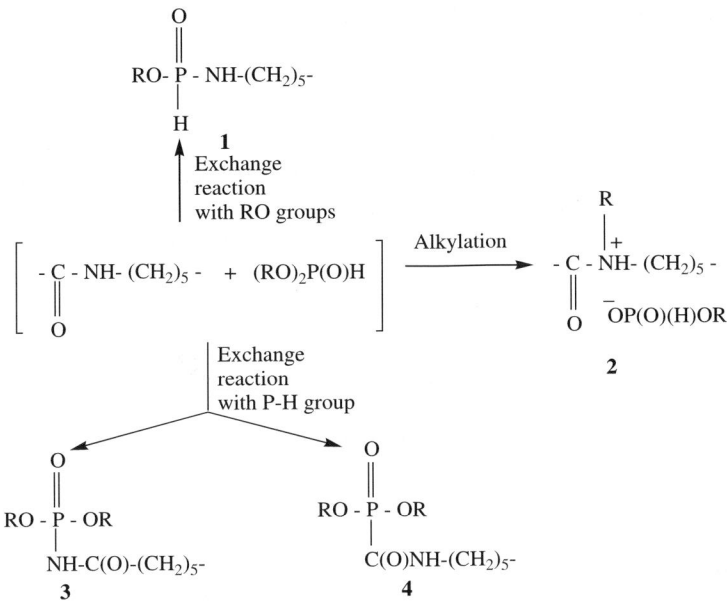

Scheme 5.3 Chemical degradation of polyamide 6 with diesters of the H-phosphonic acid.

5.4 HEAT, LIGHT, AND UV STABILIZERS

Heat stabilizers are chemical additives used to protect polymers, especially those that contain regular repeating units, against degradation resulting from heat or exposure to UV radiation. Complexes of CuI and CuCl with dioctyl H-phosphonate or bis(2-chloroethyl) H-phosphonate are efficient thermal stabilizers for polyamide [122]. Polyamide, containing 0.012% of Cu and 0.011% of P, shows a weak reduction in its viscosity measured in 1000 h.

$$(RO)_2\underset{H}{P} = O \longrightarrow CuX \quad X = I; Cl$$

The zinc derivatives of monoalkyl H-phosphonates are effective UV stabilizers for cellulose, polyesters, and other polymeric materials when added in quantities from 0.1 to 5% [123].

Mixtures of metal salts (M = Mg, Ca, Zn, Ba, Cd, Sn, and Pb) of different monoalkyl H-phosphonates (R = alkyl [C_1 to C_{20}], alkylaryl, cycloalkyl) are used as stabilizers for polyvinyl chloride in quantities ranging from 0.1 to 10 parts by weight [124]. They efficiently protect the color, transparency, and compatibility of these plastic materials when heated up to 175 °C.

5.5 CATALYSTS

Only dibutyl H-phosphonate has been reported to be used directly (together with other cocatalysts) for the preparation of isotactic polypropylene [125]. Most of the other catalysts based on dialkyl H-phosphonates comprise of their metal and ammonium derivatives. Sodium dialkyl phosphite is used as a catalyst for the cyclotrimerization of 2,4-toluenediisocyanate with phenyl, *m*-chlorophenylene or α-naphthylisocyanates, forming oligomeric diisocyanates containing triisocyanurate cycles [126].

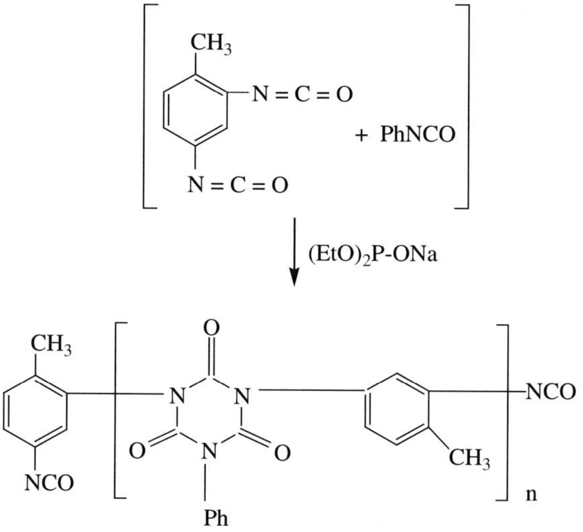

Terephthalic acid polymerizes with ethylene oxide in the presence of tetraalkylammonium salt of monoalkyl H-phosphonate to poly(ethylene terephthalate) [127].

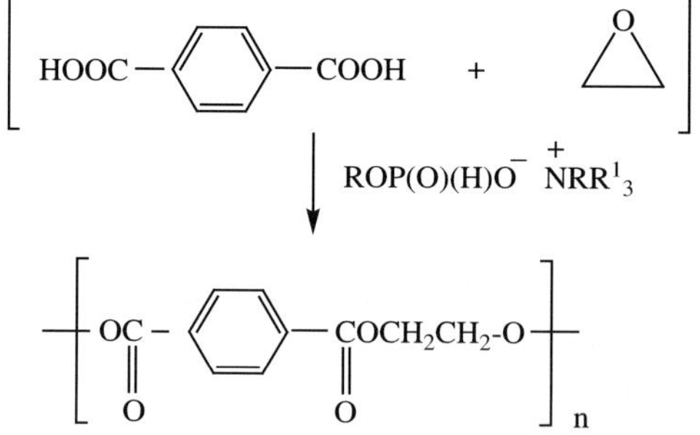

The complex $Ir(H)(Cl)[P(O)(OCH_3)_2][P(OH)(OCH_3)_2]_3$ obtained from $[IrCl(C_8H_{14})_2]_2$ and dimethyl H-phosphonate catalyses the stereoselective reduction of 4-*tert*-butylcyclohexanone to 97/3 *cis/trans* 4-*tert*-butylcyclohexanol [128]. This complex also catalyzes the

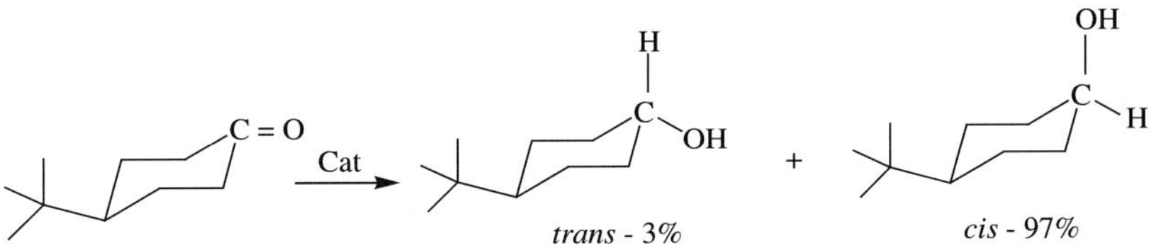

5.5 Catalysts

[Reaction scheme: ketone → trans-3% alcohol + cis-97% alcohol, via Cat]

hydrogenation of olefins [129].

High molecular weight polycarbamides and polythiocarbamides are formed by the condensation of carbon dioxide or carbon disulfide with aromatic diamines in the presence of diaryl H-phosphonates and tertiary amines [130].

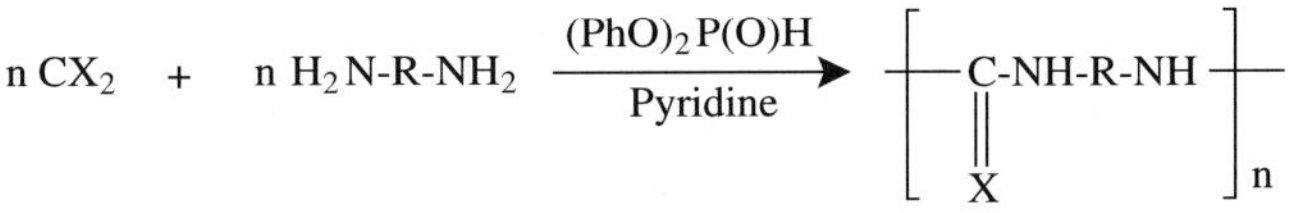

$$n\,CX_2 + n\,H_2N\text{-}R\text{-}NH_2 \xrightarrow[\text{Pyridine}]{(PhO)_2P(O)H} \left[\!\!\begin{array}{c} \text{-C-NH-R-NH-} \\ \| \\ X \end{array}\!\!\right]_n$$

X = O, S

The diphenyl H-phosphonate/pyridine mixture has been found to be the most effective catalyst for this reaction. Carbamides and thiocarbamides are formed in high yields when diamines are substituted for aniline in the above scheme [131].

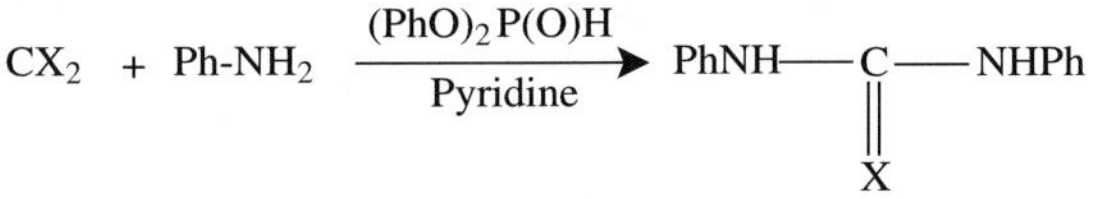

$$CX_2 + Ph\text{-}NH_2 \xrightarrow[\text{Pyridine}]{(PhO)_2P(O)H} PhNH\text{-}\underset{\underset{X}{\|}}{C}\text{-}NHPh$$

X = O, S

Phosphorus- and metal-containing oligomers having the general formula **1**, obtained

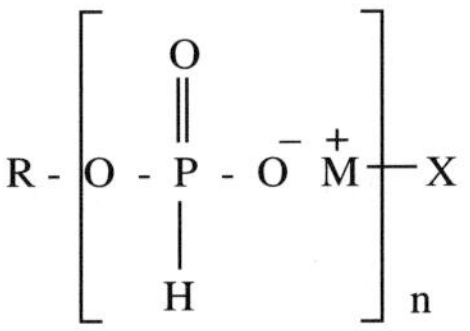

$$R\text{-}\left[O\text{-}\underset{\underset{H}{|}}{\overset{\overset{O}{\|}}{P}}\text{-}\overset{-}{O}\ \overset{+}{M}\right]_n\!\!X$$

1

where R is Et or Me; M = Ca, Cd, Zn; and X = OCOMe and Cl
by reacting dialkyl H-phosphonates with metal salts, act as catalysts of the polycondensation stage of the poly(ethylene terephthalate) synthesis [95–97]. These catalysts are much more effective than cobalt acetate, which is widely used in industry for this purpose. Thus, a resin with a relative viscosity of 1.409 is obtained after 30 min of polycondensation in the presence of the phosphorus- and zinc-containing oligomers, while the same results are obtained after 120–140 min. with $Co(OCOMe)_2$ as a catalyst under the same reaction conditions. The polycondensation of bis(2-hydroxyethyl) terephthalate in the presence of

Co(OCOMe)$_2$ occurs via complex formation where the terminal hydroxyl groups are activated by complexation to the metal atom. A similar catalytic mechanism was proposed for the phosphorus- and metal-containing catalysts.

$$\begin{array}{c} \text{—OCH}_2\text{CH}_2\text{OH} \quad \text{HOCH}_2\text{CH}_2\text{O—} \\ \searrow \downarrow \downarrow \swarrow \\ \text{—O—P(=O)(H)—O—M—O—P(=O)(H)—O—} \\ \nearrow \uparrow \uparrow \nwarrow \\ \text{—OCH}_2\text{CH}_2\text{OH} \quad \text{HOCH}_2\text{CH}_2\text{O—} \end{array}$$

The higher activity of the phosphorus- and metal-containing oligomers as compared to cobalt acetate is explained in terms of easier access of the terminal hydroxyl groups to the coordination sphere of the metal atom in the presence of the phosphonate ligands.

Dimethyl H-phosphonate is used as a cocatalyst in the synthesis of propylene carbonate from propylene oxide and carbon dioxide [132].

$$H_3C-H_2C-\overset{O}{\overset{\triangle}{C}H_2} + CO_2 \xrightarrow{\text{cat.}} H_3C-H_2C-\underset{O\diagdown\underset{\underset{O}{\|}}{C}\diagup O}{CH_2}$$

The actual catalyst is prepared in a preceding step from CaCl$_2$, NEt$_3$, and (MeO)$_2$P(O)H and is assumed to contain calcium phosphonate and triethyl methylammonium chloride.

5.6 CORROSION INHIBITORS

The corrosion inhibitors appear to possess properties that impart to metals resistance to attack by a variety of corrosive agents, such as brines, weak inorganic acids, organic acids, CO$_2$, H$_2$S, etc. The method of carrying out this process is relatively simple in principle: the corrosion preventive reagent is dissolved in the liquid corrosive medium in small amounts and is thus kept in contact with the metal surface to be protected. Alternatively, the corrosion inhibitor may be applied first to the metal surface, either as it is or as a solution in some carrier liquid or paste. Continuous application, as in the corrosive solution, is the preferred method, however. The concentration of the corrosion inhibitors varies widely, but the preferable concentrations are 15–250 ppm.

The quinoline phosphonates with the general formula given below, prepared by reacting quinoline wherein the nitrogen atom is in the form of a quaternary alkoxy derivative with

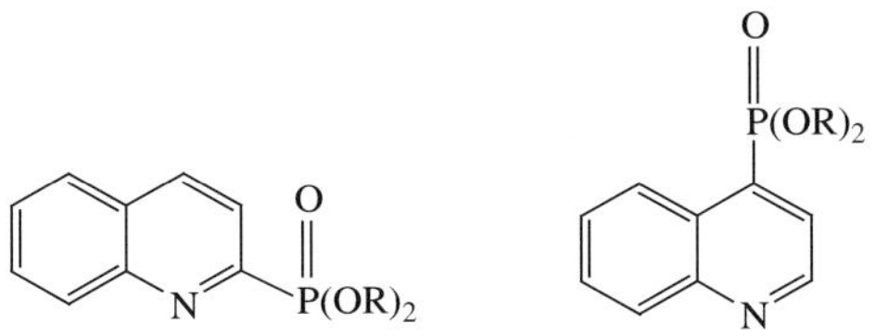

5.7 Scale Inhibitors

dialkyl H-phosphonate salt, were found to possess corrosion-inhibiting properties [133]. These compounds can be used as biocides, plastics, flocculants, and agents in the textile industry such as mercerizing assistants, wetting agents, rewetting agents, penetrating agents, dispersing agents, softening agents, dyeing assistants, antistatic agents, etc.

Pyridine derivatives having phosphonate group substituents in para or ortho positions with respect to the nitrogen atom are used as corrosion inhibitors [134,135]. These compounds effectively protect metal surfaces against the action of air humidity, soil humidity, and water, water steam, hot water, weak inorganic acids, organic acids, CO_2, H_2S, and others.

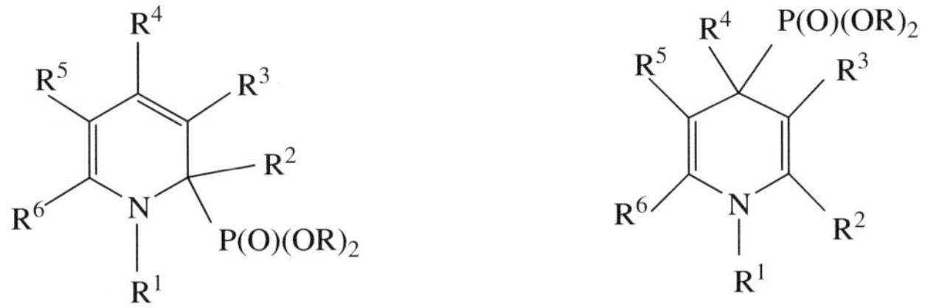

5.7 SCALE INHIBITORS

Most commercial water contains alkaline earth metal cations such as calcium, barium, magnesium, etc., and anions such as bicarbonate, carbonate, sulfate, oxalate, phosphate, etc. When combinations of these anions and cations are present in concentrations that exceed the solubility of their reaction products, precipitates form until their product solubility concentrations are no longer exceeded. For example, when the concentrations of calcium ion and carbonate ion exceed the solubility of the calcium carbonate reaction product, a solid phase of calcium carbonate will form as a precipitate. As these reaction products precipitate on the surfaces of the water-carrying system, they form scale. The scale prevents effective heat transfer, interferes with fluid flow, facilitates corrosive processes, and harbors bacteria. Scale-forming compounds can be prevented from precipitating by inactivating their cations with chelating of sequestering agents, so that the solubility of their reaction products is not exceeded.

Quaternary aminomethylphosphonates [136] with the general formula given below obtained via Kabachnik–Fields reaction and subsequent alkylation of the amine are used as scale inhibitors. They are active at room temperature, and also effective at elevated

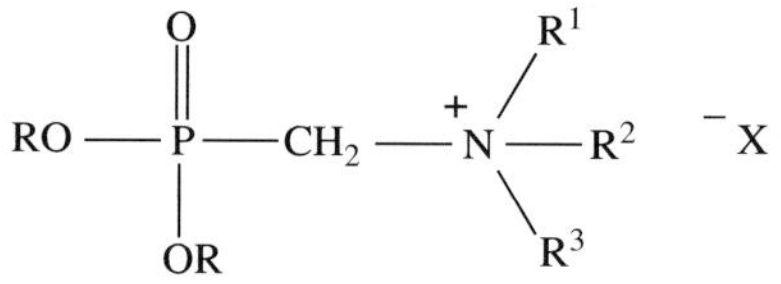

$$RO-\underset{\underset{OR}{|}}{\overset{\overset{O}{\|}}{P}}-CH_2-\overset{+}{\underset{R^3}{\overset{R^1}{N}}}-R^2 \quad ^-X$$

temperatures in amounts ranging from 0.1 to 100 ppm of the total aqueous system.

5.8 LUBRICANTS

Antiwear and load-carrying additives

A lubricant is any material that can be placed between surfaces to lessen friction. The purpose of a lubricant is to reduce the frictional resistance between two contacting surfaces forced to slide over one another, to minimize wear, and prevent corrosion. Antiwear agents produce a surface film either by a chemical or by a physical adsorption mechanism to minimize friction and wear under boundary lubrication conditions.

A product obtained from dialkyl maleate, dialkyl H-phosphonate, and diamines is an

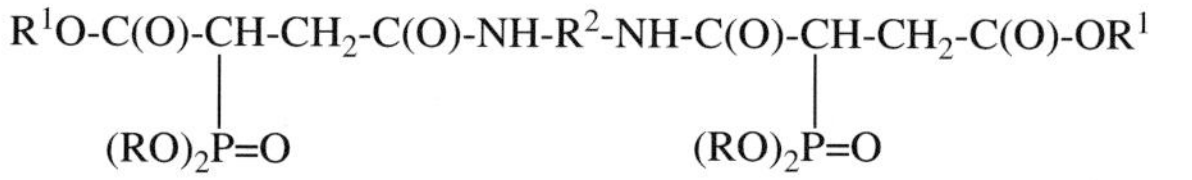

$$R^1O\text{-}C(O)\text{-}\underset{(RO)_2P=O}{\overset{|}{CH}}\text{-}CH_2\text{-}C(O)\text{-}NH\text{-}R^2\text{-}NH\text{-}C(O)\text{-}\underset{(RO)_2P=O}{\overset{|}{CH}}\text{-}CH_2\text{-}C(O)\text{-}OR^1$$

excellent additive used to impart improved load-carrying and antiwear characteristics to lubricants [137]. This additive comprises of 0.5% to 2.0% of the general weight of the lubricant.

Good antiwear and load-carrying characteristics are also imparted to the lubricants by addition of 0.5% to 2.0% of the bisaminomethylphosphonate, obtained by the reaction between dialkyl H-phosphonate, formaldehyde, and diamine (Kabachnik–Fields reaction) [138].

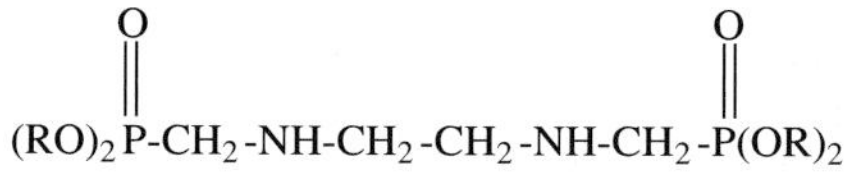

$$(RO)_2\overset{\overset{O}{\|}}{P}\text{-}CH_2\text{-}NH\text{-}CH_2\text{-}CH_2\text{-}NH\text{-}CH_2\text{-}\overset{\overset{O}{\|}}{P}(OR)_2$$

and /or

$$\begin{array}{c}(RO)_2\overset{\overset{O}{\|}}{P}\text{-}CH_2\diagdown \\ \phantom{(RO)_2P\text{-}CH_2}N\text{-}CH_2\text{-}CH_2\text{-}N \\ (RO)_2\underset{\underset{O}{\|}}{P}\text{-}CH_2\diagup\end{array}\begin{array}{c}\diagup CH_2\text{-}\overset{\overset{O}{\|}}{P}(OR)_2\\ \\ \diagdown CH_2\text{-}\underset{\underset{O}{\|}}{P}(OR)_2\end{array}$$

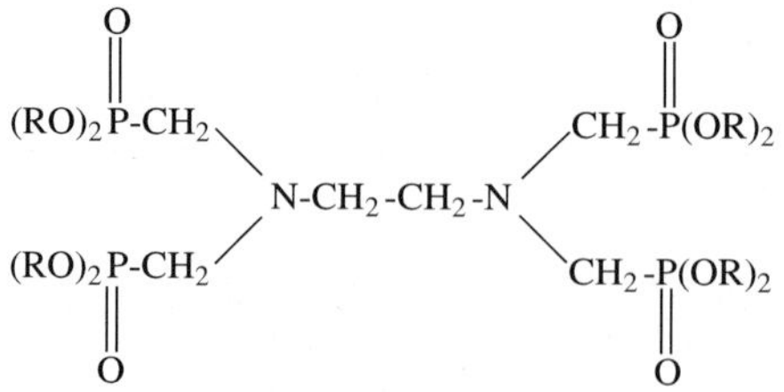

Aminomethylphosphonates are added to lubricants for aircraft gas turbines in amounts varying from 1% to 5% in order to improve load characteristics, anticorrosive properties, and viscosity indices [139].

REFERENCES

1. H. Mel'nikov, *Khimia i Tehnologia Pestitsidov*, Moskva, Mir, **1974**, 766.
2. C. Fest, K. Schmidt, *The Chemistry of Organophosphorus Pesticides*, Springer, Berlin, **1973**, 339.
3. T. Turpaev, T. Putnitseva, *Farmakol. i Toksikol.*, **1957**, 20, 22.
4. M. Kabachik, A. Brestkin, M. Mihel'son, *O Mekhanizme Fiziol. Deistvie Fosfororganicheskih soed.*, Moskva, Mir, **1965**.
5. K. Kaynon, H. Huston, *Res. Rev.*, **1973**, 47, 55.
6. K. Kacmmerer, S. Buntenkotter, *Res. Rev.*, **1973**, 46, 1.
7. D. Dull, *Res. Rev.*, **1972**, 43, 11.
8. DDR Pat. 247,825 (**1987**).
9. Eur. pat. 249566 (**1987**); C. A., 108, 70628 (**1988**).
10. L. Divinskaya, V. Limanov, E. Skrotsova, *Zh. Obshch. Khim.*, **1966**, 36, 1244.
11. V. Boev, U. Volianskii, *Khim. Farm. Zh.*, **1977**, 11, 39.
12. Eur. Pat. 156729 (**1985**).
13. Fr. Pat. 2,457,873 (**1980**).
14. E. Nifantief, A. Zavalishina, *Zh. Obshch. Khim.*, **1967**, 37, 1854.
15. D. D. Baird, R. P. Upchurch, W. B. Homesley, J. E. Franz, *Proc. North Cent. Weed Control Conf.*, **1971**, 26, 64.
16. *Herbicide Handbook of the Weed Science Society of America*, Fourth Edition, WSSA, Champaign, Illinois, **1979**.
17. U.S. Pat. 3,799,758 (**1974**).
18. U.S. Pat. 4,315,765 (**1980**).
19. P. B. Chykaliuk, J. R. Abernathy, J. R. Gipson, *Bibliography of Glyphosate*, The Texas Agricultural Experimental Station, Lubbock, **1980**.
20. P. Sprankle, W. F. Meggitt, D. Penner, *Weed Sci.*, **1975**, 23, 235.
21. C. L. Sandberg, W. F. Meggitt, D. Penner, *Weed Res.*, **1980**, 20, 195.
22. P. C. Lolas, H. D. Coble, *Weed Res.*, **1980**, 20, 267.
23. L. A. Sherp, *Proc. N. Z. Weed Pest Control Conf.*, **1975**, 28, 165.
24. J. J. Kells, C. E. Rieck, Proc. *South Weed Sci. Soc.*, **1978**, 31, 243.
25. M. L. Rueppel, B. B. Brightwell, J. Schaefer, J. T. Marvel, *J. Agric. Food Chem.*, **1977**, 25, 517.
26. *The Herbicide Glyphosate* (eds. E. Grossbard, D. Atkinson), Butterworths, London, **1985**.
27. Ed. C. D. S. Tomlin, *The Pesticide Manual – a World Compendium*, 11th edn., British Crop Protection Council, Farnham, Surrey, **1997**.
28. N. Amrhein, B. Deus, P. Gehrke, H. C. Steinrücken, *Plant Physiol.*, **1980**, 66, 830.
29. E. Haslam, *The Shikimate Pathway*, Wiley, New York, **1974**.
30. J. E. Franz, K. K. Mao, J. A. Sirorski, *Glyphosate: A Unique Global Herbicide*, ACS Monograph 189, American Chemical Society, Washington, DC, **1997**.
31. W. E. Blondiell, J. Vnek, P. F. Knowles, M. Sprecher, D. B. J. Sprinson, *Biol. Chem.*, **1971**, 246, 6191.
32. J. A. Sikorski, K. S. Anderson, D. G. Cleary, M. J. Miller, P. D. Pansegrau, J. E. Ream, R. D. Sammons, K. A. Johnson. In: *Chemical Aspects of Enzyme Biotechnology Fundamentals, Proc. 8 th Annual Industrial Univ. Ciiop. Chem. Progs. Symp.* (eds. T. O. Baldwin, F. M. Rauschel, A. I. Scott), Plenum Press, New York, **1990**, 23–39.

33. K. S. Anderson, K. A. Johnson, *Chem. Rev.*, **1990**, 90, 1131.
34. H. C. Steinrücken, N. Amrhein, *Eur. J. Biochem.*, **1984**, 143, 351.
35. E. K. Merabet, M. C. Walker, H. K. Yuen, J. A. Sikorski, *Biochim. Biphys., Acta*, **1993**, 1161, 272.
36. S. Castellino, G. C. Leo, R. D. Sammons, J. A. Sikorski, *Biochemistry*, **1989**, 28, 3856.
37. W. F. Campbell, S. O. Evans, S. C. Reed, *Weed Sci.*, **1976**, 24, 22.
38. H. M. Hull, C. A. Bleckmann, H. L. Morton, *Proc. West Weed Sci. Soc.*, **1977**, 30, 18.
39. L. M. Kitchen, W. W. Witt, C. E. Rieck, *Weed Sci.*, **1981**, 29, 513.
40. T. T. Lee, *Weed Res.*, **1981**, 21, 161.
41. N. S. Nomura, H. W. Hilton, *Weed Res.*, **1977**, 17, 113.
42. E. A. Bababunmi, O. O. Olorunsogo, O. Bassir, *Toxicol. Appl. Pharmacol.*, **1978**, 45, 319.
43. R. E. Hoagland, *Weed Sci.*, **1980**, 28, 393.
44. Belg. Pat. 840114 (**1980**).
45. Belg. Pat. 906023 (**1987**).
46. C. de Duve, *Biochem. Pharmacol.*, **1974**, 23, 2495.
47. E. Karanov, D. Trendafilova, M. Georgieva, V. Aleksieva, V. Vasileva, K. Troev, *Dokl. Bulg. Akad. Nauk.*, **1991**, 44, 11.
48. U. S. Pat. 4,475,943 (**1984**).
49. R. Rawls, *Chem. Eng. News*, **1998**, June 29, 13.
50. F. Roberts, C. Roberts, J. Johnson, D. Kyle, T. Krell, J. Coggins, G. Coombs, W. Milhous, S. Tzipori, D. Freguson, D. Chakrabarti, R. Mcleod, *Nature*, **1998**, 393, 801.
51. T. A. Bandurina, V. N. Konyuhov, O. A. Ponomareva, A. S. Baribin, E. V. Pushkareva, *Pharm. Chem.*, **1978**, 12, 35.
52. J. B. Camden, Procter & Gamble Company, (a) U.S. Pat. 5,665,713 (**1997**); (b) U.S. Pat. 5,656,615 (**1997**); (c) U.S. Pat. 5,854,231 (**1998**); (d) U.S. Pat. 5,902,804 (**1999**); (e) U.S. Pat. 6,090,796 (**2000**).
53. Ru-Yu Chen, Li-J. Mao, *Phosphorus, Sulfur, Silicon*, **1994**, 89, 97.
54. H. Mitsuya, K. Weinhold, P. A. Furman, M. H. St. Clair, S. N. Lehrman, R. C. Gallo, D. Bolognesi, D. W. Barry, S. Broder, *Proc. Natl. Acad. Sci., USA*, **1985**, 82, 7096.
55. P. A. Furman, J. A. Fyfe, M. H. St. Clair, J. Rideout, G. A. Freeman, S. N. Lehrman, D. P. Bolognesi, S. Broder, H. Mitsuya, D. W. Barry, *Proc. Nat. Acad. Sci., USA*, **1986**, 83, 8333.
56. H. Mitsuya, R. F. Jarrett, M. Matsukura, F. D. M. Veronese, A. L. DeVica, M. G. Sarngadharan, D. G. Johns, M. S. Reitz, S. Broder, *Proc. Natl. Acad. Sci., USA*, **1987**, 84, 2033.
57. C. F. Perno, R. Yarchoan, D. A. Cooney, N. R. Hartman, D. S. Webb, Z. Hao, H. Mitsuya, D. G. Johns, S. Broder, *J. Exp. Med.*, **1989**, 169, 933.
58. T. S. Lin, R. F. Schinazi, W. H. Prusoff, *Biochem. Pharmacol.*, **1987**, 36, 2713.
59. (a) J. Machado, H. Salomon, M. Olivera, *Nucleosides Nucleotides*, **1999**, 18, 901; (b) N. Tarussova, M. K. Kukhanova, A. A. Krayevsky, E. K. Karamov, V. V. Lukashov, G. B. Kornilayeva, M. A. Rodina, G. A. Galegov, *Nucleos. Nucleot.*, **1991**, 10, 351.
60. O. Yurin, A. Kravtchenko, L. Serebrovskaya, E. Golochvastova, N. Burova, E. Voronin, V. Pokrovsky, *AIDS*, **1998**, 12, 240.
61. (a) J. H. Boal, R. P. Iyer, W. Egan, *Nucleosides Nucleotides*, **1993**, 12, 1075; (b) Q. Xiao, J. Sun, Y. Ju, Yu-fen Zhao, Yu-xin Cui, *Tetrahedron Lett.*, **2002**, 43, 5281.
62. G. Gosselin, C. Perigaud, I. Lefebvre, A. Pompou, A. M. Aubertin, A. Kern, T. Szabo, J. Stawinski, J. L. Imbach, *Antiviral Res.*, **1993**, 22, 143.
63. A. A. Krayevsky, N. B. Tarussova, Q. Y. Zhu, P. Vidal, T. C. Chou, P. Baron, *Nucleosides Nucleotides*, **1992**, 11,177.
64. C. Perigaud, J. L. Giaradet, G. Gosselin, J. L. Imbach. In: *Advances in Antiviral Drug Design* (ed. E. De Clercq), JAI Pres Inc., Ct, **1996**, 2, 147–172.

References

65. E. V. Karamov, V. V. Lukashov, A. P. Gorbacheva, T.V. Makarova, G. V. Kornilaeva, N. B. Tarusova, A. A. Ktrevskii, *Mol. Biol.* (Mosk.), **1992**, 26, 3192.
66. C. Meier, L. Habel, W. Laux, E. De Clercq, Jan Balzarini, *Nucleosides Nucleotides*, **1995**, 14 (3–5), 759.
67. R. M. Francis, *Curr. Ther. Res.*, **1997**, 58, 656.
68. H. Fleisch, R. Russel, M. D. Francis, *Science*, **1969**, 165, 1264.
69. P. N. Sambrook, P. Geusens, C. Ribot, J. A. Solimano, J. Ferrer-Barriendos, K. Gains, N. Verbruggen, M. E. Melton, *J. Intern. Med.*, **2004**, 255, 503.
70. S. Vasireddy, A. Talwakar, H. Muller, R. Mehan, D. R. Swinson, *Clin. Rheumatol.*, **2003**, 22, 376.
71. N. A. Dawson, *Expert. Opin. Pharmacother.*, **2003**, 4, 705.
72. L. S. Rosen, D. H. Gordon, W. Jr. Dugan, P. Major, P. D. Eisenberg, L. Provencher, M. Kaminski, J. Simeone, J. Seaman, B. L. Chen, R. E. Coleman, *Cancer*, **2004**, 100, 36.
73. T. H. Cromartie, K. J. Fisher, J. N. Grossman, *Pesticide Biochem. Phys.*, **1999**, 63, 114.
74. E. van Beek, E. Pieterman, L. Cohen, C. Lowik, S. Papapoulos, *Biochem. Biophys. Res. Commun.*, **1999**, 265, 491; **1999**, 264, 108.
75. R. K. Keller, S. J. Fliesler, *Biochem. Biophys. Res. Commun.*, **1999**, 266, 560.
76. J. D. Bergstrom, R. G. Bostedor, P. J. Masarachia, A. A. Reszka, G. Rodan, *Arch. Biochem. Biophys.*, **2000**, 373, 231.
77. J. E. Grove, R. J. Brown, D. J. Watts, *J. Bone Miner. Res.*, **2000**, 15, 971.
78. J. E. Dunford, K. Thompson, F. P. Coxon, S. P. Luckman, F. M. Hahan, C. D. Poulter, F. H. Ebetino, M. J. Rogers, *J. Pharmacol. Exp. Ther.*, **2001**, 296, 235.
79. S. P. Luckman, D. E. Hughes, F. P. Coxon, R. Graham, G. Russell, M. J. Rogers, *J. Bone Miner. Res.*, **1998**, 13, 581.
80. J. E. Fisher, M. J. Rogers, J. M. Halasy, S. P. Luckman, D. E. Hughes, P. J. Masarachia, G. Wesolowski, R. G. Russell, G. A. Rodan, A. A. Reszka, *Proc. Natl. Acad. Sci. USA*, **1999**, 96, 133.
81. E. van Beek, C. Lowik, G. van der Pluijm, S. Papapoulos, *J. Bone Miner. Res.*, **1999**, 14, 722.
82. M. Neves, L. Gano, N. Pereira, M. C. Costa, M. R. Costa, M. Chandia, M. Rosado, R. Fausto, *Nucl. Med. Biol.*, **2002**, 29, 329.
83. GB Pat. 2,118,042 (**1983**).
84. J. M. Sanders, Y. Song, J. M. W. Chan, Y. Zhang, S. Jennings, T. Kosztowski, S. Odeh, R. Flessner, C. Schwerdtfeger, E. Kotsikorou, G. A. Meints, A. O. Gomez, D. Gonzalez-Pacanowska, A. M. Raker, H. Wang, E. R. van Beek, S. E. Papapoulos, C. T. Morita, E. Oldfield, *J. Med. Chem.*, **2005**, 48, 2957.
85. C. J. Hilado, *Flammability Handbook for Plastic*, Technomic, Stamford, CT, **1969**.
86. P. C. Warren. In: *Stabilization Against Burning in Polymer Stabilization* (ed. W. L. Hawkins), Wiley-Interscience, NY, **1972**.
87. Brit. Pat. 862539; C.A., 56, 117584 (**1962**).
88. K. Leland, R. Drogin, J. Shewmaker, *Prod. Res. Dev.*, **1963**, 2, 145.
89. K. Troev, R. Ilieva, G. Borisov, *Commun. Dept. Chem.*, **1977**, 9, 540.
90. G. Borisov, K. Troev, At. Grozeva, T. Troev, R. Georgieva, *Commun. Dept. of Chemistry*, **1976**, 9, 545.
91. K. Troev, At. Grozeva, G. Borisov, *Eur. Pol. J.*, **1979**, 15, 437.
92. K. Troev, At. Grozeva, G. Borisov, *Eur. Pol. J.*, **1979**, 15, 1143.
93. K. Troev, At. Grozeva, G. Borisov, *Eur. Pol. J.*, **1981**, 17, 27.
94. K. Troev, At. Grozeva, G. Borisov, *Eur. Pol. J.*, **1981**, 17, 31.
95. K. Troev, Ts. Kisyova, At. Grozeva, G. Borisov, *Eur. Polym. J.*, **1993**, 29, 1499.
96. K. Troev, Ts. Kisyova, At. Grozeva, G. Borisov, *J. Appl. Polym. Sci.*, **1993**, 49, 777.
97. K. Troev, Ts. Kisyova, At. Grozeva, G. Borisov, *Eur. Polym. J.*, **1993**, 29, 1211.
98. K. Troev, Ts. Kisyova, At. Grozeva, G. Borisov, *Eur. Polym. J.*, **1993**, 29, 1205.

99. U.S. Pat. 3,979,533 (**1976**).
100. U.S. Pat. 3,235,517 (**1966**); 3,294,710 (**1966**).
101. U.S. Pat. 3,855,363 (**1974**); C.A., 82, 112161c (**1975**).
102. Jpn. Pat, 57-163388 (**1981**); C.A., 98, 107535d (**1983**).
103. U.S. Pat. 3,821,263 (**1974**); C.A., 81, 170208r (**1974**).
104. Jpn. Pat, 89798 (**1974**); C. A., 83, 60605z (**1975**).
105. E. Tashev, S. Shenkov, K. Troev, G. Borisov, L. ZAbski, Z. Edlinski, *Eur. Polym. J.*, **1988**, 24, 1101.
106. U.S. Publ. Pat. Appl. 530174 (**1976**); C.A., 84, 151476v (**1976**).
107. U.S Pat. 3,183,214 (**1965**); C.A., 63, 3129g (**1965**).
108. Jpn. Pat. 57-163,388; C.A., 98, 107, 535 d (**1980**).
109. P. Zavlin, M. Sokolovskii, R. Tenisheva, *Z. Prikl. Khim.*, **1964**, 37, 928.
110. B. Agre, E. Vasilenko, I. Gvozdukovich, V. Rumiantseva, E. Nifantiev, *Plast. Massi*, **1977**, 2, 59.
111. C. Azuma, K. Sanui, K. Koshiishi, K. Ogata, *J. Pol. Sci., Pol. Chem. Ed.*, **1979**, 17, 287.
112. (a) I. Kraicheva, B. I. Liogonkii, R. Stefanova, G. Borisov, *Eur. Polym. J.*, **1988**, 24, 1167; (b) I. Kraicheva, S. Varbanov, G. Borisov, *Eur. Polym. J.*, **1992**, 28, 795; (c) I. Kraicheva, *Phosphorus, Sulfur, Silicon*, **1996**, 118, 21; (d) I. Kraicheva, *Phosphorus, Sulfur, Silicon*, **1998**, 134/135, 287; (e) I. Kraicheva, *Phosphorus, Sulfur, Silicon*, **1999**, 155, 127.
113. K. Troev, Vl. Atanasov, R. Tsevi, G. Grancharov, A. Tsekova, *Polym. Degr. Stab.*, **2000**, 67, 159.
114. K. Troev, A. Tsekova, R. Tsevi, *Polym. Degr. Stab.*, **2000**, 67, 397.
115. K. Troev, G. Grancharov, R. Tsevi, *Polym. Degr. Stab.*, **2000**, 70, 43.
116. K. Troev, A. Tsekova, R. Tsevi, *J. Appl. Polym. Sci.*, **2000**, 78, 2565.
117. K. Troev, G. Grancharov, R. Tsevi, St. Shenkov, At. Topliyska, I. Gitsov, *Eur. Polym. J.*, in press, **2005**.
118. St, Vassileva, N. Stefanova, K. Troev, P., *J. Appl. Polym. Sci.*, submitted **2005**.
119. K. Troev, R. Tsevi, I. Gitsov, *Polymer*, **2001**, 42, 39.
120. K. Troev, N. Stefanova, V. Mitova, St. Vassileva, I. Gitsov, *Polym. Degr. Stab.*, **2005,** 91, 778.
121. (a) K. Troev, D. M. Roundhill, *Phosphorus Sulfur*, **1988**, 37, 247; (b) K. Troev, *Heteroatom Chem.*, **2000**, 11, 205.
122. Ger. Pat. 1,259,094 (**1968**); C.A., 68, 87905z (**1968**).
123. U.S. Pat. 3,274,014 (**1966**).
124. U.S. Pat. 4,224,218 (**1980**).
125. Ger. Offen, 2,602,182 (**1977**); C.A., **87**, 136642q (**1977**).
126. D. Urkovich, M. Bahitov, E. Kuznetsov, *Izv. VUZ, Khim. i Khim. Teknol.*, **1977**, 8, 1249.
127. Jpn. Pat; 74-29317 (**1974**); C.A., 82, 140802f (**1975**).
128. M. Benuett, T. Mitchell, *J. Org. Chem*, **1974**, 70, 630.
129. E. Nifantief, T. Kuhareva, *Zh. Obshch. Khim.*, **1981**, 51, 2146.
130. N. Yamazaki, T. Iguchi, F. Higushi, *J. Pol. Sci., Pol. Chem. Ed.*, **1975**, 13, 785.
131. N. Yamazaki, T. Iguchi, F. Higushi, *Tetrahedron*, **1975**, 31, 3031.
132. Jpn. Kokai, 77-58796 (**1977**); C.A., 87, 118427u (**1977**).
133. (a) U.S. Pat. 3,786,055 (**1974**); C.A., 80, 96154d (**1974**); (b) U.S. Pat. 3,888,627 (**1975**); C.A., 83, 97570k (**1975**).
134. U.S. Pat. 4,557,845 (**1985**); C.A., 104, 152169r (**1986**).
135. U.S. Pat. 4,529,528 (**1985**); C.A., 103, 126320x (**1985**).
136. U.S. Pat. 4,420,399 (**1983**).
137. U.S. Pat. 4,267,063 (**1981**); C.A., 95, 45844f (**1981**).
138. U.S. Pat. 4,563,299 (**1986**); C.A., 104, 20991m (**1986**).
139. Ger. Pat. 1,126,056 (**1962**).

Index

1

1,1-cyclopropanediylbis(phosphonic acid), 157
1,3,2-dioxaphospholanes, 34
1,3,2-dioxaphosphorinanes, 34

2

2,3-epoxyalkylphosphonates, 185
2-amino-1-hydroxyethylene-1,1-bisphosphonic acid, 148
2-chloro-4-methyl-1,3,2-dioxaphospholane, 2
2-hydro-2-oxo-1,3,2-dioxaphosphorinane, 35, 194

4

4-methyl-2-hydro-2-oxo-1,3,2-dioxaphospholane
    tautomeric forms, 36
4-methyl-2-oxo-2-hydro-1,3,2-dioxaphospholane, 2

5

5′-O-hydrogen phospholipids, 179, 262
    synthesis of, 179
5-fluorouracil, 156, 210

A

ab initio conformational analyses, 13
Abramov reaction, 59, 103, 109, 269
    catalysts
        cesium fluoride, 62
        potassium fluoride, 62
    mechanism, 59
acetylene, 55
acetylenes, 151
acidity, 15, 23, 24, 46, 58, 107, 109, 110, 125, 164
acidolysis, 29

activation energy, 17
AIDS, 66, 160, 260, 261
alcohol, 1–4, 8, 25, 27, 30, 32, 33, 37, 46, 48–50, 52, 71, 77, 80, 116, 167, 179, 188
aldehydes, 59, 61, 62, 64, 108, 114–116, 122, 123, 134, 183, 217, 246
alkenes, 151, 157
alkyl halides, 87, 187, 212, 214, 225, 227
alkylammonium salt of phosphonic acid monomethyl ester
    structure, 80
alkylating agents, 83, 264, 271
aminoalkylbisphosphonates, 153
    from nitriles and dialkyl H-phosphonates, 153
    from nitriles, H_3PO_3 and PBr_3, 153
aminophosphonates
    from dialky H-phosphonates and nitriles, 132
    from dialkyl H-phosphonates and 1,3,5-trisubstitutedhexahydro-s-triazines, 133
    from dialkyl H-phosphonates and methyl-(N-diethyl)aminomethyl ether, 131
    from dialkyl H-phosphonates and methylenediamines, 130
    from dialkyl H-phosphonates and α-alkylamino dialkyl mercaptenes, 131–132
    from imines and H-phosphonic acid, 126
    from Schiff base and dialkyl H-phosphonates, 124
    methods for preparation, 108
aminophosphonic acids, 107, 108, 116, 126, 137–139, 141, 246, 260
antioxidants, 264
aryl nucleoside phosphorothioate, 173
arylselenophosphates, 76
asymmetric dialkyl H-phosphonates, 4, 25
Atherton-Todd reaction, 41–52, 102, 103, 188, 196, 198, 203, 208, 250, 268
    N-phosphorylated proteins, 48
    N-phosphorylation, 48

O-phosphorilation, 48
reaction mechanism, 42
synthetic application, 48
azidation, 52
AZT, 206, 261
AZT-5′-H-phosphonate
anti-HIV prodrug, 261

B

Beckmann rearrangement, 150
bis-(2-chloroethyl H-phosphonate)
reaction with triethylamine, 80
bis-(2chloroethyl) phosphite, 80
bis(ω-hydroxyalkyl) phosphonates, 34, 105
bisphosphonate conjugates, 155
bisphosphonates, 144
aledronate (Fosamax), 263
anti-resorption drugs, 263
biological activities, 145
from acetic anhydride and H_3PO_3, 146
from acetylenes, 151
from amides, H_3PO_3 and PCl_3, 146
from carboxylic acid, H_3PO_3 and PCl_3, 146
from dibromo-diflouromethane, 156
from N-alkyl lactams, 154
from oximes, 150
from terminal alkynes, 151
from tri-Et orthoformate, 146
from α-amino-α-alkoxyalkanes, 145
inhibitors of osteoclastic bone resorption, 144
P–C–P backbone, 144
pamidronate (Aredia), 263
risendronate (Actonel), 263
zolendronate (Zometa), 263
bisphosphonic acids, 145, 263
bone resorption, 263, 264
borohydride, 47, 102
bromomethylenebis(phosphonates), 157

C

calcium dichloride, 88
camptothecin, 202
carbon disulfide, 216, 277
carboxylic acid anhydride, 3, 8
catalysts, 30, 55, 63, 64, 91, 115, 125, 210, 214, 264, 276–278
dibutyl H-phosphonate, 276, 277

monoalkyl H-phosphonate ammonium salt, 276
phosphorus- and metal-containing oligomers, 276
sodium dialkyl phosphite, 276
chemical bonds, 12
chiral center, 4, 115, 151
chloral, 60
chloroalkylbisphosphonates, 154
copper dichloride, 47
corrosion inhibitors, 210, 264, 278, 279
quinoline phosphonates, 279
cyanation, 52
cyclic chlorophosphites, 2, 8
cysteamine hydrochloride
immobilization, 204

D

Darzens reaction, 182
dealkylation, 1, 87–89, 91–93, 204, 211, 220, 223, 225–227
degrading agents, 264
deoxyribonucleosides, 159
dialkyl 1,2-epoxyalkylphosphonates
from dialkyl chloromethylphosphonates and carbonyl compounds, 182, 183
from dialkyl vinylphosphonates, 185
from sodium dialkyl phosphonates and α-halo ketones, 180
reaction with alkyl halides, 187
reaction with amines, 186
reaction with aqueous H_2SO_4, 186
reaction with ethanol, 186
reaction with methanol, 186
thermal rearrangement, 185
dialkyl 1-alkyl-1,2-epoxyethylphosphonates
reaction with thiourea, 187
dialkyl 1-alkylvinylphosphonates, 187
dialkyl 2-chloroethyl phosphates, 47
dialkyl acylphosphonates, 69, 71
dialkyl arylphosphonates
synthesis, 214
dialkyl epoxyalkylphosphonates, 180
dialkyl H-phosphonate
complexes, 222
dialkyl H-phosphonates
addition reactions, 53
addition to acetylen, 55

Index

addition to aldehydes, 59
addition to chloral, 60
addition to isocyanates, 73
addition to ketones, 59
addition to paraformaldehyde, 61
addition to unsaturated compounds, 54
addition to vinyl carbonate, 56
alkylating agents, 271
alkylation of amines, 18
amines, 77
aminoalcohols, 81
azidation, 52
by hydrolysis of cyclic chlorophosphites, 2
cobalt complexes, 232
complexes
    P(III) type ligands, 227
    P(V) type ligands, 226
complexes with f-elements, 224
complexes with transition metals, 226
cyanation, 52
degrading agents, 264
from H-phosphonic acid, 3
from phosphorus trichloride and
    alcohols, 1
from white phosphorus and alcohols, 4
homolytic addition, 75
hydrolysis, 25
Jackson–Meizenheymer complex, 223
metal salts, 210
molecular complexes
    with Lewis acid, 222
nickel complexes, 231
phosphorus atom
    electronic structure, 11
phosphoryl group
    chemical bonds, 12
physiologically-active substances
    fungicides and bactericides, 254
proton abstraction, 18
reaction with ammonium halides, 91
reaction with chlorosilanes, 92
reaction with halocarbons, 52
reaction with hydrogen halides, 87
reaction with metal phosphites, 212
reaction with metal salts, 88
reduction, 25
thiocyanation, 52
transesterification
    side reactions, 32

dialkyl phosphinous acids, 25
diaryl H-phosphonates
    reaction with amines, 82
dibromo-diflouromethane, 156
diesters of H-phosphonic acid, 1, 15, 18, 19,
    23, 39, 41, 44, 52, 180, 264, 271
    ^{13}C-NMR spectra, 20
    ^{17}O-NMR spectra, 20
    ^{1}H-NMR spectra, 18
    ^{31}P-NMR spectra, 19
    acidity, 23
    disproportionation, 25
    infrared spectra, 18
    physical properties, 4
    pK_a values, 23, 24
    tautomerization, 15
diethyl 1,2-dihydroxyethylphosphonate, 186
diethyl 1-hydroxy-2-alkoxyethylphosphonates,
    186
diethyl chlorophosphate, 216
diethyl H-phosphonate
    addition to acrylonitrile, 56
    palladium complexes, 228
diethyl α-lithioepoxyethylphosphonate, 187
dimethyl acetylphosphonate
    P–C(O) bond cleavage, 69
dimethyl H-phosphonate
    ab initio conformational analyses, 13
    acidolysis, 29
    from carboxylic acid anhydride and
        alcohol, 3
    iridium complexes, 233
    molecular structure, 13
    pyrolysis, 6
    reaction with amides, 85
    reaction with ethyl carbamate, 85
    rhodium complexes, 233
    tautomerization
        activation energy, 17
    tautomerization energy, 16
    thermal stability, 5
    transesterification reaction, 29
dimethyl methylphosphonate, 5
dimethyl phosphite
    ab-initio computations, 15
dinucleoside H-phosphonates, 166
    preparation, 166
dinucleoside phosphorotioates, 173, 174
diphenyl diselenide, 76

diphenyl H-phosphonate, 3
  from H-phosphonic acid and triphenyl phosphite, 3
  palladium complexes, 228
  reaction with mercaptans, 37
  transesterification, 31
disproportionation, 25
DNA, 128, 159, 170, 171, 174, 175, 206, 208, 209, 261
doxorubicin, 202

E

electronic shielding, 80
electronic structure, 11
epoxide resin
  phosphorus-containing, 271
esterification, 3, 8
ethenylidenebis(phosphonic acid), 158
ethyl 2-bromoacetate, 157
exchange reaction
  between urethane and methoxy groups, 85, 86

F

flame retardants, 145, 158, 264, 265
fosfomycin, 184
fungicides and bactericides, 254
Fyrol 6, 267, 268

G

glyphosate
  derivatives
    plant growth regulators, 258
  LD_{50}, 258
  monoisopropylammonium salt
    LD_{50}, 258

H

halocarbons, 52
heat and light stabilizers, 264
heat stabilizers
  dioctyl H-phosphonate metal complexes, 275
heretrobimetallic multifunctional catalysts, 64
HIV, 59, 66, 67, 160, 206, 261, 262

Horner–Emons reaction, 218
H-phosphonic acid, 3
  esterification, 3
  sodium salts
    ^{31}P-NMR data, 222
hybridization probe, 179
hydrogels, 209, 210
hydrolysis, 25–29
hydroxyalkylbisphosphonates, 153
hydroxyapatite, 144, 263

I

immobilization, 38, 204–206, 250, 251, 259
Infrared spectra, 15, 18

K

Kabachnik–Fields reaction, 108, 110–113, 121–123, 245, 268, 279, 280
ketones, 39, 59, 61, 62, 91, 108, 183

L

LDA, 187, 248
Lewis acid, 117, 219, 222, 223
ligands, 28, 152, 156, 222, 226–228, 278
lithium diisopropylamine, 187
lubricants, 264, 280, 281

M

Mannich-type reaction, 128, 130
metal phosphites
  reaction with alkyl halides, 212, 214
metal salts
  phosphite structure, 210, 211
  phosphonate structure, 211
  synthesis, 220
Michaelis–Becker reaction, 212, 214, 215
molecular structure, 13
monoalkyl H-phosphonate anion, 18, 43–45, 79–81, 84
monoalkyl H-phosphonate salts
  addition to ketones, 91
  addition to unsaturated compounds, 91
monoalkyl H-phosphonic acids
  alkylammonium salts
    plant growth regulator, 259

Index

monoethyl H-phosphonate
  metal salts
    ^{1}H-NMR data, 221
    ^{31}P-NMR data, 221
monomethyl ester of H-phosphonic acid
  thermal rearrangement, 5
monomethyl ester of the methyl phosphonic acid, 5

N

N-(phosphonomethyl) glycine
  drugs
    against cancer, 260
    against tumor, 260
    against viruses, 260
  enzymic cleavage, 257
  herbicide, 256, 257
  mechanism of action, 257
  methods for preparation, 134–137
N-alkyl lactams, 154
naturally occurring phosphonate, 54
nitric oxide, 150
N-phosphonomethyl glycine derivatives, 137
N-phthaloylglycine, 148
N-protected amino acids, 50
nucleoside 3′-H-phosphonates, 159, 160, 162, 171
nucleoside 5′-H-phosphonates
  methods of preparation, 164
nucleoside H-phosphonates
  methods of preparation, 160
  reactivity, 166
nucleoside H-phosphorothioates, 173
nucleoside α-hydroxyphosphonates, 66

O

O-alkylation, 213
oligonucleotides, 159, 167, 170, 171, 174, 175, 179
  synthesis of, 174
osteoporosis, 145, 263
oxidation, 4, 6, 39–41, 69, 104, 135, 137, 143, 154, 171, 176, 178, 180, 184, 188, 195, 196, 204, 253, 265
oximes, 150, 216

P

paclitaxel, 203
P-alkylation, 212, 213
pamidronate, 145, 263
P–C bond, 53, 54, 63, 64, 67, 68, 70, 107, 138–140, 217
  hydrolytic cleavage, 138
peptidic α-hydroxphosphonates, 65, 66
P–H group, 15, 19, 38, 39, 52, 53, 61, 76, 82, 89, 188, 195, 199, 210
  oxidation, 39
  protection, 53
  reaction with trialkyl orthoformates, 52
phosphine, 6
phospho-aldol reaction, 59, 63
phospholipids, 177, 178
  synthesis of, 177
phosphonate-phosphate rearrangement, 60, 67, 68, 111, 165, 218
phosphonic acid, 24, 25, 28, 29, 44, 54
  pK_a values, 24
phosphonic end groups
  treatment with diazomethane, 193
phosphonopeptides, 141
phosphor(isothiocyanatidate)s, 52
phosphorazidates, 52
phosphorcyanidates, 52
phosphoric acid, 6
phosphorus atom, 4, 11
phosphorus trichloride, 1, 2, 8, 9, 146, 147
phosphoryl group, 11, 12, 15, 23, 29, 49, 50, 65, 7, 142, 194
physiologically-active substances, 83, 253
pK_a values, 23, 24, 27
poly(alkylene H-phosphonate)s, 34, 51, 188
  addition to carbonyl group, 199
  application, 201
  methods of preparation
    by polycondensation, 189
    by polymerization, 194
  reactivity
    addition to C=C double bond, 199
    hydrolysis, 195
    oxidation, 196
    reaction with amines, 198
poly(alkylene phosphoramidate), 198
poly(ethylene terephthalate)
  phosphorus-containing, 265

poly(ethylene)
  phosphorus-containing, 270
poly(N-vinylpyrrolidone-co-vinylamine), 51, 83
poly(oxyethylene H-phosphonate)s
  ^{13}C{H}NMR spectrum, 191, 192
  ^{1}H NMR spectrum, 190
  ^{31}P{H}NMR spectrum, 190, 191
  degree of polymerization, 191
polyacroleine
  phosphorus-containing, 269
polyamide-6
  degrading by dimethyl H-phosphonate, 274, 275
polycarbonate
  degrading by dimethyl H-phosphonate, 273
polycondensation, 188, 189, 193, 249, 265, 277
polymer additives, 264
polymer-drug conjugates, 201
polymerization, 2, 37, 159, 188, 189, 194, 209
polystyrene
  phosphorus-containing, 269
polyurethanes
  degrading by dimethyl H-phosphonate, 271
  phosphorus-containing, 38, 269
protection, 48
Pudovik reaction, 54, 103
pyrolysis, 6
pyrophosphates, 144, 263, 264
  P–O–P backbone, 144
p_π–d_π conjunction, 12

R

reduction, 25, 73
ribonucleoside 3′-H-phosphonates, 160, 162
ribonucleosides, 159, 160
ring-opening polymerization, 194
Ringsdorf, 201
RNA, 159, 170, 171, 174, 175, 206
rubbers
  phosphorus-containing, 269, 270

S

scale inhibitors, 264, 279
  aminomethylphosphonates, 279
Schiff bases, 39, 47, 117, 124, 125
side reactions, 32

sodium dialkyl phosphites
  reaction with carbon disulfide, 216
  reaction with N-alkoxypyridinium salts, 218
  reaction with N-alkylpyridinium salts, 218
  reaction with oximes, 216
sodium diethyl phosphite
  reactioin with aromatic aldehydes, 219
  reaction with diethyl chlorophosphate, 216
  reaction with fluorinated alkylketones, 218
sodium salt of monoethyl H-phosphonate
  ^{31}C{^{1}H} NMR, 90
spectral characteristics, 18
stable conformers, 13, 14, 15
structure
  of ammonium salt of monomethyl H-phosphonate, 90
  of sodium salt of monoethyl H-phosphonate, 90

T

tautomerization, 15
tautomerization energy, 16
teichoic acids, 194
terminal alkynes, 151
tetramethyl pyrophosphonate, 5
tetravalent phosphorus atom, 15
thermal stability, 5
thiocyanation, 52
three-coordinated phosphite form, 15
transesterification
  synthetic applications, 33
transesterification reaction, 29
trialkyl orthoformates, 52
trialkyl phosphite, 1, 15, 69, 156, 224
triethyl methylammonium salt of monomethyl H-phosphonate
  ^{31}C{^{1}H} NMR, 90
trinuclear heterometallic complexes, 233
triphenyl phosphite, 3, 82
triphosgene, 156, 161, 169
trivalent phosphorus atom, 15

U

unsaturated polyesters
  phosphorus-containing, 270
urethane, 85, 86, 271, 272

Index

W

white phosphorus, 4

Z

zoledronic acid, 145

α

α-carbon atom, 19, 23, 32, 33, 59, 77, 80, 81, 87, 88, 90–92, 154, 187, 193, 204, 212
α-hydroxy alkanephosphonic esters, 59
α-hydroxyalkylphosphonates
   dehydration, 73
   halogenation, 71
   herbicide, 260
   hydrolysis, 68
   oxidation, 69
   phosphonate-phosphate rearrangement, 67
   reduction, 73
α-ketophosphonates, 69
   from α-hydroxyalkylphosphonates, 69
α-aminophosphonate nucleosides, 143
α-ethyl-N-(phosphonomethyl) glycine, 137, 138

β

β- and γ-hydroxyphosphonate, 65